The Golden Century of Oil 1950–2050

The GeoJournal Library

Volume 19

The titles published in this series are listed at the end of this volume.

The Golden Century of Oil 1950–2050

The Depletion of a Resource

by

C. J. CAMPBELL

Consultant Geologist,
Milhac, Gourdon, France

SPRINGER SCIENCE+BUSINESS MEDIA, B.V.

ISBN 978-94-010-5548-2 ISBN 978-94-011-3500-9 (eBook)
DOI 10.1007/978-94-011-3500-9

Printed on acid-free paper

for
Bobbins
Julia & Jack
Simon & Oddny
Anne & Patrick

TABLE OF CONTENTS

PREFACE

Oil is the lifeblood of the World's economy. It was a critical element in two World Wars and in the Cold War, and, as recent events in the Middle East confirm, people are willing to fight for it.

The cheap energy it provides, especially for transport and agriculture, was one of the main factors that made possible the economic prosperity and growth that the World has enjoyed for the past fifty years and more. People rely on it everywhere, and in many forms, and they have become so accustomed to its ready availability that they take it for granted. To conceive of a world without traffic jams and airliners is unthinkable, and while not so obvious, oil lies behind every supermarket shelf, fuelling the tractor that ploughs the field and the delivery van that brings the consumer his food.

Yet everyone knows that it is a finite and irreplaceable commodity, formed long ago in the geological past. What no one knows is just how finite it is.

This book is an effort to try to answer that question : not in detail, but at least in orders of magnitude. More useful than the figures themselves is the discussion of the elements involved in addressing the subject. While it is impossible to predict the precise pattern of future production, which will be affected by many unforeseeable factors, one can at least begin to think in terms of resource constraint instead of an ever expanding supply of oil.

The scenarios, developed here, are therefore intended more to draw attention to the general limits of the underlying resource base than to claim to depict the precise shape of future production. They are also not intended to be didactic. There is an inevitability about the depletion of oil : the World will adjust to using less oil, much less oil, when it has to ; and it is probably unrealistic to press for conservation before necessity provides the motivation. By 2050, at the end of what has been called "The Golden Century of Oil", production will have fallen to about what it was in 1950. That sounds calamitous, but some people look back with a certain nostalgia for the low-consumeristic days of the 1950s. It was not that bad, so we may even come to benefit from the depletion of oil if it leads us to live in a better relationship with our environment.

Even so, it is somewhat staggering to realize that we will

have to face a radical new situation so soon : probably well before the end of the century, less than a decade away.

It may be asked upon what qualifications the author presumes to address this important subject. It is a valid question, because he does not have access to the global files of an international oil company or a government agency. Furthermore, it is some years since he was directly concerned with worldwide exploration that could give an up-to-date insight into the possibilities. So, to that extent, he is not qualified. On the other hand, it may be that he is qualified by default : insofar as one should never trust an expert, least of all one who has something to sell. The exploration community has a strong vested interest in there being a large undiscovered potential; and the oil company executive knows that investors will not favour a company that cannot claim to replace its reserves. Governments, too, are under political pressures, and may have reason to exaggerate their oil reserve base, which strengthens national collateral and in some cases even provides a certain brute power.

This book is offered as an expression of opinion rather than a scientific treatise in which every statement requires its authority and reference. Attention is nevertheless drawn to two authorities : first, L.F.Ivanhoe, who has made an outstanding contribution to the subject in many articles; and, second, to a splendid book, entitled "The Prize", in which Daniel Yergin tells the story of the growth of oil during the first half of the "Golden Century", when production was expanding.

It is necessary to touch on the subject of professional bias. The three professions most concerned with the production of oil are Geologists, Engineers and Economists, each of whom views the subject of World Resources from a very different vantage point. The Economist tends to think that the flow of oil is determined by economic factors such as supply and demand. If you need more oil, you invest in exploration and secure what you need, albeit at a price. By contrast, the Engineer tends to think that it is determined by technological factors. If you need more oil, you drill in deeper water, or use new technology. Turning to the Geologist, he can do no more than describe the nature and distribution of the oil-bearing rocks, clearly being unable to influence them in any way. It should follow that he is especially well qualified to express an objective opinion of where the conditions for oil accumulation and production are met. But, in practice, he can rarely do so because he is under enormous commercial pressure to be an Optimist: a pressure that is likely to become ever more extreme, as the viable prospects of the World are depleted. His work has become a sort of Act of Faith, in which he is ceasing at heart to be a believer. So, it is hard to come by a realistic assessment no matter which breed of expert is consulted.

Before going any further, please note carefully the following addition to the well-known catalogue of Murphy Laws, which most certainly applies to all World Reserve Studies.

ALL FIGURES ARE WRONG

It means, indeed, that without reliable statistics, there can be no real experts anyway, and the door is open for informed speculation by whoever cares to address the problem. We can at least try to understand the patterns and trends, and above all, to study carefully the implications of successive revisions.

It also means that

ALL FIGURES ARE TO BE GENEROUSLY ROUNDED.

Indeed in almost all cases, the quoted figures should be regarded as being, at best, no more than close to the mid-point of a range. We will return to the subject of rounding.

This study is divided into three parts. Part 1 comprises a general discussion of the World situation and the methods of analysis ; Part 2 is a region by region assessment ; and Part 3 gives production data and forecasts for each country, together with a list of giant fields.

It is based on 1989 Production and Reserve figures, mainly as reported by the Oil & Gas Journal. The invasion of Kuwait on August 2nd 1990 and the ensuing war, which was in progress at the time of writing, make both 1990 and 1991 anomalous. Even so, 1990 figures, which became available during the study, do provide the first, albeit qualified, test of the model. A great deal is, in fact, to be learnt from the trend of progressive revision, an essential procedure for a study of this sort. It might be an idea to attempt a revision in 1993 to see what changes are called for. The fewer the changes, the greater will be the evolving confidence.

The study is concerned only with oil as is produced in today's oilfields. It does not concern itself with oil from tar or oil shales, nor with the proceeds of unconventional enhanced recovery techniques, although they are likely to become more important in the future.

Natural Gas is also not covered here. It is an increasingly important source of hydrocarbons, being much more widely distributed in nature than is oil.

Many oilfields contain a gas-cap, which is termed Associated Gas, and in general terms its depletion follows the same pattern as the related oil, although it is normal practice to defer the production of Associated Gas for as long as possible for economic and technical reasons.

In addition, there are many accumulations of what is termed Dry Gas, which consists mainly of methane, with only minor liquid components. Such accumulations are produced where there is a market and a pipeline connection : many remain undeveloped in remote parts of the world.

A third, and increasingly important, category is Gas-Condensate, which is a gas with a high liquid content that is normally found at greater depths where the liquid hydrocarbons have been cracked due to higher temperatures.

The world reserves of gas are colossal and at a relatively early stage of depletion in most territories. Accordingly, its development is primarily controlled by economic factors, especially those related to pipeline investment. Ultimately, it will also be affected by the same resource constraints as now apply to oil as discussed in these pages but, except perhaps in the case of countries such as the United States, United Kingdom, and the Netherlands, the depletion of gas resources is unlikely to become a serious issue until well into the next century.

ACKNOWLEDGMENTS

I am very grateful to L.F.Ivanhoe for several years of always helpful correspondence and much insight into this and other subjects. I express my appreciation to Farouk al-Kasim, Oystein Kristiansen, Ole Preben Berget and Tøve Rødder of the Norwegian Petroleum Directorate for their interest and support in this study.

My wife, Bobbins, kindly proof-read and offered comments, not all favourable.

I also take the opportunity to record my gratitude to those friends, teachers and colleagues who have contributed in many different ways to my knowledge and enjoyment of geology and the oil business. They include, chronologically, I.Higginbottom, Dr Fleet, N.Goodman, K.S.Sandford, J.V.Harrison, L.R.Wager, W.S.McKerrow, T.A.Chavasse, G.H.Pickles, H.J.C.Kirk, G.E.Wilford, J.M.Bowen, H.G.Kugler, K.W.Barr, G.E.Higgins, J.B.Saunders, H.Bürgl, H.Tanner, H.Cely, R.Canfield, I.Poulson, D.Henao, F.C.R.Martin, J.Spence, J.Harrison, D.B.Walker, R.F.P.Hardman, R.A.Barry, R.Quintero, F.Barrientos, W.J.George, R.C.Herrera, W.E.Humphrey, N.J.Sander, G.R.Collomb, C.H.Miller, C.Baker, R.Blanton, H.Loser, M.C.Coffield, H.B.Zaremba, C.R.Carr, J.C.Teran, X.Espinosa, J.Leach, W.B.Pace, R.Stoneley, B.J.Kellenberger, O.J.Buis, W.K.Hall, D.E.Welch, A.R.Baker, K.G.Finstad, P.G.Gulbrandsen, R.A.Halpin, J.W.Bausch van Bertsberg, M.Parkinson, P.H.Sykes, F.Wellhauser, S.Cambi, H.R.Warman, S.Cancar, M.J.Whelan, J.J.Stafford, J.Holloway, T.Jamison, O.Skree, A.Saebøe, P.Songstad, R.Kihle, K.A.Shepard, J.G.C.Fuller, G.Dow, G.E.Bennett, G.Flores, G.Orbell, A.Morland, F.G.Stumm, L.J.Graves, R.C.Leonard, J.Munns, B.Noryani, O.Beyer, K.Jakobsen, A.Johansen, R.Humphreys, G.Knag, S.Johnson, A.R.Ellis, J.Rivertz, O.A.Lindseth, G.Gjerde, L.Dons, G.Toffelordi, A.Ianiello, M.Melli, W.Walton, A.Allen, A.Armitage, A.N.Spencer, J.Berg, I.Aarseth, C.E.Needham, G.Wardle, F.Caselli, O.P.Hansen, G.Fortems, L.J.Laurent, P.Dechanxhe, M.Hara, B.Bienaime, R.Thiers, A.Salamon, K.Lea, D.van Roye, W.H.Ziegler, M.Bjoroy, R.Koenig, R.Peacock, J.Junca, A.Robinson, T.Lund, O.J.Aga, T.Sund, E.Bergsager, P.T.Smedvig.

Lastly, I thank those who have typed, drafted, surveyed, driven, carried, and cut paths through forests in support of the endeavours. They include : D.Guyu, Njambong, Gaja, "The White Rajah", Duncan, H.Cifuentes, A.Prieto, A.Albaracin, H.Patino, D. Cano, K.McGowan, M.Vargas, M.Luz Gutierrez, J.de Chacon, E.P.Murray, A.Duffy, H.Godson, A.Ericson, J.Neish, Obando, C.Tufino, L.Lopez, G.Lopez, G.Horve, S.Lea, T.A.Brun, Julio, Neftali.

UNITS OF MEASUREMENT AND CONVERSION FACTORS

The following units of measurement and conversion factors are useful to addressing the subject. The barrel remains the most commonly used unit of measurement.

Oil Units

By Volume :
Barrel, 1 = 42 U.S.gals. = 159 litres = 0.159 Sm^3
Cubic meter, 1 = 6.29 barrels
By Weight : Tonne (metric), 1 = 2,205 lbs
Conversion (assuming Arabian Light Oil with 33.5° API)

Barrel	Sm^3	Tonne
1	0.159	0.136
6.29	1	0.855
7.33	1.17	1

Reserves are commonly measured in billions of barrels (10^9) BBO = Billion barrels of oil, although the metric system is applied in many countries.

Production is usually measured in thousands of barrels per day Mb/d (1 b/d = 49.8 tonnes per year)

Gas Units

Gas is usually measured in cubic feet or cubic meters
1 Sm^3 = 35.55 cu ft.
1 cu ft. = 0.0283 Sm^3
Reserves are commonly measured in trillion cubic feet (10^{12}) or cubic meters

Energy Equivalents

In cases where it is not practicable to differentiate oil from gas, the gas is sometimes quoted as an oil equivalent, based on its calorific content
t.o.e (tonne of oil equivalent
b.o.e (barrel of oil equivalent)
1000 Sm^3 gas = 1 t.o.e. = 7.168 b.o.e.

PART 1

WORLD ASSESSMENT

FIGURES

TABLES

THE OCCURRENCE OF PETROLEUM

There are many textbooks on petroleum geology, which explain the complex processes that are responsible for the generation and entrapment of petroleum : one of the best is "Petroleum Geology" by F.K.North. It will therefore be sufficient here to touch on only a few aspects of petroleum geology to set the scene for the debate on the state of the World's reserves. We must remember that oil is a natural resource, whose occurrence, distribution, and eventual production, are subject to geological constraints. Each discovery depends on a unique combination of geological circumstances.

SOURCE-ROCKS

In earlier years, the explorer had no very clear idea of what exactly oil was, nor where it came from. Any dark-looking sediment was considered a promising source, especially so if it smelt or looked oily, as is sometimes the case. The petroleum geologist tended, instead of worrying about source, to give most of his attention to where reservoirs and traps might be found, which are certainly also vital elements in the search for oil.

The fundamental question of petroleum source-rock was not indeed that important in the established producing provinces, such as for example the southern United States, here there was ample evidence of generation. Many petroleum geologists, working the established oil patches, could have successful careers without ever having to seriously address the problems of source-rock, and some later found themselves running the international exploration efforts of large companies. It is not surprising that, for many of them, the World offered limitless opportunities, being perceived almost as a huge extension of Texas, burdened only by grasping foreign governments, who owned the oil rights.

However, in the past decade, enormous progress has been made in the science of Geochemistry, so that now it is possible to relate the oil flowing from a well with the mother-rock from which it came, and to understand the temperature constraints that control its generation.

Something strange happened to the Adriatic in the summer of 1988: the beaches became inundated with an evil-smelling glutinous material ; and offshore, the fishermen found their nets clogged with the same stuff, which was poisoning the fish. The Adriatic is a land-locked sea with negligible currents or tides, and its depths tend to hold stagnant water.

What had happened to the Adriatic in 1988 was due to an exceptional proliferation of algae, a minute marine organism, that occurred because of exceptional climatic conditions.

It provides a present day example of the conditions required for the generation of petroleum. The algal material, together with the remains of the poisoned marine life, sinks to the sea-floor, where it is progressively covered and sealed by layers of sediment washed into the sea by the rivers that flow into it. As the sea-floor subsides, partly under the weight of the sediment deposited on it, the organic material is buried, and heated by the Earth's heat-flow. Chemical changes take place that convert it into proto-petroleums, called ***kerogens,*** some more oil-prone than others.

As burial proceeds over geological time, the ***kerogen*** is further heated and converted into oil and gas, depending on its type. The oil once formed may be in turn cracked into gas with still more burial. The temperature range for oil generation generally occurs at depths of between 2000 and 5000m, depending on the geothermal gradient of the area in question; the precise nature of the ***kerogen***; and to a lesser extent, on the duration of its exposure to heat. A low temperature over a long time can give the same result as a high temperature over a short time.

Appropriate conditions for the generation of oil occurred only rarely in the history of the Earth, and conditions for the ***prolific*** generation of oil, sufficient to support a major producing province, occurred only ***extremely*** rarely in both time and place. In effect, it had to be necessary for the conditions of the Adriatic in 1988 to be repeated for millions of years, and for very little other sediment to enter the sea, so that the organic material could become concentrated.

Just such circumstances occurred, for example, about 150 million years ago in the North Sea when a few hundred meters of highly organic, Upper Jurassic clay were deposited. It is the source of almost all the oil in the North Sea. Its distribution has now been accurately mapped, and the depth zone for oil generation has been clearly identified. For a structure to be productive, it must be in communication with this belt of generation.

This knowledge has not expanded the scope for viable exploration in the North Sea, for it has effectively condemned large areas that are not in communication with the essential source, which might otherwise have been considered prospective. It shows that successful research does not necessarily improve the chance of finding oil : more often than not, better knowledge confirms a negative assessment.

The truly exceptional nature of prolific petroleum source-rocks cannot be emphasized too strongly in addressing World Resource Studies : the greater part of the globe is barren of petroleum for very good geological reasons, which are now well understood. Generally, the World is very definitely not an extension of Texas.

In very simplistic terms, most of the World's oil comes from two major productive belts (see Figure 1). The main

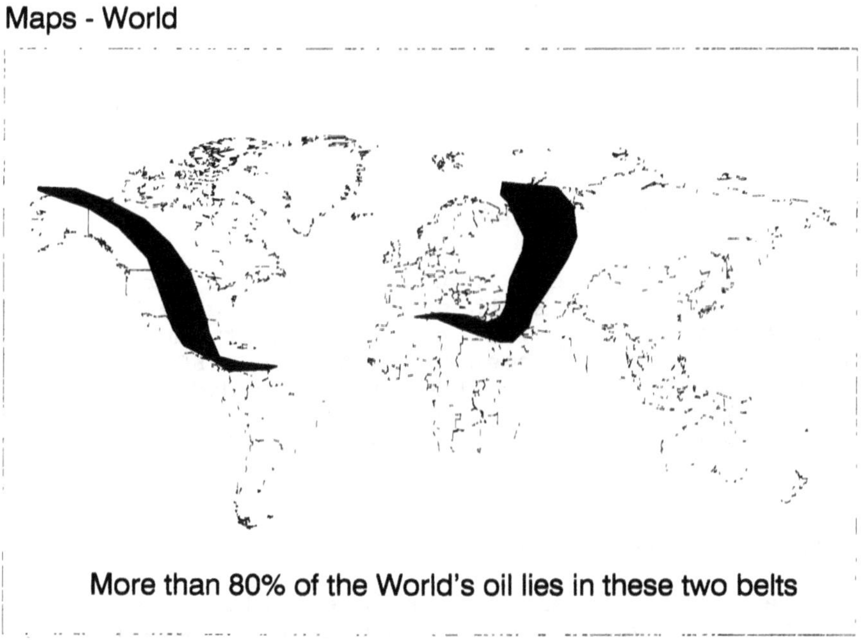

Figure 1. Most of the World's oil comes from two trends of prolific generation.

reason for such concentration is that it was here that conditions for the ***prolific*** generation of oil were particularly favourable.

Klemme & Ulmishek have studied the distribution of oil source-rocks : there were only certain, rather brief, periods in the geological history of the earth when the special conditions needed for the formation of source-rocks occurred (see Figure 2).

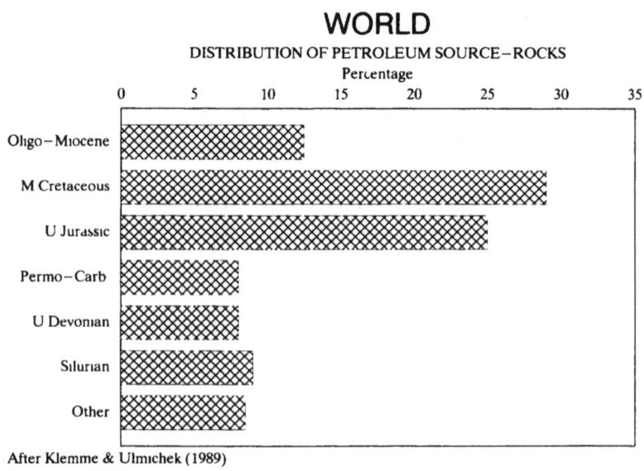

Figure 2. Distribution of hydrocarbon source-rocks.

Secondly, much of the oil that was generated during the earlier periods of the Earth's history has gone because the traps that once held it lost their integrity during later earth movements. It has been estimated that 80% of the oil so far discovered was generated since the Middle Cretaceous, namely in the last 100 million years, with 50% of it being generated and trapped since the Oligocene in the last 25 million years. Much of the World is made up of older rocks, with far less oil potential.

OIL IN THE RESERVOIR

By far the most important factor in petroleum geology is this question of whether or not prolific oil was formed in the first place. In the established producing areas, the question can obviously be answered positively, and the geologist can turn his attention to address its movement and entrapment, which are themselves very challenging subjects. But in new provinces and in areas outside the productive trends of producing basins, the question of source is paramount.

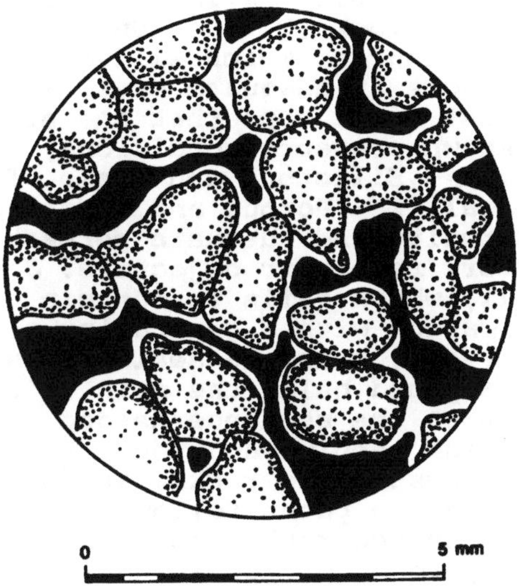

Figure 3. Idealized oil traps

Explanation of Figure 3
This idealized section illustrates the geological conditions for productive (F) and unproductive (G and J) prospects in a hypothetical offshore area, close to land. The structure has been based on seismic surveys, and the wellbore information allows the sequence of rocks to be identified and correlated with onshore outcrops.

A - Silt and clay. **B** - Salt acting as an effective seal to the underlying trap. **C** - Black organic shale with oil indications, observed at outcrop onshore, and found to be oil-prone by geochemical analysis. **D** - Same as "C" but sufficiently buried to generate oil. **E** - Sandstone reservoir with porosity and permeability. The insert shows the same rock under the microscope. Oil fills the space between the grains. Communication (permeability) between the pores is good, allowing the oil to flow through it. This formation could form a conduit allowing oil to flow from source-rocks farther out in the basin into any other traps that may develop along the margin. **F** - Oil accumulation in a porous sandstone beneath a salt seal (B) in an anticlinal trap, which is in communication with a generating oil source-rock (D). **G** - Untested prospect with reservoir, trap and source, which will fail, if tested by a well, because it lacks an adequate seal, although an optimistic geologist might hope that the silt and clay above (A), or unknown pressure conditions, might just hold the oil in place. **H** - Siltstone and Shale, forming a possible deeper seal. **I** - Porous and fractured limestone with excellent porosity and adequate permeability which forms a satisfactory reservoir objective. **J** - Prospect, tested by the well, which fails to contain oil, because it is not in communication with a generating source-rock, although the reservoir, trap and seal are satisfactory. The well was wisely deepened below the main objective (F) to obtain information on the sequence needed for regional evaluation. **K** - Basal pebbly sandstone. **L** - Ancient indurated and non-prospective rocks, known as "Basement". **M** - Arrows depicting the paths followed by the oil migrating from the source-rock into the reservoir. **N** - Arrows depict path of oil escaping from the trap due to absence of seal and flowing to the surface.

The minute droplets of oil that are generated in the source-rocks, when heated to the necessary temperatures, are subject to high pressures, caused partly by the chemical reactions that form them and partly by the weight of the several thousands of meters of overburden that lie above them. As the oil is formed, it is forced to flow towards zones of lower pressure, which are normally found at shallower depths. It searches out the paths of least resistance through the rocks in which it is embedded, such as are provided by thin zones of porosity or by fractures. A network of veinlets develops which conveys the oil (and gas) to such larger arteries as are present. The latter, often comprising porous sandstones, provide major conduits through which the oil can migrate. If there is no obstacle, the conduits may lead to the surface, allowing the oil to escape and dissipate. But more normally, folds and faults due to earth movements that occurred over long periods of geological time, form barriers that trap the migrating oil, see Figure 3.

Each trap has a catchment area from which the oil filling

it came. With the help of geochemical functions and other geological data, it is possible to calculate the amount of oil generated. Only a small fraction is available to accumulate as an exploitable deposit. Much is held by capillary attraction in the pore-space, and in constrictions in the tortuous migration paths. The traps themselves are often not perfectly sealed, so that the oil and gas may gradually leak out of them.

Reservoir rocks, mostly comprising sandstone or limestone, have porosity, commonly in the range of 15-25%, provided by the spaces between the grains of which they are formed. Rising damp in old buildings illustrates how fluids can flow through seemingly solid stone. The porosity of the reservoir and the volume of the trap determine the amount of oil that a field contains, although in some cases traps are only partly filled. It is termed Oil-in-Place.

The amount of oil that can be extracted from the reservoir depends on many factors : the permeability of the rock and the viscosity of the oil determine the rate of flow, as does the pressure differential between the formation and the well-bore. The pressure itself may be due to the existence of an expanding gas-cap above the accumulation or to the hydrostatic head of water below it, which are known as drive mechanisms. Pressure can be artificially maintained by injecting gas (gas injection), and recovery can be improved by injecting water to sweep the oil through the reservoir towards the producing wells (waterflood). Other techniques, such as artificially fracturing the reservoir by injecting fluids into it at high pressure, or of treating it with acid, can stimulate production. The management of a reservoir is a highly sophisticated business covering the selection of the best locations for wells; as well as determining the optimal production rates, the number of wells and the scope for stimulation. It falls within the domain of Reservoir Engineering. It is easy to damage a reservoir, often irreparably, by over-production and bad management. The damage done to Kuwait's reservoirs by firing the wells in the recent war has probably not yet been fully assessed, but it could well prove to be one of the major costs of the war.

The ratio of producible oil to Oil-in-Place is termed the Recovery Factor, which is normally in the range of 20-50%. It is often a major unknown during the early stages of development, and one of the causes of uncertainty in calculating reserves, as discussed later.

SEAL

The question of seal is more important than is often appreciated: some oil provinces depend on seal almost as much as source.

The reservoir in which the oil is held is commonly sandwiched between less permeable strata, such as clay or shale, that prevent or limit the seepage of oil, and many traps rely on a lateral seal caused by a fault placing the reservoir in juxtaposition with the less permeable strata.

The most effective seal is salt, but few seals are absolutely impermeable, and pressure conditions commonly contribute to their integrity, often in ways that are not fully understood. The Persian Gulf province, the most prolific of all, owes much to having a very satisfactory seal.

CRITICAL TIMING

The question of timing is critical to the formation of a viable oil bearing trap. Peak oil generation occurs under specific conditions which develop over only relatively short periods of geological time. The structural trap has to be in existence at the time the oil migrated : structures that developed before or after migration are barren. Furthermore, the accumulation once formed must be preserved from further movement : regional tilting can often cause oil to spill out of it ; or radical subsequent uplift may lead to such a pressure differential that the seal ruptures - like bringing a deepwater fish to the surface. In addition, there are more subtle factors such as over-pressuring that can be of critical importance.

The analysis of the geological history of the prospect is therefore most important : not only must all the ingredients be present, but they must have occurred in the right order. It takes only one flawed link in the chain to destroy the prospect, which is one of the reasons why normal statistical methods of analysis give such limited results.

Considering the many factors that have to come together in the right way, it is easy to see why the discovery of a new oil province is such an exceptional event. The accumulation of oil in the Ekofisk fields of Norway, for example, depends on a remarkable combination of circumstances : the prolific generation of Upper Jurassic oil during the late Tertiary ; the absence of a lateral conduit which meant that the oil had to migrate upwards; the distribution of

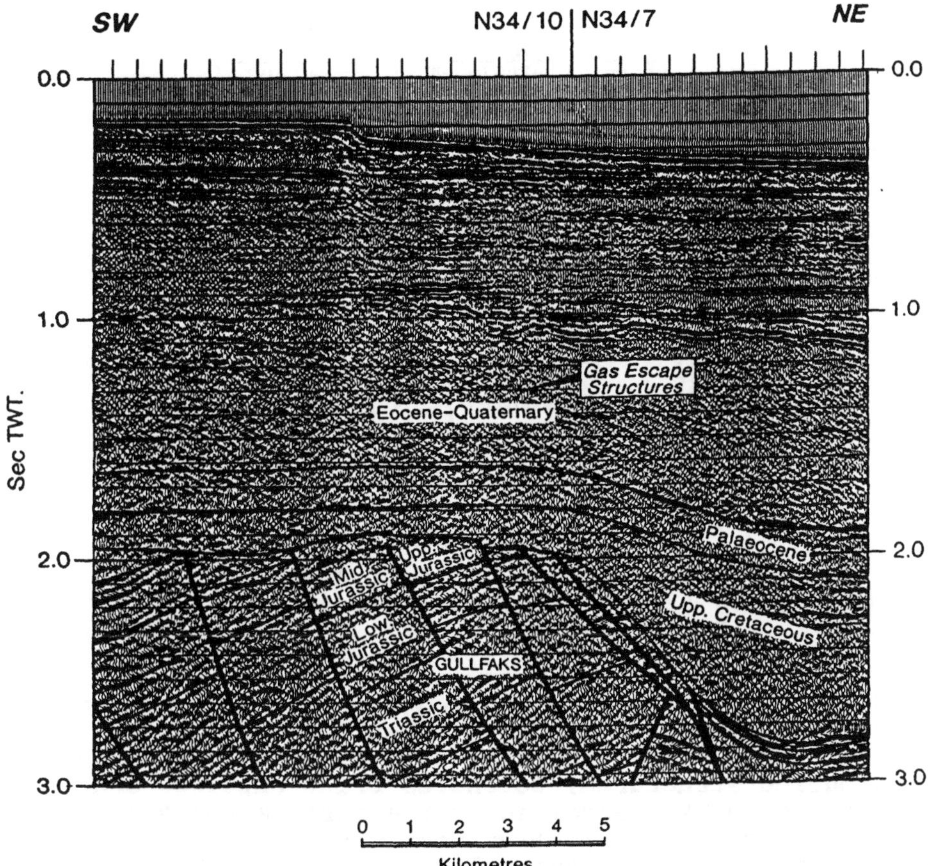

Figure 4. Interpreted seismic line across the Gullfaks Field in the North Sea (From Miles A.J., 1990, AAPG 74/11 p.1733)

slumps in the chalk sea which together with over-pressuring provided local reservoirs in what is normally an unsatisfactory formation; and the development of pressure regimes to provide just enough seal. The coincidence of all these conditions being met in one area makes Ekofisk almost one of the Wonders of the World.

MEANS OF ANALYSIS

The geologist is able to study the geological conditions far beneath the Earth's surface with the help of seismic surveys and boreholes. Seismic surveys release energy at the surface from an explosive charge or in other ways, and

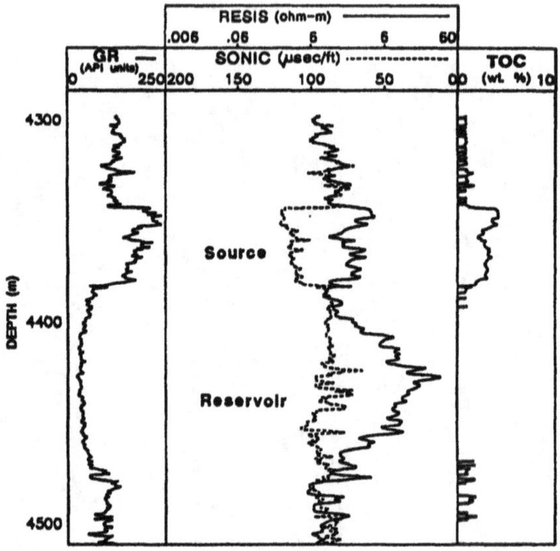

Figure 5. Example of a borehole log (From Passey Q.R. et al, 1990 AAPG 74/12 1790

record the reflections from buried geological formations. Computations, based on the time taken for the echo to return to the surface, indicate the depth and configuration of the buried structures. Boreholes bring cuttings and cores of the rocks themselves to the surface for direct study, and instruments lowered down them also provide a means of analyzing the physical properties of the rocks and their fluid content. It is as if the geologist were a doctor using seismic surveys as an X-ray to find the tumour, and the borehole as a scalpel to determine if it is benign or not. Examples of a seismic record and a borehole log are shown in respectively Figure 4 and 5.

Great technological advances in seismic and borehole surveys have been made in recent years, and the data-base from world-wide exploration is now colossal.

This is an important factor to remember in World Resource Studies. The means are now available to investigate in extraordinary detail the depth-zone in which most oil occurs, and indeed, it is already very well understood in most productive areas. There are, as a consequence, much fewer unknowns than was the case even a few years ago. But, despite the advances, the amount of oil held in a reservoir can be estimated only within broad limits. This has very important implications for World Resource Studies: reserve figures are no more than estimates, whose validity depends on unbiased professional judgment, knowledge and experience. In fact the size of an oilfield can be known absolutely only on the day when it is finally abandoned after all its oil has been produced. Until then, some measure of uncertainty will always attach to the size of its reserves.

Despite the many uncertainties discussed, exploration has become a remarkably efficient business in technical terms. It is often possible now to predict the depth and nature of the formations to be penetrated by the borehole as much as five kilometers beneath the North Sea with almost uncanny accuracy. As discussed later, the failure to find oil in a particular well does not necessarily imply technical weakness or inefficiency : the result is all too often entirely predictable in technical terms.

This brief, indeed very brief, outline sets a scene from which to address the subject of World Oil Resources itself.

UNRELIABLE DATA-BASE

In the past, when the bulk of the World's oil production was controlled by the international oil companies, most of the figures for Production and Reserves were subject to corporate and Stock Exchange audit and control. To-day, by contrast, the oil industry in the principal producing countries is controlled by the State, and is not in many cases subject to external audit.

The data-base on the essential figures for World Resource Studies is accordingly becoming increasingly unreliable, which poses a difficulty quite apart from the intrinsic geological uncertainties.

This study is based on statistics reported by the Oil & Gas Journal, being one of the most readily accessible and widely used sources of information. The Journal makes a valiant effort to compile official information, but the compilation cannot be more accurate than the input.

Another widely used authoritative source of information is the annual BP Statistical Review of World Energy. It carries the disclaimer:

> "Statistics published in this Review are taken from government sources and published data. No use is made of confidential information obtained by BP in the course of its business".

The Company cannot be expected to vouchsafe the validity of the figures.

There are other compilations, providing broadly similar statistics, although often disagreeing in detail. One source of discrepancy is the treatment of Natural Gas Liquids and Condensate: for example, BP includes North American NGL, whereas the Oil & Gas Journal does not.

Although the international oil companies can improve on these published data, where they have proprietary information, it is doubted that their global data-base, taken as a whole, is really much more reliable in terms of order of magnitude. Information on the Soviet Union, which is of critical importance, is not readily available, having been treated as a military secret, and the companies are now excluded from many of the major producing countries. The amateur analyst need not therefore feel greatly deprived in relation to the essential data-base at his disposal. Probably, the discrepancies are small in relation to the general uncertainties of the subject.

Attention is drawn later to anomalous statistics from certain important producing countries, which are believed to contain what we term "political reserves". Indeed, we think that these suspect figures are equivalent in size to as much as almost half of all the oil produced in the World to-date. With this large cloud of uncertainty hanging over us, we need not be too concerned about detailed discrepancies in the various data-bases available for general use.

THE ELEMENTS OF STUDY

The objective of the study is to describe in very general terms the distribution of oil discovery and production in time and place. Figure 6 shows the "Golden Century of Oil"

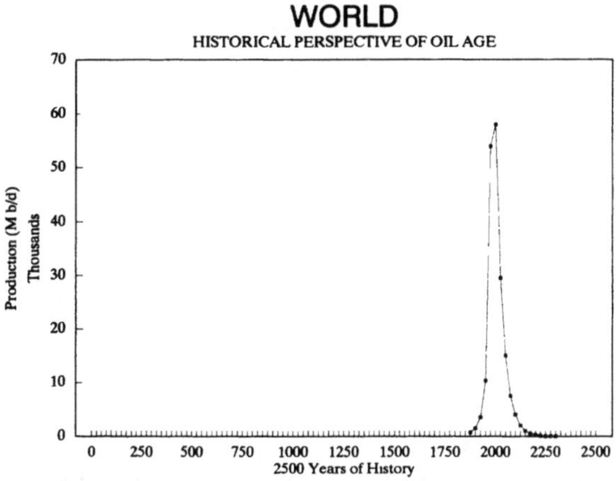

Figure 6. The age of oil in an historical perspective of 2500 years

in an historical perspective of 2500 years. Few would disagree that the age of prolific oil is but a fleeting chapter in history, when Man burns up a resource that has taken millions of years of geological time to create. Our experience over the past 50 years has witnessed the rapid growth of production : soon, we must face the other side of the coin as the progressive exhaustion of reserves leads to falling production. While this general picture is clear, the immediate future is less easily predicted.

The subject is addressed by considering the following elements:

1. CUMULATIVE PRODUCTION : the cumulative total of production to the reference date (here, end 1989).

2. RESERVES : the amount of oil remaining to be produced from discovered fields, as of the reference date.

3. TOTAL DISCOVERED : the sum of the CUMULATIVE PRODUCTION and the RESERVES (1+2).

4. UNDISCOVERED : the amount remaining to be discovered (6-3).

5. REMAINING : the sum of the RESERVES and the UNDISCOVERED (2+4).

6. ULTIMATE : the Cumulative Production when no more oil remains to be produced.

Rounding

It should be noted in passing that the Undiscovered is derived by subtracting the Total Discovered from the Ultimate. This is done to avoid the drawbacks of multiple rounding which results in figures not adding up to correct totals. The World Ultimate is the rounded starting point for the calculations. It is the element least subject to change, contrasting with the others which should change every year as the reserves are produced or new reserves found.

Although this approach gives a misleading impression of unjustified accuracy, it is the lesser of the two evils. Put in other terms, the figures quoted here should be considered as being, at best, no more than close to the mid-points of ranges, perhaps in the order of plus or minus 20%.

Furthermore, it is tiresome and repetitious, considering the many uncertainties, to qualify every figure, title or conclusion with an adjective such as "Possible", "Probable" or "Estimated", although it is true that there are few aspects of the matter which do not in reality deserve such qualification.

Figure 7 sets the scene, illustrating the division of the Ultimate into the Produced (36%), the Reserves (52%) and the Undiscovered (12%). A slice, representing 5 years of current consumption, has been removed from the Reserves to provide a sense of scale.

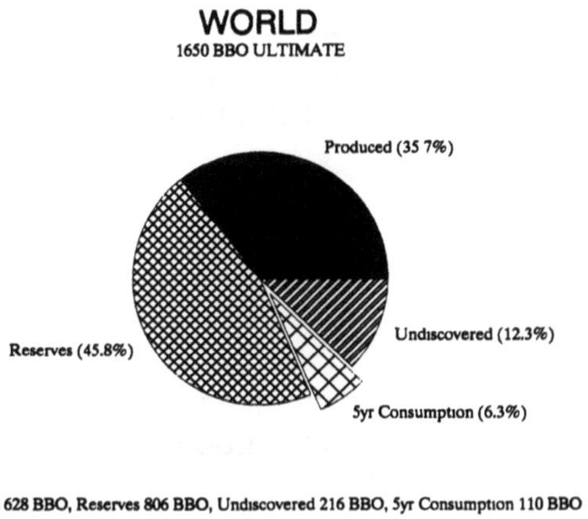

Figure 7. World's Ultimate Endowment of oil

CUMULATIVE PRODUCTION

Cumulative Production is the total amount of oil produced at the reference date (here, end 1989).

The data-base used here was not complete. It comprises data from the Oil & Gas Journal, going back to 1938 for most countries, together with 1930 data. The intervening years were extrapolated. Pre-1930 Cumulative Production for the countries that were producing then has been calculated by subtracting the 1930-89 production from the 1989 Cumulative Production.

Further research could improve the accuracy of the early production history, but it would not significantly affect the overall picture.

The figures for past production should be relatively accurate, but we note with scepticism several instances of where a particular country's reported production stays exactly stable for a number of years, even to a decimal place. For example, the reported production of Bahrain from 1948 to 1956 is shown in Table 1.

Table 1. Suspect Stable Production from Bahrain

1948	29.80 M b/d (Thousand barrels per day)
1949	30.10
1950	30.20
1951	30.10
1952	30.10
1953	29.54
1954	30.10
1955	30.10
1956	30.00

We have no way of knowing if these figures are correct or not, but production at such a constant rate seems implausible. This is the first of many doubts that arise when one begins to look into the data-base in detail.

The validity of production figures for countries which were infringing their OPEC quotas may also be in some doubt. Furthermore, the quantities sold internally or refined and exported as product may not be fully reported in some countries.

The following dates are milestones in the Cumulative Production of the World :

Table 2. Production Milestones

1930	15 BBO
1950	61
1989	624
2050	1421 Forecast (Scenario-2)

The Cumulative Production forecast of 1421 BBO for 2050 comes ominously close to the estimated Ultimate Recovery of 1650 BBO. (The basis for Scenario-2 is discussed later.)

Furthermore, some anomalies are introduced due to changes in political frontiers, such as, for example, Israel's high production from 1971-74 which presumably reflects the temporary control of Egyptian fields.

The 1989 distribution of Cumulative Production is illustrated in Figure 8. It naturally varies with time as the fortunes of the various producing countries change. Table 3 shows the early dominance of the USA, which provided 95% of all the oil produced in the World until 1930, and the growing importance of the six principal Middle East producers. They comprise Abu Dhabi, Iran, Iraq, Kuwait,

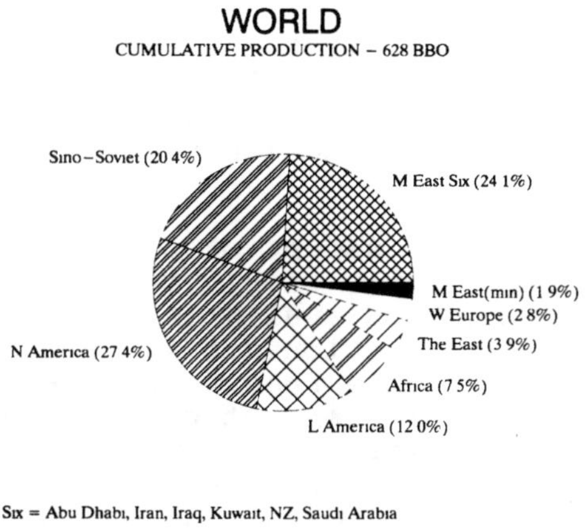

Figure 8. Distribution of Cumulative Production by Region

Neutral Zone, and Saudi Arabia. (The Neutral Zone, which is not itself such a major producer, is included because it is administered jointly by Kuwait and Saudi Arabia). For convenience we refer to these countries as the "Middle East Six"

Table 3. Cumulative Production by Region (BBO)

	N.America		USSR & China		M.East Six	
1930	14	(95%)	<1		<1	
1950	42	(69%)	5	(8%)	4	(6%)
1989	155	(25%)	106	(17%)	153	(24%)
Forecast						
2050	199	(14%)	195	(12%)	672	(47%)

Cumulative Production continues to rise as the Reserves are depleted, and will become the Ultimate when all the oil has been produced.

RESERVES

We have already described the way in which oil is generated and how it migrates generally upwards until it is trapped in a closed and sealed geological structure. We have summarized how it is held in a sponge-like reservoir, albeit a sponge made of rock, and we have touched on some of the factors that control its extraction.

The ratio of producible oil to Oil-in-Place is termed the Recovery Factor, which is normally in the range of 20-50%. The term Reserves refers only to what is producible. The oil itself changes in volume as it is extracted due to the pressure and temperature differences between reservoir and surface conditions. The first estimate, before production commences, is known as the Initial or Original Reserves, which if there is no revision would be the same as the Ultimate Recovery. As production proceeds, the reserves are inexorably depleted.

As explained above, the amount of oil contained in a reservoir can be estimated only within certain limits, based on the amount of information available. The information grows as exploration and production activities in the area proceed, which means that the reserve estimates are subject to progressive revision. The amount of information on a field, especially one with multiple reservoirs and traps, also varies from place to place within the complex, so that some Reserves are treated as Proven, whereas others are regarded as Probable. These categories are in some cases further categorized as Proven Producing, if they are being produced, or Proven Undeveloped (also termed Static), if they are awaiting development.

Proven (also termed Proved) Reserves are generally defined as

> "Those quantities which geological and engineering information indicate with reasonable certainty can be recovered in the future from known reservoirs under existing economic and operating conditions".

It is clearly not a very satisfactory definition, being subject to two important question marks. What is "reasonably certain" ? and, how far can "existing economic and operating conditions" be extended into the future ?

In practice, it is normal for the Reserves of a Prospect prior to drilling to be overstated (due to the phenomenon of Exploration Optimism already mentioned). A certain overstatement also often follows the discovery well to encourage delineation and development. On the other hand, the Initial Reserves, as formally determined when the field goes into development, often reflect over-caution, especially in the case of offshore fields when they are collateral for financing, and are commonly subject to progressive upward revision as production proceeds. It is obvious that the figures reflect something of the persona of the issuing authority : some companies work on optimistic premises; others are conservative; and State-controlled entities are sometimes under political pressures.

Greater difficulties yet are encountered in defining possible reserves for accumulations that have not yet been delineated, or indeed identified. This has led to the principle that unknown, or inadequately known, deposits, as well as those of questionable commercial viability, should be classed as Resources until they qualify for Reserve status.

This approach has drawbacks too. First, it tends to give the comforting, yet very erroneous, notion that the World is full of oil Resources which will be inexorably converted to Reserves by the diligent application of technology or under improved economic conditions. Secondly, it simply means that the term Resource has to be further divided into categories for what is more or less sure. Taken to an extreme, commercial criteria such as sanctity of contract or the financial strength of the operator, might even become logical criteria for distinguishing reserves from resources.

In World Resource Studies, it is not particularly useful to try to draw fine distinctions between Reserves and Resources, because it implies a greater knowledge than is generally available. It is also rather cumbersome.

While it is by all means true that improved economic and technological conditions can increase Reserves by making more oil commercially producible, we think that the amount of oil production dependent on such improvement is relatively small in global terms.

Enhanced recovery programmes can contribute considerable additional reserves, especially where applied to giant fields. We believe that most quoted international reserve figures already include the contributions from viable conventional enhanced recovery, such as arise from pressure maintenance, water-flood or other means.

We have not here specifically considered the contributions of unconventional enhanced recovery programmes that may be developed and applied in the far future, on the grounds that the proceeds do not at present qualify as Reserves and because that would go beyond the accuracy that can be reasonably applied to long range predictions.

There is something to be said for the traditional oil industry classification of :

Proven Reserves	:	for what is sure
Probable Reserves	:	for what is almost sure
Possible Reserves	:	for what is unsure

In this study the "Proven" and "Probable" are classed together as "Reserves". The term "Undiscovered" is used for what remains to be found and produced.

NGL and Condensate

The treatment of Natural Gas Liquids and Condensate also gives rise to some difficulties. Strictly speaking, this study is concerned with oil and not gas, but the boundary between the two phases is not always clear. Oil commonly contains some dissolved gas, and vice versa, and both separate out if the temperature and pressure change. A liquid, known as Condensate, separates naturally from certain gas accumulations on being brought to the surface from a deep reservoir, whereas Natural Gas Liquids are extracted from gas by processing.

In some cases, these substances are reported separately from oil, but in others the statistics do not differentiate them. We anticipate that the production of NGL and Condensate will grow considerably in the future as more gas fields are produced and as more deep drilling is undertaken.

In most cases, we accept the reported oil reserve figures (mainly from the Oil & Gas Journal) but, as discussed below, in some instances we have felt obliged to adjust them. We thus distinguish ***Reported*** from ***Adjusted*** Reserves.

The World's Reported Reserves by region for 1989 are

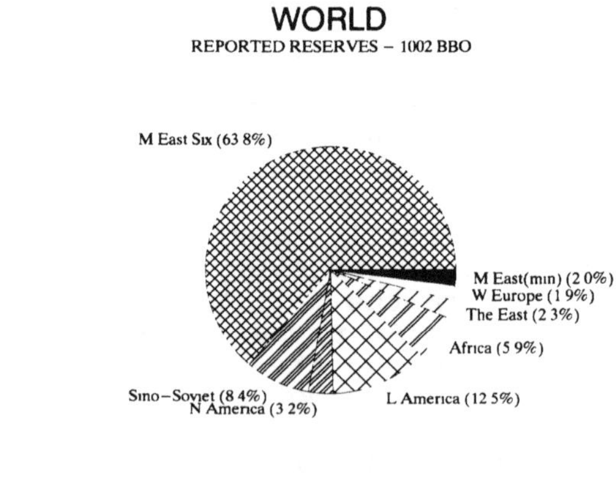

Figure 9. Reported Reserves by Region (including "political" reserves in some countries)

illustrated in Figure 9.

The study of the historical record of Reported Reserves provides a certain insight into the validity of the figures, and indeed raises some serious questions. We identify two types of anomaly : the first and most serious is termed "suspect political" ; and the second shows "suspect stability".

Table 4 is a listing of "suspect political" reserves. In addition, Mexico reported similar large increases earlier (Table 5).

Considered in greater detail, Iraq claimed an 11 BBO increase in 1983, which with some misgivings we accept, although only one field, Balad, with no more than 8 wells on it, was reported to have been found in that year (O&GJ). Then, Kuwait claimed a 26 BBO increase in 1985, which, if valid, presumably reflects a reserve revision, as no new discovery was reported in the year.

Table 4 "Suspect Political" Reserves (BBO)

	Abu Dhabi	Dubai	Iran	Iraq	Kuwait	S.Arabia	Venezuela
1980	28.00	1.40	58.00	31.00	65.40	163.35	17.87
1981	29.00	1.40	57.50	30.00	65.90	165.00	17.95
1982	30.60	1.27	57.00	29.70	64.48	164.60	20.30
1983	30.51	1.44	55.31	*41.00	64.23	162.40	21.50
1984	30.40	1.44	51.00	43.00	63.90	166.00	24.85
1985	30.50	1.44	48.50	44.50	*90.00	169.00	25.85
1986	31.00	1.40	47.88	44.11	89.77	168.80	25.59
1987	31.00	1.35	48.80	47.10	91.92	166.57	25.00
1988	*92.21	*4.00	*92.85	*100.0	91.92	166.98	*56.30
1989	92.21	4.00	92.85	100.0	91.92	169.97	58.08
1990	92.21	4.00	92.85	100.0	94.52	*257.50	59.04
Suspect Increases	61.21	2.65	44.05	52.90	26.1	87.53	31.3

__ = Discontinuity * = Suspect Increase

But these two increases, while involving huge amounts of oil, pale into insignificance when compared with what was to follow in 1988 when Abu Dhabi, Dubai, Iran, Iraq, and Venezuela all reported colossal increases, with Saudi

Arabia following suit in the next year (included in 1990 data). Together, these increases amount to 280 BBO, rising to 355 BBO if Kuwait's and Mexico's suspect figures are included.

Table 5. "Suspect" Political Reserves of Mexico (BBO)

1974	3.60
1975	13.58
1976	9.50
1977	7.00
1978	14.00
1979	16.00
1980	31.25
1981	44.00
1982	56.99
1983	48.33

Suspect increase 1977-82 = 50 BBO

Mexico has indeed some tradition in this respect, having in 1941 grossly understated its Reserves when calculating compensation following the nationalization of the assets of foreign oil companies.

These suspect increases are equivalent to more than half the total amount of oil produced in the World to-date. Even more suspect than their huge size is the fact that most occurred at almost the same time. It would be remarkable if all these countries had almost simultaneous technical reason to revise their historic reserves by so great an amount.

On the other hand, there may have been very good political reasons for them to increase their reserves in the aftermath of the 1986 "glut", when they saw their production and oil revenues decimated. The reported increase might have been intended to discourage exploration and production elsewhere, or it may have been a factor in difficult OPEC quota negotiations.

At the same time, it may also be asked if the earlier lower figures may not have reflected a certain modesty inherited from the companies, which were previously active in the countries who may have hoped to discourage sequestration.

Whatever the explanation, it is very hard to know how to treat these figures in a World Resource study. It is difficult to accept them at face value : and it is as difficult to reject them out of hand. One solution, adopted here, is compromise : we reject about 200 BBO of the Reported increase, but accept the balance as Reserve Revision, mainly reflecting higher recovery assumptions. Compromises are never satisfactory, but there seems no other way out of the dilemma : an uncertainty of this magnitude casts a heavy shadow.

These are blatant cases, but we note also that many countries hold identical figures for years on end, notwithstanding that every barrel that is produced must reduce the Reserves by an equal amount unless ***exactly*** matched by upward reserve revision and/or new discovery. There is accordingly something suspect about Reserve data from any producing country that does not change its figures from year to year. The following countries exhibit such "suspect stability".

Table 6. Countries with "suspect" stability

Argentina	1985-90
China	1977-88
Italy	1986-89
Libya	1982-90
Malaysia	1988-90
Nigeria	1981-89
Qatar	1987-89
Tunisia	1986-89
USSR	1981-85

The explanation is probably no more than inertia by the authority concerned: it is easier to leave the figures unchanged than to justify a revision. That, incidentally, is another good reason for the practice adopted here of taking the value for the Ultimate as the corner stone for oil statistics. It not only underlines that oil resources are finite, but it means that production automatically depletes the Reserves, unless specific discovery or revision indicates otherwise.

Figure 10 shows the distribution of the Reserves after Adjustment. (The term Reserve in this study always refers to the Adjusted figure, if an adjustment has been made, unless otherwise stated).

The dominance of the Middle East Six with 60% of the Reserves is immediately apparent. The Soviet Union, China, Mexico and Venezuela together hold another 20%. North America, which had produced as much as 95% of the world's oil up to 1930, now has less than 5% of the Reserves. It is

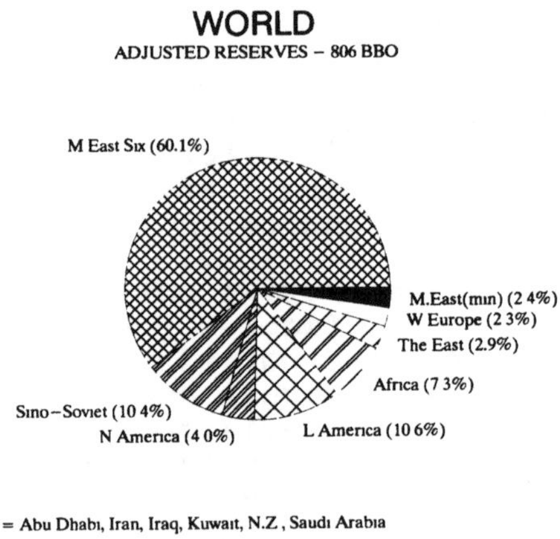

Figure 10. "Adjusted" reserves by Region (after removal of "political" reserves)

sad but true that the more efficient a country is in producing its oil, the shorter will be its producing life.

TOTAL DISCOVERED

We have discussed the uncertainties relating to the amounts of oil produced (Cumulative Production) from the discovered fields and the amounts remaining to be produced from them (Reserves). It has also been pointed out that the Reserve estimates are subject both to genuine ***revision*** on technical grounds and to ***adjustment*** to compensate for spurious assessments. The need for the latter is likely to increase as the producing countries react in political terms to the pressures of a dwindling oil resource base.

The sum of the Cumulative Production and the Reserves gives the TOTAL DISCOVERED, namely 1436 BBO. As illustrated in Figure 11, 44% of the Total Discovered lies in the Middle East Six, almost 15% in North America and as much in the Sino-Soviet region, followed by Latin America with 11%.

The concept of Total Discovered facilitates the study of the pattern of discovery since it removes the effects of the ever-changing relationship between Reserves and Cumulative Production as production proceeds.

It is particularly instructive to investigate the rate of discovery and the size distribution of the finds. It has not been possible to plot the historical development of the Total Discovered because information on early reserve estimates is not readily available. However, we come close

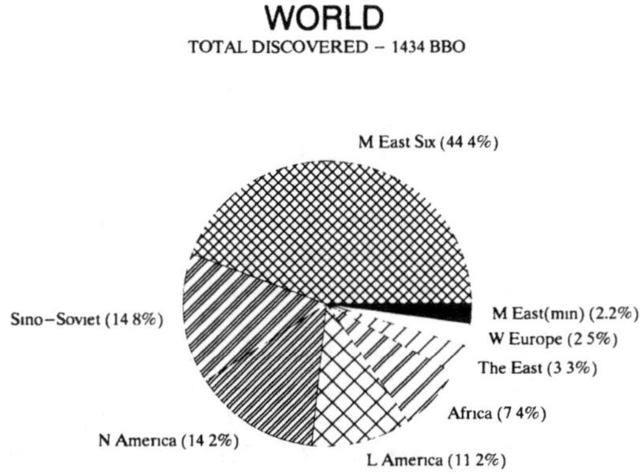

Figure 11. Total Discovered by Region (Produced + Reserves)

to achieving the same end by concentrating on the distribution of Giant Fields, which hold about 70% of the Total Discovered.

The Importance of Giant Fields

A giant field is defined in international terms as one having an Ultimate Recovery of more than 500 million barrels. In the United States, by contrast, a giant field is one with in excess of 100 million barrels.

A list of giant fields is included in Part 3. It is based on the published reports of Nehring (1979) and Roadifer (1986), and has been supplemented and adjusted on the basis of other information. We also add certain fields whose Cumulative Production, as reported by the Oil & Gas Journal, is sufficiently high to indicate that they are in fact of giant status, even if they have not been so reported. The list comprises 330 entries, which are deemed to be relatively sure, with another 139 entries that are less sure. It is also a weak and incomplete data-base. Published reserve figures may vary over wide ranges ; name-changes may in some cases cause unidentified repetitions ; there are classification difficulties depending on whether or not satellite fields are lumped together. Furthermore, the status of revisions is often uncertain. We must also note the suspect abundance of fields with reported reserves of a round 500 million barrels, just qualifying for giant status. It may be noted in passing that the list for Venezuelan giant fields seems to be particularly confused.

Attention has already been drawn to the radical upward revision of Reported Reserves in certain countries. Although some of the increase is here rejected as spurious, we are left with substantial upward revisions that are not explainable in terms of new discovery. The most likely explanation is an upward revision of giant fields based on assumed or indicated improved recovery factors. It follows that the reserves for individual giant fields in these countries are also due for upward revision.

The statistics show that about 62% of the World's Total Discovered comes from giant fields (based on the list of "sure" fields), but this is believed to be an understatement for the above reasons. The real figure may be closer to 70%.

Even this is a little misleading, because it is heavily influenced by the Middle East with its abundance of giant and supergiant fields. As shown in Table 7 the proportion of giant discovery varies greatly from region to region, depending largely on the geological conditions.

The discovery of the 330 Giants by time is plotted in Figure 12.

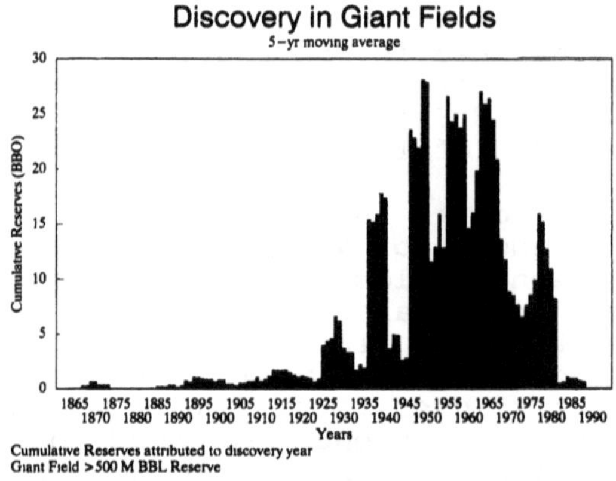

Figure 12. Giant Fields (Cumulative Reserves of all finds attributed to discovery year)

The first reported Giant was La Brea-Parinas in Peru which is attributed to 1868, although Pizzaro is said to have caulked his ships from tar seeps in the area as early as 1530. There were other isolated Giant finds around the turn of the century, but the main period of Giant discovery commenced in the 1920s and grew to a peak in the late 1950s and 1960s. For the past 30 years, there has been a decline, which has become extreme during the past decade, when, if the data-base is to be believed, less oil was found in Giants than at any other time this century. The trend is very clear, even if the data-base lacks some recent discoveries (such as Rabi-Echiza in Gabon and Avan-Bu in Nigeria) that may not yet have been reported : it has sobering implications for the future.

Table 7 : Original reserves of Giant Fields by Region

Region	Amount (BBO)	% of Discovered	% of Ultimate
North America	63	34	28
Western Europe	20	46	48
Latin America	103	64	59
Africa	58	46	48
The East	17	36	28
Sino-Soviet	109	51	43
Middle East (other)	17	51	40
Middle East Six	489	77	70

The early discovery of Giants is entirely consistent with experience in most basins. They are relatively easy to find simply because they are so large. It is, incidentally, worth bearing this in mind in connection with assessing the potential of the Soviet Union. Some commentators have suggested that the application of modern technology will bring in substantial new discoveries, but experience from the early days in the West shows that Giants can be found, at any rate onshore, without very sophisticated technology. Likewise, we wonder if the abundance of early Giants in the Middle East may not be distorting our view of this region's future potential.

We may introduce here what may be termed the Moment-of-Truth concept. It holds that exploration in any province requires a preliminary period, termed the Lead-Time, during which the essential geological knowledge is gathered, partly by drilling. In earlier years, the Lead-Time could last for a decade or so, but now, especially in offshore areas, where comprehensive high quality seismic data may be acquired rapidly and cheaply, the time span has shrunk to no more than a year or two. After the Lead-Time, comes the Moment-of-Truth, when the basin reveals its secrets. Either large, if not Giant, discoveries are made, or the basin will fail to deliver, no matter how much more drilling is undertaken. In most cases, the Moment-of-Truth reflects the presence or absence of effective and prolific hydrocarbon source-rocks.

Late stage giant discoveries can occur, such as for example Canon Limon in Colombia, but they are rare, and normally reflect some extenuating circumstance, such as remoteness, adverse terrain or political restriction. (The writer can't resist mentioning that even he managed to have ranked the very remote Northern Llanos of Colombia, where Canon Limon lies, as prime acreage as long ago as 1965, 20 years before it was found, so it cannot have been such a surprise.)

The concept of the Moment-of-Truth has a critical bearing on the assessment of the future potential.

The evidence suggests that the epoch of Giant discovery is almost over. A non-giant future has to be viewed from an entirely different perspective from that built on past experience, which was heavily influenced by the contribution of Giants. Clearly, it would take a prodigious effort to find equivalent amounts of oil from non-giant fields, even if the resource base were, contrary to expectation, sufficient to make that a possibility.

It looks as if nearly 87% of the World's oil has already been discovered : it is a high percentage.

UNDISCOVERED

The weaknesses of the data-base on Cumulative Production and Reserves has already been stressed. They were man-made weaknesses due to imperfect records or spurious estimates. When it comes to assessing the Undiscovered potential, geological uncertainties of a different order of magnitude have to be faced.

It is, however, in this arena that the New Ventures Exploration Geologist works, and he is not lacking in expertise, armed as he is with the benefits of modern technology and a vastly greater data-base than was available to his predecessors. While he is not a visionary able to see clearly the finds of the future, his judgment of the potential for discovery is not to be ignored. His views can, indeed, be remarkably accurate, but he is rarely free to express them honestly.

The assessment of Undiscovered potential is central to the subject: many who have difficulty in accepting the finite nature of the resource and the pattern of its depletion, naturally turn to new discovery as a palliative. It is necessary therefore to look a little behind the scenes to see how exploration decisions are taken in order to judge the extent to which they imply any realistic anticipation of discovery. It makes strange reading.

Most of the major international oil companies owe their existence to the early discovery of oil, and for many years, when they were developing the World's major producing basins, exploration was a major and realistic part of their business. It was not unusual to find the Chief Executive having an exploration background, or for the Chief Geologist to have a respected place on the Board. Those days are over. Now, it is more usual to find engineers, financiers and representatives of the downstream in senior positions.

Companies have a strong incentive to try to replace their reserves, as their shareholders would not be encouraged if they thought they were investing in a rapidly depleting asset. The management therefore continues to express its faith in exploration, even though the record of genuine discovery often tells the opposite story. Many do indeed invest large amounts of their shareholders' (or more often the taxpayers') money in exploration. The reason that there is not more to show for it is not that the companies have become less efficient than previously, quite the contrary, but that they have already located most of the oil that is to be found in the areas accessible to them.

It almost seems as if the relegation of exploration ironically facilitates its prosecution in ever less promising areas: the management can more easily do it if they deny themselves real knowledge of what its chances are. They know that they need to find more oil, and they therefore hire explorers to look for it, without ever really addressing the question of how much might reasonably be left to find. In many cases the explorers are required to do the best they can with the areas presented to them by the negotiators without really expressing an objective view of the prospects.

Boone Pickens in his admirable autobiography, "Boone", gives insight into the style of things and the attitudes found amongst major oil company senior executives. It is easy to understand how they are attracted by the imagery of exploration. It is the one area in the business where a failure, namely a dry hole, can be readily excused, if not almost deemed a successful, indeed valiant, endeavour, despite the outcome, and despite the often flawed judgment that led to it.

So, it has become a certain world of unreality. The Exploration Community has a strong vested interest in justifying exploration; and the management, now no longer constrained by their own professional exploration qualifications and judgment, has a strong vested interest in hearing word not only of the general exploration potential, but of the exceptional expertise of its explorers. It looks almost as if an elaborate self-deception is being practised, in some cases through tier upon tier of management committee, each progressively further removed from reality. Success in these circumstances is measured by massaging the system rather than by expressing good judgment : nothing is more distressing to it than a negative recommendation by a knowledgeable geologist. Not only burdened by committee, some companies find themselves exposed to a certain authoritarianism in which the Chief Executive or his delegate makes arbitrary, even whimsical, decisions about which well location or concession is to be preferred.

Perpetual rotation helps: the first team will score by their initiative in taking the company into a new territory; and the replacement team will score by their wisdom in taking it out again. This is often repeated time and time again, although the basic prospects, or lack of same, have not changed at all. Rotation helps them to deny themselves knowledge of the bad news.

It does make strange reading. It would be unfair to describe all companies in these terms, but few wholly escape the virus, with its many mutations.

Several factors encourage this state of affairs. First, is the application of scoping economics, which endeavours to define the hypothetical return in success against alternative risk factors. Our friend, the Optimistic Geologist, can be relied upon to furnish the required "technical" input : if reserves of 100 million barrels do not make the prospect viable, he will find good reason to double them by increasing the anticipated reservoir thickness or by improving the assumed porosity. However, what can not be easily changed are the contract terms imposed by the government or the partners. Therefore, emphasis in the management debate soon passes to consideration of the deal on which everyone can pronounce : and nearly everyone likes to picture himself as a shrewd negotiator ; even lawyers speak. The geological aspects are glossed over, which is not surprising since all prospects tend to look alike after the assiduous application of technical massage.

It is obvious to most explorers that the running of hypothetical economics on virgin exploration projects is almost entirely fatuous, but the figures provide a useful shield. Second, is the influence of the government in the host-country. Companies often seek to ingratiate themselves with host-governments, partly to protect their existing investments, and to do so by committing to exploration projects, which commonly carry heavy drilling obligations. Furthermore, holding acreage with the right to explore and produce is often deemed to be an opportunity asset, although it only too easily becomes a liability. The geologist is again under pressure to justify the drilling needed to fulfil the commitment, even if the prospect, perhaps never strong, has in the meantime been down-graded by new knowledge.

Probably, only a small fraction of the exploration wells drilled worldwide are justifiable on strictly geological criteria : those that do meet such criteria often find oil.

All of this goes forward on the premise of the well worn adage that "Exploration is the Business of Surprise". Of course, there often are surprises, but to describe it as a Business of Surprise is to unjustifiably denigrate the available, if misused, geological expertise.

It is hard to understand how enterprises conducting their affairs in this style stay in business. One explanation is the enormous hidden subsidy that is available to exploration in countries with high marginal tax rates on historic production. In Norway, for example, the marginal tax rate for most operators is 85%, which means that all but 15% of the cost of a dry exploration well is written off. The operation of PRT in the United Kingdom is said to have an even greater effect : the taxpayer pays.

It is not therefore strange that companies, risking only a small fraction of the cost of well, are ready to invest in "Surprises".

Such is the environment in the international oil companies, but what must it be like in the State-run agencies ? Although they lack the direct tax subsidy, some in fact do receive an indirect subsidy, in cases where the concession terms provide for the foreign companies in joint ventures to carry the State's share of the cost of exploration, claiming same as a tax deduction. Any lack of direct subsidy is probably more than made up for by bureaucratic momentum. In the Soviet Union, for example, they apparently had a Ministry of Geology which drilled well after well under five-year programmes, irrespective of the outcome.

Since it is very difficult for management in either private or state-owned companies to come to terms with the finite nature of oil, it is difficult for them to curtail exploration whatever the prospects or results. It would be too much like admitting defeat. Some companies do, however, periodically purge their staffs, many with experience and knowledge, before recruiting new blood with which to start all over again.

This mild exposé explains some of the attitudes and pressures that affect the conduct of exploration. It is not criticism : who but the shareholder shall judge how things are to be done ? It does not imply that there is reason to suspend the subsidy nor curb the activity : perhaps after all it is the best approach. But it is very important to understand that, in these circumstances, a willingness to look for oil does not necessarily imply any realistic chance of finding it. That has important implications in assessing the Undiscovered potential.

While a large number of the exploration wells are drilled with flawed technical justification for the above reasons, it is not to say that all are spurious. Some wells have to be drilled to investigate the geology of new areas, during the Lead-Time of exploration, and genuine prospects continue to be found that deserve testing for good technical reasons.

Despite the fact that it is becoming increasingly difficult to obtain objective assessments of exploration potential, many scientific and statistical attempts have been made to define the World's endowment. Calculations have been made of the comparative yield of rock-volumes, and statistical approaches have been applied to risking all the parameters essential to success.

None of the procedures is wholly satisfactory, and in the end it is very much a matter of pragmatic judgment based on results of past effort and a geological gut-feeling as to the untested possibilities.

In Figure 13 are listed 47 estimates of the World's Ultimate Recovery that have been published by qualified authorities. A particularly useful assessment is that of Masters (1987), itself based partly on Ivanhoe's work. They demonstrate a clear trend. In 1942, when operations were mainly limited to the onshore, a figure of 600 BBO was proposed : it was evidently an understatement as more than that has now been produced. Then, in the 1950s and 1960s the estimates grew rapidly on the strength of the record of giant discovery and with the opening of the offshore, which was held to offer great promise. An all-time high estimate of 3550 BBO was published in 1969. Since then, the trend has been downwards, reflecting the decline in giant discovery and the disappointments from many offshore areas.

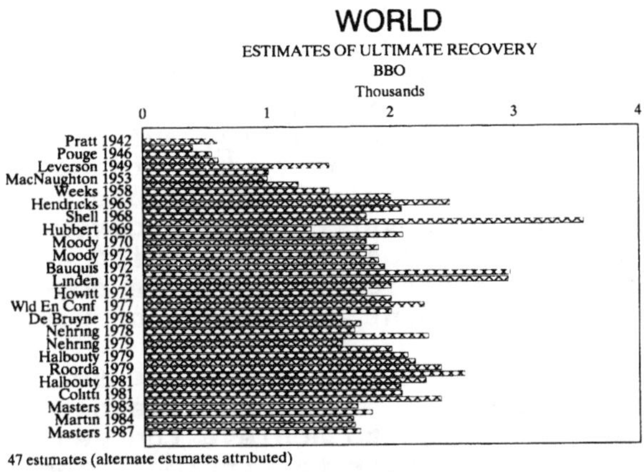

Figure 13. Expert estimates of the World's Ultimate endowment (alternate estimates attributed for reasons of space)

The present consensus seems to be about 1700 BBO.

We believe that the collapse in Giant discovery in recent years is reason to reduce the figure further, and here propose 1650 BBO as a good rounded figure. It would have been even lower but for the large upward revision of existing reserves, already discussed. It is, of course, to be treated as no more than a First Approximation or a Working Hypothesis rather than as an exact figure. Nevertheless, we believe that it is correct for order of magnitude.

It has been taken as the foundation stone upon which this study is built. In view of the many uncertainties, it does not seem particularly profitable to consider sensitivities to a higher or lower figure, but it would be very useful to subject the study to critical review every few years in the light of developments and new knowledge.

Starting with this global figure of 1650 BBO Ultimate Recovery, we subtract the Total Discovered of 1434 BBO to derive a figure of 216 BBO for the Undiscovered potential. Rounded, the figures become 1450 discovered and 200 undiscovered.

Despite the scepticism regarding exploration expressed above, this figure does imply a considerable future exploratory success. Put in other terms, it means the discovery of more than 2000 hundred million barrel fields, which would be no mean achievement.

Starting with this global figure, the problem is reduced to allocating a reasonable share to each of the producing and potentially producing countries. That has been achieved by viewing the production record, judging the maturity and geological prospects of each country, and making some, hopefully intelligent, guess as to the Undiscovered potential. Certain patterns and relationships emerge from comparing the ratios of Discovered to Undiscovered and of Cumulative Production to Reserve, and build a degree of confidence. Even so, the figure should be considered as, at best, close to the mid-point of a range of possibility.

In fact, in undertaking this exercise, we fell short of reaching the 216 BBO total. It would have been possible to revise the figures upwards to achieve the desired total, but instead we felt it better to attribute the balance to what is termed "New Provinces". By this is meant some new prolific provinces that remain to be discovered. They could comprise an as yet untested deepwater trend, or a territory such as the Barents Sea (although that is now seriously downgraded by drilling), East or North Greenland, the Falklands Arc, Antarctica or parts of the Soviet Union, including for example its Arctic offshore.

In addition to assessing the amount of oil to be discovered in New Provinces, it is necessary to make some assumptions as to when it might be found. In this study it is arbitrarily assumed that three major discoveries will be made at five year intervals between 1995 and 2005, leading to a peak production from new provinces of 2.7 million barrels per day by 2023.

Not many major new provinces have been found in recent years, and we anticipate a future downward revision of this category.

Figure 14 shows the distribution of the Undiscovered potential resulting from this allocation.

The highest potential is attributed to the Middle East Six, which is held to have about 30% of the total. We have some misgivings about this because we fear that the history of Giant and Super-Giant discovery may be distorting judgment. As has been stressed, Giants are found early. Most of the Middle East has been extensively explored, and much of it is characterized by rather straightforward geology. Take for example the case of Saudi Arabia : almost half of its Ultimate comes from only two fields, Safaniya and Ghawar,

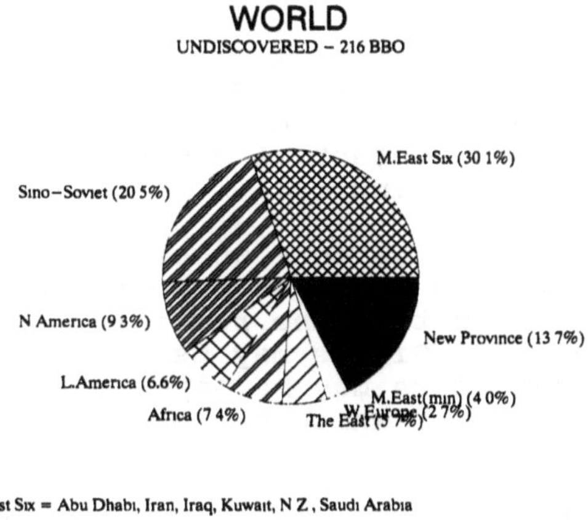

Figure 14. Undiscovered potential by Region.

found more than 40 years ago on an exceptionally prospective and easily identifiable geological arch. The remaining untested areas are much less prospective. No doubt, more discoveries will be made, but they are likely to be smaller by several orders of magnitude.

The greatest potential seems to be along the mountain front of Iran and Iraq, where more complex geology may still hold relatively large secrets. Furthermore, boundary disputes offshore may have left some tracts under-evaluated.

Attention has already been drawn to the huge reserve revisions, claimed by some of these countries, which seem suspect or partly suspect. It is asking a great deal to accept that not only have the historical reserves been increased by huge amounts but that there is also a colossal undiscovered potential.

But, notwithstanding these reservations, we do for the present attribute a large proportion of the World's Undiscovered to the Middle East.

The second most important region with perceived Undiscovered potential is the Sino-Soviet bloc, to which is attributed 21% of the World's Undiscovered, with 35 BBO coming from the Soviet Union and 6.5 BBO from China (respectively 18% and 15% of their assessed Ultimate endowments). This region is a great Unknown. Some commentators have high hopes for new exploration in these territories, which are vast, and we must take note of the remarkable Tenghis Field with reported reserves of 10-18 BBO. It was found as recently as 1979, although it is not yet under development partly because it is burdened by a high H_2S content. The Soviet Arctic is another promising area, with large undeveloped discoveries reported on the Yamal Peninsula, although burdened by a hostile environment, and a high gas content. A recent paper by Khain and others (1991) gives an indication of Soviet potential.

Despite this promise, the more accessible and prospective tracts in the Soviet Union have been thoroughly and systematically explored, so that the remaining potential for the critical discovery of Giants may be limited. Some commentators suggest that there may be large discovered but undeveloped accumulations which are not included in the official reserve figures.

China seems on the whole to be less promising : certainly, the offshore which was opened to foreign companies has been a disappointment.

No doubt, further discoveries will continue to be made in other parts of the World, and there may be some good surprises. We think that the deepwater potential is generally limited for geological rather than technical reasons, but Brasil is probably not the only deepwater basin to yield giant discovery. The deep Gulf of Mexico, for example, looks promising in geological terms.

REMAINING OIL

The sum of the Reserves and the Undiscovered gives what remains to be produced, namely 1019 BBO, of which 21% is from the Undiscovered category. The Remaining Oil amounts to 62% of the Ultimate, so the World has yet to reach the Mid-Point of its depletion. That would be comforting, were it not for its most uneven distribution.

The distribution of Remaining Oil is given in Figure 15. 54% lies in the Middle East Six, 13% in the Sino-Soviet bloc, and 10% in Latin America, mainly in Venezuela and Mexico.

It is useful here to introduce the concept of Depletion Rate, which is the percentage of what remains that is produced each year. It is a more useful concept than the commonly used Production to Reserve (P/R) ratio, which does not address at all the finite nature of the resource, and is also somewhat misleading because it tends to remain stable because falling reserves are usually matched by falling production.

We do not specifically predict when the Undiscovered will

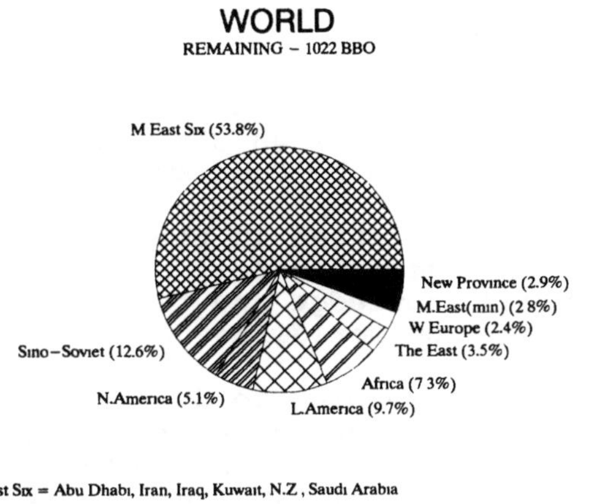

Figure 15. Remaining Oil (Reserves + Undiscovered) by Region.

be brought in, but assume that most will be found early.

ULTIMATE

The term Estimated Ultimate Recovery, or EUR, is commonly used for the total endowment of an area. In other words, it is the Cumulative Production when nothing more remains to be produced.

The question of definition has already been discussed, and it was pointed out that here we refer only to conventional oil as produced in oilfields under conditions as we know them to-day. To speak of Ultimate leads us to the far future, when conditions very different from those we know to-day will almost certainly obtain. So, in a way, we beg the question of definition when speaking of the Ultimate.

The basis for determining the Ultimate has already been reviewed in connection with the Undiscovered : namely, by using the trend of expert estimates that have been published over the years.

Many people discount the validity of Ultimate estimates with the comment :

> "My father said we were running out of oil 30 years ago, and look, we are producing more than ever".

Indeed, there have been many false alarms in the past when supply shortfalls were anticipated. Such fears prompted the

U.S. companies to move overseas, and they were of particular concern in the context of national defense.

Theories have even been proposed to explain the observed lag-time between the date of an Ultimate Estimate and the date when it is overtaken by Cumulative Production.

This scepticism was justified in the past, but more weight can now be given to the estimates, as they have the benefit of greatly improved knowledge and technology, and a hugely expanded data base from worldwide drilling.

Furthermore, it has something to do with the stage of depletion. The depletion of any finite resource follows a bell-shaped curve. Production expands before the Mid-Point and declines afterwards, as discussed further below. This curve may be indirectly reflected in the estimates of Ultimate as there is a parallel curve of knowledge: first of the expanding resource base but later of its depletion. The lag-time after the Mid-Point may be in the opposite sense : over-stating rather than under-stating the Ultimate.

The component parts of the Ultimate were illustrated in Figure 7, and the distribution is shown in Figure 16. As

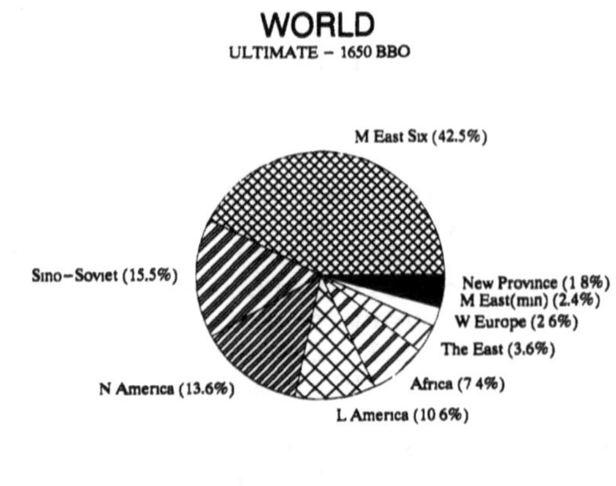

Figure 16. The Ultimate recovery by Region.

much as 42% lies in the Middle East Six countries; 16% in the Sino-Soviet Region ; 14% in North America ; and 11% in Latin America, with no more than 17% in the rest of the World. The extremely uneven distribution is self-evident : it is largely due to the exceptional and rare geological conditions responsible for prolific oil generation.

This concludes the discussion of the six elements : Cumulative Production, Reserves, Total Discovered, Undiscovered, Remaining and Ultimate. They are the ingredients of World Resource studies.

Now, we turn to the record of production, and to the forecasts of production to come, which in turn leads to considerations of price.

PRODUCTION

Attention has already been drawn to the bell-shaped depletion curve, a concept which was applied to oil production by the distinguished American petroleum geologist, M. King Hubbert, some fifty years ago. It provides a mathematical model that explains how the production of a new resource expands until a peak is reached, when about half has been consumed, whereupon it contracts due to supply constraints. North (1985) gives a good review of this and other procedures.

The half-life or Mid-Point of the full cycle of depletion is therefore of critical importance marking a fundamental change in the pattern of production as increase is replaced by decrease.

In the real world, the inflexion point does not necessarily come exactly when 50% of the Ultimate has been produced, the peak being determined by all sorts of short-term factors, but it is nevertheless a valid approximation. Most countries are to-day too close to the inflexion point for it to be evident, but several countries are now sufficiently mature to confirm the underlying pattern.

The Mid-Point inflexion point is used here as a basis for predicting future production, applying different procedures for different circumstances, as follows:

Post Mid-Point Countries

In countries that are at an advanced state of depletion, way past their Mid-Point, it is assumed that their Remaining Oil will be depleted at the present (1989) Depletion Rate. The procedure is well illustrated by the case of Germany. It had produced 50% of its 2.25 BBO Ultimate already in 1974, with 73% having been produced by the 1989 reference date. 1989 production, at 75.9 M b/d, is 4.36% of the 0.599 BBO Remaining. So, future production is assumed to decline at a rate that reduces the Remaining by 4.36% each year.

We may note in this connection the different patterns of countries with onshore, as opposed to offshore, production. Onshore production tends to rise and decline gradually, whereas offshore, the curves are much steeper, with generally high terminal Depletion Rates. The United Kingdom is an extreme example with a 1989 Depletion Rate of 11.3% : 62% of its Ultimate having been produced, no more than 15 years after production was established. It reflects the nature of offshore infrastructure and economics, where the high cost of platforms gives an incentive to achieve a plateau of high production as soon as possible and to hold it for as long as possible.

Pre Mid-Point Countries

It is more difficult to predict production for countries that have not yet reached their Mid-Point, and have the capacity, at least in resource terms, to increase their production until their Mid-Point is reached. There are wider options in these cases. After having addressed the problem in several countries in this study it became evident that it was reasonable to assume a 5% annual increase for countries approaching their Mid-Point. Higher annual increases apply to countries commencing production, and they have to be treated on an individual basis. Once the Mid-Point is reached, production is expected to decline at the Depletion Rate then in effect.

Naturally, there will be short term departures from these ideal production curves for any number of reasons, but the important point to remember is that for any period when production lies above the line there must be a corresponding period below the line, given the constraint of a finite Ultimate figure.

The special case of the Middle East Six

Lastly is the special case of the Middle East Six (Abu Dhabi, Iran, Iraq, Kuwait, Neutral Zone and Saudi Arabia) which are at such an early stage of their Depletion that they will not be resource constrained at all until far into the next century.

These countries are treated as swing producers making up the difference between World demand and what is produced by the other countries. It is further assumed that each of the Middle East Six countries will continue to produce the same proportion of the region's total as was the case in 1989 until it reaches the Mid-Point of its Depletion. Thereafter production is assumed to decline at the then Depletion Rate in that country, with the balance being made up by Saudi

Arabia. The consequences of the recent war will almost certainly impose important departures from the 1989 distribution, with Iraq and Kuwait not being expected to return to full production for at least a year. Assumptions about the distribution of production within the Middle East Six are thus due for revision when the situation stabilizes.

To derive the contribution of the Middle East Six, it is, accordingly, first necessary to address the subject of World demand.

Predicting World demand is essentially outside the scope of this study, but we believe that the actual will probably be within the confines of three alternative scenarios that are outlined below:

Scenario-1

Scenario-1 assumes that World production will rise at 2% per annum, the average of the past five years, until the World reaches the Mid-Point of its depletion, when it will fall at the then Depletion Rate due to resource constraints. This scenario basically assumes the continuation of the price environment of the past five years that created the current trend of demand. That in turn implies that the Middle East Six will not use their increasing share of the world market to impose radically higher prices.

Scenario-2

Scenario-2 assumes that the Middle East Six will raise prices when they are in a position to do so, which is likely to occur when they supply more than about 30% of the World's needs. Accordingly, Scenario-2 matches Scenario-1 with a 2% annual increase until the 30% threshold is reached. Price rises are then expected to hold production approximately level until the World's Mid-Point is reached, when it declines at the then Depletion Rate due to resource constraints.

Scenario-2 is considered the more likely one, and unless otherwise stated, the figures and discussion relate to it.

Scenario-3

A third Scenario assumes that production stays at the 1989 level until the World Mid-Point is reached, and thereafter declines at the then Depletion Rate. Such a scenario might apply if the present recession is prolonged or if the Middle East situation deteriorates further.

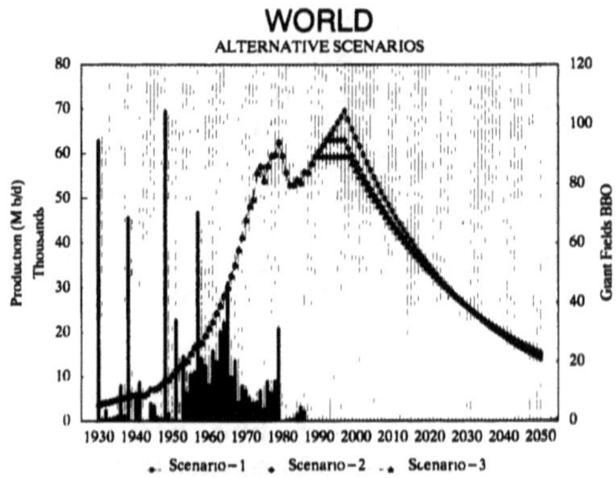

Figure 17. Production Forecasts : alternative scenarios.

Scenarios-1,-2 and -3 are illustrated on Figure 17, which also shows the distribution of giant discovery (1930 includes the Cumulative giant discovery to that year). Figure 18 shows the production by region, based on Scenario-2

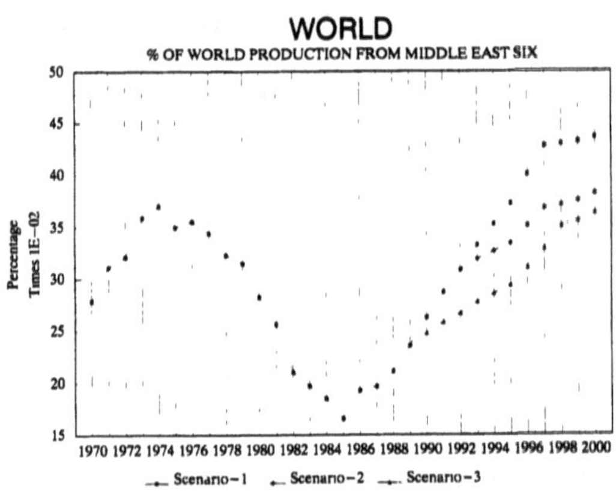

Figure 18. Production profiles by Region (Scenario-2)

THE IMPORTANCE OF THE MIDDLE EAST SIX

As already discussed, the Middle East Six countries effectively control the World's destiny in terms of oil production. We therefore give more emphasis to it than to the political grouping of countries making up OPEC. That organization has endeavoured to control oil prices. It tends to lack self-discipline when there is excess supply,

and when prices are weak, but it may well regain strength in times of shortage and higher prices. The position of the Middle East Six, and especially Saudi Arabia, is in any event critical, and in large measure controls the direction of OPEC. While the Middle East Six is identified as a geopolitical region in terms of its high resource base, it is far from having economic cohesion, due to the disparate ethnic, religious and political composition of its members.

An alternative approach would have been to include Venezuela, Mexico, Libya and Nigeria with the Middle East Six because they are also far from the mid-points of their depletion curves. But on balance, it was considered preferable to treat the Middle East Six separately from the other major producers.

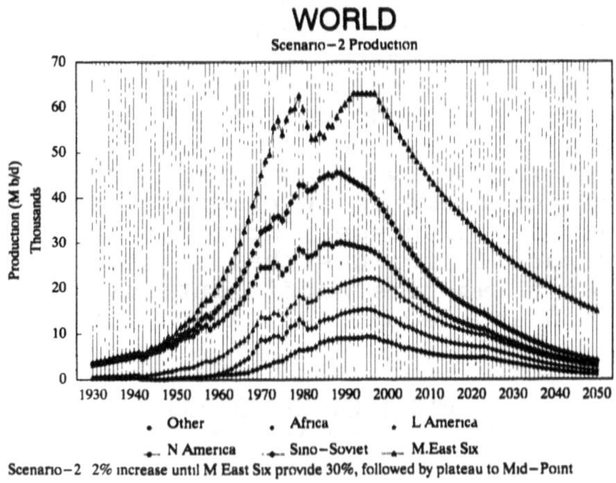

Figure 19. Percentage of World production from Middle East Six

Figure 19 shows the percentage of World supply coming from the Middle East Six under the three scenarios. In historical terms their share rose from 16% in 1950 to a peak of 37% in 1974, before sinking to a low of 17% in 1985. Since then the share has been rising, and is expected to continue to do so in the absence of any new alternative source of major supply. The following table summarizes the three scenarios, proposed here:

Table 8. Middle East Six share of World Production

Scenario	-1	-2	-3
30%	1992	1992	1996
35%	1994	1996	1998
40%	1996	2002	2003
At World Depletion Mid-Point	1997	1997	1998

It will be recalled that production is assumed to decline at the Mid-Point of the World's depletion at the Depletion Rate then in effect, due to resource constraints.

The long term historical price of oil is given in Figure 20, and the correlation with the percentage coming from the Middle East Six is illustrated in Figure 21. The First Oil Shock in 1973 shows a clear correlation. But the Middle East Six's share was already declining at the time of the Second Oil Shock in 1979, and it is easy to see why the high prices then imposed were not sustainable. Now a new trend of increasing share is in progress. The resource constraints, already discussed, strongly suggest that it is likely to continue. It follows that prices are likely to reflect it after a lag-time sufficient for the position to become understood and recognized. A 30% threshold may well be critical.

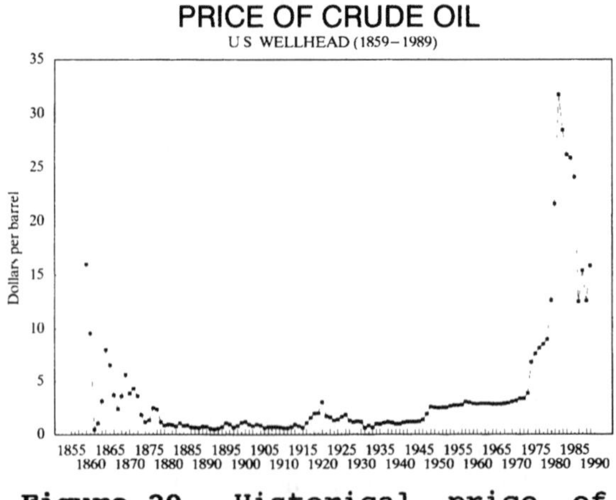

Figure 20. Historical price of oil.

Scenarios are no more than hypotheses that consider the consequences of various assumptions, and history rarely is so accommodating as to unfold in accordance with any preconceived scenario.

The scenarios proposed here have both strengths and weaknesses. Their weakness is that they are generalized, and do not take into account the details of actual planned field developments in each of the countries. There is certainly scope to improve the near-term forecasts by adding such information. Their strength, on the other hand, is that they are resource constrained. Thus, if higher than predicted short-term production levels are achieved, that will simply advance the Mid-Point and lead to a steeper subsequent decline. Likewise, present production levels cannot be maintained for very long without giving rise to an unsustainably steep subsequent decline.

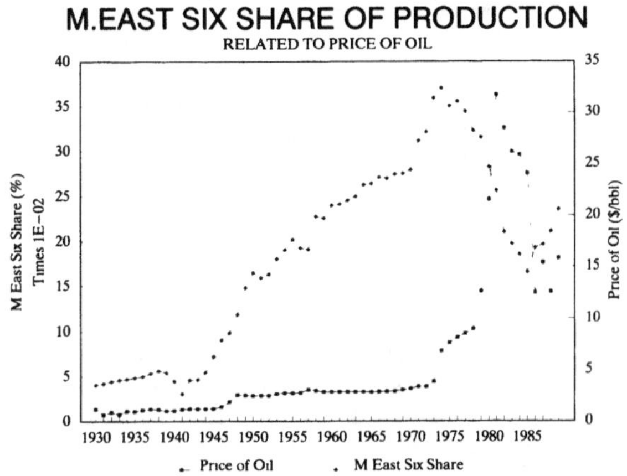

Figure 21. Price of oil compared with share of production coming from the Middle East Six

This is precisely the value of the model that touches the heart of the subject. With a finite resource the faster production rises and the longer it is maintained, the steeper will be the subsequent decline and the shorter its life. Production is governed by a fundamental bell-shaped curve : square bells do not ring true ! There is a price to be paid for excess.

These are very important constraints to build into any economic model of the future. Indeed, they dwarf the recognized imperfections in the data-base and the uncertainties of the scenarios.

WORLD DISTRIBUTION

Having explained the ingredients of the analysis, we may now summarize the distribution of oil between the World's regions.

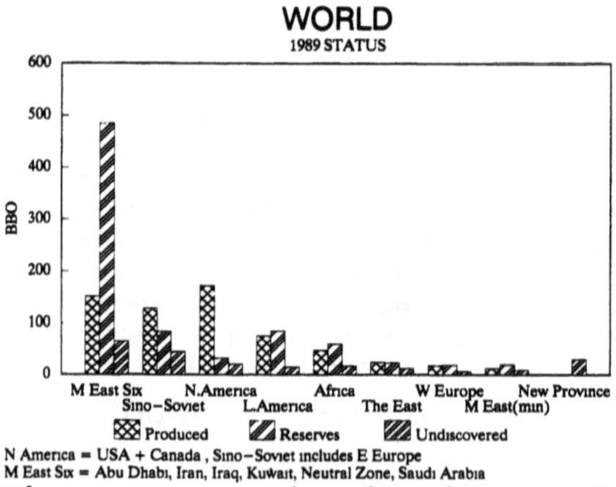

Figure 22. Distribution of Produced, Reserves and Undiscovered by Region.

BASIC ELEMENTS

Figure 22 shows the distribution of the three basic elements : Produced, Reserves and Undiscovered. The enormous reserves of the Middle East Six are immediately evident, as by contrast is the fact that North America has already produced most of its oil. The relatively small contribution of the Undiscovered is also noteworthy.

Remaining Oil and Depletion Rate

Figure 23 shows the distribution of Remaining Oil, namely Reserves plus Undiscovered. Again, the dominance of the Middle East Six is very evident, contrasting with the position of North America.

Depletion Rates show the opposite pattern : the lowest rate is in the Middle East Six, at less than 1% ; and the highest are in North America and Western Europe, where the Rates are as much as five times higher.

Maturity and Undiscovered Potential

Figure 24 illustrates Maturity, as measured by the proportion of the Ultimate that remains to be discovered. North America is at the most mature level. Most of the

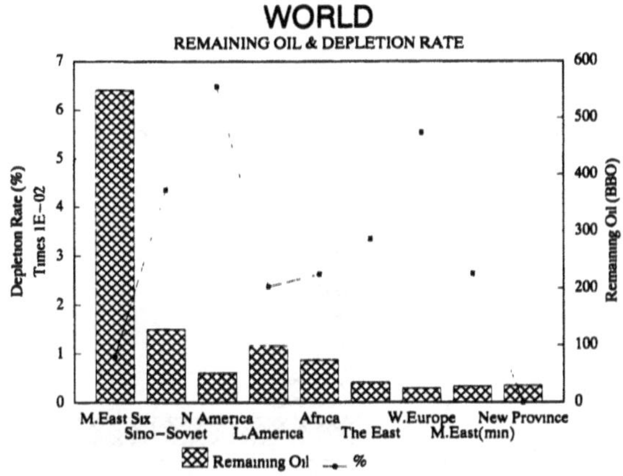

Figure 23. Remaining Oil (Reserves+Undiscovered) and Depletion Rates by Region

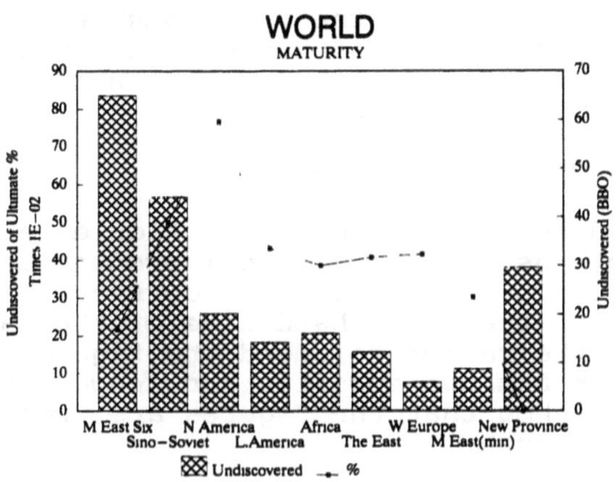

Figure 24. Maturity and Undiscovered Potential

other regions are broadly comparable.

Years to Depletion Mid-Point

Figure 25 illustrates the number of years to or from the Depletion Mid-Point for the several regions. The contrast between the Middle East Six and North America is striking. As discussed, the recognition of the Depletion Mid-Point is an important element of the study.

Lastly, the overall position is summed up in two tables : Table 9 is a Summary Data Sheet showing the distribution of

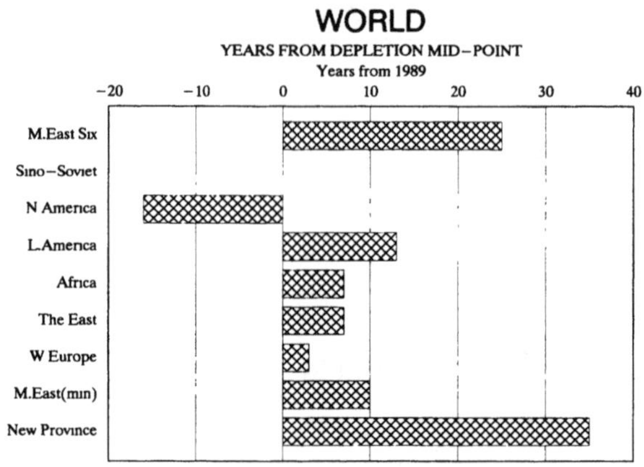

Figure 25. Number of years to or from the Mid-Point of Depletion.

the several parameters and the relationships between them for the nine regions into which the World is divided ; and Table 10 is a production profile for Scenario-2, which is considered the more likely case.

THE IMPORTANCE OF REVISION ANALYSIS

As was stressed in the Preface, "all figures are wrong". It is obvious that the forecasts of oil production given in Part 3 will not prove to be exactly correct. Indeed, the "Notes" column on the right hand side of the tables is there to allow the reader to amend the figures either from new data or from his own specific knowledge. The important thing is not to show that the figures are wrong but to determine by how much.

In the same way as important conclusions were drawn from the trend in the historical evolution of estimates of the Ultimate Recovery, already discussed, so this model too is subject to evolution as the patterns of production and discovery unfold in the future. The trends of revision are likely to mean more than the figures themselves. Thus, if a country has consistently higher production than anticipated, it might be reason to increase its assumed Ultimate, unless there is some other short term explanation.

1990 figures became available during the course of the study, and provide the first test of the model. Figure 26 illustrates the variance between the prediction and the actual. World Production was 0.44% below prediction. Lower than anticipated production from the Middle East Six was

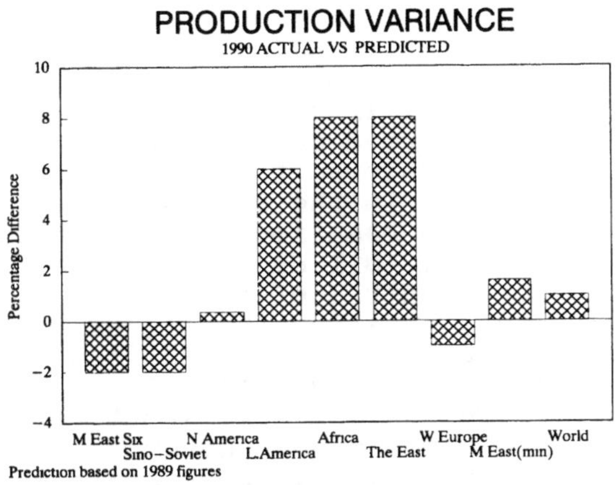

Figure 26. Production variance : 1990 actual vs. forecast.

offset by higher than anticipated production from Latin America, Africa and the East. This very likely reflects the Iraq conflict. Sino-Soviet production was also lower than predicted, perhaps due to the difficult politico-economic situation in the U.S.S.R.

CONCLUSIONS

The distribution of oil in time and place has been discussed in the preceding pages. The weakness of the data-base has been stressed: there are difficulties in definit-ion ; and considerable uncertainties affect the estimates of the Ultimate and Undiscovered Potential.

The pattern of Depletion is also open to discussion.

Despite these uncertainties, we are reasonably confident that the resource base depicted here is correct in order of magnitude. Of critical importance is recognition of the fundamental validity of the bell-shaped full cycle deplet-ion curve. One can argue about the exact size and shape of the curve, but the general picture can now be clearly seen. It shows that the World is indeed approaching the Mid-Point in the depletion of its oil resources : the epoch of increasing production is almost over ; and the epoch of declining production is about to begin.

Oil has been a story of boom and bust, which indirectly reflects the geological distribution of oil and the natural laws of its depletion. The problems of marketing flush production from new giant discoveries led the industry to structure itself in a way that solved that difficulty but it could not at the same time meet the needs of conservation that would inevitably follow the flush production. Efforts were made to regulate supply and provide an orderly market : the Standard Oil Trust ; the Achnacarry agreement of 1929 ; the Texas Railroad Commission and OPEC. But control was no sooner achieved than it became decried as monopolistic, especially in the United States. Strategic considerations have also been important as nations feared for the security of their oil supply. As the producing countries gained more control of their oil, they pressed for increased revenue, sometimes by higher prices, but often by higher production. Lastly oil itself created its own demand by changing the way people lived : for example commuting suburbia. The essentially short-term landlord-tenant relationships between host-government and concessionaire was not conducive to conservation, nor was the competitive environment between the companies themselves.

All of these things were made possible by the fact that the resource base was growing during the first half of the "Golden Century". As the World passes the Mid-Point of depletion during this decade, it will come to face an entirely new situation characterized by a dwindling resource base. That must surely provide an incentive for slowing the pace of depletion by curbing consumption.

Future generations will likely look back and see this inflexion point on the curve as one of the great turning points in history, considering the dominant position that oil based energy has held in the economy of the remarkable Twentieth Century.

REFERENCES

British Petroleum Co, 1989, BP Statistical Review of World Energy.
--------------------, 1990, BP Statistical Review of World Energy.
Hubbert, M. King, 1982, Techniques of prediction as applied to the production of oil and gas. In NBS Special Publication 631 - Oil & Gas Modelling, Ed. Saul I. Gass, US Dept. of Commerce, Nat. Bureau of Standards
Ivanhoe, L.F., 1990, Soviet oil production drop may be major, lasting global supply woe. Oil & Gas Journ., Sept 3 1990 p. 69-71.
------- L.F., 1988, Future crude oil supply and prices. Oil & Gas Journ., July 25, 1988, p.111-112.
------- L.F., 1980, World's giant petroleum provinces. Oil & Gas Journ. June 30, 1989. p.146-148
Khain, V.E., Sokolov, B.A., Kleshchev, K.A., Shein V.S., 1991, Tectonic and geodynamic setting of Oil and Gas basins of the Soviet Union. AAPG bull. v.75/2 p.313-325
Klemme, D. & Ulmishek,G., 1989, Depositional controls, distribution and effectiveness of World's petroleum source rocks. AAPG bull. v 73/3 pp.372-373
Masters, C, 1987, Potential of known, possible major oil provinces. OPEC Review, Summer 1987.
Nehring, R, 1979, The Outlook for Conventional Petroleum Resources, Paper P-6413, 21p. Rand Corp.
North, F.K., 1985, Petroleum Geology. 607p. Publ. Allen & Unwin, Winchester, Mass. USA
Oil & Gas Journal, 1989, Worldwide oil and gas at a glance. OGJ Special Dec. 25, 1989
-----------------, 1990, Worldwide oil and gas at a glance. OGJ Special Dec. 31, 1990
Pickens, T. Boone, Jr., 1987, Boone. Publ. Houghton Miffin, Boston.
Roadifer, R.E.,1986, Size distribution of world's largest, known oil, tar accumulations. Oil & Gas Journ., Feb. 24, 1986, p.93-98.
Yergin, D., 1990, The Prize - The epic quest for oil, money and power. 877p. Publ. Simon and Schuster, New York.

SUMMARY DATA SHEET WORLD
REFERENCE DATE 1989

Country	Production M b/d	BBO/ year	Cumulative Production	Reported Reserves	Adjusted Reserves	Discovered	Undiscovered	Ultimate	Remaining	Und. of Ult.	Prod of Ult.	Rem. of Ult.	Prod of Disc	Disc. in Giants	Dep. Rate
A	B	C	D	E	F	G	H	I	J	K	L	M	N	O	P
M.East Six	13998	5.109	152	639	484	636	65	701	549	9%	22%	78%	24%	77%	0.93%
Sino-Soviet	15279	5.577	128	84	84	212	44	256	128	17%	50%	50%	60%	55%	4.35%
N.America	9264	3.381	172	32	32	204	20	224	52	9%	77%	23%	84%	31%	6.48%
L.America	6451	2.355	76	125	85	161	14	175	99	8%	43%	57%	47%	64%	2.37%
Africa	5372	1.961	47	59	59	106	16	122	75	13%	39%	61%	44%	57%	2.62%
The East	3234	1.180	24	23	23	47	12	60	35	21%	41%	59%	51%	43%	3.35%
W.Europe	3764	1.374	18	19	19	37	6	42	25	14%	42%	58%	48%	54%	5.54%
M.East(min)	2033	0.742	12	20	19	32	9	40	28	21%	30%	70%	39%	52%	2.64%
New Province	0	0	0	0	0	0	30	30	30	100%	0%	100%	ERR	ERR	0.00%
Region	59395	21.679	628	1002	806	1434	216	1650	1022	13%	38%	62%	44%	62%	2.12%

CUM.PROD	628.1 BBO	GIANT FIELDS			WORLD	
RESERVES	805.9 BBO	TOTAL BBO	887		TOTAL	
DISCOVERED	1434.0 BBO	% OF DISC.	62%		Scenario-2	
UNDISCOVERED	216.0 BBO	Last Giant	1986			
REMAINING	1021.9 BBO				Variance	Trend
ULTIMATE	1650.0 BBO				1990	+1%
Rept.Reserve	1001.5	Ref.date(end)	1989		Future	n/a

Milestones	Date	Prod Mb/d	Dep. Rate	% Produced	Forecast n/a
					A=Exponential decline
1st Peak	1979	62638	1.83%	26%	B=5% incr. to Mid-Point
2nd Peak	1998	61271	2.67%	51%	C=Other
Mid-Point	1998	61271	2.67%	51%	Pre-1930 Cum. 18.977
Final Decline		n/a			(Added to 1930 prod)

Date	Prod. M.b/d	Prod. BBO/yr	Cum. Prod.	Remain	Deplet. Rate	% Produced	Giant	Notes & Names
1930	3712	1.355	20	1630	ERR	1%	94.53	
1931	4017	1.466	22	1628	0.09%	1%	0.76	
1932	4190	1.529	23	1627	0.09%	1%	3.71	
1933	4379	1.598	25	1625	0.10%	2%	0.81	
1934	4646	1.696	27	1623	0.10%	2%	1.27	
1935	4855	1.772	28	1622	0.11%	2%	1.79	
1936	5066	1.849	30	1620	0.11%	2%	12.17	
1937	5258	1.919	32	1618	0.12%	2%	1.96	
1938	5418	1.977	34	1616	0.12%	2%	68.59	
1939	5697	2.079	36	1614	0.13%	2%	0.00	
1940	5860	2.139	38	1612	0.13%	2%	8.32	
1941	6070	2.215	41	1609	0.14%	2%	13.30	
1942	5734	2.093	43	1607	0.13%	3%	0.00	
1943	6169	2.252	45	1605	0.14%	3%	0.00	
1944	7076	2.583	48	1602	0.16%	3%	6.27	
1945	7101	2.592	50	1600	0.16%	3%	5.23	
1946	7518	2.744	53	1597	0.17%	3%	1.50	
1947	8253	3.012	56	1594	0.19%	3%	1.03	
1948	9324	3.403	59	1591	0.21%	4%	104.53	
1949	9251	3.377	63	1587	0.21%	4%	2.68	
1950	10343	3.775	66	1584	0.24%	4%	0.52	
1951	11620	4.241	71	1579	0.27%	4%	34.30	
1952	12232	4.465	75	1575	0.28%	5%	0.00	
1953	13163	4.805	80	1570	0.31%	5%	22.18	
1954	13563	4.951	85	1565	0.32%	5%	9.51	
1955	15243	5.564	90	1560	0.36%	5%	15.80	
1956	16585	6.053	96	1554	0.39%	6%	16.62	
1957	17462	6.373	103	1547	0.41%	6%	70.41	
1958	17625	6.433	109	1541	0.42%	7%	21.39	
1959	19258	7.029	116	1534	0.46%	7%	19.11	
1960	20927	7.638	124	1526	0.50%	8%	12.30	
1961	22365	8.163	132	1518	0.53%	8%	23.96	
1962	24276	8.861	141	1509	0.58%	9%	20.38	
1963	26001	9.490	150	1500	0.63%	9%	30.34	
1964	28257	10.314	161	1489	0.69%	10%	33.48	
1965	30088	10.982	172	1478	0.74%	10%	46.93	
1966	32679	11.928	184	1466	0.81%	11%	15.10	
1967	35122	12.820	197	1453	0.87%	12%	20.64	
1968	38500	14.052	211	1439	0.97%	13%	6.41	
1969	41229	15.049	226	1424	1.05%	14%	12.11	
1970	45088	16.457	242	1408	1.16%	15%	10.52	
1971	48167	17.581	260	1390	1.25%	16%	7.91	
1972	49612	18.108	278	1372	1.30%	17%	6.09	
1973	55671	20.320	298	1352	1.48%	18%	6.66	
1974	57206	20.880	319	1331	1.54%	19%	10.32	
1975	54001	19.710	339	1311	1.48%	21%	4.29	
1976	57306	20.917	360	1290	1.60%	22%	13.50	
1977	59505	21.719	381	1269	1.68%	23%	9.50	
1978	59816	21.833	403	1247	1.72%	24%	13.61	
1979	62638	22.863	426	1224	1.83%	26%	31.53	
1980	59636	21.767	448	1202	1.78%	27%	0.00	
1981	55930	20.415	468	1182	1.70%	28%	0.55	
1982	53092	19.378	488	1162	1.64%	30%	0.71	
1983	53034	19.358	507	1143	1.67%	31%	1.00	
1984	54242	19.798	527	1123	1.73%	32%	2.60	
1985	53483	19.521	546	1104	1.74%	33%	4.85	
1986	55938	20.417	567	1083	1.85%	34%	3.49	
1987	55745	20.347	587	1063	1.88%	36%	0.00	
1988	57738	21.074	608	1042	1.98%	37%	0.00	

1989	59395	21.679	630	1020	2.08%	38%
1990	60583	22.113	652	998	2.17%	40%
1991	61795	22.555	674	976	2.26%	41%
1992	63030	23.006	697	953	2.36%	42%
1993	63030	23.006	720	930	2.42%	44%
1994	63030	23.006	743	907	2.47%	45%
1995	63030	23.006	766	884	2.54%	46%
1996	63030	23.006	789	861	2.60%	48%
1997	63030	23.006	812	838	2.67%	49%
1998	61271	22.364	835	815	2.67%	51%
1999	59635	21.767	857	793	2.67%	52%
2000	58042	21.185	878	772	2.67%	53%
2001	56493	20.620	898	752	2.67%	54%
2002	54984	20.069	918	732	2.67%	56%
2003	53516	19.533	938	712	2.67%	57%
2004	52087	19.012	957	693	2.67%	58%
2005	50697	18.504	976	674	2.67%	59%
2006	49343	18.010	994	656	2.67%	60%
2007	48026	17.529	1011	639	2.67%	61%
2008	46743	17.061	1028	622	2.67%	62%
2009	45495	16.606	1045	605	2.67%	63%
2010	44281	16.162	1061	589	2.67%	64%
2011	43098	15.731	1077	573	2.67%	65%
2012	41948	15.311	1092	558	2.67%	66%
2013	40828	14.902	1107	543	2.67%	67%
2014	39737	14.504	1121	529	2.67%	68%
2015	38676	14.117	1135	515	2.67%	69%
2016	37644	13.740	1149	501	2.67%	70%
2017	36639	13.373	1163	487	2.67%	70%
2018	35660	13.016	1176	474	2.67%	71%
2019	34708	12.669	1188	462	2.67%	72%
2020	33782	12.330	1201	449	2.67%	73%
2021	32880	12.001	1213	437	2.67%	73%
2022	32002	11.681	1224	426	2.67%	74%
2023	31147	11.369	1236	414	2.67%	75%
2024	30316	11.065	1247	403	2.67%	76%
2025	29506	10.770	1257	393	2.67%	76%
2026	28718	10.482	1268	382	2.67%	77%
2027	27952	10.202	1278	372	2.67%	77%
2028	27205	9.930	1288	362	2.67%	78%
2029	26479	9.665	1298	352	2.67%	79%
2030	25772	9.407	1307	343	2.67%	79%
2031	25084	9.156	1316	334	2.67%	80%
2032	24414	8.911	1325	325	2.67%	80%
2033	23762	8.673	1334	316	2.67%	81%
2034	23128	8.442	1342	308	2.67%	81%
2035	22510	8.216	1351	299	2.67%	82%
2036	21909	7.997	1359	291	2.67%	82%
2037	21324	7.783	1366	284	2.67%	83%
2038	20755	7.576	1374	276	2.67%	83%
2039	20201	7.373	1381	269	2.67%	84%
2040	19661	7.176	1388	262	2.67%	84%
2041	19136	6.985	1395	255	2.67%	85%
2042	18625	6.798	1402	248	2.67%	85%
2043	18128	6.617	1409	241	2.67%	85%
2044	17644	6.440	1415	235	2.67%	86%
2045	17173	6.268	1422	228	2.67%	86%
2046	16715	6.101	1428	222	2.67%	87%
2047	16268	5.938	1434	216	2.67%	87%
2048	15834	5.779	1439	211	2.67%	87%
2049	15411	5.625	1445	205	2.67%	88%
2050	15000	5.475	1450	200	2.67%	88%

Note
This scenario assumes World production rises at 2% (last 5 yr av.) until 50% of World ultimate is depleted. The Middle East Six producers make up difference between World demand and other countries' production. In 1998, when Middle East Six has produced 46% of World production, it is assumed to decline at 2.63% (then Depletion Rate), having by then produced 50% of its Ultimate.

PART 2

REGIONAL ASSESSMENT

KEY TO LAYOUT

The following regions are recognized, with a reference abbreviation shown:

Middle East Six	ME6
Middle East (Minor)	MEm
Sino-Soviet	S-S
North America	NA
Latin America	LA
Africa	AF
The East	E
Western Europe	WE
New Provinces	NP

Each Figure carries a suffix according to the key indicated above:

Explanation of Tables and Figures

The illustrations follow a standard format for each region as follows :

Table 1 - Production Profile

This is an important element of the study whose elements comprise the following :

1.CUM.PROD : Cumulative Production.
2.RESERVES : Reserves (as adjusted).
3.DISCOVERED : (1+2)
4.UNDISCOVERED : (6-3)
5.REMAINING : (2+4)
6.ULTIMATE : Estimate of Cumulative Production at the end of production.
7.Rept.Reserves : Reported Reserves (mainly from O&G Journal).
8.Milestones
  8a. 1st Peak : Date of first peak in production.
  8b. 2nd Peak : Date of second peak (if any).
  8c. Mid-Point : Date when 50% of the Ultimate has been produced.
  8d. Final Decline : Depletion Rate after Mid-Point.
  8e. Dep.Rate : Annual Production as a percentage of Remaining Oil at the end of the previous year.
  8f. % Produced : Percentage of Ultimate produced.
9.GIANT FIELDS : Fields with Ultimate Reserves of >500 M Barrels.
  9a. % OF DISC. : Percentage of Discovered Oil (3) in Giant Fields.
  9b. Last Giant : Date of last giant discovery.
10. Ref. date (end) : Reference Date (end 1989) to which the figures apply.

11. COUNTRY
12. REGION
13. Scenario-1 : Reference to Production Scenario.
14. Variance Trend
    14a.1990 : Percentage difference between actual and predicted production (thus +4% means actual 4% higher than predicted).
    14b.Future : Anticipated trend of future revision.
15. Forecast
    15a. "A" : Exponential production decline at (1989) depletion rate.
    15b. "B" : Production increases at 5% to Depletion Mid-Point and then declines at then Depletion Rate.
    15c. "C" : Other profiles (mainly in case of countries with new production).
16. Pre-1930 CUM : Pre-1930 Cumulative Production.

The columns list the following :

A. Date : Date (1930-2050)
B. Prod M.b/d : 1930-1989 Production in M b/d as reported.
 : 1990-2050 Column C divided by 0.000365.
C. Prod BBO/yr : 1930-1989 Column B multiplied by 0.000365.
 : 1990-2050 Previous Year's Remaining in BBO (Column E) multiplied by Depletion Rate (Column F) or increased at 5% to the Mid-Point or as otherwise determined.
D. Cum.Prod. : Previous year's cumulative production (including pre-1930 figure) plus this year's production in BBO.
E. Remaining : Ultimate (6) less Cum. Prod (Column D).
F. Dep.Rate : Depletion Rate - Annual Production as a percentage of Remaining Oil at the end of the previous year. (Also fixed at Depletion Mid-Point, or at 1990 if that post-dates Mid-Point).
G. % Produced : Cum Prod (Column C) as a Percentage of Ultimate.
H. Giant : Ultimate recovery of giant fields attributed to date of discovery.
I. Notes & Names : see base of column for notes referenced by #1: names of giants, Milestones etc.

Table 2 - Summary Data Sheet

This table requires no special explanation, except to note that the figures in Columns K to O are percentages : thus, "Und. of Ult." means the percentage of the Ultimate that is Undiscovered.

INTRODUCTION

Each of the eight regions into which the World has been divided, together with the "New Provinces", is reviewed in this section. The regions are in some regards artificial groupings, made up of diverse countries often having little in common with each other. All the same, it is convenient to break up the world into more manageable units.

The North America Region is defined here as the United States and Canada, two countries which have had a similar oil experience. Mexico, which geographically is part of North America, is here included in Latin America with the countries in South and Central America, as well as the Caribbean.

The Sino-Soviet Region covers China, the USSR and the East European countries with the exception of the former East Germany, which is included in Western Europe. Although there are great and increasing differences between these countries, it remains convenient to treat them together in view of their Communist background, which has undoubtedly had a considerable influence on their oil development.

The Middle East is divided into two regions. The principal producers are termed the Middle East Six, which comprises Abu Dhabi, Iran, Iraq, Kuwait, Neutral Zone and Saudi Arabia. (The Neutral Zone is included, not because it is a major producer, but because it is administered jointly by Kuwait and Saudi Arabia). The rest of the Middle East countries are included in what is termed the Middle East (Minor) region.

Western Europe comprises countries from Norway to Turkey, and includes the important North Sea area.

The Africa Region, as the name implies, covers the African continent.

The East Region comprises the remaining countries, lying geographically between India and Japan, and including Australasia, with very diverse geo-political backgrounds.

The category "New Provinces" covers such new important producing areas as may be found, either in countries with existing production elsewhere within their frontiers or in countries, not yet producing oil, including Antarctica.

A brief regional summary of the salient points introduces a number of graphs and tables.

MIDDLE EAST SIX
(Prefix - "ME6")

SUMMARY

The principal producing countries of the Middle East, which occupy a unique position in the World of Oil are here distinguished as The Middle East Six. They comprise Abu Dhabi (part of the United Arab Emirates), Iran, Iraq, Kuwait and Saudi Arabia, to which the Neutral Zone, a territory administered by Kuwait and Saudi Arabia, may be conveniently added.

The Region dwarfs all others having 42% of the World's oil endowment.

Oil exploration commenced in Iran in the early years of the century with the establishment of what is now BP. The remaining countries had been part of the Ottoman Empire, which was broken up following the First World War and divided between French and British spheres of influence, with the United States later successfully exerting its claims. British and American interests have dominated the oil industry, often with political influence. Iraq was developed by the Iraq Petroleum Co (BP, Shell, CFP, Mobil, Exxon, and Gulbenkian) ; Kuwait by BP and Gulf, Saudi Arabia by Chevron (later joined by Texaco, Exxon and Mobil) through Aramco ; and the Neutral Zone by Getty and others. BP, Shell and CFP were also responsible for Abu Dhabi developments. The foundations were laid before the Second World War, but the main growth of production was between 1950 and 1970.

National companies have now taken over the rights of the foreign companies in all of these countries, except Abu Dhabi and the Neutral Zone, although Saudi Arabia still accords certain privileges to the foreign companies that previously owned Aramco.

The story of Middle East oil is well described by Yergin (1990). The most recent chapter involves the Iraqi invasion of Kuwait in August 1990, and the subsequent successful war to liberate Kuwait, which was led by the United States under a United Nations' mandate. Probably this will have a colossal impact on the future of oil in ways that are not yet easy to predict. Serious damage may have been done to Kuwait's reservoirs when the retreating Iraqis fired the wells, which are expected to burn for months, if not years, before they can be finally extinguished.

The Region's share of World oil is as follows :

Cumulative Production	24%
Reserves	60%
Discovered	44%
Undiscovered	30%
Remaining	54%
Ultimate	42%

The Region lies within the Persian Gulf province, which is blessed with a remarkable combination of geological circumstances that makes it the most prolific hydrocarbon basin on earth. In short, it forms a wedge of sediment between the Arabian Shield and the Zagros Mountains. This wedge has had a long history of gentle subsidence in which thick sequences of hydrocarbon source-rock and both carbonate and sandstone reservoirs were deposited and capped by effective evaporitic seals. In the west and central areas, huge gentle growth anticlines and domes developed, partly over salt cores, and were fed with hydrocarbons over long periods of time, whereas in the east more complex and markedly disharmonic structures developed in the mountain front of Iran and Iraq.

The main reservoirs in the west are Jurassic and Cretaceous limestones, with important Albian sandstones in Kuwait. In Iran and Iraq production comes mainly from Tertiary limestones. Most of the oil was sourced in the Upper Jurassic, although the Middle Cretaceous and the Lower Tertiary also contain important source-rocks.

In addition, there are deeper prospects in the Palaeozoic strata, principally for gas.

In geotectonic terms, the Persian Gulf lies on the contact between the African and Asian continents. The Arabian Plate, which belongs to the African continent, is being subducted below the Zagros Mountains. It is, in geological terms, a comparatively recent subduction, with the result that most of the prolific foredeep has not yet been consumed in the mountain building.

In short, the God of Oil, if there is one, has built an immaculate edifice to a grand scale. There is nothing to compare with it.

Ultimate

Saudi Arabia is by far the most important oil country in the region with an Ultimate endowment of 275 BBO. Kuwait (120 BBO), Iran (115 BBO) and Iraq (100 BBO) make up the second tier, followed by Abu Dhabi (80 BBO) and the Neutral Zone (11 BBO).

Reserves

It has been considered necessary to make major downward adjustments to the reported reserve figures for all the countries in the Region, with the exception of the Neutral Zone, for the reasons already discussed. (It is perhaps significant in this respect that the Neutral Zone is not a discrete political entity, and its reserve figures may not therefore be subject to political influence).

While it seems clear that the reported reserve figures require adjustment, it is very difficult to know by how much. The adjustments proposed here are very uncertain, but if accepted show that Saudi Arabia is the most important country with 200 BBO of reserves, followed by Kuwait with 94 BBO. Abu Dhabi with 65 BBO, and Iran and Iraq with 60 BBO each come next. The Neutral Zone's 11 BBO completes the list.

Undiscovered

Again, it is very difficult to know how to allocate the Undiscovered potential to this province. As discussed, it is at an advanced stage of maturity, and its prolific past may not be matched by comparable future discovery. Here, the greatest potential is attributed to Iraq with 21 BBO followed by Saudi Arabia and Iran with 17 BBO each. Kuwait, Abu Dhabi and the Neutral Zone are considered to have less than 10 BBO each.

To attribute more reserves or undiscovered potential to the Middle East Six is almost like saying that there is more water in the sea. Although the range of uncertainty involves huge amounts of oil, it does not affect the dominance of the region in World terms.

Production

Production commenced by the early years of the century in Iran, but it was not until the 1950s that the region began to assume its present dominance. The first peak (21,193 M b/d) occurred in 1974, which was followed by a decline to 8857 M b/d in 1985. Production is now expected to rise to a second peak in 1998 of 22,711 M b/d. The World will then have reached the Mid-Point of its depletion, and production is expected to decline thereafter, although the Middle East Six will not reach their Mid-Point until 2014.

The Scenarios behind these assumptions are discussed in Part 1. The region is not effectively resource constrained until well into the next century.

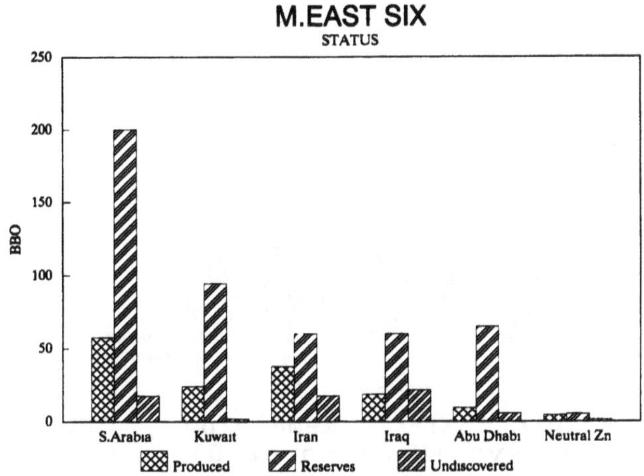

Adjusted Reserve (484 BBO) used in place of Reported Reserves (639 BBO)

Figure 1-ME6. Produced, Reserves and Undiscovered

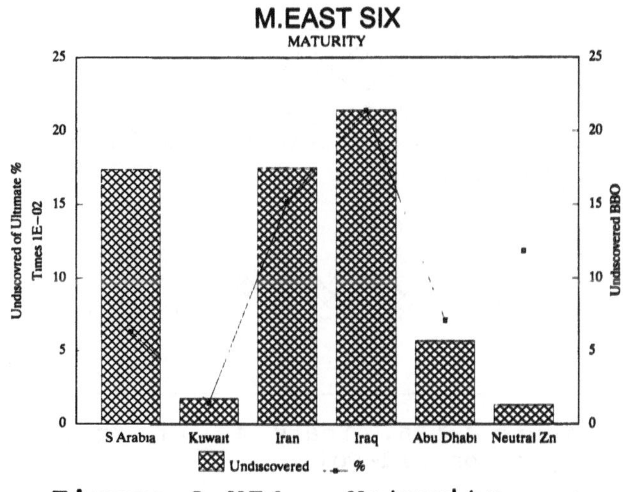

Figure 2-ME6 Maturity : Undiscovered as a percentage of Ultimate

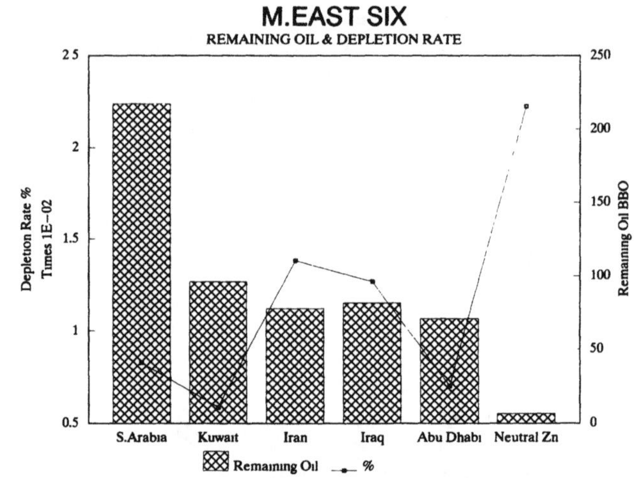

Figure 3-ME6 Remaining Oil (Reserves+Undiscovered) and Depletion Rates

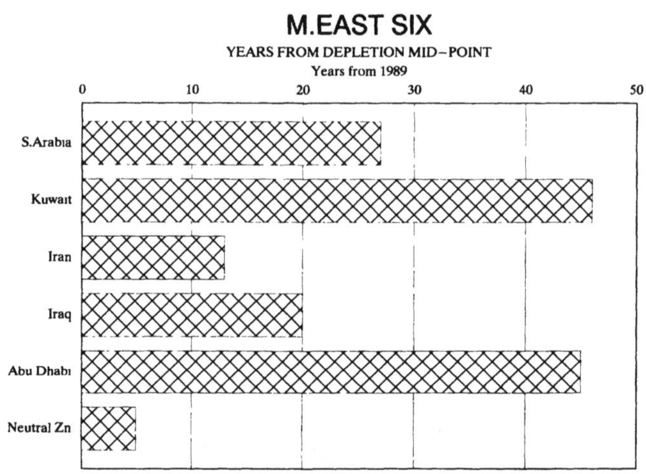

Figure 4-ME6. Years to or from Depletion Mid-Point.

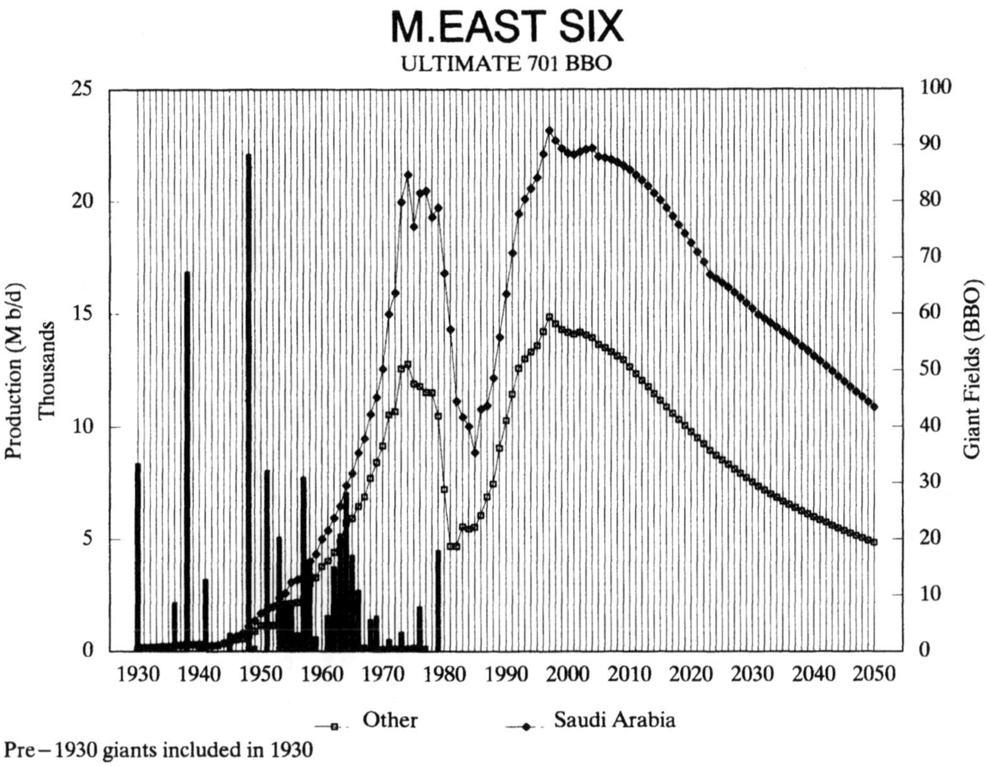

Figure 5-ME6. Production Profile 1930-2050 with Giant Fields

CUM.PROD	151.60 BBO	GIANT FIELDS		M.EAST (MAJOR)	
RESERVES	484.45 BBO	TOTAL BBO	489	REGION	
DISCOVERED	636.05 BBO	% OF DISC.	77%	Scenario-1	
UNDISCOVERED	64.95 BBO	Last Giant	1979		
REMAINING	549.40 BBO			Variance Trend	
ULTIMATE	701.00 BBO			1990	-2%
Rept.Reserve	639.27	Ref.date(end)	1989	Future	n/a

Milestones	Date	Prod Mb/d	Dep. Rate	% Produced	Forecast n/a
1st Peak	1974	21193	1.22%	11%	A=Exponential decline
2nd Peak	1998	29018	2.20%	33%	B=5% incr. to Mid-Point
Mid-Point	2011	23744	2.44%	51%	C=Other
Final Decline		N/A			Pre-1930 Cum. 1.960 (Added to 1930 prod)

Date	Prod. M.b/d	Prod. BBO/yr	Cum. Prod.	Remain	Deplet. Rate	% Produced	Giant	Notes & Names
1930	151	0.055	2.02	698.98	ERR	0%	33.53	
1931	168	0.061	2.08	698.92	0.01%	0%	0.00	
1932	185	0.068	2.14	698.86	0.01%	0%	0.00	
1933	202	0.074	2.22	698.78	0.01%	0%	0.00	
1934	219	0.080	2.30	698.70	0.01%	0%	0.00	
1935	236	0.086	2.38	698.62	0.01%	0%	1.00	
1936	253	0.092	2.48	698.52	0.01%	0%	8.70	
1937	280	0.102	2.58	698.42	0.01%	0%	0.00	
1938	306	0.112	2.69	698.31	0.02%	0%	67.55	
1939	309	0.113	2.80	698.20	0.02%	0%	0.00	
1940	261	0.095	2.90	698.10	0.01%	0%	1.76	
1941	185	0.068	2.97	698.03	0.01%	0%	12.80	
1942	264	0.097	3.06	697.94	0.01%	0%	0.00	
1943	287	0.105	3.17	697.83	0.01%	0%	0.00	
1944	386	0.141	3.31	697.69	0.02%	0%	0.00	
1945	512	0.187	3.49	697.51	0.03%	0%	3.20	
1946	680	0.248	3.74	697.26	0.04%	1%	0.00	
1947	813	0.297	4.04	696.96	0.04%	1%	0.00	
1948	1108	0.404	4.44	696.56	0.06%	1%	88.37	
1949	1372	0.501	4.94	696.06	0.07%	1%	0.95	
1950	1703	0.622	5.57	695.43	0.09%	1%	0.00	
1951	1850	0.675	6.24	694.76	0.10%	1%	32.30	
1952	1997	0.729	6.97	694.03	0.10%	1%	0.00	
1953	2372	0.866	7.84	693.16	0.12%	1%	20.38	
1954	2581	0.942	8.78	692.22	0.14%	1%	9.00	
1955	3086	1.126	9.90	691.10	0.16%	1%	7.88	
1956	3191	1.165	11.07	689.93	0.17%	2%	3.30	
1957	3330	1.215	12.28	688.72	0.18%	2%	31.10	
1958	4016	1.466	13.75	687.25	0.21%	2%	15.66	
1959	4346	1.586	15.34	685.66	0.23%	2%	2.60	
1960	5022	1.833	17.17	683.83	0.27%	2%	0.00	
1961	5395	1.969	19.14	681.86	0.29%	3%	6.40	
1962	5956	2.174	21.31	679.69	0.32%	3%	15.10	
1963	6493	2.370	23.68	677.32	0.35%	3%	20.96	
1964	7417	2.707	26.39	674.61	0.40%	4%	28.30	
1965	7938	2.897	29.29	671.71	0.43%	4%	17.14	
1966	8852	3.231	32.52	668.48	0.48%	5%	10.90	
1967	9468	3.456	35.97	665.03	0.52%	5%	1.20	
1968	10562	3.855	39.83	661.17	0.58%	6%	5.71	
1969	11332	4.136	43.97	657.03	0.63%	6%	6.40	
1970	12579	4.591	48.56	652.44	0.70%	7%	0.90	
1971	15008	5.478	54.03	646.97	0.84%	8%	2.13	
1972	15950	5.822	59.86	641.14	0.90%	9%	1.03	
1973	19989	7.296	67.15	633.85	1.14%	10%	3.43	
1974	21193	7.735	74.89	626.11	1.22%	11%	1.00	1ST PEAK
1975	18900	6.899	81.79	619.21	1.10%	12%	1.17	
1976	20375	7.437	89.22	611.78	1.20%	13%	8.00	
1977	20485	7.477	96.70	604.30	1.22%	14%	1.00	
1978	19320	7.052	103.75	597.25	1.17%	15%	0.00	
1979	19740	7.205	110.96	590.04	1.21%	16%	18.00	
1980	16830	6.143	117.10	583.90	1.04%	17%	0.00	
1981	14340	5.234	122.33	578.67	0.90%	17%	0.00	
1982	11167	4.076	126.41	574.59	0.70%	18%	0.00	
1983	10450	3.814	130.22	570.78	0.66%	19%	0.00	
1984	10024	3.659	133.88	567.12	0.64%	19%	0.00	
1985	8857	3.233	137.12	563.88	0.57%	20%	0.00	
1986	10790	3.938	141.05	559.95	0.70%	20%	0.00	
1987	10956	3.999	145.05	555.95	0.71%	21%	0.00	
1988	12178	4.445	149.50	551.50	0.80%	21%	0.00	

1989	13998	5.109	154.61	546.39	0.93%	22%	0.00	
1990	15900	5.803	160.41	540.59	1.06%	23%	0.00	
1991	17720	6.468	166.88	534.12	1.20%	24%	0.00	
1992	19461	7.103	173.98	527.02	1.33%	25%	0.00	
1993	21371	7.800	181.78	519.22	1.48%	26%	0.00	
1994	23124	8.440	190.22	510.78	1.63%	27%	0.00	
1995	24915	9.094	199.32	501.68	1.78%	28%	0.00	
1996	27302	9.965	209.28	491.72	1.99%	30%	0.00	
1997	29740	10.855	220.14	480.86	2.21%	31%	0.00	
1998	29018	10.592	230.73	470.27	2.20%	33%	0.00	2ND PEAK
1999	28297	10.328	241.06	459.94	2.20%	34%	0.00	
2000	27735	10.123	251.18	449.82	2.20%	36%	0.00	
2001	27311	9.968	261.15	439.85	2.22%	37%	0.00	
2002	27121	9.899	271.05	429.95	2.25%	39%	0.00	
2003	26922	9.826	280.87	420.13	2.29%	40%	0.00	
2004	26682	9.739	290.61	410.39	2.32%	41%	0.00	
2005	26010	9.494	300.11	400.89	2.31%	43%	0.00	
2006	25692	9.378	309.48	391.52	2.34%	44%	0.00	
2007	25343	9.250	318.73	382.27	2.36%	45%	0.00	
2008	24967	9.113	327.85	373.15	2.38%	47%	0.00	
2009	24582	8.972	336.82	364.18	2.40%	48%	0.00	
2010	24174	8.823	345.64	355.36	2.42%	49%	0.00	
2011	23744	8.667	354.31	346.69	2.44%	51%	0.00	50% PRODUCED
2012	23296	8.503	362.81	338.19	2.45%	52%	0.00	
2013	22830	8.333	371.15	329.85	2.46%	53%	0.00	
2014	22350	8.158	379.30	321.70	2.47%	54%	0.00	
2015	21853	7.976	387.28	313.72	2.48%	55%	0.00	
2016	21343	7.790	395.07	305.93	2.48%	56%	0.00	
2017	20822	7.600	402.67	298.33	2.48%	57%	0.00	
2018	20293	7.407	410.08	290.92	2.48%	58%	0.00	
2019	19753	7.210	417.29	283.71	2.48%	60%	0.00	
2020	19204	7.010	424.30	276.70	2.47%	61%	0.00	
2021	18646	6.806	431.10	269.90	2.46%	61%	0.00	
2022	18079	6.599	437.70	263.30	2.44%	62%	0.00	
2023	17402	6.352	444.05	256.95	2.41%	63%	0.00	
2024	17115	6.247	450.30	250.70	2.43%	64%	0.00	
2025	16817	6.138	456.44	244.56	2.45%	65%	0.00	
2026	16508	6.026	462.46	238.54	2.46%	66%	0.00	
2027	16190	5.909	468.37	232.63	2.48%	67%	0.00	
2028	15863	5.790	474.16	226.84	2.49%	68%	0.00	
2029	15527	5.667	479.83	221.17	2.50%	68%	0.00	
2030	15207	5.550	485.38	215.62	2.51%	69%	0.00	
2031	14867	5.426	490.81	210.19	2.52%	70%	0.00	
2032	14618	5.336	496.14	204.86	2.54%	71%	0.00	
2033	14365	5.243	501.39	199.61	2.56%	72%	0.00	
2034	14108	5.149	506.54	194.46	2.58%	72%	0.00	
2035	13847	5.054	511.59	189.41	2.60%	73%	0.00	
2036	13585	4.958	516.55	184.45	2.62%	74%	0.00	
2037	13321	4.862	521.41	179.59	2.64%	74%	0.00	
2038	13055	4.765	526.18	174.82	2.65%	75%	0.00	
2039	12790	4.668	530.84	170.16	2.67%	76%	0.00	
2040	12525	4.572	535.42	165.58	2.69%	76%	0.00	
2041	12260	4.475	539.89	161.11	2.70%	77%	0.00	
2042	11996	4.379	544.27	156.73	2.72%	78%	0.00	
2043	11734	4.283	548.55	152.45	2.73%	78%	0.00	
2044	11474	4.188	552.74	148.26	2.75%	79%	0.00	
2045	11216	4.094	556.83	144.17	2.76%	79%	0.00	
2046	10960	4.000	560.83	140.17	2.77%	80%	0.00	
2047	10707	3.908	564.74	136.26	2.79%	81%	0.00	
2048	10457	3.817	568.56	132.44	2.80%	81%	0.00	
2049	10210	3.727	572.29	128.71	2.81%	82%	0.00	
2050	9966	3.638	575.92	125.08	2.83%	82%	0.00	

CUM.PROD	151.60 BBO	GIANT FIELDS			M.EAST (MAJOR)	
RESERVES	484.45 BBO	TOTAL BBO	489		REGION	
DISCOVERED	636.05 BBO	% OF DISC.	77%		Scenario-2	
UNDISCOVERED	64.95 BBO	Last Giant	1979			
REMAINING	549.40 BBO				Variance	Trend
ULTIMATE	701.00 BBO				1990	-2%
Rept.Reserve	639.27	Ref.date(end)	1989		Future	n/a

Milestones	Date	Prod Mb/d	Dep. Rate	% Produced	Forecast n/a
1st Peak	1974	21193	1.22%	11%	A=Exponential decline
2nd Peak	1998	22711	1.70%	31%	B=5% incr. to Mid-Point
Mid-Point	2014	20371	2.06%	50%	C=Other
Final Decline		N/A			Pre-1930 Cum. 0.980 (Added to 1930 prod)

Date	Prod. M.b/d	Prod. BBO/yr	Cum. Prod.	Remain	Deplet. Rate	% Produced	Giant	Notes & Names
1930	151	0.055	1.04	699.96	ERR	0%	33.53	
1931	168	0.061	1.10	699.90	0.01%	0%	0.00	
1932	185	0.068	1.16	699.84	0.01%	0%	0.00	
1933	202	0.074	1.24	699.76	0.01%	0%	0.00	
1934	219	0.080	1.32	699.68	0.01%	0%	0.00	
1935	236	0.086	1.40	699.60	0.01%	0%	1.00	
1936	253	0.092	1.50	699.50	0.01%	0%	8.70	
1937	280	0.102	1.60	699.40	0.01%	0%	0.00	
1938	306	0.112	1.71	699.29	0.02%	0%	67.55	
1939	309	0.113	1.82	699.18	0.02%	0%	0.00	
1940	261	0.095	1.92	699.08	0.01%	0%	1.76	
1941	185	0.068	1.99	699.01	0.01%	0%	12.80	
1942	264	0.097	2.08	698.92	0.01%	0%	0.00	
1943	287	0.105	2.19	698.81	0.01%	0%	0.00	
1944	386	0.141	2.33	698.67	0.02%	0%	0.00	
1945	512	0.187	2.51	698.49	0.03%	0%	3.20	
1946	680	0.248	2.76	698.24	0.04%	0%	0.00	
1947	813	0.297	3.06	697.94	0.04%	0%	0.00	
1948	1108	0.404	3.46	697.54	0.06%	0%	88.37	
1949	1372	0.501	3.96	697.04	0.07%	1%	0.95	
1950	1703	0.622	4.59	696.41	0.09%	1%	0.00	
1951	1850	0.675	5.26	695.74	0.10%	1%	32.30	
1952	1997	0.729	5.99	695.01	0.10%	1%	0.00	
1953	2372	0.866	6.86	694.14	0.12%	1%	20.38	
1954	2581	0.942	7.80	693.20	0.14%	1%	9.00	
1955	3086	1.126	8.92	692.08	0.16%	1%	7.88	
1956	3191	1.165	10.09	690.91	0.17%	1%	3.30	
1957	3330	1.215	11.30	689.70	0.18%	2%	31.10	
1958	4016	1.466	12.77	688.23	0.21%	2%	15.66	
1959	4346	1.586	14.36	686.64	0.23%	2%	2.60	
1960	5022	1.833	16.19	684.81	0.27%	2%	0.00	
1961	5395	1.969	18.16	682.84	0.29%	3%	6.40	
1962	5956	2.174	20.33	680.67	0.32%	3%	15.10	
1963	6493	2.370	22.70	678.30	0.35%	3%	20.96	
1964	7417	2.707	25.41	675.59	0.40%	4%	28.30	
1965	7938	2.897	28.31	672.69	0.43%	4%	17.14	
1966	8852	3.231	31.54	669.46	0.48%	4%	10.90	
1967	9468	3.456	34.99	666.01	0.52%	5%	1.20	
1968	10562	3.855	38.85	662.15	0.58%	6%	5.71	
1969	11332	4.136	42.99	658.01	0.62%	6%	6.40	
1970	12579	4.591	47.58	653.42	0.70%	7%	0.90	
1971	15008	5.478	53.05	647.95	0.84%	8%	2.13	
1972	15950	5.822	58.88	642.12	0.90%	8%	1.03	
1973	19989	7.296	66.17	634.83	1.14%	9%	3.43	
1974	21193	7.735	73.91	627.09	1.22%	11%	1.00	1ST PEAK
1975	18900	6.899	80.81	620.19	1.10%	12%	1.17	
1976	20375	7.437	88.24	612.76	1.20%	13%	8.00	
1977	20485	7.477	95.72	605.28	1.22%	14%	1.00	
1978	19320	7.052	102.77	598.23	1.17%	15%	0.00	
1979	19740	7.205	109.98	591.02	1.20%	16%	18.00	
1980	16830	6.143	116.12	584.88	1.04%	17%	0.00	
1981	14340	5.234	121.35	579.65	0.89%	17%	0.00	
1982	11167	4.076	125.43	575.57	0.70%	18%	0.00	
1983	10450	3.814	129.24	571.76	0.66%	18%	0.00	
1984	10024	3.659	132.90	568.10	0.64%	19%	0.00	
1985	8857	3.233	136.14	564.86	0.57%	19%	0.00	
1986	10790	3.938	140.07	560.93	0.70%	20%	0.00	
1987	10956	3.999	144.07	556.93	0.71%	21%	0.00	
1988	12178	4.445	148.52	552.48	0.80%	21%	0.00	

1989	13998	5.109	153.63	547.37	0.92%	22%	0.00	
1990	15900	5.803	159.43	541.57	1.06%	23%	0.00	
1991	17720	6.468	165.90	535.10	1.19%	24%	0.00	
1992	19461	7.103	173.00	528.00	1.33%	25%	0.00	
1993	20110	7.340	180.34	520.66	1.39%	26%	0.00	
1994	20577	7.511	187.85	513.15	1.44%	27%	0.00	
1995	21057	7.686	195.54	505.46	1.50%	28%	0.00	
1996	22105	8.068	203.61	497.39	1.60%	29%	0.00	
1997	23179	8.460	212.07	488.93	1.70%	30%	0.00	
1998	22711	8.289	220.36	480.64	1.70%	31%	0.00	2ND PEAK
1999	22361	8.162	228.52	472.48	1.70%	33%	0.00	
2000	22154	8.086	236.60	464.40	1.71%	34%	0.00	
2001	22070	8.055	244.66	456.34	1.73%	35%	0.00	
2002	22205	8.105	252.76	448.24	1.78%	36%	0.00	
2003	22317	8.146	260.91	440.09	1.82%	37%	0.00	
2004	22374	8.167	269.08	431.92	1.86%	38%	0.00	
2005	21986	8.025	277.10	423.90	1.86%	40%	0.00	
2006	21940	8.008	285.11	415.89	1.89%	41%	0.00	
2007	21851	7.976	293.09	407.91	1.92%	42%	0.00	
2008	21722	7.929	301.01	399.99	1.94%	43%	0.00	
2009	21574	7.874	308.89	392.11	1.97%	44%	0.00	
2010	21391	7.808	316.70	384.30	1.99%	45%	0.00	
2011	21178	7.730	324.43	376.57	2.01%	46%	0.00	
2012	20935	7.641	332.07	368.93	2.03%	47%	0.00	
2013	20665	7.543	339.61	361.39	2.04%	48%	0.00	
2014	20371	7.436	347.04	353.96	2.06%	50%	0.00	50% PRODUCED
2015	20053	7.319	354.36	346.64	2.07%	51%	0.00	
2016	19712	7.195	361.56	339.44	2.08%	52%	0.00	
2017	19352	7.063	368.62	332.38	2.08%	53%	0.00	
2018	18976	6.926	375.55	325.45	2.08%	54%	0.00	
2019	18583	6.783	382.33	318.67	2.08%	55%	0.00	
2020	18173	6.633	388.97	312.03	2.08%	55%	0.00	
2021	17747	6.478	395.44	305.56	2.08%	56%	0.00	
2022	17305	6.316	401.76	299.24	2.07%	57%	0.00	
2023	16747	6.113	407.87	293.13	2.04%	58%	0.00	
2024	16573	6.049	413.92	287.08	2.06%	59%	0.00	
2025	16382	5.979	419.90	281.10	2.08%	60%	0.00	
2026	16175	5.904	425.80	275.20	2.10%	61%	0.00	
2027	15953	5.823	431.63	269.37	2.12%	62%	0.00	
2028	15716	5.736	437.36	263.64	2.13%	62%	0.00	
2029	15467	5.645	443.01	257.99	2.14%	63%	0.00	
2030	15227	5.558	448.57	252.43	2.15%	64%	0.00	
2031	14964	5.462	454.03	246.97	2.16%	65%	0.00	
2032	14788	5.398	459.43	241.57	2.19%	66%	0.00	
2033	14603	5.330	464.76	236.24	2.21%	66%	0.00	
2034	14410	5.260	470.02	230.98	2.23%	67%	0.00	
2035	14210	5.187	475.20	225.80	2.25%	68%	0.00	
2036	14004	5.111	480.31	220.69	2.26%	69%	0.00	
2037	13793	5.034	485.35	215.65	2.28%	69%	0.00	
2038	13578	4.956	490.30	210.70	2.30%	70%	0.00	
2039	13359	4.876	495.18	205.82	2.31%	71%	0.00	
2040	13138	4.795	499.98	201.02	2.33%	71%	0.00	
2041	12914	4.714	504.69	196.31	2.34%	72%	0.00	
2042	12688	4.631	509.32	191.68	2.36%	73%	0.00	
2043	12461	4.548	513.87	187.13	2.37%	73%	0.00	
2044	12234	4.465	518.33	182.67	2.39%	74%	0.00	
2045	12006	4.382	522.72	178.28	2.40%	75%	0.00	
2046	11778	4.299	527.02	173.98	2.41%	75%	0.00	
2047	11551	4.216	531.23	169.77	2.42%	76%	0.00	
2048	11325	4.134	535.37	165.63	2.43%	76%	0.00	
2049	11099	4.051	539.42	161.58	2.45%	77%	0.00	
2050	10875	3.969	543.39	157.61	2.46%	78%	0.00	

CUM.PROD	151.60 BBO	GIANT FIELDS		M.EAST (MAJOR)		
RESERVES	484.45 BBO	TOTAL BBO	489	REGION		
DISCOVERED	636.05 BBO	% OF DISC.	77%	Scenario-3		
UNDISCOVERED	64.95 BBO	Last Giant	1979			
REMAINING	549.40 BBO			Variance Trend		
ULTIMATE	701.00 BBO			1990	-2%	
Rept.Reserve	639.27	Ref.date(end)	1989	Future	n/a	

Milestones	Date	Prod Mb/d	Dep. Rate	% Produced	Forecast	n/a
1st Peak	1974	21193	1.22%	11%	A=Exponential decline	
2nd Peak	1998	20835	1.53%	30%	B=5% incr. to Mid-Point	
Mid-Point	2016	19311	1.95%	50%	C=Other	
Final Decline		N/A			Pre-1930 Cum.	3.040
					(Added to 1930 prod)	

Date	Prod. M.b/d	Prod. BBO/yr	Cum. Prod.	Remain	Deplet. Rate	% Produced	Giant	Notes & Names
1930	151	0.055	3.10	697.90	ERR	0%	33.53	
1931	168	0.061	3.16	697.84	0.01%	0%	0.00	
1932	185	0.068	3.22	697.78	0.01%	0%	0.00	
1933	202	0.074	3.30	697.70	0.01%	0%	0.00	
1934	219	0.080	3.38	697.62	0.01%	0%	0.00	
1935	236	0.086	3.46	697.54	0.01%	0%	1.00	
1936	253	0.092	3.56	697.44	0.01%	1%	8.70	
1937	280	0.102	3.66	697.34	0.01%	1%	0.00	
1938	306	0.112	3.77	697.23	0.02%	1%	67.55	
1939	309	0.113	3.88	697.12	0.02%	1%	0.00	
1940	261	0.095	3.98	697.02	0.01%	1%	1.76	
1941	185	0.068	4.05	696.95	0.01%	1%	12.80	
1942	264	0.097	4.14	696.86	0.01%	1%	0.00	
1943	287	0.105	4.25	696.75	0.02%	1%	0.00	
1944	386	0.141	4.39	696.61	0.02%	1%	0.00	
1945	512	0.187	4.57	696.43	0.03%	1%	3.20	
1946	680	0.248	4.82	696.18	0.04%	1%	0.00	
1947	813	0.297	5.12	695.88	0.04%	1%	0.00	
1948	1108	0.404	5.52	695.48	0.06%	1%	88.37	
1949	1372	0.501	6.02	694.98	0.07%	1%	0.95	
1950	1703	0.622	6.65	694.35	0.09%	1%	0.00	
1951	1850	0.675	7.32	693.68	0.10%	1%	32.30	
1952	1997	0.729	8.05	692.95	0.11%	1%	0.00	
1953	2372	0.866	8.92	692.08	0.12%	1%	20.38	
1954	2581	0.942	9.86	691.14	0.14%	1%	9.00	
1955	3086	1.126	10.98	690.02	0.16%	2%	7.88	
1956	3191	1.165	12.15	688.85	0.17%	2%	3.30	
1957	3330	1.215	13.36	687.64	0.18%	2%	31.10	
1958	4016	1.466	14.83	686.17	0.21%	2%	15.66	
1959	4346	1.586	16.42	684.58	0.23%	2%	2.60	
1960	5022	1.833	18.25	682.75	0.27%	3%	0.00	
1961	5395	1.969	20.22	680.78	0.29%	3%	6.40	
1962	5956	2.174	22.39	678.61	0.32%	3%	15.10	
1963	6493	2.370	24.76	676.24	0.35%	4%	20.96	
1964	7417	2.707	27.47	673.53	0.40%	4%	28.30	
1965	7938	2.897	30.37	670.63	0.43%	4%	17.14	
1966	8852	3.231	33.60	667.40	0.48%	5%	10.90	
1967	9468	3.456	37.05	663.95	0.52%	5%	1.20	
1968	10562	3.855	40.91	660.09	0.58%	6%	5.71	
1969	11332	4.136	45.05	655.95	0.63%	6%	6.40	
1970	12579	4.591	49.64	651.36	0.70%	7%	0.90	
1971	15008	5.478	55.11	645.89	0.84%	8%	2.13	
1972	15950	5.822	60.94	640.06	0.90%	9%	1.03	
1973	19989	7.296	68.23	632.77	1.14%	10%	3.43	
1974	21193	7.735	75.97	625.03	1.22%	11%	1.00	1ST PEAK
1975	18900	6.899	82.87	618.13	1.10%	12%	1.17	
1976	20375	7.437	90.30	610.70	1.20%	13%	8.00	
1977	20485	7.477	97.78	603.22	1.22%	14%	1.00	
1978	19320	7.052	104.83	596.17	1.17%	15%	0.00	
1979	19740	7.205	112.04	588.96	1.21%	16%	18.00	
1980	16830	6.143	118.18	582.82	1.04%	17%	0.00	
1981	14340	5.234	123.41	577.59	0.90%	18%	0.00	
1982	11167	4.076	127.49	573.51	0.71%	18%	0.00	
1983	10450	3.814	131.30	569.70	0.67%	19%	0.00	
1984	10024	3.659	134.96	566.04	0.64%	19%	0.00	
1985	8857	3.233	138.20	562.80	0.57%	20%	0.00	
1986	10790	3.938	142.13	558.87	0.70%	20%	0.00	
1987	10956	3.999	146.13	554.87	0.72%	21%	0.00	
1988	12178	4.445	150.58	550.42	0.80%	21%	0.00	

1989	13998	5.109	155.69	545.31	0.93%	22%	0.00	
1990	14712	5.370	161.06	539.94	0.98%	23%	0.00	
1991	15320	5.592	166.65	534.35	1.04%	24%	0.00	
1992	15826	5.776	172.43	528.57	1.08%	25%	0.00	
1993	16475	6.013	178.44	522.56	1.14%	25%	0.00	
1994	16942	6.184	184.62	516.38	1.18%	26%	0.00	
1995	17422	6.359	190.98	510.02	1.23%	27%	0.00	
1996	18470	6.742	197.72	503.28	1.32%	28%	0.00	
1997	19544	7.134	204.86	496.14	1.42%	29%	0.00	
1998	20835	7.605	212.46	488.54	1.53%	30%	0.00	2ND PEAK
1999	20602	7.520	219.98	481.02	1.54%	31%	0.00	
2000	20506	7.485	227.47	473.53	1.56%	32%	0.00	
2001	20528	7.493	234.96	466.04	1.58%	34%	0.00	
2002	20765	7.579	242.54	458.46	1.63%	35%	0.00	
2003	20974	7.655	250.19	450.81	1.67%	36%	0.00	
2004	21124	7.710	257.90	443.10	1.71%	37%	0.00	
2005	20826	7.601	265.50	435.50	1.72%	38%	0.00	
2006	20865	7.616	273.12	427.88	1.75%	39%	0.00	
2007	20858	7.613	280.73	420.27	1.78%	40%	0.00	
2008	20807	7.595	288.33	412.67	1.81%	41%	0.00	
2009	20734	7.568	295.90	405.10	1.83%	42%	0.00	
2010	20623	7.527	303.42	397.58	1.86%	43%	0.00	
2011	20477	7.474	310.90	390.10	1.88%	44%	0.00	
2012	20300	7.409	318.31	382.69	1.90%	45%	0.00	
2013	20092	7.334	325.64	375.36	1.92%	46%	0.00	
2014	19858	7.248	332.89	368.11	1.93%	47%	0.00	
2015	19597	7.153	340.04	360.96	1.94%	49%	0.00	
2016	19311	7.048	347.09	353.91	1.95%	50%	0.00	50% PRODUCED
2017	19002	6.936	354.03	346.97	1.96%	51%	0.00	
2018	18676	6.817	360.84	340.16	1.96%	51%	0.00	
2019	18329	6.690	367.53	333.47	1.97%	52%	0.00	
2020	17964	6.557	374.09	326.91	1.97%	53%	0.00	
2021	17580	6.417	380.51	320.49	1.96%	54%	0.00	
2022	17179	6.270	386.78	314.22	1.96%	55%	0.00	
2023	16660	6.081	392.86	308.14	1.94%	56%	0.00	
2024	16522	6.031	398.89	302.11	1.96%	57%	0.00	
2025	16365	5.973	404.86	296.14	1.98%	58%	0.00	
2026	16191	5.910	410.77	290.23	2.00%	59%	0.00	
2027	16000	5.840	416.61	284.39	2.01%	59%	0.00	
2028	15793	5.764	422.37	278.63	2.03%	60%	0.00	
2029	15571	5.684	428.06	272.94	2.04%	61%	0.00	
2030	15359	5.606	433.66	267.34	2.05%	62%	0.00	
2031	15121	5.519	439.18	261.82	2.06%	63%	0.00	
2032	14968	5.463	444.65	256.35	2.09%	63%	0.00	
2033	14805	5.404	450.05	250.95	2.11%	64%	0.00	
2034	14633	5.341	455.39	245.61	2.13%	65%	0.00	
2035	14453	5.275	460.67	240.33	2.15%	66%	0.00	
2036	14265	5.207	465.87	235.13	2.17%	66%	0.00	
2037	14072	5.136	471.01	229.99	2.18%	67%	0.00	
2038	13873	5.064	476.07	224.93	2.20%	68%	0.00	
2039	13670	4.989	481.06	219.94	2.22%	69%	0.00	
2040	13462	4.914	485.98	215.02	2.23%	69%	0.00	
2041	13252	4.837	490.81	210.19	2.25%	70%	0.00	
2042	13039	4.759	495.57	205.43	2.26%	71%	0.00	
2043	12823	4.681	500.25	200.75	2.28%	71%	0.00	
2044	12607	4.601	504.85	196.15	2.29%	72%	0.00	
2045	12389	4.522	509.38	191.62	2.31%	73%	0.00	
2046	12170	4.442	513.82	187.18	2.32%	73%	0.00	
2047	11951	4.362	518.18	182.82	2.33%	74%	0.00	
2048	11732	4.282	522.46	178.54	2.34%	75%	0.00	
2049	11514	4.203	526.67	174.33	2.35%	75%	0.00	
2050	11296	4.123	530.79	170.21	2.37%	76%	0.00	

SUMMARY DATA SHEET
REFERENCE DATE 1989

REGION MIDDLE EAST (Major)

Country	Production M b/d	BBO/ year	Cumulative Production	Reported Reserves	Adjusted Reserves	Discovered	Undiscovered	Ultimate	Remaining	Und. of Ult.	Prod of Ult.	Rem. of Ult.	Prod of Disc	Disc. in Giants	Dep. Rate
A	B	C	D	E	F	G	H	I	J	K	L	M	N	O	P
S.Arabia	4936	1.801	57.64	254.96	200.00	257.64	17.36	275.00	217.36	6%	21%	79%	22%	81%	0.83%
Kuwait	1543	0.563	24.02	94.25	94.25	118.28	1.72	120.00	95.98	1%	20%	80%	20%	72%	0.59%
Iran	2934	1.071	37.52	92.86	60.00	97.52	17.48	115.00	77.48	15%	33%	67%	38%	52%	1.38%
Iraq	2830	1.033	18.59	100.00	60.00	78.59	21.41	100.00	81.41	21%	19%	81%	24%	96%	1.27%
Abu Dhabi	1359	0.496	9.33	92.00	65.00	74.33	5.67	80.00	70.67	7%	12%	88%	13%	84%	0.70%
Neutral Zn	397	0.145	4.50	5.20	5.20	9.70	1.30	11.00	6.50	12%	41%	59%	46%	57%	2.23%
Region	13998	5.109	151.60	639.27	484.45	636.05	64.95	701.00	549.40	9%	22%	78%	24%	77%	0.93%

SINO-SOVIET ("S-S")

SUMMARY

The Sino-Soviet Region comprises the Soviet Union, China and the countries of Eastern Europe, except the former East Germany. They are classed as a region because of their Communist background, although they are rapidly losing their connections.

It is the most difficult region to assess. Information is limited, and the record of past results is difficult to interpret because the industry was conducted under such a different economic environment.

The oil industry has, however, a long history pre-dating the Communist revolution by several decades. Russian oil was a very important element in the early days of the century, with production from Baku being exported widely. Such disparate characters as Joseph Stalin and Calouste Gulbenkian, one of the fathers of western oil, spent their early years in Baku. Romania had great importance prior to and during the Second World War, and Polish drillers were once famous.

The Region, as here assessed, has 15.5% of the World's Ultimate oil endowment, slightly more than North America.

A number of giant discoveries were made before 1930, although most were made in the 1950s and 1960s.

The Region's share of World oil is as follows :

Cumulative Production	20.3%
Reserves	8.4%
Discovered	14.8%
Undiscovered	20.4%
Remaining	12.5%
Ultimate	15.5%

There is a very large number of sedimentary basins containing oil and gas in this huge territory which we can hope to classify only in the most general sense. It is now a huge single continental land mass, but its geotectonic setting was very different in the past, and seaways developed across it at various times.

Rift Basins

The Urals form an important northerly trending tectonic element that came into existence in late Palaeozoic times and bisects the continent.

A series of rift basins occurs to the west of the Urals from the Arctic region, including the offshore, to the Caspian. Important oil and gas fields are found in both Palaeozoic and Mesozoic reservoirs. The principal source-rock in this province is the prolific Domanik Shale of Upper Devonian age which yields either oil or gas, depending on the precise depositional environments in the many different basins that make up the province. Gas-prone sources predominate and account for the enormous gas accumulations, some surprisingly sealed beneath permafrost. But there are also major oil accumulations, with several giant discoveries, including the famous Samotlor Field, the country's largest. It is at a mature stage of development.

To the east of the Urals lies the great West Siberian Basin, which contains as many as twelve Mesozoic grabens, bearing a certain resemblance to the North Sea rifts. As in the North Sea, the principal source-rock is an Upper Jurassic shale, the Bazhenov Shale, which feeds overlying Lower Cretaceous deltaic sandstone reservoirs. The province becomes gas-prone northwards due to deep burial. The basin is also at a fairly mature stage of development, so it can be assumed that most of the giants have been found.

Other rift basins develop on the eastern margin of the continent, extending into China, and are also present, although much less well known, along the exceedingly remote Arctic margin of the northeastern Soviet Union.

A third development connects to the southern North Sea and runs southeastward through the eastern Baltic and Poland, where some small finds were made in early years in Lower Palaeozoic strata.

Foreland Basins

A series of basins developed between the Alpine-Carpathian-Himalayan orogenic belt and the shield. They are characterised by fold structures, and have Mesozoic and Tertiary reservoirs and source-rocks. These basins are more oil prone, with local depocenters yielding prolific generation. The Romanian and Baku fields lie in this category. The bordering mountain chains have had long tectonic histories, and several phases of basin formation are represented. Furthermore, transcurrent faulting and orogenic collisions complicate the pattern. The province has been extensively explored, although for example the offshore Caspian remains to be investigated.

Marginal Basins

The eastern margin of the continent is flanked by basins at the interface of the Asian continent and the Pacific plates. Such are represented by Sakhalin where oil has been produced from earliest times from an enormously thick sequence of late Tertiary sediments, bearing perhaps some similarity with California. Other similar, less well known basins lie offshore in the area.

Generally speaking, this continental mass, relies heavily on Palaeozoic source-rocks laid down in marine conditions under regional tectonic conditions, very different from those obtaining today. The Silurian and Devonian Domanik facies is particularly important. These deep oils have migrated upwards to collect in a wide variety of reservoirs, but much must have been lost. With the important exception of the West Siberian Basin with its analogies to North Sea geology, the interior rifts are predictably gas prone, although some contain more oil than would be at first anticipated. The foredeep and marginal basins are characterised by structural complexity, giving numerous, yet individually often small, fields. Lastly, there is a huge cratonic basin, the Lena-Tungusta Basin in Eastern Siberia, where some early Palaeozoic and Pre-Cambrian finds have been made beneath a thick seal of Cambrian salt. This province also contains one of the world's largest bitumen deposits in the Khatanga Basin.

Ultimate

With 256 BBO, the Region has the World's second largest endowment. Most of it comes from the Soviet Union with 200 BBO and from China with 45 BBO. There are only four other countries with in excess of 1 BBO : Romania (7 BBO), Yugoslavia (1.25 BBO), Hungary (1.25 BBO) and Albania (1 BBO)

Reserves

It is said that the Soviet Union has large proved undeveloped reserves, but the extent to which they are incorporated in the Reserve figures is uncertain. Here, the reserves figures are accepted as reported.

Again, the Soviet Union and China dominate the Region with respectively 58.4 BBO and 24 BBO. None of the other countries have reserves exceeding 1 BBO.

Undiscovered

Despite some already expressed reservations, the Soviet Union is considered to hold considerable undiscovered potential. While the more accessible regions have already been explored sufficiently to find most of their giant potential, the rift zones along the Arctic margin as well as other huge tracts in the far east, together with the offshore potential of the Okhotsk Sea have received only limited attention. The undiscovered potential is here assessed at 35 BBO. This could well be due for upward revision when more information emerges. China seems to be much less promising, but still is thought to have potential for about 6 BBO, in view of its vast territory. None of the other countries is thought to have an undiscovered potential in excess of 1 BBO.

Production

China has been producing oil for hundreds of years, and Russian production commenced in Baku and Sakhalin before the turn of the century. Production grew rapidly in the Soviet Union from the late 1950s to a peak in 1988. The country reached its Mid-Point in the same year and has now commenced its inexorable decline. China is at a less mature stage, not reaching its Mid-Point until the end of the century. The production profiles of the USSR and China given in Part 3 have a normal pattern. The profiles of the East European countries are too small and uncertain to merit plotting.

The Soviet Union has relied heavily on oil exports for its hard currency, and these are likely to reduce considerably in scale as production falls and domestic demand increases. It may be an increasingly important factor affecting the World price of oil.

The Region was the world's largest producer in 1989 with 15,279 M b/d (26% of the World total). Of this, 12,140 M b/d (20% of the World total) came from the Soviet Union. In terms of Cumulative Production, the region is in third place behind the Middle East Six and North America with 128 BBO, of which 106 BBO came from the Soviet Union.

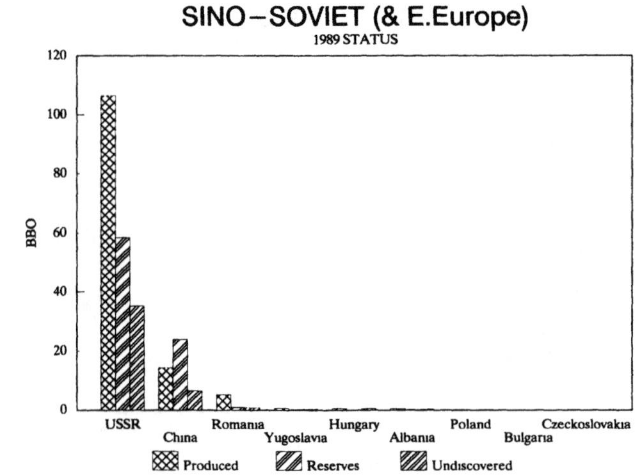

Figure 1-SS. Produced, Reserves and Undiscovered

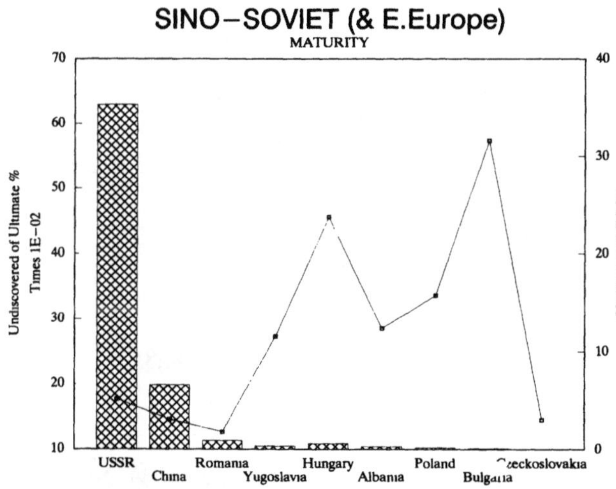

Figure 2-SS. Maturity: Undiscovered and a percentage of Ultimate

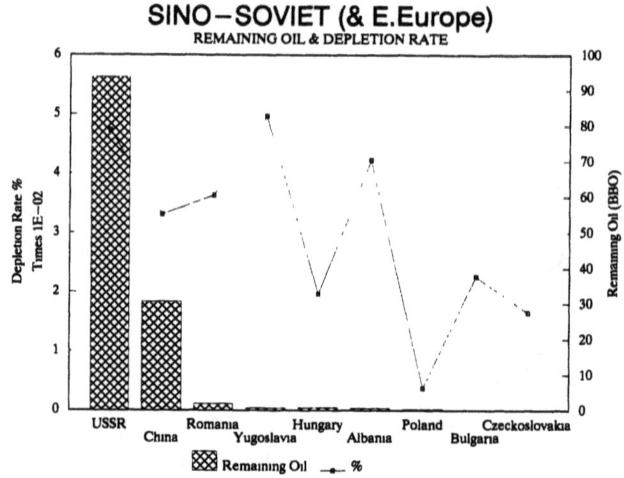

Figure 3-SS. Remaining Oil (Reserves+Undiscovered) and Depletion Rates.

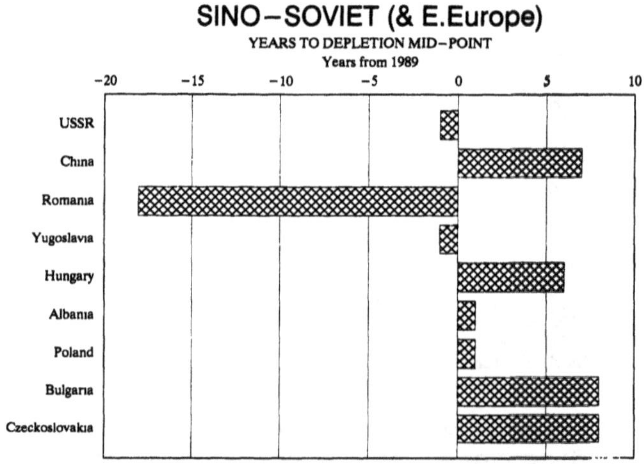

Figure 4-SS. Years to or from Depletion Mid-Point

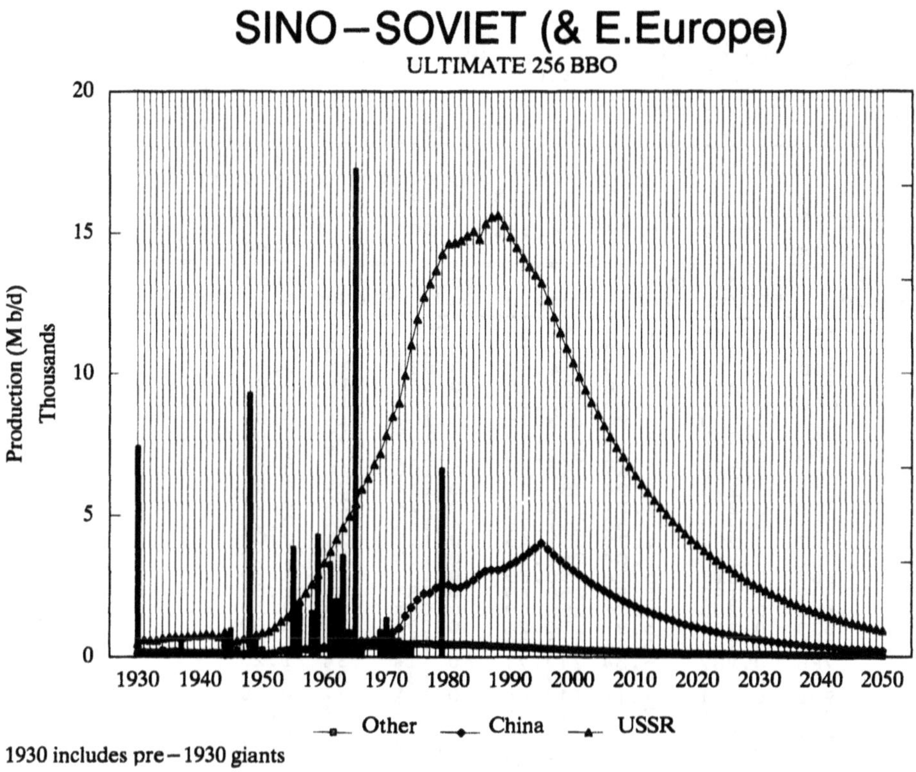

Figure 5-SS. Production Profile 1930-2050 with Giant Fields.

CUM.PROD	127.60	BBO	GIANT FIELDS			SINO-SOVIET REGION
RESERVES	84.08	BBO	TOTAL BBO	109		
DISCOVERED	212.07	BBO	% OF DISC.	51%		
UNDISCOVERED	44.18	BBO	Last Giant	1979		Variance Trend
REMAINING	128.26	BBO				1990 -2%
ULTIMATE	256.25	BBO				Future n/a
Rept.Reserve	83.999		Ref.date(end)	1989		

Milestones	Date	Prod Mb/d	Dep. Rate	% Produced	Forecast n/a
1st Peak	1988	15613	4.07%	48%	A=Exponential decline
2nd Peak	-	-	-	-	B=5% incr. to Mid-Point
					C=Other
Mid-Point	1989	15279	4.15%	50%	Pre-1930 Cum. 1.576
Final Decline	n/a				(Added to 1930 prod)

Date	Prod. M.b/d	Prod. BBO/yr	Cum. Prod.	Remain	Deplet. Rate	% Produced	Giant	Notes & Names
1930	480	0.175	1.75	254.50	ERR	1%	11.17	
1931	603	0.220	1.97	254.28	0.09%	1%	0.00	
1932	590	0.215	2.19	254.06	0.08%	1%	0.00	
1933	591	0.216	2.40	253.85	0.08%	1%	0.00	
1934	668	0.244	2.65	253.60	0.10%	1%	0.50	
1935	686	0.250	2.90	253.35	0.10%	1%	0.00	
1936	705	0.257	3.15	253.10	0.10%	1%	0.00	
1937	695	0.254	3.41	252.84	0.10%	1%	1.00	
1938	718	0.262	3.67	252.58	0.10%	1%	0.00	
1939	747	0.273	3.94	252.31	0.11%	2%	0.00	
1940	749	0.273	4.21	252.04	0.11%	2%	0.00	
1941	792	0.289	4.50	251.75	0.11%	2%	0.00	
1942	777	0.284	4.79	251.46	0.11%	2%	0.00	
1943	702	0.256	5.04	251.21	0.10%	2%	0.00	
1944	867	0.317	5.36	250.89	0.13%	2%	1.25	
1945	539	0.197	5.56	250.69	0.08%	2%	1.53	
1946	556	0.203	5.76	250.49	0.08%	2%	0.50	
1947	633	0.231	5.99	250.26	0.09%	2%	0.00	
1948	729	0.266	6.26	249.99	0.11%	2%	14.00	
1949	786	0.287	6.54	249.71	0.11%	3%	1.23	
1950	861	0.314	6.86	249.39	0.13%	3%	0.52	
1951	908	0.331	7.19	249.06	0.13%	3%	0.00	
1952	1051	0.384	7.57	248.68	0.15%	3%	0.00	
1953	1261	0.460	8.03	248.22	0.19%	3%	0.00	
1954	1416	0.517	8.55	247.70	0.21%	3%	0.51	
1955	1673	0.610	9.16	247.09	0.25%	4%	5.85	
1956	1958	0.715	9.88	246.37	0.29%	4%	2.90	
1957	2251	0.822	10.70	245.55	0.33%	4%	0.00	
1958	2556	0.933	11.63	244.62	0.38%	5%	2.46	
1959	2914	1.063	12.69	243.56	0.43%	5%	6.50	
1960	3322	1.212	13.91	242.34	0.50%	5%	0.60	
1961	3715	1.356	15.26	240.99	0.56%	6%	5.06	
1962	4145	1.513	16.77	239.48	0.63%	7%	3.08	
1963	4562	1.665	18.44	237.81	0.70%	7%	5.43	
1964	4959	1.810	20.25	236.00	0.76%	8%	1.45	
1965	5402	1.972	22.22	234.03	0.84%	9%	25.90	
1966	5920	2.161	24.38	231.87	0.92%	10%	0.60	
1967	6296	2.298	26.68	229.57	0.99%	10%	0.00	
1968	6779	2.474	29.15	227.10	1.08%	11%	0.00	
1969	7161	2.614	31.77	224.48	1.15%	12%	1.44	
1970	7810	2.851	34.62	221.63	1.27%	14%	2.10	
1971	8496	3.101	37.72	218.53	1.40%	15%	1.10	
1972	8965	3.272	40.99	215.26	1.50%	16%	1.00	
1973	9933	3.625	44.62	211.63	1.68%	17%	0.50	
1974	11009	4.018	48.64	207.61	1.90%	19%	0.50	
1975	11926	4.353	52.99	203.26	2.10%	21%	0.00	
1976	12719	4.642	57.63	198.62	2.28%	22%	0.00	
1977	13204	4.819	62.45	193.80	2.43%	24%	0.00	
1978	13674	4.991	67.44	188.81	2.58%	26%	0.00	
1979	14255	5.203	72.64	183.61	2.76%	28%	10.00	
1980	14627	5.339	77.98	178.27	2.91%	30%	0.00	
1981	14651	5.348	83.33	172.92	3.00%	33%	0.00	
1982	14734	5.378	88.71	167.54	3.11%	35%	0.00	
1983	14902	5.439	94.15	162.10	3.25%	37%	0.00	
1984	15068	5.500	99.65	156.60	3.39%	39%	0.00	
1985	14797	5.401	105.05	151.20	3.45%	41%	0.00	
1986	15335	5.597	110.65	145.60	3.70%	43%	0.00	
1987	15566	5.682	116.33	139.92	3.90%	45%	0.00	
1988	15613	5.699	122.03	134.22	4.07%	48%	0.00	

1989	15279	5.577	127.60	128.65	4.15%	50%	0.00	50% PRODUCED
1990	14872	5.428	133.03	123.22	4.22%	52%	0.00	
1991	14483	5.286	138.32	117.93	4.29%	54%	0.00	
1992	14125	5.156	143.47	112.78	4.37%	56%	0.00	
1993	13798	5.036	148.51	107.74	4.47%	58%	0.00	
1994	13501	4.928	153.44	102.81	4.57%	60%	0.00	
1995	13227	4.828	158.27	97.98	4.70%	62%	0.00	
1996	12600	4.599	162.86	93.39	4.69%	64%	0.00	
1997	12002	4.381	167.25	89.00	4.69%	65%	0.00	
1998	11433	4.173	171.42	84.83	4.69%	67%	0.00	
1999	10890	3.975	175.39	80.86	4.69%	68%	0.00	
2000	10374	3.787	179.18	77.07	4.68%	70%	0.00	
2001	9883	3.607	182.79	73.46	4.68%	71%	0.00	
2002	9415	3.436	186.22	70.03	4.68%	73%	0.00	
2003	8969	3.274	189.50	66.75	4.68%	74%	0.00	
2004	8545	3.119	192.62	63.63	4.67%	75%	0.00	
2005	8141	2.971	195.59	60.66	4.67%	76%	0.00	
2006	7756	2.831	198.42	57.83	4.67%	77%	0.00	
2007	7390	2.697	201.12	55.13	4.66%	78%	0.00	
2008	7041	2.570	203.69	52.56	4.66%	79%	0.00	
2009	6708	2.448	206.13	50.12	4.66%	80%	0.00	
2010	6392	2.333	208.47	47.78	4.66%	81%	0.00	
2011	6090	2.223	210.69	45.56	4.65%	82%	0.00	
2012	5803	2.118	212.81	43.44	4.65%	83%	0.00	
2013	5530	2.018	214.83	41.42	4.65%	84%	0.00	
2014	5269	1.923	216.75	39.50	4.64%	85%	0.00	
2015	5021	1.833	218.58	37.67	4.64%	85%	0.00	
2016	4785	1.746	220.33	35.92	4.64%	86%	0.00	
2017	4560	1.664	221.99	34.26	4.63%	87%	0.00	
2018	4345	1.586	223.58	32.67	4.63%	87%	0.00	
2019	4141	1.511	225.09	31.16	4.63%	88%	0.00	
2020	3946	1.440	226.53	29.72	4.62%	88%	0.00	
2021	3761	1.373	227.90	28.35	4.62%	89%	0.00	
2022	3585	1.308	229.21	27.04	4.62%	89%	0.00	
2023	3416	1.247	230.46	25.79	4.61%	90%	0.00	
2024	3256	1.189	231.65	24.60	4.61%	90%	0.00	
2025	3104	1.133	232.78	23.47	4.60%	91%	0.00	
2026	2958	1.080	233.86	22.39	4.60%	91%	0.00	
2027	2820	1.029	234.89	21.36	4.60%	92%	0.00	
2028	2688	0.981	235.87	20.38	4.59%	92%	0.00	
2029	2562	0.935	236.80	19.45	4.59%	92%	0.00	
2030	2442	0.891	237.70	18.55	4.58%	93%	0.00	
2031	2328	0.850	238.55	17.70	4.58%	93%	0.00	
2032	2219	0.810	239.36	16.89	4.58%	93%	0.00	
2033	2116	0.772	240.13	16.12	4.57%	94%	0.00	
2034	2017	0.736	240.86	15.39	4.57%	94%	0.00	
2035	1923	0.702	241.57	14.68	4.56%	94%	0.00	
2036	1833	0.669	242.24	14.01	4.56%	95%	0.00	
2037	1748	0.638	242.87	13.38	4.55%	95%	0.00	
2038	1666	0.608	243.48	12.77	4.55%	95%	0.00	
2039	1589	0.580	244.06	12.19	4.54%	95%	0.00	
2040	1515	0.553	244.61	11.64	4.54%	95%	0.00	
2041	1444	0.527	245.14	11.11	4.53%	96%	0.00	
2042	1377	0.503	245.64	10.61	4.52%	96%	0.00	
2043	1313	0.479	246.12	10.13	4.52%	96%	0.00	
2044	1252	0.457	246.58	9.67	4.51%	96%	0.00	
2045	1194	0.436	247.02	9.23	4.51%	96%	0.00	
2046	1139	0.416	247.43	8.82	4.50%	97%	0.00	
2047	1086	0.396	247.83	8.42	4.49%	97%	0.00	
2048	1035	0.378	248.21	8.04	4.49%	97%	0.00	
2049	988	0.360	248.57	7.68	4.48%	97%	0.00	
2050	942	0.344	248.91	7.34	4.47%	97%	0.00	

SUMMARY DATA SHEET   REGION SINO-SOVIET
REFERENCE DATE 1989

Country	Production M b/d	BBO/ year	Cumulative Production	Reported Reserves	Adjusted Reserves	Discovered	Undiscovered	Ultimate	Remaining	Und. of Ult.	Prod. of Ult.	Rem. of Ult.	Prod. of Disc	Disc. in Giants	Dep. Rate
A	B	C	D	E	F	G	H	I	J	K	L	M	N	O	P
USSR	12140	4.431	106.33	58.40	58.40	164.73	35.27	200.00	93.67	18%	53%	47%	65%	59%	4.73%
China	2770	1.011	14.44	24.00	24.00	38.44	6.56	45.00	30.56	15%	32%	68%	38%	26%	3.31%
Romania	180	0.066	5.19	0.88	0.92	6.11	0.89	7.00	1.81	13%	74%	26%	85%	13%	3.63%
Yugoslavia	80	0.029	0.66	0.23	0.25	0.91	0.34	1.25	0.59	27%	53%	47%	73%	0%	4.96%
Hungary	38	0.014	0.54	0.13	0.14	0.68	0.57	1.25	0.71	46%	43%	57%	79%	0%	1.96%
Albania	60	0.022	0.48	0.24	0.24	0.71	0.29	1.00	0.52	29%	48%	52%	67%	0%	4.22%
Poland	3	0.001	0.25	0.08	0.09	0.33	0.17	0.50	0.25	34%	49%	51%	74%	0%	0.38%
Bulgaria	6	0.002	0.06	0.00	0.00	0.06	0.09	0.15	0.09	57%	40%	60%	94%	0%	2.27%
Czeckoslovak	3	0.001	0.04	0.04	0.04	0.09	0.01	0.10	0.06	15%	42%	58%	50%	0%	1.65%
Region	15279	5.577	127.60	84.00	84.08	212.07	44.18	256.25	128.26	17%	50%	50%	60%	51%	4.35%

NORTH AMERICA ("NA")

SUMMARY

North America has 13.6% of the World's Ultimate oil endowment. As here defined it is made up of the United States and Canada, Mexico being included in Latin America.

Most of the Region's giant discoveries were made before 1930. A combination of geological and economic conditions, including the private ownership of mineral rights in the USA, has tended to encourage the development of small accumulations, which have an exceptionally low commercial threshold. Consequently, giant fields are responsible for no more than 29% of the Ultimate.

North America's share of World oil is as follows :

Cumulative Production	27.3%
Reserves	4.0%
Discovered	14.2%
Undiscovered	9.2%
Remaining	5.0%
Ultimate	13.6%

North America's main productive provinces are :

United States

The South-West and Mid-Continent

Many of the early large finds were made in an extensive province, made up of a large number of discrete basins in Louisiana, Texas and Oklahoma with reservoirs and source-rocks occurring in Palaeozoic, Mesozoic and Tertiary strata.

Gulf of Mexico

Substantial production comes from a thick Tertiary succession in the Gulf of Mexico.

California

The Tertiary is productive onshore and offshore California, which also saw some of the country's important early developments.

Alaska

Large finds have been made in the Mesozoic of the foothills of the Brooks Range in Alaska, including the giant and immensely important Prudhoe Bay Field. The area was opened in the 1960s.

Other established areas

Lesser finds have been made elsewhere : in Pennsylvania and the Mid-West in Palaeozoic strata, in the Rocky Mountain thrust belt, and in the Cook Inlet of Alaska.

Atlantic Margin

The Atlantic shelf seems to have been substantially condemned as an oil province.

Arctic Waters

The Alaskan offshore offers attractive prospects under harsh environmental conditions for Mesozoic and Tertiary traps.

Deep Gulf of Mexico

The Deep Gulf of Mexico offers promising prospects, including the very interesting Perdido Trend in 11000 feet of water, when technology and economic factors permit.

Canada

Western Canada Basin

Almost all of Canada's oil comes from the Western Canada basin, mainly in Alberta. It is essentially an extension of the main producing trend of the United States, with production coming from a wide range of reservoirs in front of the Rocky Mountains.

Atlantic Shelf

Mesozoic rift basins, bearing some comparison with the North Sea, have yielded the Hibernia giant discovery off Newfoundland. More finds can be expected although the results so far have been rather disappointing.

Arctic Islands and offshore

The Arctic islands and offshore form a huge province, with analogies to the Barents Sea and Svalbard. Some gas discoveries have been made, but it has been severely downgraded by drilling.

McKenzie Delta and Beaufort Sea

The McKenzie Delta and Beaufort Sea cover a Tertiary deltaic basin, very different from the adjoining Alaskan productive zone. Most of the discoveries have contained gas, which speaks of adverse source-rock conditions, but there have also been some important oil discoveries, including the near-giant Amauligak Field in 1984.

Ultimate

With 224 BBO, North America has the third largest endowment of all the regions.

Reserves

It has not been considered necessary to adjust the reported reserves of North America, although it should be noted that the definition of reserves in North America is perhaps more restrictive than is generally the case internationally. The two countries have a total of 32 BBO : 26 BBO in the USA and 6 BBO in Canada.

Undiscovered
Both countries are considered to be at a very mature stage of development onshore and in the productive offshore basins of California and the Gulf of Mexico. The Mid-Point of depletion was passed in 1973, and it is thought that very few important discoveries are left to be made. There is, however, an enormous potential for minor reserve revisions from infill and outstep drilling, because of the favourable commercial climate. A figure of 17 BBO is attributed to the USA, somewhat less than the official figure, and 3 BBO for Canada. The Arctic offshore and the very deep Gulf of Mexico seem to offer the best potential. It is claimed that important discoveries could be made if various tracts, which are closed for environmental reasons, were to be opened, but that is far from certain. Possibly, Canada has a greater potential than is here considered likely.

Production
Production commenced in 1859 in the United States, which was effectively the birthplace of the industry. It maintained its dominance until 1950, when it was overtaken by the Middle East. Canada did not yield significant production until 1950. Peak production was reached in the Region in 1972, falling subsequently to a low in 1976, before staging a partial recovery to a second peak in 1985. Production is now thought to be set for its final decline : both countries being way past the Mid-Point of their depletion, as illustrated in Part 3.

North America previously provided the Western World with an important strategic reserve through the mechanism of the Texas Railroad Commission that restricted production to maintain prices in the main producing states, but those days are over.

North America provided almost 16% of World production in 1989.

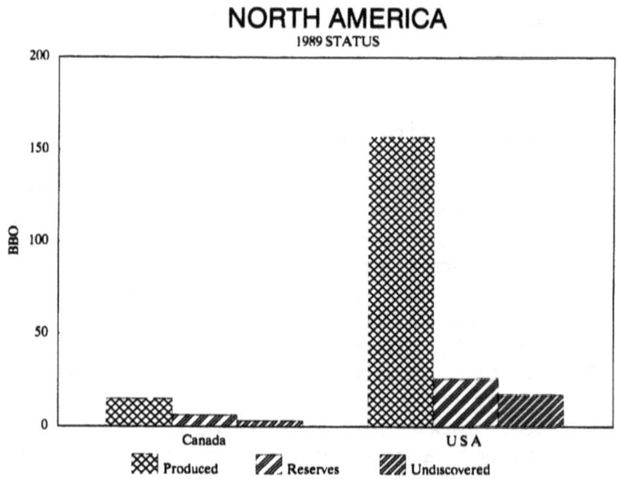

Figure 1-NA. Produced, Reserves and Undiscovered

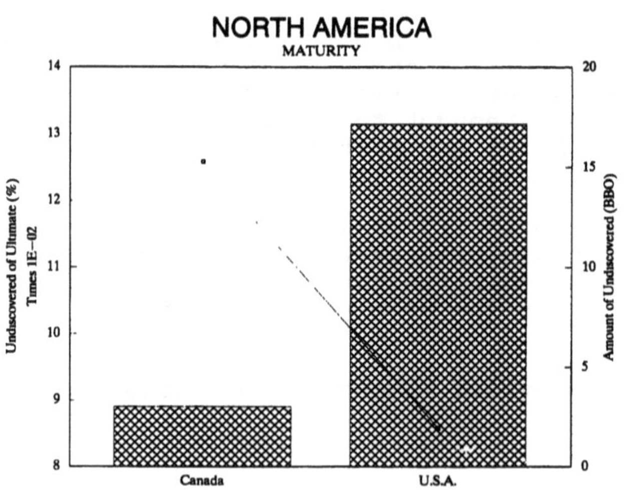

Figure 2-NA. Maturity: Undiscovered as a percentage of Ultimate

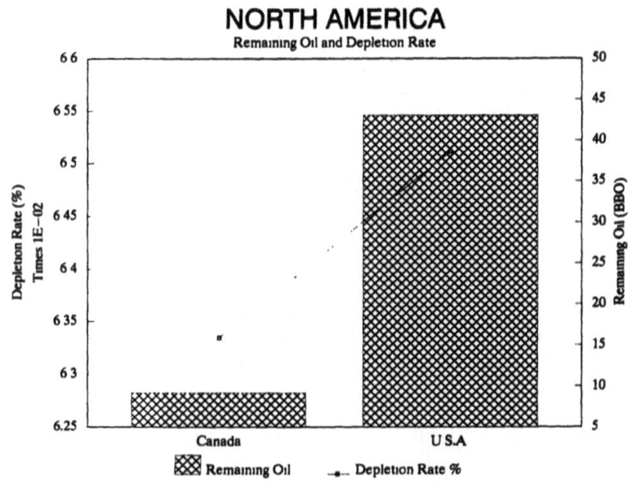

Figure 3-NA Remaining Oil (Reserves+Undiscovered) and Depletion Rate.

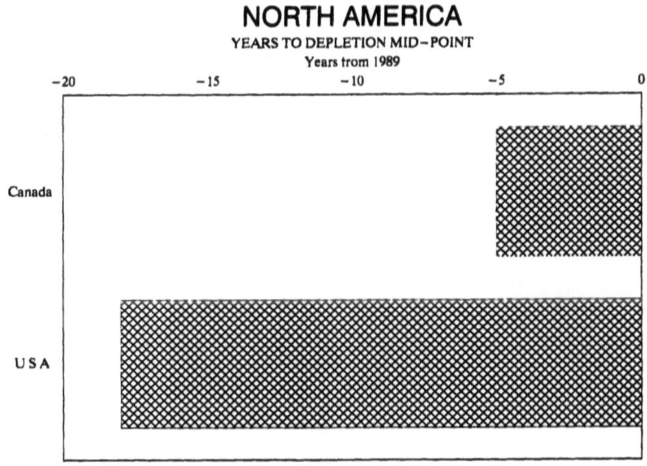

Figure 4-NA. Years to or from Depletion Mid-Point.

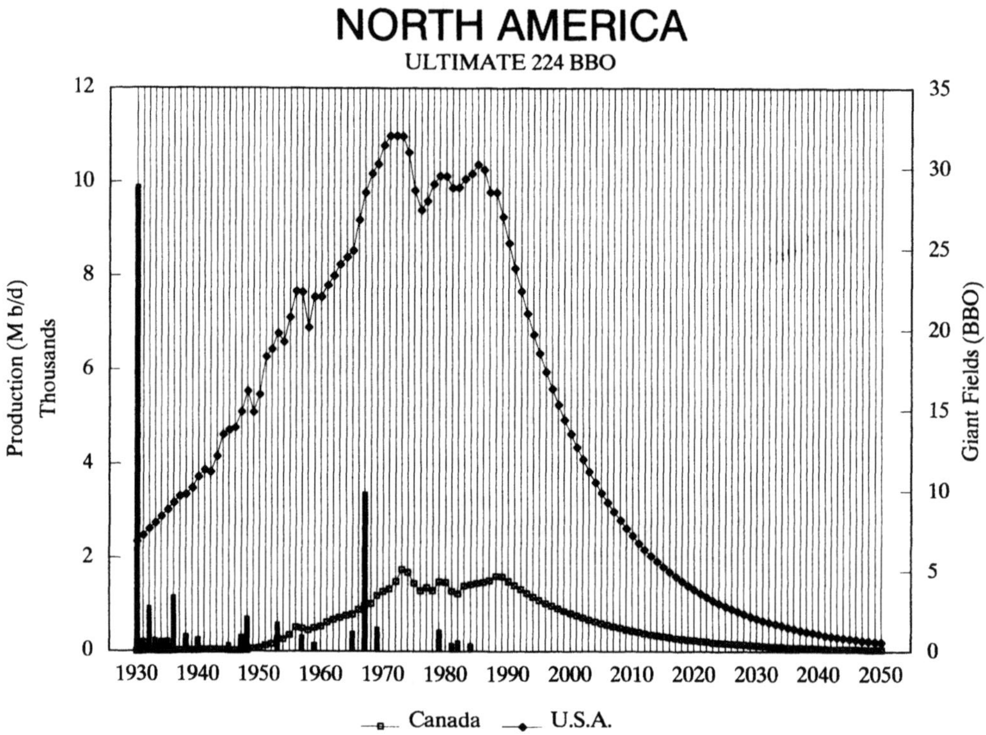

Pre – 1930 giants included in 1930

Figure 5-NA. Production Profile 1930-2050 with Giant Fields.

CUM.PROD	171.82 BBO	GIANT FIELDS		N.AMERICA
RESERVES	31.99 BBO	TOTAL BBO	63	REGION
DISCOVERED	203.81 BBO	% OF DISC.	31%	
UNDISCOVERED	20.19 BBO	Last Giant	1984	
REMAINING	52.18 BBO			Variance Trend
ULTIMATE	224.00 BBO			1990 +0.36%
Rept.Reserve	31.99	Ref.date(end)	1989	Future UP

Milestones	Date	Prod Mb/d	Dep. Rate	% Produced	Forecast n/a
					A=Exponential decline
1st Peak	1972	10990	3.39%	49%	B=5% incr. to Mid-Point
2nd Peak	1985	10372	5.39%	70%	C=Other
Mid-Point	1973	10975	3.51%	51%	Pre-1930 Cum. 13.313
Final Decline	n/a				(Added to 1930 prod)

Date	Prod. M.b/d	Prod. BBO/yr	Cum. Prod.	Remain	Deplet. Rate	% Produced	Giant	Notes & Names
1930	2351	0.858	14.17	209.83	ERR	6%	28.93	
1931	2482	0.906	15.08	208.92	0.43%	7%	0.76	
1932	2616	0.955	16.03	207.97	0.46%	7%	2.79	
1933	2752	1.005	17.04	206.96	0.48%	8%	0.81	
1934	2890	1.055	18.09	205.91	0.51%	8%	0.77	
1935	3028	1.105	19.20	204.80	0.54%	9%	0.79	
1936	3167	1.156	20.35	203.65	0.56%	9%	3.47	
1937	3306	1.207	21.56	202.44	0.59%	10%	0.00	
1938	3346	1.221	22.78	201.22	0.60%	10%	1.04	
1939	3487	1.273	24.05	199.95	0.63%	11%	0.00	
1940	3721	1.358	25.41	198.59	0.68%	11%	0.86	
1941	3870	1.412	26.82	197.18	0.71%	12%	0.00	
1942	3827	1.397	28.22	195.78	0.71%	13%	0.00	
1943	4153	1.516	29.74	194.26	0.77%	13%	0.00	
1944	4612	1.683	31.42	192.58	0.87%	14%	0.00	
1945	4717	1.722	33.14	190.86	0.89%	15%	0.50	
1946	4771	1.742	34.88	189.12	0.91%	16%	0.00	
1947	5108	1.864	36.75	187.25	0.99%	16%	1.03	
1948	5552	2.027	38.77	185.23	1.08%	17%	2.16	
1949	5104	1.863	40.64	183.36	1.01%	18%	0.00	
1950	5481	2.000	42.64	181.36	1.09%	19%	0.00	
1951	6280	2.292	44.93	179.07	1.26%	20%	0.00	
1952	6438	2.350	47.28	176.72	1.31%	21%	0.00	
1953	6782	2.475	49.76	174.24	1.40%	22%	1.80	
1954	6595	2.407	52.16	171.84	1.38%	23%	0.00	
1955	7117	2.598	54.76	169.24	1.51%	24%	0.00	
1956	7668	2.799	57.56	166.44	1.65%	26%	0.00	
1957	7655	2.794	60.35	163.65	1.68%	27%	1.01	
1958	6909	2.522	62.87	161.13	1.54%	28%	0.00	
1959	7551	2.756	65.63	158.37	1.71%	29%	0.54	
1960	7551	2.756	68.39	155.61	1.74%	31%	0.00	
1961	7797	2.846	71.23	152.77	1.83%	32%	0.00	
1962	8005	2.922	74.15	149.85	1.91%	33%	0.00	
1963	8254	3.013	77.17	146.83	2.01%	34%	0.00	
1964	8403	3.067	80.23	143.77	2.09%	36%	0.00	
1965	8541	3.118	83.35	140.65	2.17%	37%	1.24	
1966	9195	3.356	86.71	137.29	2.39%	39%	0.00	
1967	9773	3.567	90.28	133.72	2.60%	40%	9.92	
1968	10176	3.714	93.99	130.01	2.78%	42%	0.00	
1969	10378	3.788	97.78	126.22	2.91%	44%	1.50	
1970	10784	3.936	101.71	122.29	3.12%	45%	0.00	
1971	10986	4.010	105.72	118.28	3.28%	47%	0.00	
1972	10990	4.011	109.74	114.26	3.39%	49%	0.00	1st PEAK
1973	10975	4.006	113.74	110.26	3.51%	51%	0.00	& 50% PRODUCED
1974	10627	3.879	117.62	106.38	3.52%	53%	0.00	
1975	9820	3.584	121.20	102.80	3.37%	54%	0.00	
1976	9405	3.433	124.64	99.36	3.34%	56%	0.00	
1977	9600	3.504	128.14	95.86	3.53%	57%	0.00	
1978	9960	3.635	131.78	92.22	3.79%	59%	0.00	
1979	10130	3.697	135.47	88.53	4.01%	60%	1.33	
1980	10120	3.694	139.17	84.83	4.17%	62%	0.00	
1981	9875	3.604	142.77	81.23	4.25%	64%	0.55	
1982	9888	3.609	146.38	77.62	4.44%	65%	0.71	
1983	10065	3.674	150.05	73.95	4.73%	67%	0.00	
1984	10180	3.716	153.77	70.23	5.02%	69%	0.50	
1985	10372	3.786	157.56	66.44	5.39%	70%	0.00	2nd PEAK
1986	10265	3.747	161.30	62.70	5.64%	72%	0.00	
1987	9785	3.571	164.87	59.13	5.70%	74%	0.00	
1988	9771	3.566	168.44	55.56	6.03%	75%	0.00	

1989	9264	3.381	171.82	52.18	6.09%	77%	0.00
1990	8697	3.174	175.00	49.00	6.08%	78%	0.00
1991	8168	2.981	177.98	46.02	6.08%	79%	0.00
1992	7671	2.800	180.78	43.22	6.08%	81%	0.00
1993	7204	2.629	183.41	40.59	6.08%	82%	0.00
1994	6766	2.470	185.88	38.12	6.08%	83%	0.00
1995	6354	2.319	188.20	35.80	6.08%	84%	0.00
1996	5968	2.178	190.37	33.63	6.08%	85%	0.00
1997	5604	2.046	192.42	31.58	6.08%	86%	0.00
1998	5264	1.921	194.34	29.66	6.08%	87%	0.00
1999	4943	1.804	196.14	27.86	6.08%	88%	0.00
2000	4643	1.695	197.84	26.16	6.08%	88%	0.00
2001	4360	1.591	199.43	24.57	6.08%	89%	0.00
2002	4095	1.495	200.93	23.07	6.08%	90%	0.00
2003	3846	1.404	202.33	21.67	6.08%	90%	0.00
2004	3612	1.318	203.65	20.35	6.08%	91%	0.00
2005	3392	1.238	204.89	19.11	6.08%	91%	0.00
2006	3186	1.163	206.05	17.95	6.08%	92%	0.00
2007	2992	1.092	207.14	16.86	6.08%	92%	0.00
2008	2810	1.026	208.17	15.83	6.08%	93%	0.00
2009	2639	0.963	209.13	14.87	6.08%	93%	0.00
2010	2478	0.905	210.03	13.97	6.08%	94%	0.00
2011	2328	0.850	210.88	13.12	6.08%	94%	0.00
2012	2186	0.798	211.68	12.32	6.08%	95%	0.00
2013	2053	0.749	212.43	11.57	6.08%	95%	0.00
2014	1928	0.704	213.13	10.87	6.08%	95%	0.00
2015	1811	0.661	213.80	10.20	6.08%	95%	0.00
2016	1701	0.621	214.42	9.58	6.08%	96%	0.00
2017	1597	0.583	215.00	9.00	6.08%	96%	0.00
2018	1500	0.548	215.55	8.45	6.08%	96%	0.00
2019	1409	0.514	216.06	7.94	6.08%	96%	0.00
2020	1323	0.483	216.54	7.46	6.08%	97%	0.00
2021	1243	0.454	217.00	7.00	6.08%	97%	0.00
2022	1167	0.426	217.42	6.58	6.08%	97%	0.00
2023	1096	0.400	217.82	6.18	6.08%	97%	0.00
2024	1029	0.376	218.20	5.80	6.08%	97%	0.00
2025	967	0.353	218.55	5.45	6.08%	98%	0.00
2026	908	0.331	218.88	5.12	6.08%	98%	0.00
2027	853	0.311	219.19	4.81	6.08%	98%	0.00
2028	801	0.292	219.49	4.51	6.08%	98%	0.00
2029	752	0.275	219.76	4.24	6.08%	98%	0.00
2030	706	0.258	220.02	3.98	6.08%	98%	0.00
2031	663	0.242	220.26	3.74	6.08%	98%	0.00
2032	623	0.227	220.49	3.51	6.08%	98%	0.00
2033	585	0.214	220.70	3.30	6.08%	99%	0.00
2034	550	0.201	220.90	3.10	6.08%	99%	0.00
2035	516	0.188	221.09	2.91	6.08%	99%	0.00
2036	485	0.177	221.27	2.73	6.08%	99%	0.00
2037	455	0.166	221.43	2.57	6.08%	99%	0.00
2038	428	0.156	221.59	2.41	6.08%	99%	0.00
2039	402	0.147	221.74	2.26	6.08%	99%	0.00
2040	377	0.138	221.87	2.13	6.08%	99%	0.00
2041	354	0.129	222.00	2.00	6.08%	99%	0.00
2042	333	0.121	222.13	1.87	6.08%	99%	0.00
2043	312	0.114	222.24	1.76	6.08%	99%	0.00
2044	293	0.107	222.35	1.65	6.08%	99%	0.00
2045	276	0.101	222.45	1.55	6.08%	99%	0.00
2046	259	0.094	222.54	1.46	6.08%	99%	0.00
2047	243	0.089	222.63	1.37	6.08%	99%	0.00
2048	228	0.083	222.71	1.29	6.08%	99%	0.00
2049	214	0.078	222.79	1.21	6.08%	99%	0.00
2050	201	0.073	222.87	1.13	6.08%	99%	0.00

SUMMARY DATA SHEET
Reference Date 1989

NORTH AMERICA

Country	Production M b/d	BBO/ year	Cumulative Production	Reported Reserves	Adjusted Reserves	Discovered	Undiscovered	Ultimate	Remaining	Und. of Ult.	Prod. of Ult.	Rem. of Ult.	Prod. of Disc	Disc. in Giants	Dep. Rate
A	B	C	D	E	F	G	H	I	J	K	L	M	N	O	P
Canada	1588	0.580	14.85	6.13	6.13	20.98	3.02	24.00	9.15	13%	62%	38%	71%	30%	6.33%
U.S.A.	7676	2.802	156.97	25.86	25.86	182.83	17.17	200.00	43.03	9%	78%	22%	86%	31%	6.51%
Region	9264	3.381	171.82	31.99	31.99	203.81	20.19	224.00	52.18	9%	77%	23%	84%	31%	6.48%

LATIN AMERICA ("LA")

SUMMARY

Latin America has 10.6% of the World's Ultimate oil endowment. Fourteen oil countries make up the region.

Oil exploration commenced in the last century in Peru and Trinidad, and most of the important provinces were discovered during the early years of this century. The two largest producing countries are Venezuela and Mexico, both of which were developed in earlier years by the international companies. Mexico expropriated the assets of the foreign companies in 1938, and Venezuela followed suit in 1976. State companies have exclusive positions in these two countries and in Brasil. They also have dominant, but not exclusive, positions in Argentina, Colombia, Ecuador, Peru and Bolivia.

Latin America's share of World oil is as follows :

Cumulative Production	12.1%
Reserves	10.5%
Discovered	11.2%
Undiscovered	6.5%
Remaining	9.7%
Ultimate	10.6%

Latin America's main productive provinces are :

Northern Sub-Andean Province

A number of productive basins from Trinidad to northern Peru separate the Central Andes from the Guayana Shield. (The Maracaibo and Magdalena basins belong to this province although now separated as a result of the late Tertiary uplift of the Eastern Andes). The province depends on prolific generation of oil from the Upper Cretaceous, La Luna Formation. The trend dies out to the south in Peru, where the source-rock becomes less prolific and gas prone. Apart from Lake Maracaibo, most important production is onshore, although there is some offshore production in Trinidad.

Southern Sub-Andean Province

A similar, but much less prolific, trend develops in the

southern part of the continent from Bolivia through Argentina and Chile. The Cretaceous is again an important source-rock, although production in northern Argentina and Bolivia partly depends on Palaeozoic source-rocks.

Mexican Basin

The western margin of the Gulf of Mexico contains prolific source-rocks, primarily of late Jurassic age, which have charged mainly carbonate reservoirs both offshore and onshore Mexico, forming what was called the "Golden Lane'. Carbonate reservoirs tend to yield flush early production, but require careful attention in their latter stages to avoid water invasion. Furthermore, some of the Mexican fields produce from irregular stratigraphic traps, whose size is not easily predicted. There are therefore technical as well as political reasons for questioning Mexico's reserve estimates.

Offshore Brasil

Oil was discovered on the Atlantic coast of Brasil many years ago, and a large number of mainly small fields have been produced. Of much greater importance are the deep-water developments of recent years, which have yielded several giant fields producing from Tertiary turbidites and contourites, sourced in the underlying Cretaceous. Petrobras, the Brasilian state-owned company, is to be credited for its skill and initiative in the technological innovations in pioneering these deepwater projects.

Proto-Amazon Delta

It seems that the Amazon flowed westwards until the uplift of the Andes in late Tertiary times. It gave rise to a productive Tertiary delta straddling the Ecuador-Peru border and running offshore.

Ultimate

Venezuela and Mexico are by far the most important oil countries in the region with an Ultimate endowment of respectively 83 BBO and 60 BBO. They are followed by Brasil (10 BBO), Argentina (6.0 BBO), Colombia (5.5 BBO), Ecuador (3.5 BBO), Trinidad (3.3 BBO) and Peru (2.75 BBO).

Reserves

It has been considered necessary to adjust the reported reserves of the three most important countries : downwards for Venezuela and Mexico, but upwards for Brasil.

Only six countries have reserves in excess of 1 BBO : Venezuela (35 BBO c.f. 58 BBO reported), Mexico (40 BBO c.f. 56 BBO reported), Brasil (5 BBO c.f 2.8 BBO reported),

Argentina (2.28 BBO), Colombia (2.06 BBO) and Ecuador (1.51 BBO).

Giant fields are responsible for 64% of all discovered oil, with peak giant discovery occurring in 1957. The last giant was found in 1986.

Undiscovered
Only four countries are considered to have an Undiscovered potential in excess of 1 BBO : Venezuela (6.13 BBO), Mexico (3.64 BBO), Brasil (2.36 BBO) and Argentina (1.53 BBO). It is doubted that many very important discoveries will be made onshore in Latin America, but there may be new offshore potential on the Falklands arc and perhaps elsewhere off the Atlantic coast.

Production
Production commenced in the early years of the century in Trinidad, Argentina, Colombia, Peru and Ecuador, followed by Mexico and Venezuela. Production grew rapidly after the war, especially from Venezuela, reaching a peak in 1985 of 6349 M b/d. It then fell slightly, but is now expected to rise to a second peak in 1999 of 7461 M b/d, due largely to predicted increases from Brasil and Mexico. According to the prediction, the Depletion Mid-Point will be reached in 1994, followed by a decline at about 3%.

Mexico, with 2618 M b/d, and Venezuela, with 1732 M b/d, were by far the most important producers in 1989, and are also responsible for most of the Cumulative, having produced respectively 16.36 and 41.87 BBO. Brasil, Argentina, Colombia, Ecuador, Trinidad and Peru are next in importance.

As illustrated in the Production Profiles in Part 3, Venezuela shows a classic pattern of advancing maturity, whereas Mexico appears to have scope for increasing its near-term production considerably, even after adjusting the reported reserves down-wards. The recent discovery of giant fields in Brasil gives that country scope for near-term increase. Bolivia seems to have good scope for new, although not large, discovery. Production in Barbados, Guatemala and Suriname is too small to merit plotting.

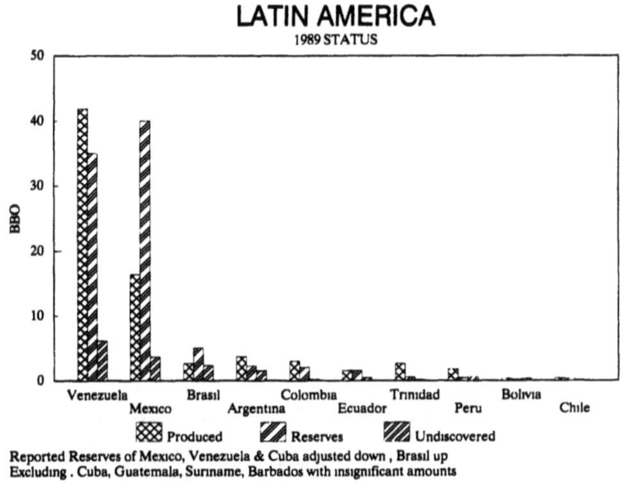

Figure 1-LA. Produced, Reserves and Undiscovered

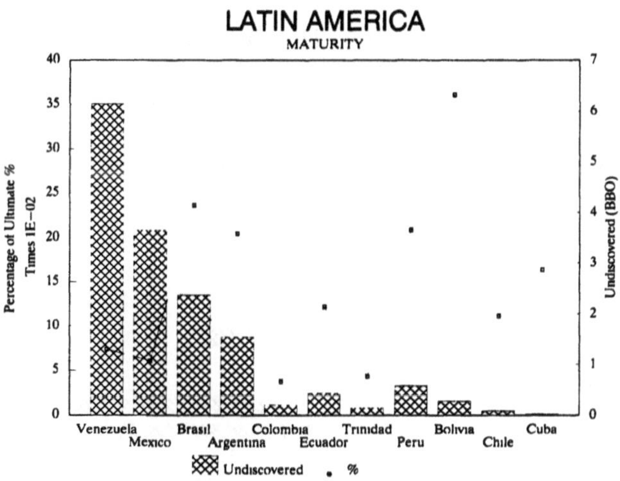

Figure 2-LA. Maturity: Undiscovered as a percentage of Ultimate

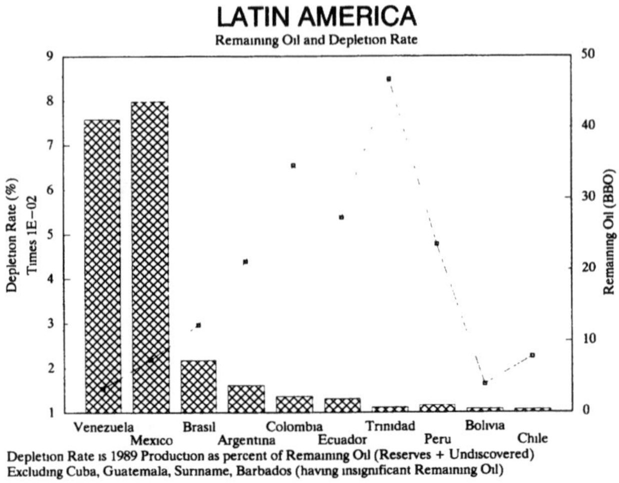

Figure 3-LA. Remaining Oil (Reserves+Undiscovered) and Depletion Rates.

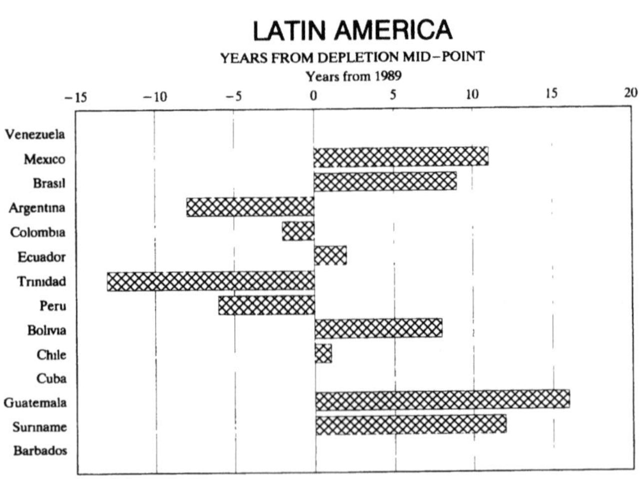

Figure 4-LA. Years to or from Depletion Mid-Point

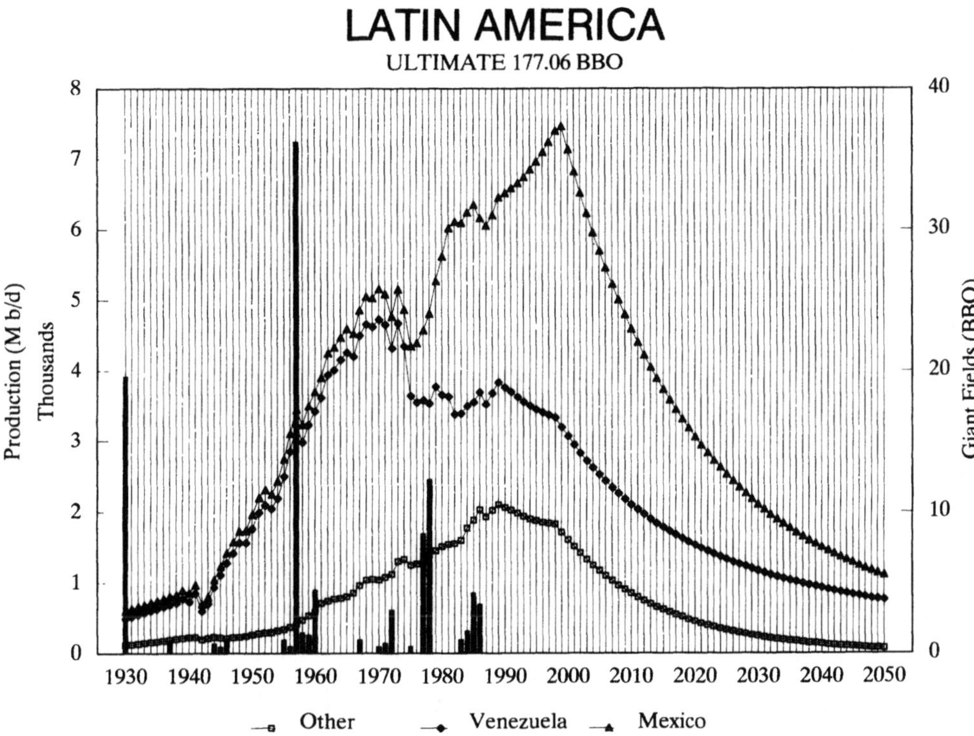

Pre – 1930 Giants included in 1930

Figure 5-LA. Production Profile 1930-2050 with Giant Fields

CUM.PROD	77.36 BBO	GIANT FIELDS		LATIN AMERICA	
RESERVES	87.39 BBO	TOTAL BBO	105	REGION	
DISCOVERED	164.75 BBO	% OF DISC.	64%		
UNDISCOVERED	12.31 BBO	Last Giant	1986		
REMAINING	99.70 BBO			Variance	Trend
ULTIMATE	177.06 BBO			1990	+6%
Rept.Reserves	58.50 BBO	Ref. Date (end)	1989	Future	UP

Milestones	Date	Prod Mb/d	Dep. Rate	% Produced
1st Peak	1999	7461	3.53%	58%
2nd Peak				
Mid-Point	1994	6845	2.78%	51%
Final Decline	n/a			

Forecast N/A
A=Exponential decline
B=5% inc. to Mid-Point
C=Other
Pre-1930 Cum 2.223
(added to 1930 Prod)

Date	Prod. M.b/d	Prod. BBO/yr	Cum Prod.	Remain	Dep. Rate	% Produced	Giant	Names & Notes
1930	608	0.222	2.45	174.61	ERR	1%	19.60	
1931	634	0.231	2.68	174.38	0.1%	2%	0.00	
1932	662	2.000	4.68	172.38	1.1%	3%	0.00	
1933	689	0.252	4.93	172.13	0.1%	3%	0.00	
1934	717	0.262	5.19	171.87	0.2%	3%	0.00	
1935	745	0.272	5.46	171.60	0.2%	3%	0.00	
1936	773	0.282	5.74	171.32	0.2%	3%	0.00	
1937	801	0.292	6.04	171.02	0.2%	3%	0.96	
1938	826	0.301	6.34	170.72	0.2%	4%	0.00	
1939	897	0.328	6.66	170.40	0.2%	4%	0.00	
1940	856	0.313	6.98	170.08	0.2%	4%	0.00	
1941	968	0.353	7.33	169.73	0.2%	4%	0.00	
1942	699	0.255	7.59	169.47	0.2%	4%	0.00	
1943	800	0.292	7.88	169.18	0.2%	4%	0.00	
1944	1046	0.382	8.26	168.80	0.2%	5%	0.72	
1945	1233	0.450	8.71	168.35	0.3%	5%	0.50	
1946	1414	0.516	9.23	167.83	0.3%	5%	1.00	
1947	1572	0.574	9.80	167.26	0.3%	6%	0.00	
1948	1730	0.632	10.43	166.63	0.4%	6%	0.00	
1949	1737	0.634	11.07	165.99	0.4%	6%	0.00	
1950	1960	0.715	11.78	165.28	0.4%	7%	0.00	
1951	2200	0.803	12.58	164.48	0.5%	7%	0.00	
1952	2312	0.844	13.43	163.63	0.5%	8%	0.00	
1953	2248	0.820	14.25	162.81	0.5%	8%	0.00	
1954	2427	0.886	15.13	161.93	0.5%	9%	0.00	
1955	2747	1.003	16.14	160.92	0.6%	9%	1.00	
1956	3107	1.134	17.27	159.79	0.7%	10%	0.52	
1957	3450	1.259	18.53	158.53	0.8%	10%	36.20	
1958	3229	1.179	19.71	157.35	0.7%	11%	1.47	
1959	3500	1.278	20.99	156.07	0.8%	12%	1.30	
1960	3702	1.351	22.34	154.72	0.9%	13%	4.50	
1961	3902	1.424	23.76	153.30	0.9%	13%	0.00	
1962	4250	1.551	25.31	151.75	1.0%	14%	0.00	
1963	4331	1.581	26.89	150.17	1.0%	15%	0.00	
1964	4470	1.631	28.52	148.54	1.1%	16%	0.00	
1965	4596	1.678	30.20	146.86	1.1%	17%	0.00	
1966	4529	1.653	31.86	145.20	1.1%	18%	0.00	
1967	4858	1.773	33.63	143.43	1.2%	19%	1.00	
1968	5044	1.841	35.47	141.59	1.3%	20%	0.00	
1969	5031	1.836	37.31	139.75	1.3%	21%	0.00	
1970	5159	1.883	39.19	137.87	1.3%	22%	0.50	
1971	5084	1.856	41.04	136.02	1.3%	23%	0.75	
1972	4763	1.739	42.78	134.28	1.3%	24%	3.06	
1973	5150	1.880	44.66	132.40	1.4%	25%	0.00	
1974	4865	1.776	46.44	130.62	1.3%	26%	0.00	
1975	4352	1.588	48.03	129.03	1.2%	27%	0.50	
1976	4400	1.606	49.63	127.43	1.2%	28%	0.00	
1977	4575	1.670	51.30	125.76	1.3%	29%	8.50	
1978	4807	1.755	53.06	124.00	1.4%	30%	12.29	
1979	5266	1.922	54.98	122.08	1.6%	31%	0.00	
1980	5620	2.051	57.03	120.03	1.7%	32%	0.00	
1981	6023	2.198	59.23	117.83	1.8%	33%	0.00	
1982	6113	2.231	61.46	115.60	1.9%	35%	0.00	
1983	6093	2.224	63.68	113.38	1.9%	36%	1.00	
1984	6241	2.278	65.96	111.10	2.0%	37%	1.60	
1985	6349	2.317	68.28	108.78	2.1%	39%	4.30	
1986	6163	2.249	70.53	106.53	2.1%	40%	3.49	
1987	6058	2.211	72.74	104.32	2.1%	41%	0.00	
1988	6207	2.266	75.01	102.05	2.2%	42%	0.00	

1989	6451	2.355	77.36	99.70	2.3%	44%	0.00	
1990	6511	2.376	79.74	97.32	2.4%	45%	0.00	
1991	6587	2.404	82.14	94.92	2.5%	46%	0.00	
1992	6658	2.430	84.57	92.49	2.6%	48%	0.00	
1993	6744	2.462	87.03	90.03	2.7%	49%	0.00	
1994	6845	2.498	89.53	87.53	2.8%	51%	0.00	50% PRODUCED
1995	6960	2.540	92.07	84.99	2.9%	52%	0.00	
1996	7090	2.588	94.66	82.40	3.0%	53%	0.00	
1997	7236	2.641	97.30	79.76	3.2%	55%	0.00	
1998	7394	2.699	100.00	77.06	3.4%	56%	0.00	
1999	7461	2.723	102.72	74.34	3.5%	58%	0.00	1ST PEAK
2000	7131	2.603	105.32	71.74	3.5%	59%	0.00	
2001	6815	2.488	107.81	69.25	3.5%	61%	0.00	
2002	6515	2.378	110.19	66.87	3.4%	62%	0.00	
2003	6231	2.274	112.46	64.60	3.4%	64%	0.00	
2004	5960	2.175	114.64	62.42	3.4%	65%	0.00	
2005	5703	2.082	116.72	60.34	3.3%	66%	0.00	
2006	5458	1.992	118.71	58.35	3.3%	67%	0.00	
2007	5226	1.907	120.62	56.44	3.3%	68%	0.00	
2008	5004	1.827	122.45	54.61	3.2%	69%	0.00	
2009	4794	1.750	124.20	52.86	3.2%	70%	0.00	
2010	4594	1.677	125.87	51.19	3.2%	71%	0.00	
2011	4404	1.607	127.48	49.58	3.1%	72%	0.00	
2012	4223	1.541	129.02	48.04	3.1%	73%	0.00	
2013	4051	1.478	130.50	46.56	3.1%	74%	0.00	
2014	3887	1.419	131.92	45.14	3.0%	75%	0.00	
2015	3731	1.362	133.28	43.78	3.0%	75%	0.00	
2016	3582	1.308	134.59	42.47	3.0%	76%	0.00	
2017	3441	1.256	135.85	41.21	3.0%	77%	0.00	
2018	3306	1.207	137.05	40.01	2.9%	77%	0.00	
2019	3178	1.160	138.21	38.85	2.9%	78%	0.00	
2020	3056	1.115	139.33	37.73	2.9%	79%	0.00	
2021	2939	1.073	140.40	36.66	2.8%	79%	0.00	
2022	2828	1.032	141.43	35.63	2.8%	80%	0.00	
2023	2722	0.994	142.43	34.63	2.8%	80%	0.00	
2024	2621	0.957	143.38	33.68	2.8%	81%	0.00	
2025	2525	0.922	144.30	32.76	2.7%	82%	0.00	
2026	2433	0.888	145.19	31.87	2.7%	82%	0.00	
2027	2345	0.856	146.05	31.01	2.7%	82%	0.00	
2028	2261	0.825	146.87	30.19	2.7%	83%	0.00	
2029	2181	0.796	147.67	29.39	2.6%	83%	0.00	
2030	2105	0.768	148.44	28.62	2.6%	84%	0.00	
2031	2031	0.741	149.18	27.88	2.6%	84%	0.00	
2032	1962	0.716	149.90	27.16	2.6%	85%	0.00	
2033	1895	0.692	150.59	26.47	2.5%	85%	0.00	
2034	1831	0.668	151.26	25.80	2.5%	85%	0.00	
2035	1770	0.646	151.90	25.16	2.5%	86%	0.00	
2036	1711	0.625	152.53	24.53	2.5%	86%	0.00	
2037	1656	0.604	153.13	23.93	2.5%	86%	0.00	
2038	1602	0.585	153.72	23.34	2.4%	87%	0.00	
2039	1551	0.566	154.28	22.78	2.4%	87%	0.00	
2040	1502	0.548	154.83	22.23	2.4%	87%	0.00	
2041	1455	0.531	155.36	21.70	2.4%	88%	0.00	
2042	1409	0.514	155.87	21.19	2.4%	88%	0.00	
2043	1366	0.499	156.37	20.69	2.4%	88%	0.00	
2044	1325	0.484	156.86	20.20	2.3%	89%	0.00	
2045	1285	0.469	157.33	19.73	2.3%	89%	0.00	
2046	1247	0.455	157.78	19.28	2.3%	89%	0.00	
2047	1210	0.442	158.22	18.84	2.3%	89%	0.00	
2048	1175	0.429	158.65	18.41	2.3%	90%	0.00	
2049	1141	0.417	159.07	17.99	2.3%	90%	0.00	
2050	1109	0.405	159.47	17.59	2.2%	90%	0.00	

SUMMARY DATA SHEET REGION LATIN AMERICA
REFERENCE DATE 1989

Country	Production M b/d	BBO/yr	Cumulative Production	Reserves Reported	Adjusted	Discovered	Undiscovered	Ultimate	Remaining	Und. of Ult.	Prod. of Ult.	Rem. of Ult.	Prod. of Disc	Disc. in Giants	Dep. Rate %
A	B	C	D	E	F	G	H	I	J	K	L	M	N	O	P
Venezuela	1732	0.632	41.87	58.50	35.00	76.87	6.13	83.00	41.13	7%	50%	50%	54%	65%	1.54%
Mexico	2618	0.956	16.36	56.37	40.00	56.36	3.64	60.00	43.64	6%	27%	73%	29%	72%	2.19%
Brasil	597	0.218	2.64	2.82	5.00	7.64	2.36	10.00	7.36	24%	26%	74%	35%	78%	2.96%
Argentina	459	0.168	3.69	2.28	2.28	5.97	1.53	7.50	3.81	20%	49%	51%	62%	34%	4.39%
Colombia	405	0.148	2.99	2.06	2.06	5.05	0.20	5.25	2.26	4%	57%	43%	59%	34%	6.55%
Ecuador	286	0.104	1.56	1.51	1.51	3.07	0.43	3.50	1.94	12%	45%	55%	51%	80%	5.38%
Trinidad	154	0.056	2.59	0.52	0.52	3.11	0.14	3.25	0.66	4%	80%	20%	83%	41%	8.47%
Peru	129	0.047	1.76	0.41	0.41	2.18	0.57	2.75	0.99	21%	64%	36%	81%	64%	4.77%
Bolivia	20	0.007	0.30	0.18	0.18	0.48	0.27	0.75	0.45	36%	40%	60%	62%	0%	1.64%
Chile	24	0.009	0.37	0.30	0.30	0.67	0.08	0.75	0.38	11%	49%	51%	55%	0%	2.25%
Cuba	19	0.007	0.08	0.10	0.05	0.13	0.02	0.15	0.07	16%	50%	50%	60%	0%	9.30%
Guatemala	4	0.001	0.02	0.04	0.04	0.06	0.04	0.10	0.08	40%	18%	82%	30%	0%	1.64%
Suriname	4	0.001	0.00	0.03	0.03	0.03	0.02	0.05	0.05	35%	9%	91%	13%	0%	2.96%
Barbados	1	0.000	0.01	0.00	0.00	0.01	0.00	0.01	0.00	16%	54%	46%	64%	0%	9.45%
Region	6451	2.355	74.22	125.13	87.39	161.61	15.45	177.06	102.84	9%	42%	58%	46%	64%	2.29%

AFRICA ("AF")

Summary
Figure-1: Status - Production, Reserves, Undiscovered
Figure-2: Maturity - Undiscovered as a % of Ultimate
Figure-3: Remaining Oil and Depletion Rate
Figure-4: Years to Depletion Mid-Point
Figure-5: Production with Giant Fields
Table 1 : Production Profile
Table 2 : Summary Data-sheet

SUMMARY

Africa has 7.3% of the World's Ultimate oil endowment. Fifteen oil countries make up the region. (Chad is due to be added when it comes on to production, but is here included in "New Provinces").

Most of the oil in the region was discovered by the international oil companies in the 1960s. Not all have been able to remain. Most were forced to leave Libya for political reasons, although some European companies remain. Algeria expelled the French companies which were active there before independence, and Nigeria expelled BP, previously in joint venture with Shell, but has now readmitted the company in partnership with the Norwegian Statoil. State companies are particularly active in Libya and Algeria, and work in cooperation with foreign companies in Egypt and Angola.

Africa's share of World oil is as follows :

Cumulative Production	7.4%
Reserves	7.3%
Discovered	7.4%
Undiscovered	7.4%
Remaining	7.3%
Ultimate	7.4%

Africa's main productive provinces are :

North Africa

A number of productive basins flank the northern boundary of the African Shield. They are mainly onshore, although there are offshore extensions in Libya, Egypt and Tunisia. The Cretaceous is the principal source-rock in this province, although there are also important older sources in the interior regions, especially of Algeria. The most important basin is the Sirte Basin in Libya, which includes some prolific reefal fields.

West Africa

The most important province in West Africa is the Niger delta region of Nigeria where numerous fields with Tertiary reservoirs have been found, both onshore and offshore, although relatively few are giants. This province may have some very deep water potential. To the south lie the rift basins of Angola, Gabon and the Congo with Cretaceous source-rocks and important salt deposits. The reservoirs both under- and overlie the salt, which acts as an effective seal to the deeper reservoirs. Production is mainly offshore. There are unconfirmed reports of recent giant discoveries in Nigeria and Gabon, which may be an important indication of promise.

Gulf of Suez

The Gulf of Suez is a small but very prolific rift basin in Egypt, where Tertiary salt deposits mask underlying structures.

Discoveries have also been made in Sudan and in Chad, the latter in an interior basin having some connection with the Nigerian basin. Two countries, Morocco and Ghana, have exhausted their resources.

Ultimate

Libya is by far the most important oil country in the region with an Ultimate endowment of 45 BBO. It is followed by Nigeria (32 BBO), Algeria (20 BBO) and Egypt (11 BBO).

A second tier of countries, of lesser but significant importance, is made up of Angola (4.24 BBO), Tunisia (3.50 BBO), Gabon (2.5 BBO), Congo (2.0 BBO) and Cameroon (1 BBO).

Reserves

Apart from a minor adjustment for Benin's reserves, the reported figures have been accepted for all the countries in the Region.

Six countries have reserves of in excess of 1 BBO : Libya 22,8 BBO, Nigeria 16 BBO, Algeria 9.2 BBO, Egypt (4.5 BBO), Angola (2.2 BBO) and Tunisia (1.25 BBO). Algeria's figure of 9.2 BBO seems excessive, but has not been adjusted downwards here.

Giant fields are responsible for 57% of all discovered oil, with peak giant discovery occurring in 1961. The last confirmed giant was found in 1990.

Undiscovered

Only four countries are considered to have an Undiscovered potential of in excess of 1 BBO : Libya (5.95 BBO), Nigeria (3.09 BBO), Algeria (2.43 BBO) and Egypt (1.85 BBO). With the possible exception of Somalia, it is doubted that any important discoveries will be made in East Africa, due to very limited source-rocks and adverse conditions for generation. There may be new very deep water potential off West Africa, similar to that off Brasil on the other side of the South Atlantic rift system.

Production

Significant production did not develop in Africa until the 1960s. It reached a peak in 1979 of 6714 M b/d, falling to a low in 1983 of 4333 M b/d. It is expected to reach a second peak in 1994. The Depletion Mid-Point should be reached in 1996, followed by a decline at just under 4%.

Nigeria (1605 M b/d) and Libya (1101 M b/d) gave by far the highest production in 1989, and are also responsible for most of the Cumulative, having produced respectively 12.91 and 16.25 BBO. Algeria, Egypt and Angola are next in importance.

As shown in the Profiles in Part 3, Algeria would have a classic mature production profile but for the somewhat suspect high figure of Remaining oil, which gives a second peak in the late 1990s. Egypt has a normal production curve following the early discovery of giant fields. Gabon has an interesting second peak following late stage discoveries. Libyan production fell radically in the late 1970s and 1980s due to political constraints and OPEC cuts, but a second peak is expected following improved conditions. Nigeria, likewise suffered seriously during the "glut', but a second peak is expected, partly following recent discoveries. Tunisia has a somewhat anomalous curve with a high second peak, due to a somewhat suspect Remaining oil figure. None of the other countries merit particular comment.

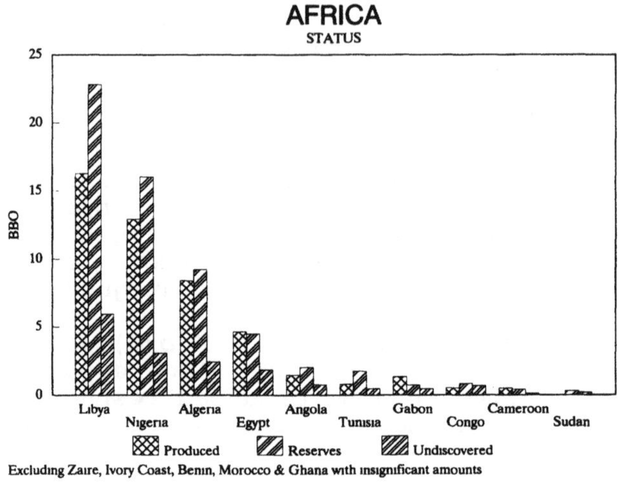

Figure 1-AF. Produced, Reserves and Undiscovered.

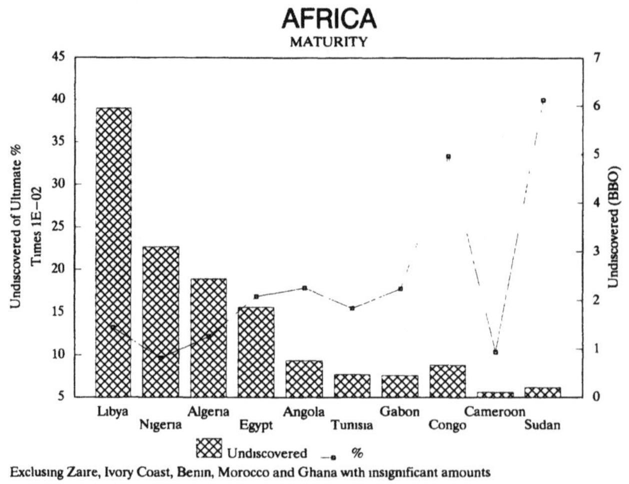

Figure 2-AF. Maturity: Undiscovered as a percentage of Ultimate

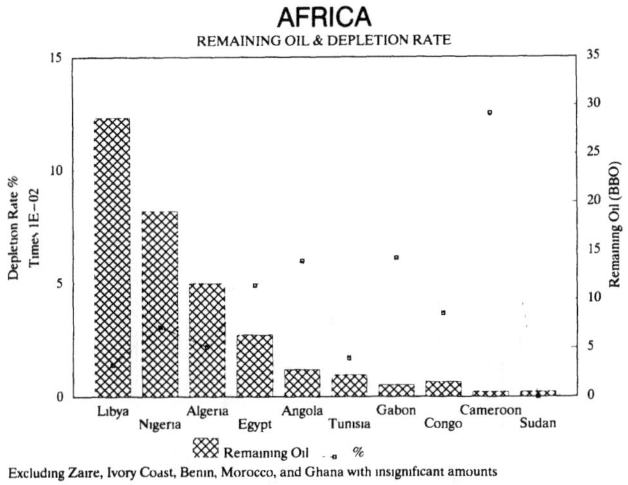

Figure 3-AF. Remaining Oil (Reserves+Undiscovered) and Depletion Rate.

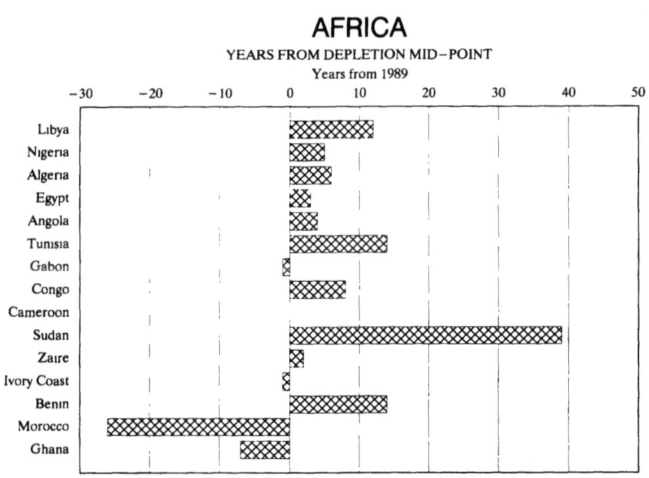

Figure 4-AF. Years to or from Depletion Mid-Point

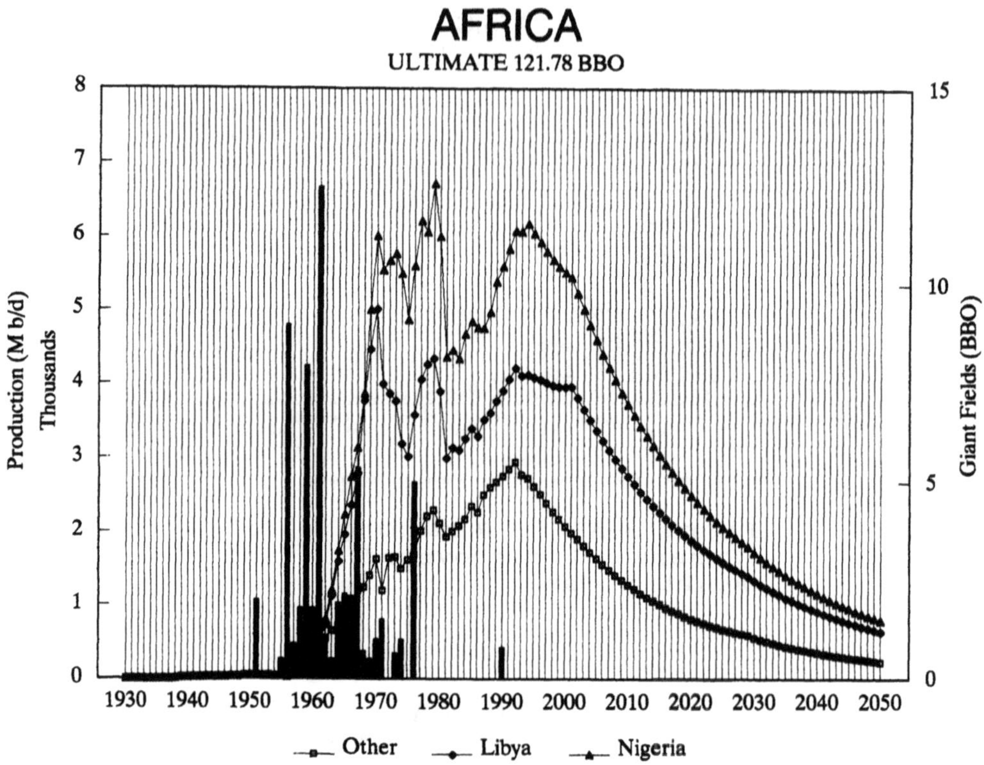

Figure 5-AF. Production Profile 1930-2050 with Giant fields.

CUM.PROD	46.97	BBO	GIANT FIELDS			AFRICA
RESERVES	58.74	BBO	TOTAL BBO	60.6		REGION
DISCOVERED	105.71	BBO	% OF DISC.	57%		
UNDISCOVERED	16.07	BBO	Last Giant	1990		Variance Trend
REMAINING	74.80	BBO				1990 +8%
ULTIMATE	121.78	BBO				
Rept.Reserves	58.79	BBO	Ref.Date (end)		1989	Future UP

Milestones	Date	Prod Mb/d	Dep Rate	% Produced
1st Peak	1979	6714	2.58%	24%
2nd Peak	1994	6173	3.40%	47%
Mid-Point	1996	5919	3.50%	51%
Final Decline	N/A			

Forecast N/A
A=Exponential decline
B=5% inc. to Mid-Point
C=Other
Pre-1930 Cum 0.00
(Added to 1930 Prod)

Date	Prod. M.b/d	Prod. BBO/yr	Cum Prod	Remain	Dep. Rate	% Produced	Giant	Names & Notes
1930	0	0.000	0.00	121.78	ERR	0%		
1931	0	0.000	0.00	121.78	0.00%	0%	0.000	
1932	0	0.000	0.00	121.78	0.00%	0%	0.000	
1933	0	0.000	0.00	121.78	0.00%	0%	0.000	
1934	0	0.000	0.00	121.78	0.00%	0%	0.000	
1935	0	0.000	0.00	121.78	0.00%	0%	0.000	
1936	0	0.000	0.00	121.78	0.00%	0%	0.000	
1937	0	0.000	0.00	121.78	0.00%	0%	0.000	
1938	4	0.002	0.00	121.78	0.00%	0%	0.000	
1939	13	0.005	0.01	121.77	0.00%	0%	0.000	
1940	18	0.006	0.01	121.76	0.01%	0%	0.000	
1941	23	0.009	0.02	121.76	0.01%	0%	0.000	
1942	23	0.008	0.03	121.75	0.01%	0%	0.000	
1943	25	0.009	0.04	121.74	0.01%	0%	0.000	
1944	26	0.009	0.05	121.73	0.01%	0%	0.000	
1945	26	0.009	0.06	121.72	0.01%	0%	0.000	
1946	25	0.009	0.07	121.71	0.01%	0%	0.000	
1947	36	0.013	0.08	121.70	0.01%	0%	0.000	
1948	36	0.013	0.09	121.68	0.01%	0%	0.000	
1949	44	0.016	0.11	121.67	0.01%	0%	0.000	
1950	45	0.017	0.13	121.65	0.01%	0%	0.000	
1951	46	0.017	0.14	121.63	0.01%	0%	2.000	
1952	47	0.017	0.16	121.62	0.01%	0%	0.000	
1953	50	0.018	0.18	121.60	0.02%	0%	0.000	
1954	42	0.015	0.19	121.58	0.01%	0%	0.000	
1955	38	0.014	0.21	121.57	0.01%	0%	0.500	
1956	35	0.013	0.22	121.56	0.01%	0%	9.000	
1957	44	0.016	0.24	121.54	0.01%	0%	0.900	
1958	91	0.033	0.27	121.51	0.03%	0%	1.800	
1959	108	0.039	0.31	121.47	0.03%	0%	7.950	
1960	276	0.101	0.41	121.37	0.08%	0%	1.800	
1961	480	0.175	0.58	121.19	0.14%	0%	12.500	
1962	794	0.290	0.87	120.90	0.24%	1%	1.000	
1963	1194	0.436	1.31	120.47	0.36%	1%	0.500	
1964	1733	0.633	1.94	119.84	0.53%	2%	1.930	
1965	2226	0.813	2.75	119.02	0.68%	2%	2.150	
1966	2741	1.001	3.75	118.02	0.84%	3%	2.100	
1967	3132	1.143	4.90	116.88	0.97%	4%	5.370	
1968	3865	1.411	6.31	115.47	1.21%	5%	0.700	
1969	5000	1.825	8.13	113.64	1.58%	7%	0.500	
1970	6003	2.191	10.32	111.45	1.93%	8%	1.000	
1971	5539	2.022	12.35	109.43	1.81%	10%	1.500	
1972	5666	2.068	14.41	107.36	1.89%	12%	0.000	
1973	5764	2.104	16.52	105.26	1.96%	14%	0.645	
1974	5487	2.003	18.52	103.26	1.90%	15%	1.000	
1975	4864	1.775	20.30	101.48	1.72%	17%	0.000	
1976	5598	2.043	22.34	99.44	2.01%	18%	5.000	
1977	6206	2.265	24.60	97.17	2.28%	20%	0.000	
1978	6058	2.211	26.82	94.96	2.28%	22%	0.000	
1979	6714	2.451	29.27	92.51	2.58%	24%	0.000	1ST PEAK
1980	5997	2.189	31.46	90.32	2.37%	26%	0.000	
1981	4360	1.591	33.05	88.73	1.76%	27%	0.000	
1982	4455	1.626	34.67	87.10	1.83%	28%	0.000	
1983	4333	1.582	36.25	85.52	1.82%	30%	0.000	
1984	4673	1.706	37.96	83.82	1.99%	31%	0.000	
1985	4839	1.766	39.73	82.05	2.11%	33%	0.000	
1986	4757	1.736	41.46	80.31	2.12%	34%	0.000	
1987	4755	1.736	43.20	78.58	2.16%	35%	0.000	
1988	4967	1.813	45.01	76.77	2.31%	37%	0.000	

1989	5372	1.961	46.97	74.80	2.55%	39%	0.000	
1990	5592	2.041	49.01	72.76	2.73%	40%	0.800	
1991	5828	2.127	51.14	70.64	2.92%	42%	0.000	
1992	6072	2.216	53.36	68.42	3.14%	44%	0.000	
1993	6059	2.212	55.57	66.21	3.23%	46%	0.000	
1994	6173	2.253	57.82	63.95	3.40%	47%	0.000	2nd PEAK
1995	6038	2.204	60.03	61.75	3.45%	49%	0.000	
1996	5919	2.161	62.19	59.59	3.50%	51%	0.000	50% PRODUCED
1997	5792	2.114	64.30	57.48	3.55%	53%	0.000	
1998	5681	2.074	66.37	55.40	3.61%	55%	0.000	
1999	5585	2.039	68.41	53.36	3.68%	56%	0.000	
2000	5504	2.009	70.42	51.35	3.76%	58%	0.000	
2001	5438	1.985	72.41	49.37	3.87%	59%	0.000	
2002	5224	1.907	74.31	47.46	3.86%	61%	0.000	
2003	5003	1.826	76.14	45.64	3.85%	63%	0.000	
2004	4793	1.750	77.89	43.89	3.83%	64%	0.000	
2005	4593	1.677	79.57	42.21	3.82%	65%	0.000	
2006	4402	1.607	81.17	40.60	3.81%	67%	0.000	
2007	4220	1.540	82.71	39.06	3.79%	68%	0.000	
2008	4046	1.477	84.19	37.59	3.78%	69%	0.000	
2009	3880	1.417	85.61	36.17	3.77%	70%	0.000	
2010	3722	1.359	86.97	34.81	3.76%	71%	0.000	
2011	3572	1.304	88.27	33.51	3.75%	72%	0.000	
2012	3428	1.252	89.52	32.25	3.74%	74%	0.000	
2013	3291	1.202	90.72	31.05	3.73%	75%	0.000	
2014	3160	1.154	91.88	29.90	3.72%	75%	0.000	
2015	3036	1.109	92.99	28.79	3.71%	76%	0.000	
2016	2916	1.065	94.05	27.72	3.70%	77%	0.000	
2017	2803	1.024	95.08	26.70	3.69%	78%	0.000	
2018	2695	0.984	96.06	25.72	3.69%	79%	0.000	
2019	2591	0.947	97.01	24.77	3.68%	80%	0.000	
2020	2493	0.911	97.92	23.86	3.68%	80%	0.000	
2021	2399	0.877	98.79	22.98	3.67%	81%	0.000	
2022	2309	0.844	99.64	22.14	3.67%	82%	0.000	
2023	2224	0.813	100.45	21.32	3.67%	82%	0.000	
2024	2142	0.784	101.24	20.54	3.67%	83%	0.000	
2025	2065	0.756	101.99	19.79	3.68%	84%	0.000	
2026	1992	0.729	102.72	19.06	3.68%	84%	0.000	
2027	1922	0.704	103.42	18.35	3.69%	85%	0.000	
2028	1856	0.680	104.10	17.67	3.70%	85%	0.000	
2029	1794	0.649	104.75	17.02	3.67%	86%	0.000	
2030	1711	0.623	105.37	16.40	3.66%	87%	0.000	
2031	1644	0.599	105.97	15.80	3.65%	87%	0.000	
2032	1579	0.575	106.55	15.23	3.64%	87%	0.000	
2033	1518	0.553	107.10	14.67	3.63%	88%	0.000	
2034	1459	0.531	107.63	14.14	3.62%	88%	0.000	
2035	1402	0.511	108.14	13.63	3.61%	89%	0.000	
2036	1348	0.491	108.63	13.14	3.60%	89%	0.000	
2037	1296	0.472	109.11	12.67	3.59%	90%	0.000	
2038	1247	0.454	109.56	12.21	3.59%	90%	0.000	
2039	1199	0.437	110.00	11.78	3.58%	90%	0.000	
2040	1154	0.420	110.42	11.36	3.57%	91%	0.000	
2041	1110	0.404	110.82	10.95	3.56%	91%	0.000	
2042	1068	0.389	111.21	10.56	3.55%	91%	0.000	
2043	1028	0.375	111.59	10.19	3.55%	92%	0.000	
2044	989	0.361	111.95	9.83	3.54%	92%	0.000	
2045	952	0.347	112.29	9.48	3.53%	92%	0.000	
2046	917	0.334	112.63	9.15	3.52%	92%	0.000	
2047	883	0.322	112.95	8.83	3.52%	93%	0.000	
2048	850	0.310	113.26	8.52	3.51%	93%	0.000	
2049	818	0.298	113.56	8.22	3.50%	93%	0.000	
2050	788	0.287	113.85	7.93	3.50%	93%	0.000	

SUMMARY DATA SHEET REGION AFRICA
REFERENCE DATE 1989

Country	Production M b/d	BBO/yr	Cumulative Production	Reported Reserves	Adjusted Reserves	Discovered	Undiscovered	Ultimate	Remaining	Und. of Ult.	Prod. of Ult.	Rem. of Ult.	Prod. of Disc	Disc. in Giants	Dep. Rate %
A	B	C	D	E	F	G	H	I	J	K	L	M	N	O	P
Libya	1101	0.402	16.25	22.80	22.80	39.05	5.95	45.00	28.75	13%	36%	64%	42%	86%	1.40%
Nigeria	1605	0.586	12.91	16.00	16.00	28.91	3.09	32.00	19.09	10%	40%	60%	45%	20%	3.07%
Algeria	698	0.255	8.37	9.20	9.20	17.57	2.43	20.00	11.63	12%	42%	58%	48%	62%	2.19%
Egypt	854	0.312	4.65	4.50	4.50	9.15	1.85	11.00	6.35	17%	42%	58%	51%	52%	4.91%
Angola	458	0.167	1.47	2.02	2.02	3.49	0.76	4.25	2.78	18%	35%	65%	42%	43%	6.00%
Tunisia	102	0.037	0.79	1.75	1.75	2.54	0.46	3.00	2.21	15%	26%	74%	31%	30%	1.68%
Gabon	197	0.072	1.32	0.73	0.73	2.06	0.44	2.50	1.18	18%	53%	47%	64%	24%	6.12%
Congo	150	0.055	0.50	0.83	0.83	1.33	0.67	2.00	1.50	33%	25%	75%	38%	75%	3.66%
Cameroon	173	0.063	0.50	0.40	0.40	0.90	0.10	1.00	0.50	10%	50%	50%	55%	0%	12.51%
Sudan	0	0.000	0.00	0.30	0.30	0.30	0.20	0.50	0.50	40%	0%	100%	0%	0%	0.00%
Zaire	28	0.010	0.13	0.10	0.10	0.22	0.08	0.30	0.17	25%	43%	57%	57%	0%	5.87%
Ivory Coast	2	0.001	0.05	0.05	0.05	0.10	0.00	0.10	0.05	0%	51%	49%	51%	0%	1.50%
Benin	5	0.002	0.02	0.10	0.05	0.07	0.03	0.10	0.08	33%	17%	83%	26%	0%	1.98%
Morocco	0	0.000	0.02	0.00	0.00	0.02	0.00	0.02	0.00	0%	86%	14%	86%	0%	3.65%
Ghana	0	0.000	0.00	0.00	0.00	0.01	0.00	0.01	0.00	4%	80%	20%	83%	0%	0.00%
Region	5372	1.961	46.97	58.79	58.74	105.71	16.07	121.78	74.80	13%	39%	61%	44%	57%	2.62%

THE EAST ("E")

SUMMARY

As defined here, The East constitutes a very diverse region ranging from India to Indonesia, including Indo-China, Japan and Australasia. The grouping is a matter of convenience, as the member countries have little in common.

The region is made up of fifteen oil countries having only 3.6% of the World's Ultimate oil endowment.

Oil exploration commenced in the last century in Indonesia and Burma (now known as Myanmar). Indonesia was then a Dutch territory, and Royal Dutch, a parent of Shell, had its roots there. Indonesia remains by far the largest producer, with more than twice the production of the next largest producer, India. Most of the region, except for India and Burma, has been open to the international companies. India's production has grown rapidly in recent years. Burma has recently opened its doors to foreign companies.

The East's share of World oil is as follows :

Cumulative Production	3.8%
Reserves	2.8%
Discovered	3.2%
Undiscovered	5.5%
Remaining	3.4%
Ultimate	3.6%

The East supports a large number of relatively small productive provinces, with production mainly coming from Tertiary sandstones, although there are some reefal deposits, as in for example West Irian. They do not lend themselves to summary classification, but are here treated as follows :

East Indies Basins

There are a number of mainly Tertiary basins in the foredeep of the East Indies island arc, including the particularly prolific Sumatra basin with the giant Minas Field. Otherwise, the fields, which occur both onshore and offshore, are generally not large.

The island of Borneo supports several productive Tertiary basins, both onshore and offshore, including Brunei in the north and the Kutai Basin in the south. The Malaysian basins in the South China Sea also fall broadly in this province.

Bombay High

The Bombay High off India is a major, relatively new, productive basin with giant fields. A number of other less important oil basins have also been found in the country.

Australasian Basins

The most important basin in Australia is the offshore Tertiary Bass Strait basin, between the mainland and Tasmania, which contains several giant fields. In addition are to be mentioned the Australian NW shelf, which is mainly gas-prone, the Timor Sea, and West Irian and Papua basins, although they are not particularly important.

Other productive basins occur in Pakistan, Burma, Thailand and Viet Nam

Ultimate

Indonesia is by far the most important oil country in the region with an Ultimate endowment of 23 BBO. India comes next with 10 BBO. Malaysia (7 BBO), Australia (6 BBO), Brunei (4 BBO) and Burma (1.2 BBO) are the only other significant countries with in excess of 1 BBO.

Reserves

It has not been considered necessary to make any adjustment to the reported reserve figures.

Only five countries have reserves of in excess of 1 BBO : Indonesia (8.2 BBO), India (7.5%), Malaysia (3.0 BBO), Australia (1.7 BBO), and Brunei (1.4 BBO).

Giant Fields are responsible for no more than 37% of all discovered oil, with peak giant discovery occurring in 1944 (Minas Field). The last giant was found in 1976.

Undiscovered

Only three countries are considered to have an Undiscovered potential of in excess of 1 BBO : India (4.28 BBO), Malaysia (1.98 BBO) and Australia (1.22 BBO). The rapid growth of Indian production in a country closed to foreign companies is somewhat remarkable and is taken to offer promise of further discovery, which may or may not be realized. Papua has had some recent discoveries in extremely difficult onshore terrain, where it is difficult, if not impossible, to obtain satisfactory seismic coverage.

Its potential is not here rated highly, although some commentators are optimistic. Burma (Myanmar) has recently opened its doors to foreign companies and may have promise, perhaps greater than here assessed. Offshore Viet Nam may also offer a useful potential.

In general, the region does not offer great potential for major new discoveries.

Production

Production had already commenced by the early years of the century in Indonesia and Burma. It has grown consistently over the past 30 years, but is now thought to have reached a peak. The Depletion Mid-Point should be reached in 1994, followed by a decline at about 4%.

Indonesia was the most important producer in 1989, with 1209 M b/d, and it also has by far the largest Cumulative Production (13.9 BBO).

As shown on the Production Profiles in Part 3, Australia has a classic mature profile with early giant finds. Burma has, by contrast, a most anomalous pattern, due to adverse political factors. India is shown to be at an immature stage, with a high peak coming in the early 2000s : possibly an unduly optimistic scenario. Indonesia has a classic mature curve for a country past its Mid-Point. None of the other countries merits particular comment.

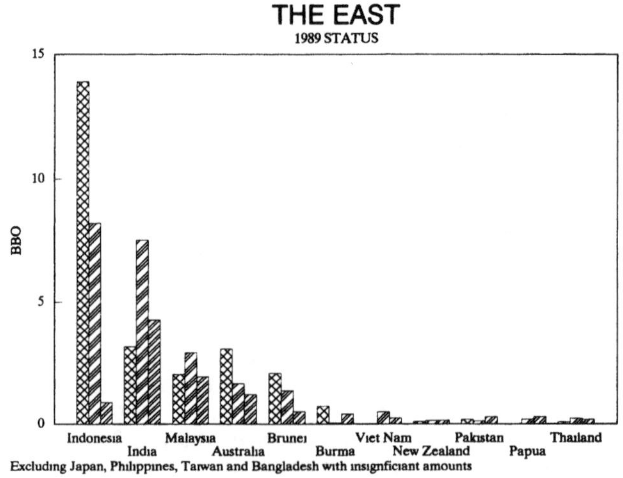

Figure 1-E. Produced, Reserves and Undiscovered.

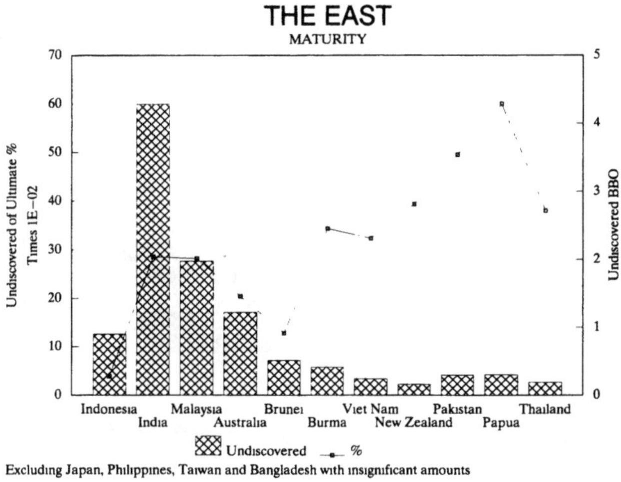

Figure 2-E. Maturity: Undiscovered as a percentage of Ultimate

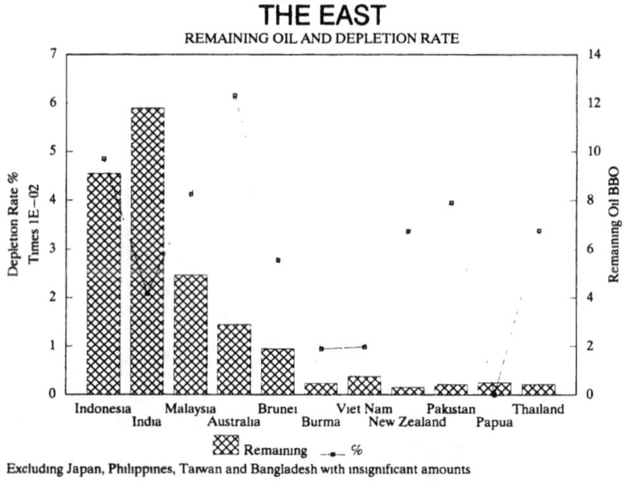

Figure 3-E. Remaining Oil (Reserves+Undiscovered) and Depletion Rate.

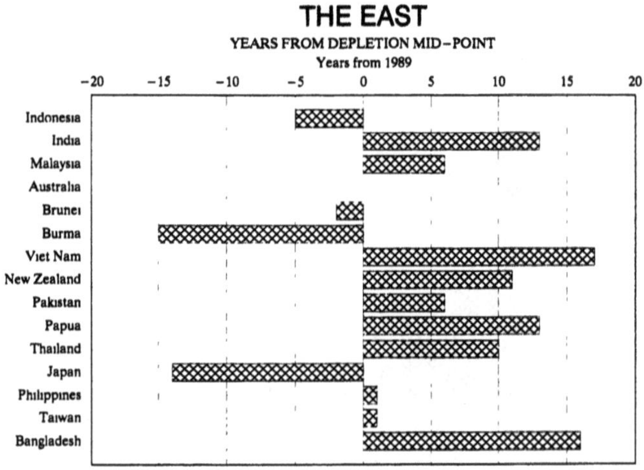

Figure 4-E. Years to or from Depletion Mid-Point.

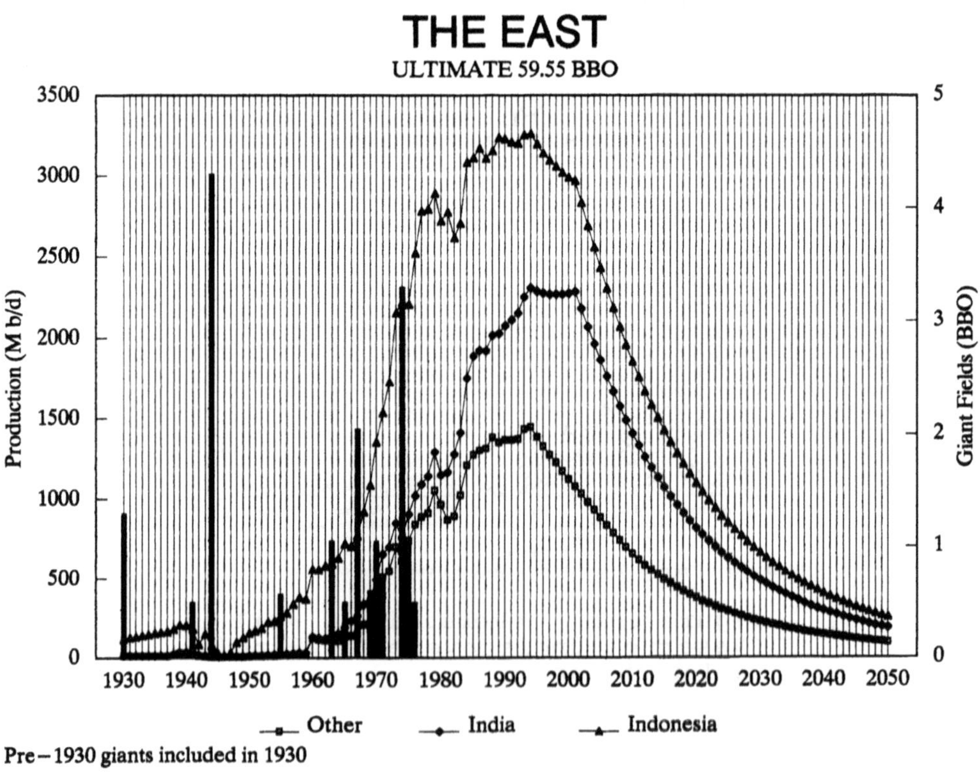

Figure 5-E. Production Profile 1930-2050 and Giant fields.

CUM.PROD	24.273 BBO	GIANT FIELDS			THE EAST	
RESERVES	23.045 BBO	TOTAL BBO	17		REGION	
DISCOVERED	47.318 BBO	% OF DISC.	37%			
UNDISCOVERED	12.027 BBO	Last Giant	1976			
REMAINING	35.072 BBO				Variance Trend	
ULTIMATE	59.345 BBO				1990	+8%
Rept.Reserve	23.046	Ref.date(end)	1989		Future	n/a

Milestones	Date	Prod Mb/d	Dep. Rate	% Produced	Forecast	n/a
1st Peak	1989	3234	3.26%	41%	A=Exponential decline	
2nd Peak	-	-	-	-	B=5% incr. to Mid-Point	
Mid-Point	1994	3291	3.96%	51%	C=Other	
Final Decline	n/a				Pre-1930 Cum.	0.880
					(Added to 1930 prod)	

Date	Prod. M.b/d	Prod. BBO/yr	Cum. Prod.	Remain	Deplet. Rate	% Produced	Giant	Notes & Names
1930	119	0.043	0.92	58.42	ERR	2%	1.300	
1931	126	0.046	0.97	58.38	0.08%	2%	0.000	
1932	132	0.048	1.02	58.33	0.08%	2%	0.000	
1933	139	0.051	1.07	58.28	0.09%	2%	0.000	
1934	145	0.053	1.12	58.22	0.09%	2%	0.000	
1935	152	0.055	1.18	58.17	0.10%	2%	0.000	
1936	159	0.058	1.23	58.11	0.10%	2%	0.000	
1937	165	0.060	1.30	58.05	0.10%	2%	0.000	
1938	181	0.066	1.36	57.98	0.11%	2%	0.000	
1939	206	0.075	1.44	57.91	0.13%	2%	0.000	
1940	205	0.075	1.51	57.83	0.13%	3%	0.000	
1941	182	0.066	1.58	57.77	0.11%	3%	0.500	
1942	85	0.031	1.61	57.74	0.05%	3%	0.000	
1943	147	0.054	1.66	57.68	0.09%	3%	0.000	
1944	78	0.028	1.69	57.65	0.05%	3%	4.300	
1945	33	0.012	1.70	57.64	0.02%	3%	0.000	
1946	17	0.006	1.71	57.64	0.01%	3%	0.000	
1947	32	0.012	1.72	57.62	0.02%	3%	0.000	
1948	97	0.035	1.76	57.59	0.06%	3%	0.000	
1949	127	0.047	1.80	57.54	0.08%	3%	0.000	
1950	156	0.057	1.86	57.49	0.10%	3%	0.000	
1951	169	0.062	1.92	57.42	0.11%	3%	0.000	
1952	186	0.068	1.99	57.36	0.12%	3%	0.000	
1953	222	0.081	2.07	57.28	0.14%	3%	0.000	
1954	230	0.084	2.15	57.19	0.15%	4%	0.000	
1955	260	0.095	2.25	57.10	0.17%	4%	0.570	
1956	282	0.103	2.35	56.99	0.18%	4%	0.000	
1957	338	0.123	2.47	56.87	0.22%	4%	0.000	
1958	382	0.139	2.61	56.73	0.25%	4%	0.000	
1959	371	0.135	2.75	56.60	0.24%	5%	0.000	
1960	556	0.203	2.95	56.39	0.36%	5%	0.000	
1961	555	0.203	3.16	56.19	0.36%	5%	0.000	
1962	579	0.211	3.37	55.98	0.38%	6%	0.000	
1963	591	0.216	3.58	55.76	0.39%	6%	1.050	
1964	631	0.230	3.81	55.53	0.41%	6%	0.000	
1965	722	0.264	4.08	55.27	0.47%	7%	0.500	
1966	706	0.258	4.33	55.01	0.47%	7%	0.000	
1967	771	0.281	4.62	54.73	0.51%	8%	2.050	
1968	920	0.336	4.95	54.39	0.61%	8%	0.000	
1969	1084	0.396	5.35	54.00	0.73%	9%	0.600	
1970	1354	0.494	5.84	53.50	0.92%	10%	1.050	
1971	1535	0.560	6.40	52.94	1.05%	11%	0.700	
1972	1725	0.630	7.03	52.31	1.19%	12%	0.000	
1973	2151	0.785	7.82	51.53	1.50%	13%	0.000	
1974	2219	0.810	8.63	50.72	1.57%	15%	3.300	
1975	2205	0.805	9.43	49.91	1.59%	16%	1.050	
1976	2522	0.920	10.35	48.99	1.84%	17%	0.500	
1977	2780	1.015	11.37	47.98	2.07%	19%	0.000	
1978	2792	1.019	12.39	46.96	2.12%	21%	0.000	
1979	2892	1.056	13.44	45.90	2.25%	23%	0.000	
1980	2719	0.992	14.43	44.91	2.16%	24%	0.000	
1981	2774	1.013	15.45	43.90	2.25%	26%	0.000	
1982	2618	0.956	16.40	42.94	2.18%	28%	0.000	
1983	2705	0.987	17.39	41.96	2.30%	29%	0.000	
1984	3084	1.126	18.52	40.83	2.68%	31%	0.000	
1985	3110	1.135	19.65	39.69	2.78%	33%	0.000	
1986	3169	1.157	20.81	38.54	2.91%	35%	0.000	
1987	3107	1.134	21.94	37.40	2.94%	37%	0.000	
1988	3153	1.151	23.09	36.25	3.08%	39%	0.000	

1989	3234	1.180	24.27	35.07	3.26%	41%	0.000	1ST PEAK
1990	3232	1.180	25.45	33.89	3.36%	43%	0.000	
1991	3221	1.176	26.63	32.72	3.47%	45%	0.000	
1992	3217	1.174	27.80	31.54	3.59%	47%	0.000	
1993	3276	1.196	29.00	30.35	3.79%	49%	0.000	
1994	3291	1.201	30.20	29.15	3.96%	51%	0.000	50% PRODUCED
1995	3234	1.180	31.38	27.97	4.05%	53%	0.000	
1996	3179	1.160	32.54	26.81	4.15%	55%	0.000	
1997	3132	1.143	33.68	25.66	4.26%	57%	0.000	
1998	3094	1.129	34.81	24.53	4.40%	59%	0.000	
1999	3057	1.116	35.93	23.42	4.55%	61%	0.000	
2000	3028	1.105	37.03	22.31	4.72%	62%	0.000	
2001	2984	1.089	38.12	21.22	4.88%	64%	0.000	
2002	2845	1.039	39.16	20.18	4.89%	66%	0.000	
2003	2700	0.986	40.15	19.20	4.88%	68%	0.000	
2004	2564	0.936	41.08	18.26	4.87%	69%	0.000	
2005	2436	0.889	41.97	17.37	4.87%	71%	0.000	
2006	2305	0.841	42.81	16.53	4.84%	72%	0.000	
2007	2182	0.796	43.61	15.74	4.82%	73%	0.000	
2008	2065	0.754	44.36	14.98	4.79%	75%	0.000	
2009	1956	0.714	45.08	14.27	4.76%	76%	0.000	
2010	1852	0.676	45.75	13.59	4.74%	77%	0.000	
2011	1755	0.640	46.39	12.95	4.71%	78%	0.000	
2012	1662	0.607	47.00	12.35	4.68%	79%	0.000	
2013	1575	0.575	47.57	11.77	4.66%	80%	0.000	
2014	1493	0.545	48.12	11.23	4.63%	81%	0.000	
2015	1415	0.517	48.64	10.71	4.60%	82%	0.000	
2016	1342	0.490	49.13	10.22	4.57%	83%	0.000	
2017	1272	0.464	49.59	9.75	4.55%	84%	0.000	
2018	1207	0.441	50.03	9.31	4.52%	84%	0.000	
2019	1145	0.418	50.45	8.90	4.49%	85%	0.000	
2020	1086	0.396	50.85	8.50	4.46%	86%	0.000	
2021	1031	0.376	51.22	8.12	4.43%	86%	0.000	
2022	978	0.357	51.58	7.77	4.40%	87%	0.000	
2023	929	0.339	51.92	7.43	4.37%	87%	0.000	
2024	882	0.322	52.24	7.11	4.33%	88%	0.000	
2025	837	0.306	52.55	6.80	4.30%	89%	0.000	
2026	795	0.290	52.84	6.51	4.27%	89%	0.000	
2027	756	0.276	53.11	6.23	4.24%	89%	0.000	
2028	718	0.262	53.37	5.97	4.20%	90%	0.000	
2029	682	0.249	53.62	5.72	4.17%	90%	0.000	
2030	648	0.237	53.86	5.49	4.14%	91%	0.000	
2031	616	0.225	54.08	5.26	4.10%	91%	0.000	
2032	586	0.214	54.30	5.05	4.07%	91%	0.000	
2033	557	0.203	54.50	4.84	4.03%	92%	0.000	
2034	530	0.193	54.69	4.65	3.99%	92%	0.000	
2035	504	0.184	54.88	4.47	3.96%	92%	0.000	
2036	480	0.175	55.05	4.29	3.92%	93%	0.000	
2037	457	0.167	55.22	4.12	3.88%	93%	0.000	
2038	435	0.159	55.38	3.97	3.85%	93%	0.000	
2039	414	0.151	55.53	3.81	3.81%	94%	0.000	
2040	394	0.144	55.67	3.67	3.77%	94%	0.000	
2041	375	0.137	55.81	3.53	3.73%	94%	0.000	
2042	358	0.131	55.94	3.40	3.69%	94%	0.000	
2043	341	0.124	56.07	3.28	3.65%	94%	0.000	
2044	325	0.119	56.18	3.16	3.61%	95%	0.000	
2045	309	0.113	56.30	3.05	3.57%	95%	0.000	
2046	295	0.108	56.41	2.94	3.53%	95%	0.000	
2047	281	0.103	56.51	2.84	3.49%	95%	0.000	
2048	268	0.098	56.61	2.74	3.45%	95%	0.000	
2049	256	0.093	56.70	2.65	3.41%	96%	0.000	
2050	244	0.089	56.79	2.56	3.37%	96%	0.000	

SUMMARY DATA SHEET REGION THE EAST
REFERENCE DATE 1989

Country	Production M b/d	BBO/ year	Cumulative Production	Reported Reserves	Adjusted Reserves	Discovered	Undiscovered	Ultimate	Remaining	Und. of Ult.	Prod of Ult.	Rem. of Ult.	Prod of Disc	Disc. in Giants	Dep. Rate
A	B	C	D	E	F	G	H	I	J	K	L	M	N	O	P
Indonesia	1209	0.441	13.90	8.20	8.20	22.10	0.90	23.00	9.10	4%	60%	40%	63%	39%	4.85%
India	673	0.246	3.20	7.52	7.52	10.72	4.28	15.00	11.80	29%	21%	79%	30%	28%	2.08%
Malaysia	557	0.203	2.07	2.95	2.95	5.02	1.98	7.00	4.93	28%	30%	70%	41%	20%	4.13%
Australia	489	0.178	3.10	1.68	1.68	4.78	1.22	6.00	2.90	20%	52%	48%	65%	53%	6.16%
Brunei	143	0.052	2.11	1.38	1.38	3.49	0.51	4.00	1.89	13%	53%	47%	61%	67%	2.77%
Burma	12	0.004	0.74	0.05	0.05	0.79	0.41	1.20	0.46	34%	61%	39%	93%	0%	0.94%
Viet Nam	20	0.007	0.01	0.50	0.50	0.51	0.24	0.75	0.74	32%	1%	99%	1%	0%	0.98%
New Zealand	28	0.010	0.10	0.15	0.15	0.24	0.16	0.40	0.30	39%	24%	76%	40%	0%	3.37%
Pakistan	45	0.016	0.18	0.12	0.12	0.30	0.30	0.60	0.42	49%	31%	69%	61%	0%	3.95%
Papua	0	0.000	0.00	0.20	0.20	0.20	0.30	0.50	0.50	60%	0%	100%	0%	0%	0.00%
Thailand	39	0.014	0.08	0.23	0.23	0.31	0.19	0.50	0.42	38%	16%	84%	26%	0%	3.38%
Japan	11	0.004	0.18	0.06	0.06	0.24	0.01	0.25	0.07	5%	71%	29%	74%	0%	5.64%
Philippines	4	0.001	0.04	0.02	0.02	0.05	0.02	0.08	0.04	32%	48%	52%	71%	0%	3.75%
Taiwan	3	0.001	0.02	0.00	0.00	0.03	0.02	0.05	0.03	43%	48%	52%	84%	0%	3.65%
Bangladesh	1	0.000	0.00	0.00	0.00	0.00	0.02	0.02	0.02	87%	10%	90%	81%	0%	1.83%
Region	3234	1.180	24.27	23.05	23.05	47.32	12.03	59.35	35.07	20%	41%	59%	51%	37%	3.37%

WESTERN EUROPE ("WE")

SUMMARY

Western Europe has no more that 2.5% of the World's Ultimate oil endowment. The Region comprises eleven oil producing countries, ranging from Norway in the north to Turkey in the south. The former East Germany is now included in Germany.

Small amounts of oil were produced in a number of countries in the region prior to the Second World War, but it did not become a significant producer until the 1970s, when giant North Sea discoveries were developed. The international oil companies were responsible for the development of the industry. In Britain, a State Company, B.N.O.C., took up a dominant position in the 1970s under a socialist government when it clawed back access to the rights previously awarded to the companies, but it was subsequently privatised on a change in government. The rights were not however returned to the original owners. In Norway, a State Company, Statoil, was formed in 1973 and retains a dominant position. Most of its successful exploration was funded by its foreign partners under carried interest provisions in the licence terms, the cost thereof being deductable against taxable income.

Western Europe's share of World oil is as follows :

Cumulative Production	2.9%
Reserves	2.4%
Discovered	2.6%
Undiscovered	2.8%
Remaining	2.4%
Ultimate	2.5%

Of the 42.2 BBO of Ultimate endowment 81% lies in the three North Sea producers : Norway, the United Kingdom and Denmark.

The reported reserve estimates are accepted as valid, and no adjustment has been called for.

Giant fields were found in Austria, Norway and the United Kingdom and between 1948 (Austria) and 1985, with a pronounced peak in 1975. No further giant discoveries are anticipated.

The attributes of the several producing countries are summarised as follows:

Austria

Pre-war production from Tertiary reservoirs in the Vienna Basin was followed by the discovery of the giant Matzen Field in 1948, and peak production was reached by 1955. Since then, production has followed a broadly exponential decline. 50% of the Ultimate had been consumed by 1973, rising to 68% by 1989. The industry is dominated by OMV, the national company, although Mobil, Shell and certain other foreign companies have been active. The Production Profile in Part 3 shows a curve of advanced maturity : a valuable confirmation of the validity of the fundamental bell-shaped Depletion Curve.

Denmark

Denmark relies on production from the Chalk of the North Sea, a particularly difficult reservoir with low recovery. The first North Sea oil discovery was in fact made in Denmark. The industry for many year was dominated by DUC, a consortium of Gulf, Shell, Caltex and others, which had exclusive rights. The Production Profile (Part 3) shows a typical offshore pattern with steep rise and decline. It is relatively immature.

Attention is turning to a second exploration campaign in the deeper Jurassic, which may lead to some important new discoveries. The present Undiscovered potential of 0.170 BBO may be due for future upward revision.

France

France relies on shallow onshore production in the Aquitaine and Paris basins, some of which was already established before the Second World War. Late, larger than anticipated, discoveries in the Paris Basin led to an unusual two-peaked production curve (see Part 3).

Germany

Germany has a perhaps surprisingly high Ultimate endowment of 2.25 BBO. Production began before the war, and comes from a large number of small fields.

The industry is a closely knit one, mainly in German hands, although some of the major foreign companies, marketing in Germany, have been active. In view of the extremely thorough exploration, no significant new discoveries are expected.

The production decline follows a classic pattern of advanced maturity with peak in 1974 at a depletion rate of 6%.(see Part 3). It, like Austria, well confirms the underlying bell-shaped curve.

Italy

Italy has supported an active oil industry since before the war, dominated by the national company, Agip. The country offers highly attractive terms, and sustains a surprisingly high level of activity both onshore and offshore in the Adriatic and around Sicily.

The Adriatic has an exceptionally low geothermal gradient, meaning that there is a viable exploration potential to great depths.

Offshore developments, especially off Sicily, have led to a rapid increase in production during the late 1980s leading to an anticipated peak in 1993. Only 29% of the estimated ultimate had been consumed by 1989 (see Part 3). An undiscovered potential of almost 1 BBO is considered justified.

Netherlands

Oil production commenced in the Netherlands during the war, coming from several small onshore Cretaceous fields, fed by the well-known North Sea Upper Jurassic source-rock.

The huge Groningen gas field together with other offshore gas fields, sustains the level of exploration in the Netherlands, which has led, almost incidentally, to some oil discoveries. Production stands close to its peak, and in view of a very high depletion rate of 16.8% is expected to decline rapidly. No significant new discoveries are expected in this very mature province.

Norway

Norway is the jewel in Europe's oil crown, having 43% of the region's Ultimate ; 61% of the Reserves ; 46% of the undiscovered potential.

Production is expected to peak in 1993, and decline at a modest depletion rate of 5%.

The production curve follows the classic pattern with the early discovery of giant fields peaking in the 1970s. No further giants are anticipated (see Part 3).

No significant further major discoveries are here predicted outside the North Sea basin. If further drilling in the Barents Sea should lead to a discovery, it would open another major province, with colossal European significance, bringing in new production for the end of the century. The chance of this occurring is not here ranked highly, and the attributed Undiscovered potential of 2.97 BBO, by far Europe's highest, is here considered realistic verging on the optimistic.

Turkey

In geological terms, Turkey lies at the northern extremity of the prolific Persian Gulf province. The structures have however been deeply breached so that much of the hydrocarbon content that they may once have contained has escaped. The terrain is rugged, and the contract terms have been generally adverse to foreign investment.

Mobil and Shell made a number of discoveries in the 1950s and 1960s which led to peak production in 1968. It has fallen since but at a relatively low depletion rate (2.7% in 1989), due to low productivities and limited infrastructure.

The Undiscovered potential of 0.22 BBO may be due for some upward revision if the industry were to dedicate a major effort to Turkey following improved contract terms.

The Production Profile (Part 3) appears somewhat anomalous, with an anticipated second peak in the late 1990s, reflecting realization of the country's assessed potential.

United Kingdom

The United Kingdom is Europe's second most important producing country after Norway. Like Norway, it has a classic offshore depletion curve with the early discovery of giant fields. It is now at an advanced stage of depletion (see Part 3).

Peak production came in 1986, and is followed by a rapid decline with an exceptionally high depletion rate of 11.3%

Some discoveries have been made onshore, but with the exception of Wytch Farm in Dorset, they are small.

The offshore was opened in the early 1960s, and most of the important finds had been made by 1975. In recent years the industry has been in a state of hyper-activity. Many statements about the glorious future are made, but the reality is that the shelf is now at a very mature stage of depletion. A late fillip was given by the discovery of several Tertiary stratigraphic traps with turbidite and contourite reservoirs : more may yet be found. There is some hope of more discoveries west of the Shetlands, but generally the picture does not seem very promising. The Reserves may be somewhat understated in view of the efforts to enhance recovery and bring in small satellites, but the Undiscovered potential of 1.4 BBO is here considered reasonable, even optimistic.

Greece, Spain and Ireland

Greece and Spain support minor production, but are not significant in a regional context. Ireland is not here treated as an oil producing country, although some small finds have been made in the Celtic Sea and the Porcupine Basin, which may be eventually developed. It has now been extensively explored, and sadly we must confess that its potential has been downgraded.

Ultimate

Norway and the United Kingdom have by far the largest Ultimate endowment in the region with 18 and 15 BBO respectively, followed by Germany, a long way behind, with no more than 2.25 BBO.

Reserves

Norway has by far the largest reserves with 11.55 BBO. The United Kingdom comes far behind with 4.26 BBO, and none of the other countries exceeds 1 BBO.

Undiscovered

Only Norway (2.97 BBO) and the United Kingdom (1.40 BBO) have an Undiscovered potential in excess of 1 BBO. But modest discoveries, albeit of considerable local importance, are anticipated in Italy (0.34 BBO), France (0.29 BBO), Austria (0.26 BBO) and Turkey (0.22 BBO).

The S.E.Barents Sea, off northern Norway, offers the greatest scope for a major positive surprise, although much of the rest of the province has been downgraded by drilling. The chances of major discovery in deepwater west of the British Isles is not here ranked highly, nor is the rest of the northern Norwegian shelf or Svalbard.

Production

North Sea production in Britain and Norway commenced in the early 1970s and rose rapidly to a peak in 1988. The United Kingdom is now past its Mid-Point, and is in sharp decline at a very high Depletion Rate of 11%, reflecting the hyperactivity with numerous satellite development projects. Norway has developed its oil much more slowly and has almost ten years to reach its Mid-Point.

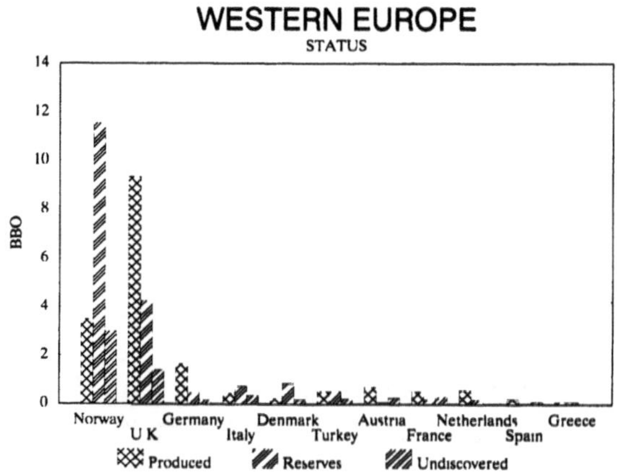

Figure 1-WE. Produced, Reserves, Undiscovered.

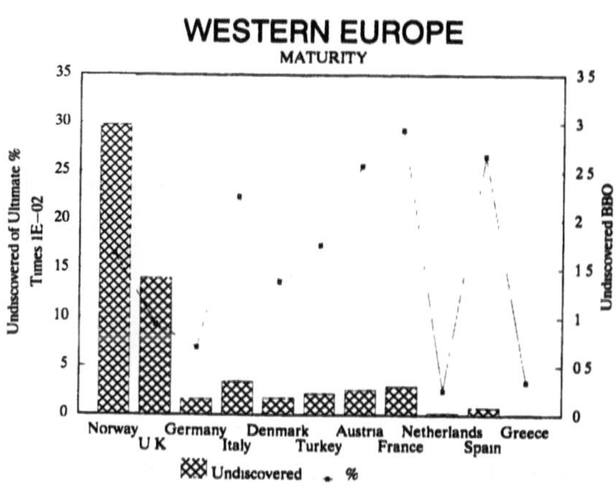

Figure 2-WE. Maturity: Undiscovered as a percentage of Ultimate.

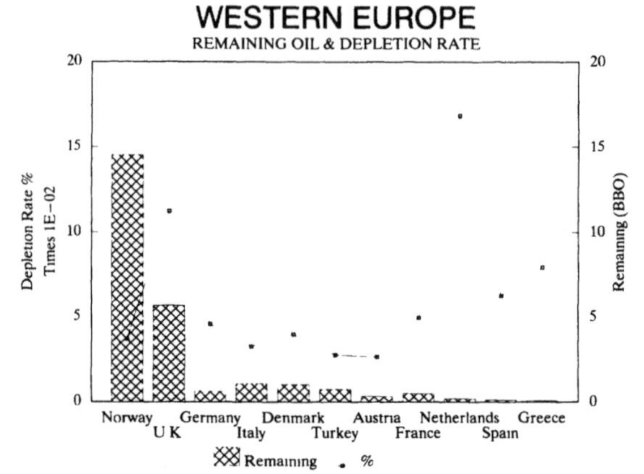

Figure 3-WE. Remaining Oil (Reserves+Undiscovered) and Depletion Rate

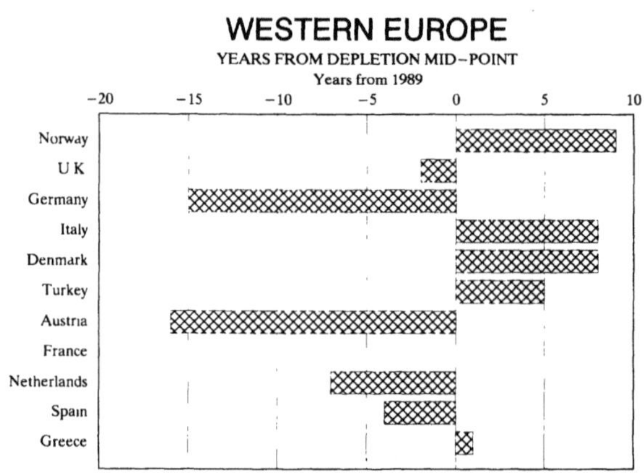

Figure 4-WE. Years to or from Depletion Mid-Point.

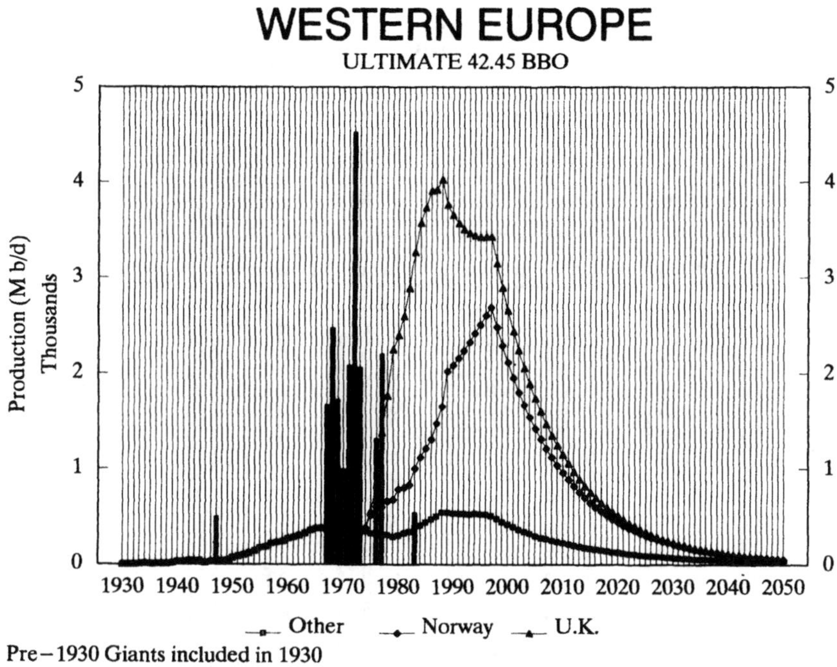

Figure 5-WE. Production Profile 1930-2050 and Giant fields.

CUM.PROD	17.637 BBO	GIANT FIELDS		W.EUROPE	
RESERVES	18.871 BBO	TOTAL BBO	20.1	REGION	
DISCOVERED	36.508 BBO	% OF DISC.	55%		
UNDISCOVERED	5.942 BBO	Last Giant	1983		
REMAINING	24.813 BBO			Variance	Trend
ULTIMATE	42.450 BBO			1990	-1%
Rept.Reserve	18.871 BBO	Ref.Date(end)	1989	Future	DOWN

Milestones	Date	Prod Mb/d	Dep. Rate	% Produced
1st Peak	1988	4027	5.32%	38%
2nd Peak				
Mid-Point	1992	3502	5.77%	51%
Final Decline	n/a			

Forecast n/a
A=Exponential decline
B=5% incr. to Mid-Point
C=Other
Pre-1930 Cum. 0.005
(Added to 1930 Prod)

Date	Prod. M.b/d	Prod. BBO/yr	Cum Prod	Remain	Dep. Rate	% Produced	Giant	Names & Notes
1930	3	0.001	0.00	42.44	ERR	0%	0.000	
1931	4	0.001	0.00	42.44	0.00%	0%	0.000	
1932	5	0.002	0.00	42.44	0.00%	0%	0.000	
1933	6	0.002	0.01	42.44	0.01%	0%	0.000	
1934	7	0.003	0.01	42.44	0.01%	0%	0.000	
1935	8	0.003	0.01	42.43	0.01%	0%	0.000	
1936	9	0.003	0.02	42.43	0.01%	0%	0.000	
1937	10	0.004	0.02	42.43	0.01%	0%	0.000	
1938	13	0.005	0.02	42.42	0.01%	0%	0.000	
1939	17	0.006	0.03	42.41	0.01%	0%	0.000	
1940	30	0.011	0.04	42.40	0.03%	0%	0.000	
1941	31	0.011	0.05	42.39	0.03%	0%	0.000	
1942	41	0.015	0.07	42.38	0.04%	0%	0.000	
1943	38	0.014	0.08	42.36	0.03%	0%	0.000	
1944	42	0.015	0.10	42.35	0.04%	0%	0.000	
1945	22	0.008	0.10	42.34	0.02%	0%	0.000	
1946	31	0.011	0.12	42.33	0.03%	0%	0.000	
1947	33	0.012	0.13	42.32	0.03%	0%	0.500	
1948	41	0.015	0.14	42.30	0.04%	0%	0.000	
1949	50	0.018	0.16	42.28	0.04%	0%	0.000	
1950	72	0.026	0.19	42.26	0.06%	0%	0.000	
1951	89	0.033	0.22	42.22	0.08%	1%	0.000	
1952	103	0.038	0.26	42.19	0.09%	1%	0.000	
1953	121	0.044	0.30	42.14	0.10%	1%	0.000	
1954	146	0.053	0.36	42.09	0.13%	1%	0.000	
1955	178	0.065	0.42	42.02	0.15%	1%	0.000	
1956	187	0.068	0.49	41.96	0.16%	1%	0.000	
1957	225	0.082	0.57	41.87	0.20%	1%	0.000	
1958	238	0.087	0.66	41.79	0.21%	2%	0.000	
1959	246	0.090	0.75	41.70	0.22%	2%	0.000	
1960	278	0.101	0.85	41.60	0.24%	2%	0.000	
1961	297	0.108	0.96	41.49	0.26%	2%	0.000	
1962	312	0.114	1.07	41.37	0.27%	3%	0.000	
1963	336	0.123	1.19	41.25	0.30%	3%	0.000	
1964	370	0.135	1.33	41.12	0.33%	3%	0.000	
1965	386	0.141	1.47	40.98	0.34%	3%	0.000	
1966	389	0.142	1.61	40.83	0.35%	4%	0.000	
1967	384	0.140	1.75	40.69	0.34%	4%	1.673	
1968	379	0.138	1.89	40.56	0.34%	4%	2.470	
1969	397	0.145	2.03	40.41	0.36%	5%	1.730	
1970	382	0.139	2.17	40.27	0.34%	5%	1.000	
1971	367	0.134	2.31	40.14	0.33%	5%	2.086	
1972	376	0.137	2.44	40.00	0.34%	6%	4.523	
1973	391	0.143	2.59	39.86	0.36%	6%	2.064	
1974	388	0.142	2.73	39.72	0.36%	6%	0.000	
1975	557	0.203	2.93	39.51	0.51%	7%	0.000	
1976	833	0.304	3.24	39.21	0.77%	8%	1.320	
1977	1376	0.502	3.74	38.71	1.28%	9%	2.199	
1978	1765	0.644	4.38	38.06	1.66%	10%	0.000	
1979	2250	0.821	5.20	37.24	2.16%	12%	0.000	
1980	2391	0.873	6.08	36.37	2.34%	14%	0.000	
1981	2593	0.946	7.02	35.42	2.60%	17%	0.000	
1982	2884	1.053	8.08	34.37	2.97%	19%	0.000	
1983	3266	1.192	9.27	33.18	3.47%	22%	0.547	
1984	3570	1.303	10.57	31.87	3.93%	25%	0.000	
1985	3733	1.363	11.93	30.51	4.28%	28%	0.000	
1986	3912	1.428	13.36	29.08	4.68%	31%	0.000	
1987	3922	1.432	14.79	27.65	4.92%	35%	0.000	
1988	4027	1.470	16.26	26.18	5.32%	38%	0.000	1st PEAK

1989	3764	1.374	17.64	24.81	5.25%	42%	0.000	
1990	3653	1.334	18.97	23.47	5.38%	45%	0.000	
1991	3566	1.302	20.27	22.17	5.55%	48%	0.000	
1992	3502	1.278	21.55	20.89	5.77%	51%	0.000	50% PRODUCED
1993	3459	1.262	22.81	19.63	6.04%	54%	0.000	
1994	3435	1.254	24.07	18.38	6.39%	57%	0.000	
1995	3423	1.249	25.32	17.13	6.80%	60%	0.000	
1996	3428	1.251	26.57	15.88	7.30%	63%	0.000	
1997	3429	1.252	27.82	14.63	7.88%	66%	0.000	
1998	3147	1.149	28.97	13.48	7.85%	68%	0.000	
1999	2888	1.054	30.02	12.42	7.82%	71%	0.000	
2000	2652	0.968	30.99	11.46	7.79%	73%	0.000	
2001	2435	0.889	31.88	10.57	7.76%	75%	0.000	
2002	2237	0.817	32.69	9.75	7.73%	77%	0.000	
2003	2056	0.750	33.45	9.00	7.70%	79%	0.000	
2004	1889	0.690	34.13	8.31	7.66%	80%	0.000	
2005	1737	0.634	34.77	7.68	7.63%	82%	0.000	
2006	1597	0.583	35.35	7.09	7.59%	83%	0.000	
2007	1469	0.536	35.89	6.56	7.56%	85%	0.000	
2008	1352	0.493	36.38	6.06	7.52%	86%	0.000	
2009	1244	0.454	36.84	5.61	7.49%	87%	0.000	
2010	1145	0.418	37.25	5.19	7.45%	88%	0.000	
2011	1054	0.385	37.64	4.81	7.41%	89%	0.000	
2012	971	0.355	37.99	4.45	7.37%	89%	0.000	
2013	895	0.327	38.32	4.13	7.34%	90%	0.000	
2014	823	0.300	38.62	3.83	7.28%	91%	0.000	
2015	759	0.277	38.90	3.55	7.24%	92%	0.000	
2016	700	0.255	39.15	3.29	7.20%	92%	0.000	
2017	646	0.236	39.39	3.06	7.16%	93%	0.000	
2018	596	0.218	39.61	2.84	7.12%	93%	0.000	
2019	550	0.201	39.81	2.64	7.07%	94%	0.000	
2020	508	0.185	39.99	2.45	7.03%	94%	0.000	
2021	469	0.171	40.16	2.28	6.98%	95%	0.000	
2022	434	0.158	40.32	2.12	6.94%	95%	0.000	
2023	401	0.146	40.47	1.98	6.89%	95%	0.000	
2024	371	0.135	40.60	1.84	6.84%	96%	0.000	
2025	343	0.125	40.73	1.72	6.80%	96%	0.000	
2026	317	0.116	40.84	1.60	6.75%	96%	0.000	
2027	294	0.107	40.95	1.49	6.70%	96%	0.000	
2028	272	0.099	41.05	1.39	6.65%	97%	0.000	
2029	252	0.092	41.14	1.30	6.60%	97%	0.000	
2030	234	0.085	41.23	1.22	6.55%	97%	0.000	
2031	217	0.079	41.31	1.14	6.49%	97%	0.000	
2032	201	0.073	41.38	1.06	6.44%	97%	0.000	
2033	186	0.068	41.45	1.00	6.39%	98%	0.000	
2034	173	0.063	41.51	0.93	6.34%	98%	0.000	
2035	161	0.059	41.57	0.88	6.28%	98%	0.000	
2036	149	0.054	41.62	0.82	6.23%	98%	0.000	
2037	139	0.051	41.68	0.77	6.17%	98%	0.000	
2038	129	0.047	41.72	0.72	6.12%	98%	0.000	
2039	120	0.044	41.77	0.68	6.06%	98%	0.000	
2040	112	0.041	41.81	0.64	6.01%	98%	0.000	
2041	104	0.038	41.84	0.60	5.95%	99%	0.000	
2042	97	0.035	41.88	0.56	5.89%	99%	0.000	
2043	90	0.033	41.91	0.53	5.84%	99%	0.000	
2044	84	0.031	41.94	0.50	5.78%	99%	0.000	
2045	79	0.029	41.97	0.47	5.72%	99%	0.000	
2046	73	0.027	42.00	0.45	5.67%	99%	0.000	
2047	69	0.025	42.02	0.42	5.61%	99%	0.000	
2048	64	0.023	42.05	0.40	5.55%	99%	0.000	
2049	60	0.022	42.07	0.38	5.50%	99%	0.000	
2050	56	0.020	42.09	0.36	5.44%	99%	0.000	

SUMMARY DATA SHEET
REFERENCE DATE 1989

REGION WESTERN EUROPE

Country	Production M b/d	BBO/yr	Cumulative Production	Reported Reserves	Adjusted Reserves	Discovered	Undiscovered	Ultimate	Remaining	Und. of Ult.	Prod. of Ult.	Rem. of Ult.	Prod. of Disc	Disc. in Giants	Dep. Rate
A	B	C	D	E	F	G	H	I	J	K	L	M	N	O	P
Norway	1473	0.538	3.48	11.55	11.55	15.03	2.97	18.00	14.52	17%	19%	81%	23%	54%	3.70%
U.K.	1743	0.636	9.35	4.26	4.26	13.60	1.40	15.00	5.65	9%	62%	38%	69%	85%	11.25%
Germany	76	0.028	1.64	0.45	0.45	2.09	0.16	2.25	0.61	7%	73%	27%	79%	0%	4.57%
Italy	95	0.035	0.44	0.73	0.73	1.16	0.34	1.50	1.06	22%	29%	71%	38%	0%	3.26%
Denmark	111	0.041	0.23	0.85	0.85	1.08	0.17	1.25	1.02	14%	18%	82%	21%	0%	3.97%
Turkey	56	0.020	0.51	0.52	0.52	1.03	0.22	1.25	0.74	17%	41%	59%	50%	0%	2.74%
Austria	23	0.009	0.68	0.07	0.07	0.74	0.26	1.00	0.32	26%	68%	32%	91%	67%	2.64%
France	65	0.024	0.52	0.19	0.19	0.71	0.29	1.00	0.48	29%	52%	48%	73%	0%	4.93%
Netherlands	87	0.032	0.56	0.17	0.17	0.73	0.02	0.75	0.19	2%	75%	25%	77%	0%	16.84%
Spain	18	0.007	0.20	0.03	0.03	0.22	0.08	0.30	0.10	27%	65%	35%	89%	0%	6.26%
Greece	17	0.006	0.07	0.07	0.07	0.14	0.01	0.15	0.08	3%	48%	52%	50%	0%	7.95%
Region	3764	1.374	17.68	18.87	18.87	36.55	5.90	42.45	24.77	14%	42%	58%	48%	55%	5.55%

MIDDLE EAST (Minor) ("MEm")

Summary
Figure-1: Status - Production, Reserves, Undiscovered
Figure-2: Maturity - Undiscovered as a % of Ultimate
Figure-3: Remaining Oil and Depletion Rate
Figure-4: Years to Depletion Mid-Point
Figure-5: Production with Giant Fields
Table 1 : Production Profile
Table 2 : Summary Data-sheet

SUMMARY

In addition to the prolific six countries that dominate Middle East production are another eleven countries of much lesser importance, which are here classified as the Middle East (Minor) Region. Two countries, Yemen North and South, have merged since the reference date, but are here still treated separately. The Region holds 2.4% of the World's Ultimate oil endowment.

Some oil had already been produced in the region prior the Second World War, but it was not until the late 1960s that production reached significant proportions, coming principally from Bahrain and Qatar. Private companies are active in all the countries. Although minor quantities of oil have been produced in Israel and Jordan, these countries lie outside the main producing trends and are of negligible importance.

The Region's share of World oil is as follows :

Cumulative Production	1.9%
Reserves	2.4%
Discovered	2.2%
Undiscovered	4.1%
Remaining	2.7%
Ultimate	2.4%

Its productive provinces are :

Western flank of the Persian Gulf Basin

Syria, Qatar, Dubai, Sharjah, Bahrain, and Ras-al-Khaimah all lie within the Persian Gulf basin, described in connection with the Middle East Six region.

Oman Basin and Yemen Rift

The Oman Mountains cut across the southern part of the Arabian sub-continent, and separate the Persian Gulf Basin from another productive belt to the south in Yemen and Oman. A large number of discoveries have been made, partly from Mesozoic reservoirs as in the Gulf, but also from Palaeozoic strata. Studies suggest that the late Pre-cambrian may even be an oil source-rock. It is a relatively new province, but most of the more important finds have probably already been made.

Ultimate

Oman and Qatar are by far the most important countries in the region with 10 BBO of Ultimate endowment each. They are followed by Dubai (5.5 BBO), Yemen South (4.5 BBO), Syria (4.0 BBO), Sharjah (2.0 BBO), Yemen North (2.0) and Bahrain (1.0 BBO). The three remaining countries, Ras-al-Khaimah, Israel and Jordan have a combined endowment of less than 1.0 BBO.

It will be noted that Yemen if treated as a single entity moves into third place.

Reserves

With the exception of Dubai, whose reported reserves have been adjusted downwards, the reported reserve figures for the region have been accepted.

Seven countries have reserves of in excess of 1 BBO : Qatar (4.5 BBO), Oman (4.25 BBO), Dubai and South Yemen (3.00 BBO each), Syria (1.73 BBO), Sharjah (1.50 BBO) and Yemen North (1.00 BBO).

Giant Fields are responsible for 52% of all discovered oil, with peak giant discovery phases occurring in 1940 and 1960. The last giant was found in 1984.

Undiscovered

Four countries are considered to have an Undiscovered potential of in excess of 1 BBO : Oman (2.73 BBO), South Yemen (1.49 BBO), Qatar (1.27 BBO) and Syria (1.00 BBO). The estimate for Qatar may be high, whereas Syria may have greater potential than realised, possibly from Palaeozoic reservoirs.

Production

Production commenced in Bahrain in 1938, one of the first Middle East countries outside Iran to be developed, followed by Qatar in 1949 and Oman in 1967. Ras-al-Khaimah and Yemen did not commence production until the 1980s. Bahrain and Syria peaked in the 1970s, whereas most of the other countries will peak within the next few years. Ras-al-Khaimah, Sharjah and Yemen are not expected to peak until well into the next century, which is responsible for the pronounced peak around 2030 seen on the production profile in Figure 5.

Oman, with 623 M b/d, Dubai (411 M b/d), Qatar (395 M b/d) and Syria (301 M b/d) were the most important producers in

1989. They are also responsible for most of the Cumulative: Qater (4.23 BBO), Oman (3.02 BBO), Dubai (2.18 BBO) and Syria (1.27 BBO).

The Production Profiles for the region are given in Part 3. Bahrain has curious periods of stable production, already commented upon, but otherwise is at an advanced state of maturity. Dubai, despite a downward adjustment of the Reported Reserves, is still relatively immature, being expected to peak in 1992, and therefore decline at a relatively high Depletion Rate. Oman is at a surprisingly immature stage, considering the giants had been discovered already in the 1960s : possibly the assessment is optimistic. Qatar has a classic mature curve, on which the effects of the 1980s "glut" are superimposed. Sharjah has a promising curve not reaching a Mid-Point until 2008. Syria has a fairly mature curve with early giant finds, but more promise.

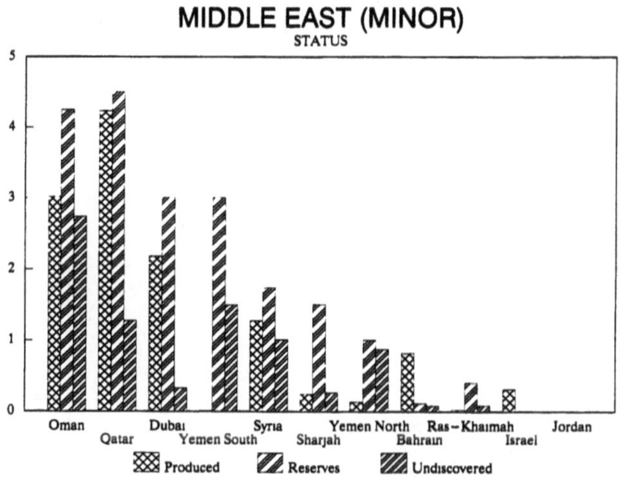

Figure 1-MEm. Produced, Reserves, Undiscovered.

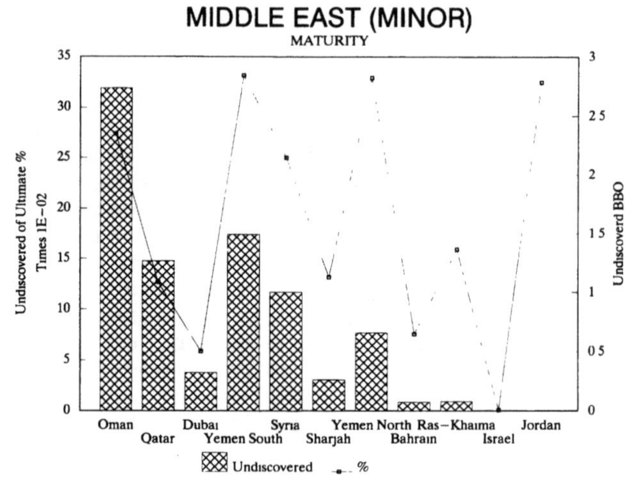

Figure 2-MEm. Maturity: Undiscovered as a percentage of Ultimate.

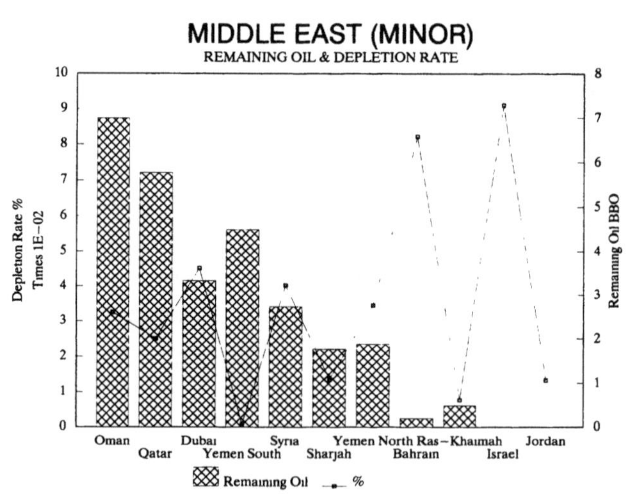

Figure 3-MEm. Remaining Oil (Reserves+Undiscovered) and Depletion Rate.

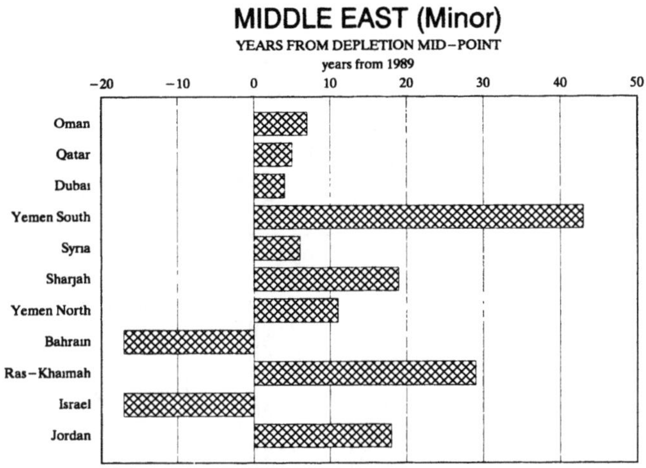

Figure 4-MEm. Years to or from Depletion Mid-Point.

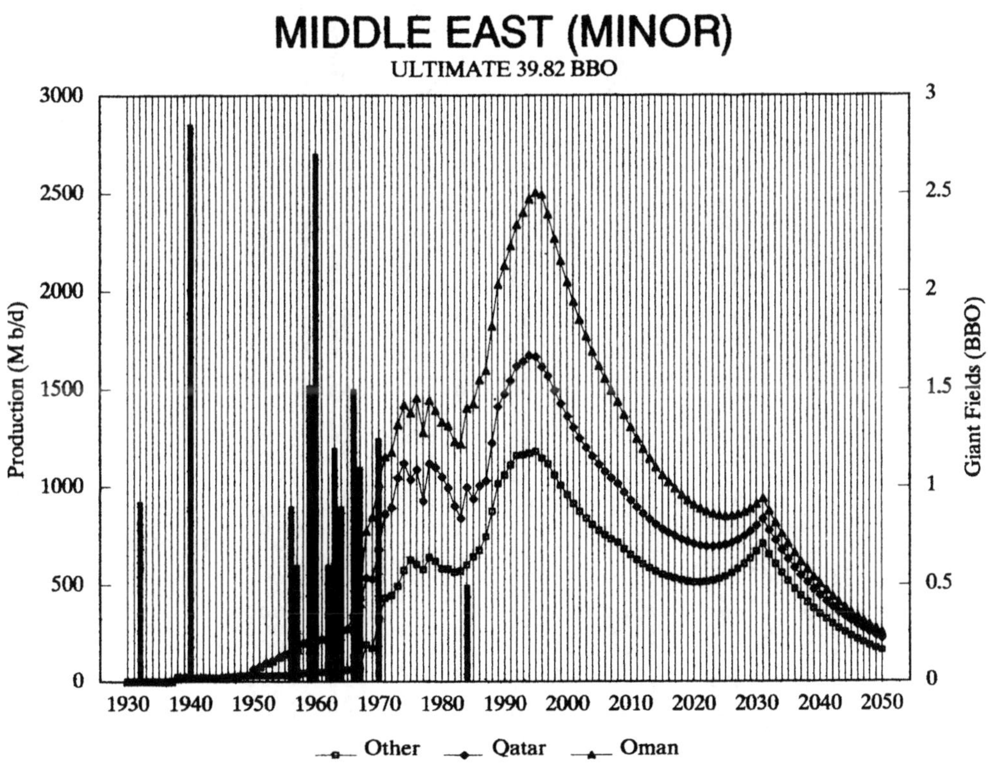

Figure 5-MEm. Production Profile 1930-2050 and Giant fields.

CUM.PROD	12.216 BBO	GIANT FIELDS		M.EAST (MINOR) REGION	
RESERVES	19.498 BBO	TOTAL BBO	17		
DISCOVERED	31.714 BBO	% OF DISC.	52%		
UNDISCOVERED	8.108 BBO	Last Giant	1984		
REMAINING	27.606 BBO			Variance Trend	
ULTIMATE	39.822 BBO			1990	+1.6%
Rept.Reserve	20.498	Ref.date(end)	1989	Future	n/a

Milestones	Date	Prod Mb/d	Dep. Rate	% Produced
1st Peak	1976	1455	1.51%	13%
2nd Peak	1995	2501	3.90%	44%
Mid-Point	1998	2268	4.00%	50%
Final Decline	n/a			

Forecast n/a
A=Exponential decline
B=5% incr. to Mid-Point
C=Other
Pre-1930 Cum. 0.000
(Added to 1930 prod)

Date	Prod. M.b/d	Prod. BBO/yr	Cum. Prod.	Remain	Deplet. Rate	% Produced	Giant	Notes & Names
1930	0	0.000	0.00	39.82	ERR	0%	0.000	
1931	0	0.000	0.00	39.82	0.00%	0%	0.000	
1932	0	0.000	0.00	39.82	0.00%	0%	0.924	
1933	0	0.000	0.00	39.82	0.00%	0%	0.000	
1934	0	0.000	0.00	39.82	0.00%	0%	0.000	
1935	0	0.000	0.00	39.82	0.00%	0%	0.000	
1936	0	0.000	0.00	39.82	0.00%	0%	0.000	
1937	0	0.000	0.00	39.82	0.00%	0%	0.000	
1938	23	0.008	0.01	39.81	0.02%	0%	0.000	
1939	21	0.008	0.02	39.81	0.02%	0%	0.000	
1940	19	0.007	0.02	39.80	0.02%	0%	2.850	
1941	19	0.007	0.03	39.79	0.02%	0%	0.000	
1942	17	0.006	0.04	39.79	0.02%	0%	0.000	
1943	18	0.007	0.04	39.78	0.02%	0%	0.000	
1944	18	0.007	0.05	39.77	0.02%	0%	0.000	
1945	20	0.007	0.06	39.77	0.02%	0%	0.000	
1946	23	0.008	0.06	39.76	0.02%	0%	0.000	
1947	26	0.009	0.07	39.75	0.02%	0%	0.000	
1948	30	0.011	0.09	39.74	0.03%	0%	0.000	
1949	32	0.012	0.10	39.73	0.03%	0%	0.000	
1950	64	0.023	0.12	39.70	0.06%	0%	0.000	
1951	79	0.029	0.15	39.67	0.07%	0%	0.000	
1952	99	0.036	0.19	39.64	0.09%	0%	0.000	
1953	108	0.039	0.22	39.60	0.10%	1%	0.000	
1954	128	0.047	0.27	39.55	0.12%	1%	0.000	
1955	144	0.053	0.32	39.50	0.13%	1%	0.000	
1956	158	0.058	0.38	39.44	0.15%	1%	0.900	
1957	169	0.062	0.44	39.38	0.16%	1%	0.600	
1958	205	0.075	0.52	39.30	0.19%	1%	0.000	
1959	222	0.081	0.60	39.22	0.21%	2%	1.519	
1960	221	0.080	0.68	39.14	0.21%	2%	2.700	
1961	224	0.082	0.76	39.06	0.21%	2%	0.000	
1962	236	0.086	0.85	38.97	0.22%	2%	0.600	
1963	241	0.088	0.93	38.89	0.23%	2%	1.200	
1964	274	0.100	1.03	38.79	0.26%	3%	0.900	
1965	276	0.101	1.14	38.69	0.26%	3%	0.000	
1966	348	0.127	1.26	38.56	0.33%	3%	1.500	
1967	441	0.161	1.42	38.40	0.42%	4%	1.100	
1968	774	0.283	1.71	38.12	0.74%	4%	0.000	
1969	847	0.309	2.02	37.81	0.81%	5%	0.000	
1970	1018	0.372	2.39	37.43	0.98%	6%	1.250	
1971	1152	0.421	2.81	37.01	1.12%	7%	0.000	
1972	1176	0.429	3.24	36.59	1.16%	8%	0.000	
1973	1319	0.481	3.72	36.10	1.32%	9%	0.000	
1974	1418	0.518	4.24	35.59	1.43%	11%	0.000	
1975	1378	0.503	4.74	35.08	1.41%	12%	0.000	
1976	1455	0.531	5.27	34.55	1.51%	13%	0.000	1ST PEAK
1977	1279	0.467	5.74	34.09	1.35%	14%	0.000	
1978	1440	0.526	6.26	33.56	1.54%	16%	0.000	
1979	1390	0.507	6.77	33.05	1.51%	17%	0.000	
1980	1332	0.486	7.26	32.57	1.47%	18%	0.000	
1981	1314	0.480	7.74	32.09	1.47%	19%	0.000	
1982	1233	0.450	8.19	31.64	1.40%	21%	0.000	
1983	1219	0.445	8.63	31.19	1.41%	22%	0.000	
1984	1402	0.512	9.14	30.68	1.64%	23%	0.500	
1985	1425	0.520	9.66	30.16	1.69%	24%	0.000	
1986	1547	0.565	10.23	29.60	1.87%	26%	0.000	
1987	1596	0.582	10.81	29.01	1.97%	27%	0.000	
1988	1822	0.665	11.47	28.35	2.29%	29%	0.000	

1989	2033	0.742	12.22	27.61	2.62%	31%	0.000	
1990	2129	0.777	12.99	26.83	2.82%	33%	0.000	
1991	2232	0.815	13.81	26.01	3.04%	35%	0.000	
1992	2339	0.854	14.66	25.16	3.28%	37%	0.000	
1993	2401	0.876	15.54	24.28	3.48%	39%	0.000	
1994	2470	0.901	16.44	23.38	3.71%	41%	0.000	
1995	2501	0.913	17.35	22.47	3.90%	44%	0.000	2ND PEAK
1996	2491	0.909	18.26	21.56	4.05%	46%	0.000	
1997	2391	0.873	19.13	20.69	4.05%	48%	0.000	
1998	2268	0.828	19.96	19.86	4.00%	50%	0.000	50% PRODUCED
1999	2154	0.786	20.75	19.07	3.96%	52%	0.000	
2000	2047	0.747	21.50	18.33	3.92%	54%	0.000	
2001	1948	0.711	22.21	17.62	3.88%	56%	0.000	
2002	1856	0.677	22.88	16.94	3.85%	57%	0.000	
2003	1771	0.646	23.53	16.29	3.82%	59%	0.000	
2004	1692	0.618	24.15	15.67	3.79%	61%	0.000	
2005	1619	0.591	24.74	15.08	3.77%	62%	0.000	
2006	1552	0.567	25.31	14.52	3.76%	64%	0.000	
2007	1491	0.544	25.85	13.97	3.75%	65%	0.000	
2008	1435	0.524	26.37	13.45	3.75%	66%	0.000	
2009	1367	0.499	26.87	12.95	3.71%	67%	0.000	
2010	1304	0.476	27.35	12.47	3.67%	69%	0.000	
2011	1245	0.455	27.80	12.02	3.64%	70%	0.000	
2012	1192	0.435	28.24	11.58	3.62%	71%	0.000	
2013	1143	0.417	28.65	11.17	3.60%	72%	0.000	
2014	1098	0.401	29.06	10.77	3.59%	73%	0.000	
2015	1058	0.386	29.44	10.38	3.59%	74%	0.000	
2016	1022	0.373	29.82	10.01	3.59%	75%	0.000	
2017	990	0.362	30.18	9.65	3.61%	76%	0.000	
2018	959	0.350	30.53	9.30	3.63%	77%	0.000	
2019	931	0.340	30.87	8.96	3.66%	78%	0.000	
2020	907	0.331	31.20	8.62	3.70%	78%	0.000	
2021	888	0.324	31.52	8.30	3.76%	79%	0.000	
2022	872	0.318	31.84	7.98	3.84%	80%	0.000	
2023	861	0.314	32.15	7.67	3.94%	81%	0.000	
2024	854	0.312	32.47	7.36	4.07%	82%	0.000	
2025	852	0.311	32.78	7.05	4.23%	82%	0.000	
2026	854	0.312	33.09	6.73	4.42%	83%	0.000	
2027	860	0.314	33.40	6.42	4.66%	84%	0.000	
2028	872	0.318	33.72	6.10	4.96%	85%	0.000	
2029	889	0.324	34.05	5.78	5.32%	85%	0.000	
2030	911	0.333	34.38	5.44	5.76%	86%	0.000	
2031	939	0.343	34.72	5.10	6.30%	87%	0.000	
2032	875	0.319	35.04	4.78	6.26%	88%	0.000	
2033	816	0.298	35.34	4.48	6.23%	89%	0.000	
2034	761	0.278	35.62	4.21	6.20%	89%	0.000	
2035	710	0.259	35.87	3.95	6.16%	90%	0.000	
2036	663	0.242	36.12	3.71	6.13%	91%	0.000	
2037	619	0.226	36.34	3.48	6.09%	91%	0.000	
2038	578	0.211	36.55	3.27	6.06%	92%	0.000	
2039	540	0.197	36.75	3.07	6.02%	92%	0.000	
2040	504	0.184	36.93	2.89	5.99%	93%	0.000	
2041	471	0.172	37.11	2.72	5.95%	93%	0.000	
2042	440	0.161	37.27	2.55	5.92%	94%	0.000	
2043	412	0.150	37.42	2.40	5.88%	94%	0.000	
2044	385	0.141	37.56	2.26	5.85%	94%	0.000	
2045	361	0.132	37.69	2.13	5.81%	95%	0.000	
2046	338	0.123	37.81	2.01	5.78%	95%	0.000	
2047	316	0.115	37.93	1.89	5.74%	95%	0.000	
2048	296	0.108	38.04	1.79	5.71%	96%	0.000	
2049	277	0.101	38.14	1.68	5.67%	96%	0.000	
2050	260	0.095	38.23	1.59	5.63%	96%	0.000	

SUMMARY DATA SHEET — REGION: MIDDLE EAST (Minor)
REFERENCE DATE: 1989

Country	Production M b/d	BBO/ year	Cumulative Production	Reported Reserves	Adjusted Reserves	Discovered	Undiscovered	Ultimate	Remaining	Und. of Ult.	Prod of Ult.	Rem. of Ult.	Prod of Disc	Disc. in Giants	Dep. Rate
A	B	C	D	E	F	G	H	I	J	K	L	M	N	O	P
Oman	623	0.227	3.02	4.25	4.25	7.27	2.73	10.00	6.98	27%	30%	70%	42%	37%	3.25%
Qatar	395	0.144	4.23	4.50	4.50	8.73	1.27	10.00	5.77	13%	42%	58%	48%	85%	2.50%
Dubai	411	0.150	2.18	4.00	3.00	5.18	0.32	5.50	3.32	6%	40%	60%	42%	50%	4.51%
Yemen South	10	0.004	0.01	3.00	3.00	3.01	1.49	4.50	4.49	33%	0%	100%	0%	0%	0.08%
Syria	301	0.110	1.27	1.73	1.73	3.00	1.00	4.00	2.73	25%	32%	68%	42%	81%	4.02%
Sharjah	65	0.024	0.24	1.50	1.50	1.74	0.26	2.00	1.76	13%	12%	88%	14%	0%	1.35%
Yemen North	177	0.064	0.13	1.00	1.00	1.13	0.87	2.00	1.87	43%	7%	93%	12%	44%	3.45%
Bahrain	42	0.015	0.81	0.11	0.11	0.92	0.08	1.00	0.19	8%	81%	19%	88%	100%	8.23%
Ras-Khaimah	10	0.004	0.02	0.40	0.40	0.42	0.08	0.50	0.48	16%	4%	96%	5%	0%	0.76%
Israel	0	0.000	0.31	0.00	0.00	0.31	0.00	0.31	0.00	0%	100%	0%	100%	0%	9.11%
Jordan	0	0.000	0.00	0.01	0.01	0.01	0.00	0.01	0.01	32%	18%	82%	26%	0%	1.33%
Region	2033	0.742	12.22	20.50	19.50	31.71	8.11	39.82	27.61	20%	31%	69%	39%	52%	2.69%

NEW PROVINCES

SUMMARY

We do not know where new prolific provinces may be found, but nevertheless we assume that some will be discovered.

Exploration is now so advanced in all areas that have been accessible to the international companies that it is very doubtful that any new important provinces remain to be discovered onshore. There is some possibility that finds might be made in areas of difficult terrain or for example beneath the lava fields of India or Brasil, but this is considered most unlikely.

Probably the most likely remaining onshore areas where discoveries could still be made are in Antarctica, the Soviet Union, or perhaps in countries such as China, Burma (Myanmar), Afganistan, Mongolia, or even eastern Europe. There is also always the possibility of turning up some unsuspected small graben having a very localised prolific source that could provide a viable new province.

Most of the offshore in the accessible world has also been so extensively explored as to make it unlikely than many, if any, new major provinces are still to be found. But potential may remain in the more remote and hostile environments. The Barents Sea could have been such a province, although much of it has now been all but condemned by drilling. There remains some reasonable possibilities on the Arctic shelves of North America, Greenland and Soviet Union, or on the Falklands Arc, although the extremely harsh operating conditions and distance from market mean that oil prices will have to have risen greatly before these areas become viable. Rather more promising might be the Sea of Okhotsk, despite its hostile environment, and adjoining areas.

Lastly, there is the deep offshore. Most deepwater areas lack satisfactory geology, but there may be other areas comparable to offshore Brasil that will yield major discoveries. The Perdido Trend in the deep Gulf of Mexico looks particularly promising, and there are leads into deep water off West Africa.

On balance, we believe that the amounts we have allocated to New Provinces are likely to be subject to downward revision. Certainly, no one should entertain any false hope for finding another Middle East.

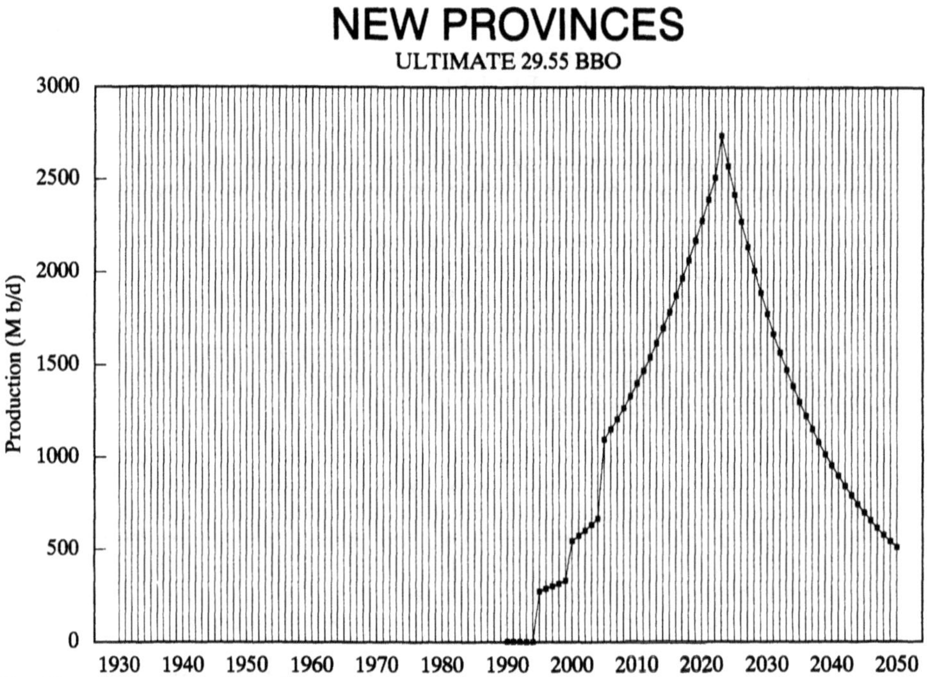

Possible major new producing provinces
(e.g. Antarctica, Barents Sea, Falklands Arc, Soviet Arctic etc,)

Figure 1-NP. Production Profile 1930-2050.

CUM.PROD	0.00	BBO	GIANT FIELDS		WORLD	
RESERVES	0.00	BBO	TOTAL BBO	0	NEW PROVINCES	
DISCOVERED	0.00	BBO	% OF DISC.	ERR		
UNDISCOVERED	29.55	BBO	Last Giant	-		
REMAINING	29.55	BBO			Variance	Trend
ULTIMATE	29.55	BBO			1990	n/a
Rept.Reserve			Ref.date(end)	1989	Future	DOWN

Milestones	Date	Prod Mb/d	Dep. Rate	% Produced	Forecast	n/a
1st Peak	2023	2739	6.01%	47%	A=Exponential decline	
2nd Peak					B=5% incr. to Mid-Point	
Mid-Point	2024	2574	6.01%	50%	C=Other	
					Pre-1930 Cum.	0
Final Decline	n/a					

Date	Prod. M.b/d	Prod. BBO/yr	Cum. Prod.	Remain	Deplet. Rate	% Produced	Giant	Notes & Names
1930		0.000	0.00	29.55	ERR	0%		
1931		0.000	0.00	29.55	0.00%	0%		
1932		0.000	0.00	29.55	0.00%	0%		
1933		0.000	0.00	29.55	0.00%	0%		
1934		0.000	0.00	29.55	0.00%	0%		
1935		0.000	0.00	29.55	0.00%	0%		
1936		0.000	0.00	29.55	0.00%	0%		
1937		0.000	0.00	29.55	0.00%	0%		
1938		0.000	0.00	29.55	0.00%	0%		
1939		0.000	0.00	29.55	0.00%	0%		
1940		0.000	0.00	29.55	0.00%	0%		
1941		0.000	0.00	29.55	0.00%	0%		
1942		0.000	0.00	29.55	0.00%	0%		
1943		0.000	0.00	29.55	0.00%	0%		
1944		0.000	0.00	29.55	0.00%	0%		
1945		0.000	0.00	29.55	0.00%	0%		
1946		0.000	0.00	29.55	0.00%	0%		
1947		0.000	0.00	29.55	0.00%	0%		
1948		0.000	0.00	29.55	0.00%	0%		
1949		0.000	0.00	29.55	0.00%	0%		
1950		0.000	0.00	29.55	0.00%	0%		
1951		0.000	0.00	29.55	0.00%	0%		
1952		0.000	0.00	29.55	0.00%	0%		
1953		0.000	0.00	29.55	0.00%	0%		
1954		0.000	0.00	29.55	0.00%	0%		
1955		0.000	0.00	29.55	0.00%	0%		
1956		0.000	0.00	29.55	0.00%	0%		
1957		0.000	0.00	29.55	0.00%	0%		
1958		0.000	0.00	29.55	0.00%	0%		
1959		0.000	0.00	29.55	0.00%	0%		
1960		0.000	0.00	29.55	0.00%	0%		
1961		0.000	0.00	29.55	0.00%	0%		
1962		0.000	0.00	29.55	0.00%	0%		
1963		0.000	0.00	29.55	0.00%	0%		
1964		0.000	0.00	29.55	0.00%	0%		
1965		0.000	0.00	29.55	0.00%	0%		
1966		0.000	0.00	29.55	0.00%	0%		
1967		0.000	0.00	29.55	0.00%	0%		
1968		0.000	0.00	29.55	0.00%	0%		
1969		0.000	0.00	29.55	0.00%	0%		
1970		0.000	0.00	29.55	0.00%	0%		
1971		0.000	0.00	29.55	0.00%	0%		
1972		0.000	0.00	29.55	0.00%	0%		
1973		0.000	0.00	29.55	0.00%	0%		
1974		0.000	0.00	29.55	0.00%	0%		
1975		0.000	0.00	29.55	0.00%	0%		
1976		0.000	0.00	29.55	0.00%	0%		
1977		0.000	0.00	29.55	0.00%	0%		
1978		0.000	0.00	29.55	0.00%	0%		
1979		0.000	0.00	29.55	0.00%	0%		
1980		0.000	0.00	29.55	0.00%	0%		
1981		0.000	0.00	29.55	0.00%	0%		
1982		0.000	0.00	29.55	0.00%	0%		
1983		0.000	0.00	29.55	0.00%	0%		
1984		0.000	0.00	29.55	0.00%	0%		
1985		0.000	0.00	29.55	0.00%	0%		
1986		0.000	0.00	29.55	0.00%	0%		
1987		0.000	0.00	29.55	0.00%	0%		
1988		0.000	0.00	29.55	0.00%	0%		

1989		0.000	0.00	29.55	0.00%	0%	
1990	3	0.001	0.00	29.55	0.00%	0%	
1991	3	0.001	0.00	29.55	0.00%	0%	
1992	3	0.001	0.00	29.55	0.00%	0%	
1993	3	0.001	0.00	29.55	0.00%	0%	
1994	3	0.001	0.01	29.54	0.00%	0%	
1995	274	0.100	0.11	29.44	0.34%	0%	Assumed major discovery
1996	288	0.105	0.21	29.34	0.36%	1%	
1997	302	0.110	0.32	29.23	0.38%	1%	
1998	317	0.116	0.44	29.11	0.40%	1%	
1999	333	0.122	0.56	28.99	0.42%	2%	
2000	548	0.200	0.76	28.79	0.69%	3%	Assumed major discovery
2001	575	0.210	0.97	28.58	0.73%	3%	
2002	604	0.221	1.19	28.36	0.77%	4%	
2003	634	0.232	1.42	28.13	0.82%	5%	
2004	666	0.243	1.66	27.89	0.86%	6%	
2005	1096	0.400	2.06	27.49	1.43%	7%	Assumed major discovery
2006	1151	0.420	2.48	27.07	1.53%	8%	
2007	1208	0.441	2.92	26.63	1.63%	10%	
2008	1269	0.463	3.39	26.16	1.74%	11%	
2009	1332	0.486	3.87	25.68	1.86%	13%	
2010	1399	0.511	4.38	25.17	1.99%	15%	
2011	1469	0.536	4.92	24.63	2.13%	17%	
2012	1542	0.563	5.48	24.07	2.29%	19%	
2013	1619	0.591	6.07	23.48	2.46%	21%	
2014	1700	0.621	6.69	22.86	2.64%	23%	
2015	1785	0.652	7.35	22.20	2.85%	25%	
2016	1874	0.684	8.03	21.52	3.08%	27%	
2017	1968	0.718	8.75	20.80	3.34%	30%	
2018	2066	0.754	9.50	20.05	3.63%	32%	
2019	2170	0.792	10.29	19.26	3.95%	35%	
2020	2278	0.832	11.13	18.42	4.32%	38%	
2021	2392	0.873	12.00	17.55	4.74%	41%	
2022	2512	0.917	12.92	16.63	5.22%	44%	
2023	2739	1.000	13.92	15.63	6.01%	47%	1ST PEAK
2024	2574	0.940	14.86	14.69	6.01%	50%	50% PRODUCED
2025	2420	0.883	15.74	13.81	6.01%	53%	
2026	2274	0.830	16.57	12.98	6.01%	56%	
2027	2137	0.780	17.35	12.20	6.01%	59%	
2028	2009	0.733	18.08	11.47	6.01%	61%	
2029	1888	0.689	18.77	10.78	6.01%	64%	
2030	1775	0.648	19.42	10.13	6.01%	66%	
2031	1668	0.609	20.03	9.52	6.01%	68%	
2032	1568	0.572	20.60	8.95	6.01%	70%	
2033	1474	0.538	21.14	8.41	6.01%	72%	
2034	1385	0.506	21.64	7.91	6.01%	73%	
2035	1302	0.475	22.12	7.43	6.01%	75%	
2036	1224	0.447	22.57	6.98	6.01%	76%	
2037	1150	0.420	22.99	6.56	6.01%	78%	
2038	1081	0.395	23.38	6.17	6.01%	79%	
2039	1016	0.371	23.75	5.80	6.01%	80%	
2040	955	0.349	24.10	5.45	6.01%	82%	
2041	898	0.328	24.43	5.12	6.01%	83%	
2042	844	0.308	24.73	4.82	6.01%	84%	
2043	793	0.289	25.02	4.53	6.01%	85%	
2044	745	0.272	25.30	4.25	6.01%	86%	
2045	700	0.256	25.55	4.00	6.01%	86%	
2046	658	0.240	25.79	3.76	6.01%	87%	
2047	619	0.226	26.02	3.53	6.01%	88%	
2048	582	0.212	26.23	3.32	6.01%	89%	
2049	547	0.200	26.43	3.12	6.01%	89%	
2050	514	0.188	26.62	2.93	6.01%	90%	

PART 3

PRODUCTION BY COUNTRY AND GIANT OILFIELDS

INTRODUCTION

An assessment of each country, listed below, is given in the following pages. A table for each country provides details of the production profile, including historical production and other information. The results are also plotted in graphic form, except for countries with only minor production. An explanation of the format and calculations was given in the introduction to Part 2.

The figures are those as generated by the computer, but it is stressed that all need to be generously rounded as the implied accuracy is misleading.

The following countries are covered (graphic plots are not provided for those shown in parenthesis)

Abu Dhabi	France	(Pakistan)
(Albania)	Gabon	(Papua)
Algeria	Germany	Peru
Angola	(Ghana)	(Philippines)
Argentina	(Greece)	(Poland)
Australia	(Guatemala)	Qatar
Austria	(Hungary)	(Ras al Khaimah)
Bahrain	India	(Romania)
(Bangladesh)	Indonesia	Saudi Arabia
(Barbados)	Iran	Sharjah
(Benin)	Iraq	(Spain)
Bolivia	(Israel)	(Sudan)
Brasil	Italy	(Suriname)
(Brunei)	(Ivory Coast)	Syria
(Bulgaria)	(Japan)	(Taiwan)
Burma	(Jordan)	(Thailand)
(Cameroon)	Kuwait	Trinidad
Canada	Libya	Tunisia
(Chile)	(Malaysia)	Turkey
China	Mexico	United Kingdom
Colombia	(Morocco)	United States
(Congo)	(Netherlands)	U.S.S.R.
(Cuba)	Neutral Zone	Venezuela
(Czeckoslovakia)	(New Zealand)	(Viet Nam)
Denmark	Nigeria	Yemen North
Dubai	Norway	Yemen South
Ecuador	Oman	(Yugoslavia)
Egypt		(Zaire)

Finally is given a list of giant fields, defined as having in excess of 500 M bbl of reported original reserves. It is made up of two parts. The first list is one on which there is a fair degree of confidence, and the second list covers reported giants on which there are some doubts.

The data-base on giant fields is weak. The reserves of many fields are probably subject to revision, in some cases upwards. There may be instances of the same field being reported under different names, and there may be some recent giant discoveries which are not yet reported (as for example off Nigeria and in Gabon).

Important note

The plots and tables give a misleading impression of greater accuracy than is intended. The figures are given as printed by the computer without individual rounding. They should be regarded as being, at best, close to the mid-point of a wide range.

ALL FIGURES ARE TO BE GENEROUSLY ROUNDED

Furthermore, clerical errors in inputting the data may have been made, and the computer itself, having done sterling work, sometimes shows that it has a mind of its own in what figures it elects to print. Some such anomalies may have escaped proof-reading and checking.

In many cases the reader will have specific knowledge by which to improve the assessments made here. The right-hand column of the tables may be used to note such corrections : the author would naturally appreciate any such information.

ABU DHABI

ULTIMATE 80 BBO

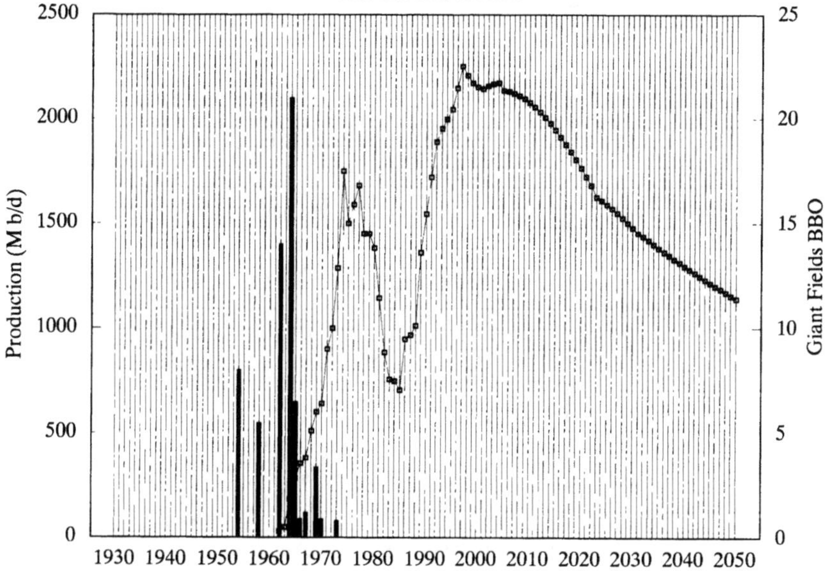

ALGERIA

ULTIMATE 20 BBO

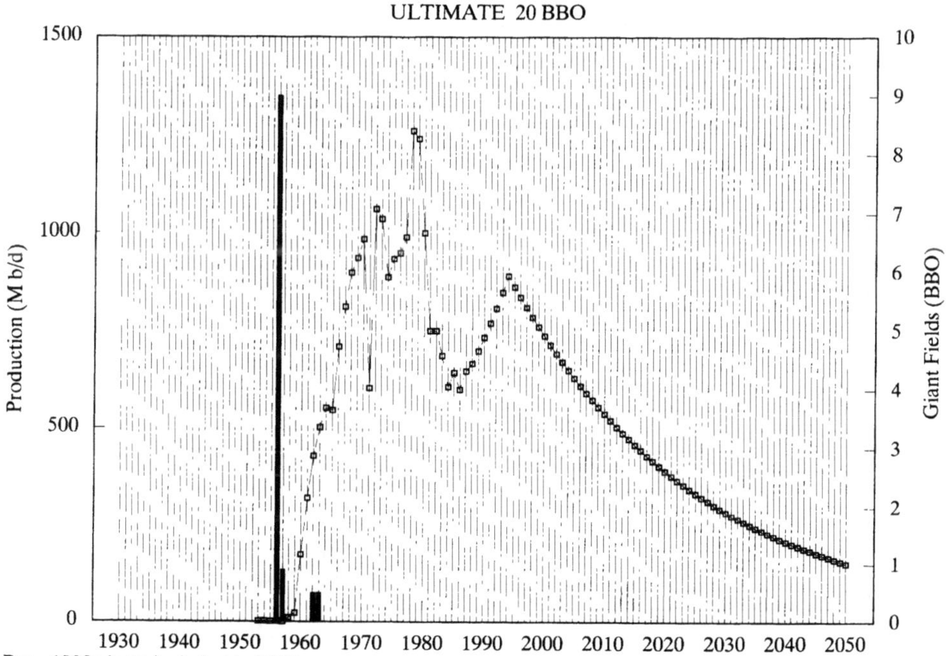

Pre–1930 giants included in 1930

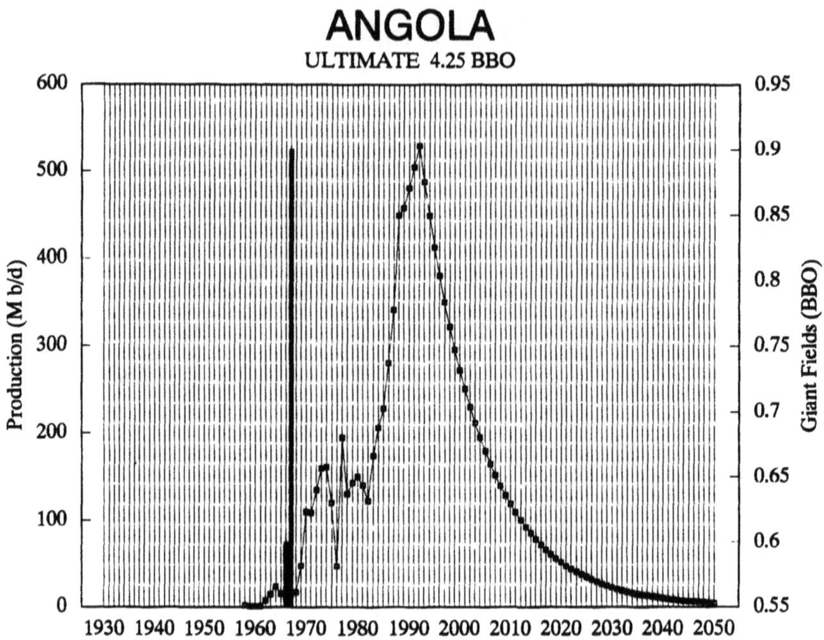

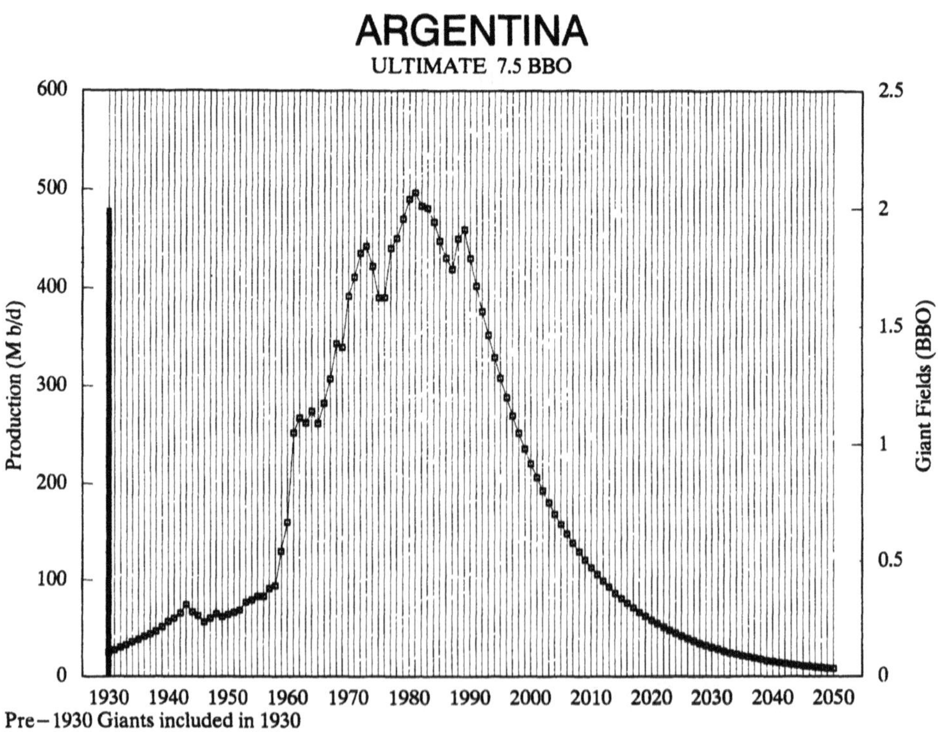

Pre–1930 Giants included in 1930

AUSTRALIA

ULTIMATE 6 BBO

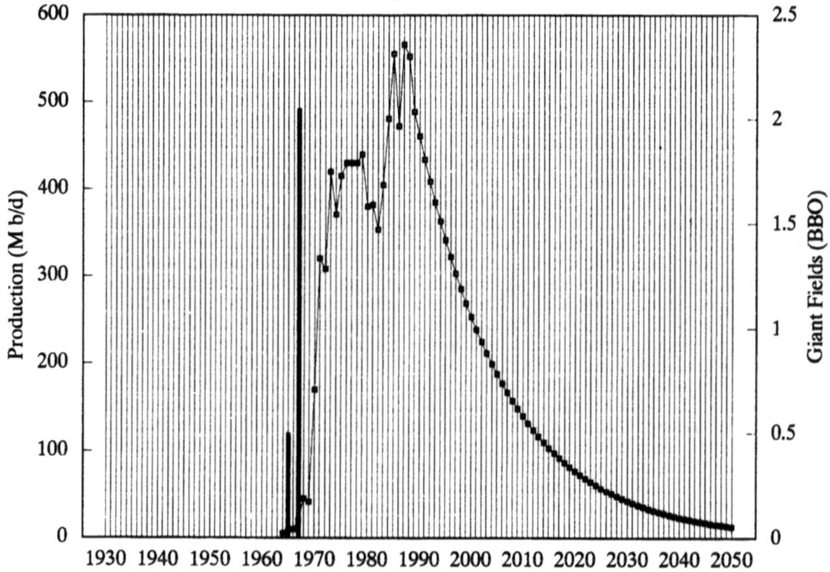

AUSTRIA

ULTIMATE 1 BBO

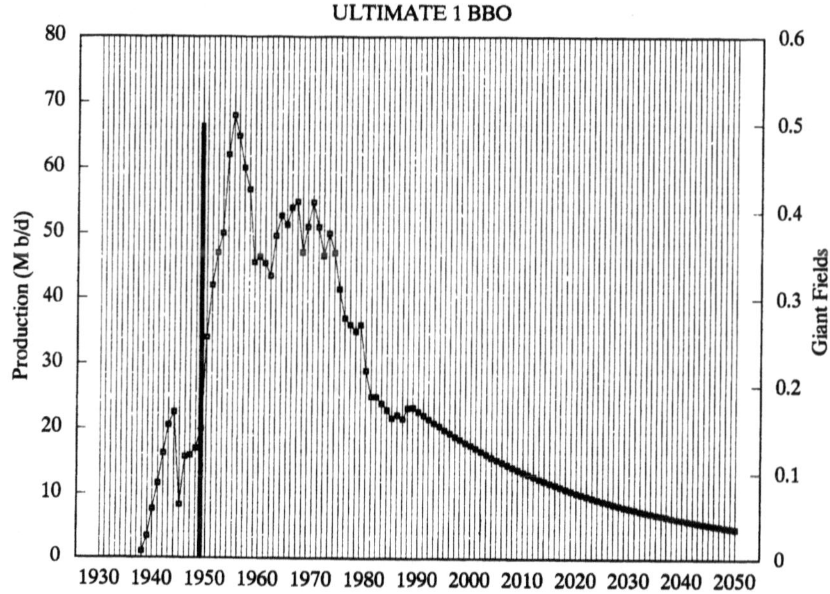

BAHRAIN

ULTIMATE 1 BBO

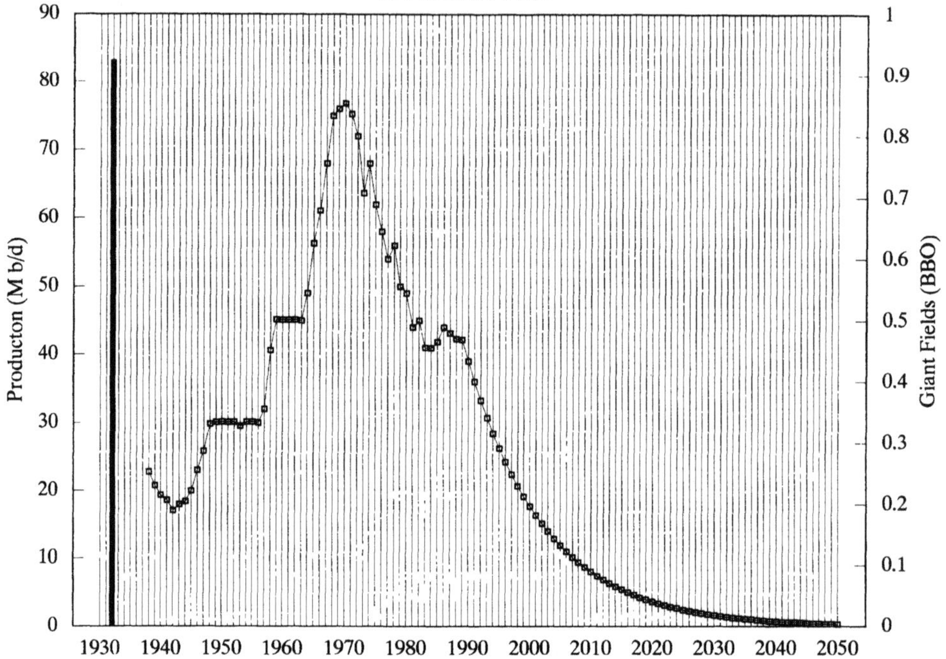

BOLIVIA

ULTIMATE 0.75 BBO

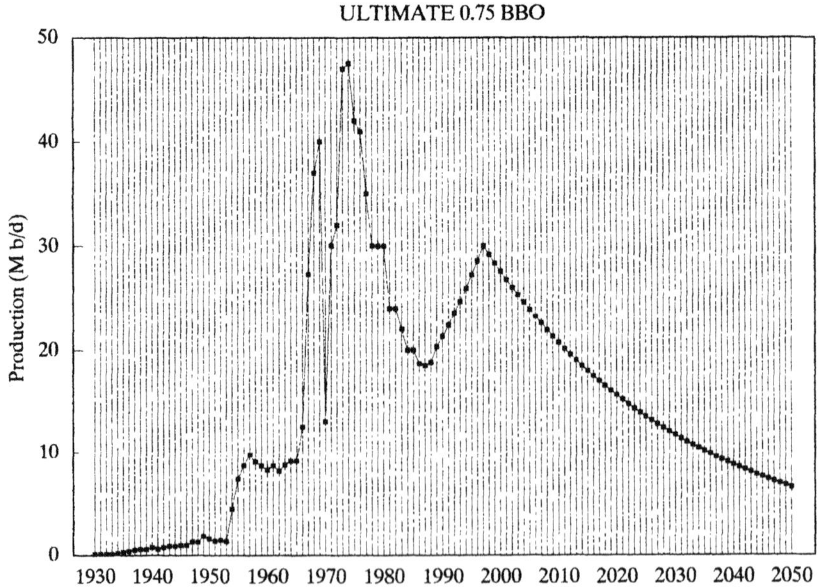

BRASIL

ULTIMATE 8.00 BBO

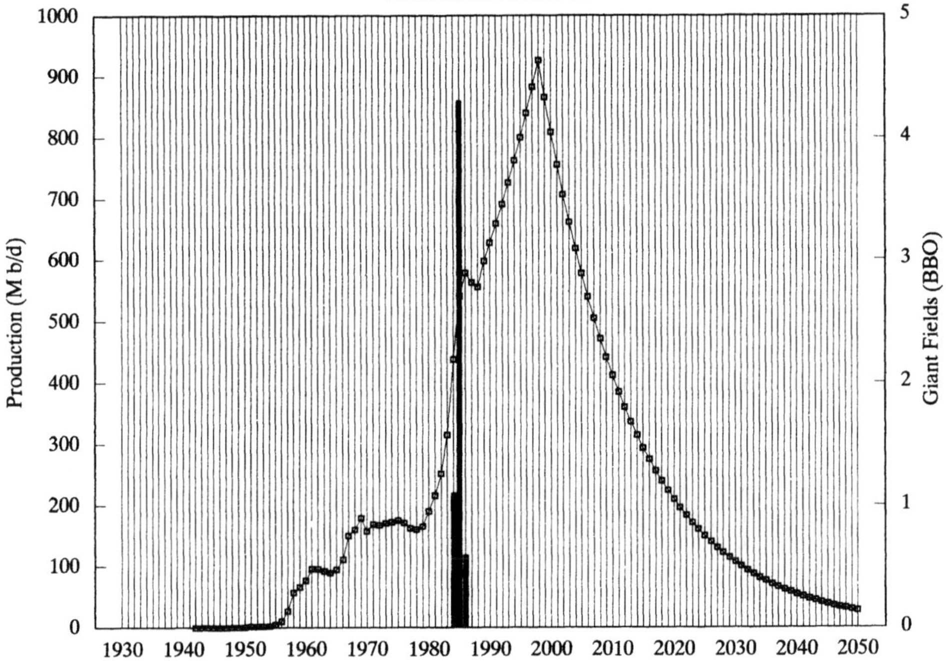

BURMA

ULTIMATE 1.2 BBO

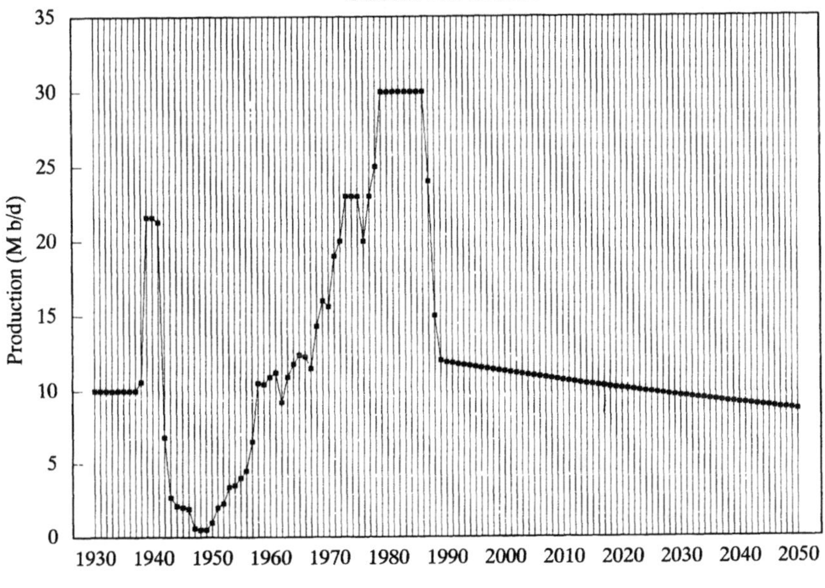

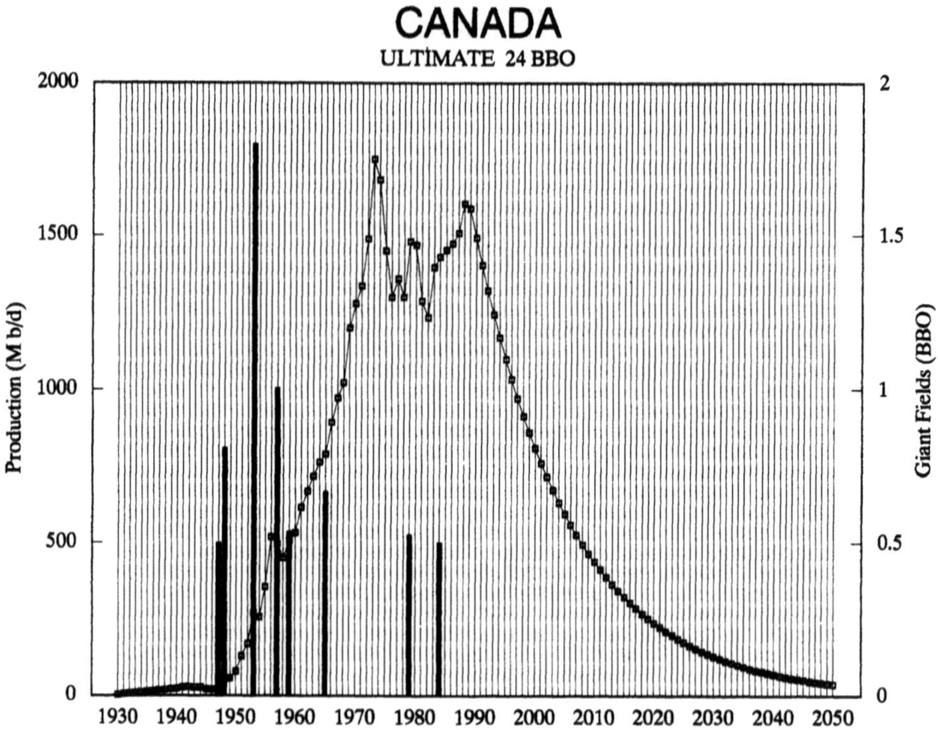
CANADA
ULTIMATE 24 BBO
Production (M b/d)
Giant Fields (BBO)
2000
1500
1000
500
0
2
1.5
1
0.5
0
1930 1940 1950 1960 1970 1980 1990 2000 2010 2020 2030 2040 2050

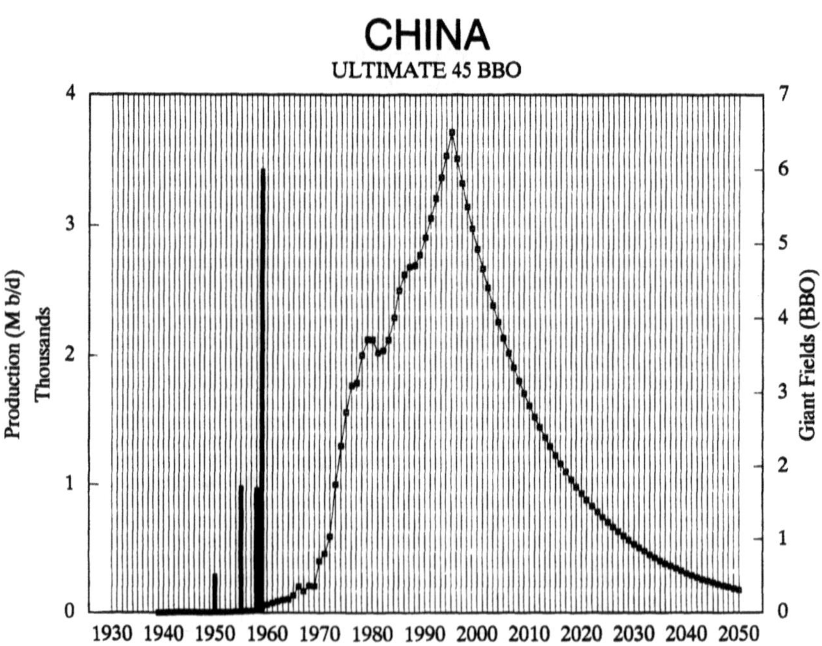
CHINA
ULTIMATE 45 BBO
Production (M b/d)
Thousands
Giant Fields (BBO)
4
3
2
1
0
7
6
5
4
3
2
1
0
1930 1940 1950 1960 1970 1980 1990 2000 2010 2020 2030 2040 2050

COLOMBIA

ULTIMATE 5.25 BBO

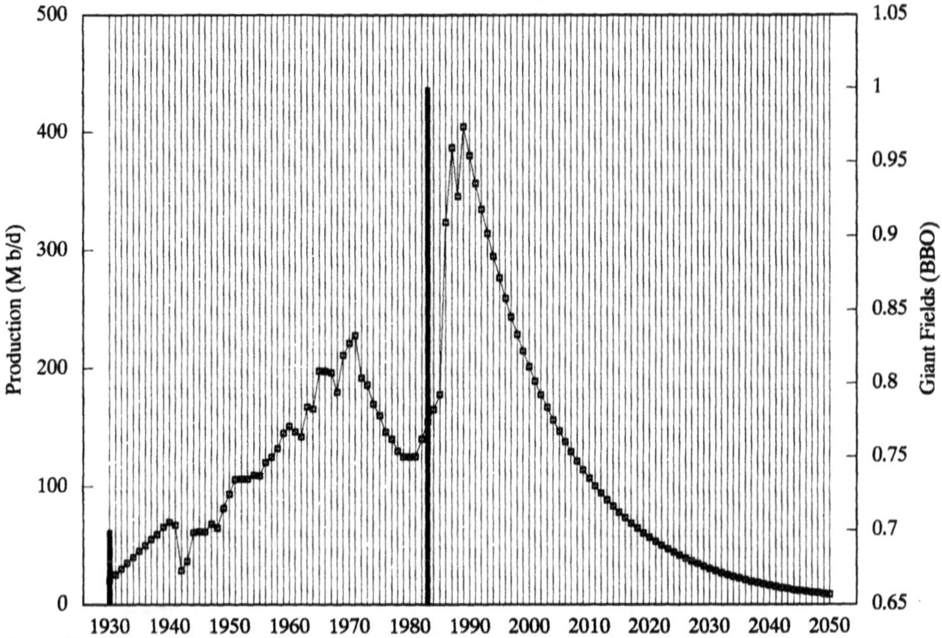

Pre – 1930 Giants included in 1930

DENMARK

ULTIMATE 1.25 BBO

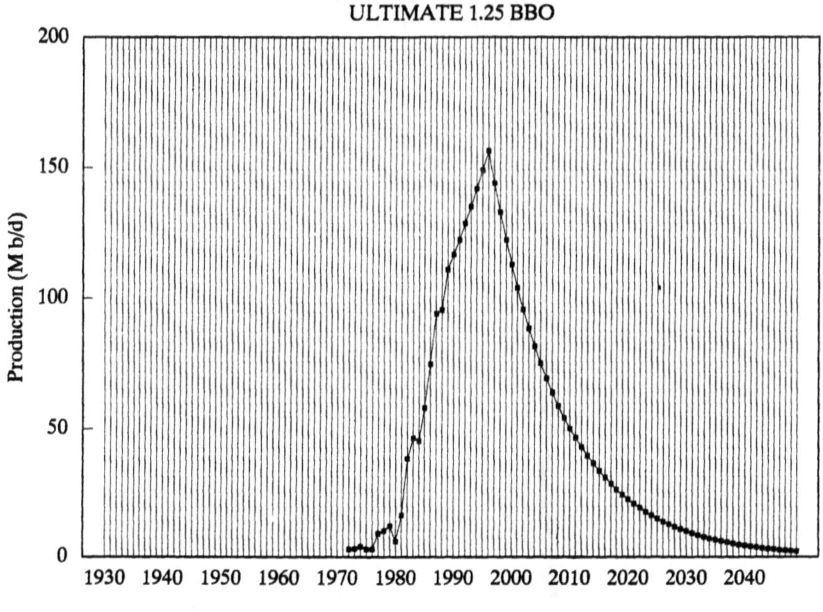

DUBAI

ULTIMATE 5.5 BBO

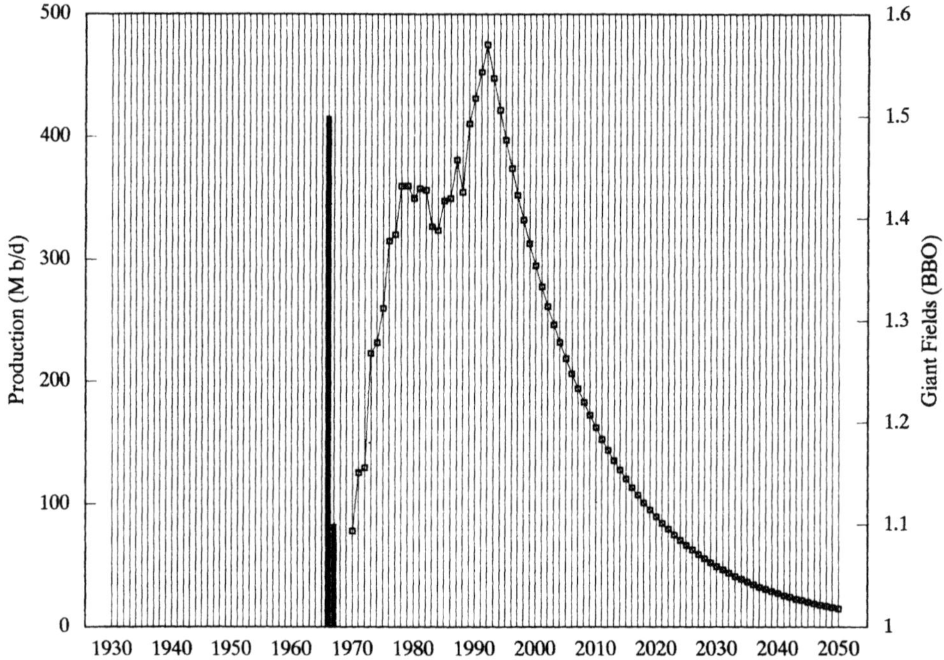

ECUADOR

ULTIMATE 3.5 BBO

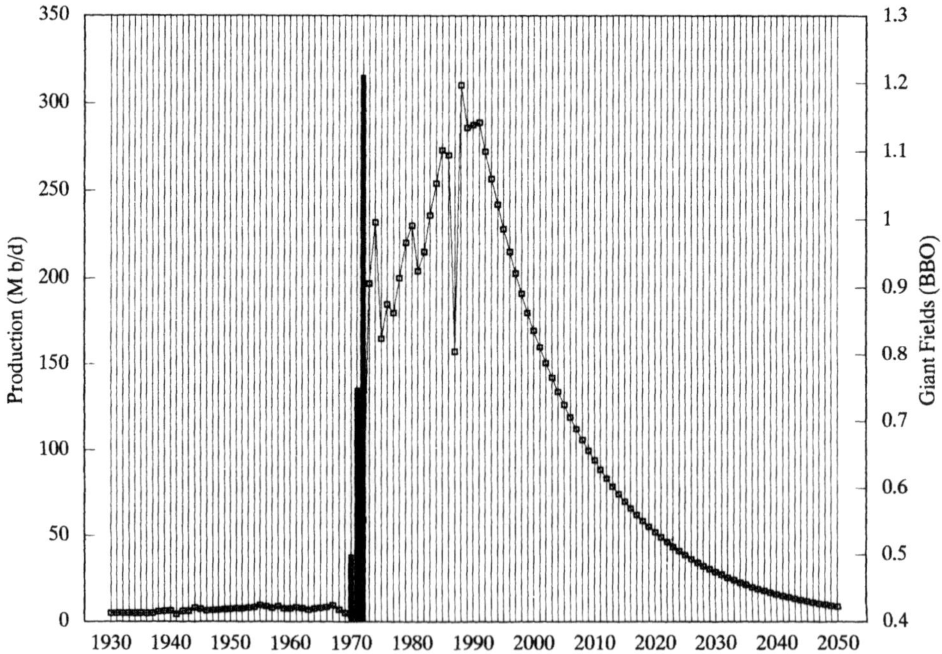

EGYPT

ULTIMATE 11 BBO

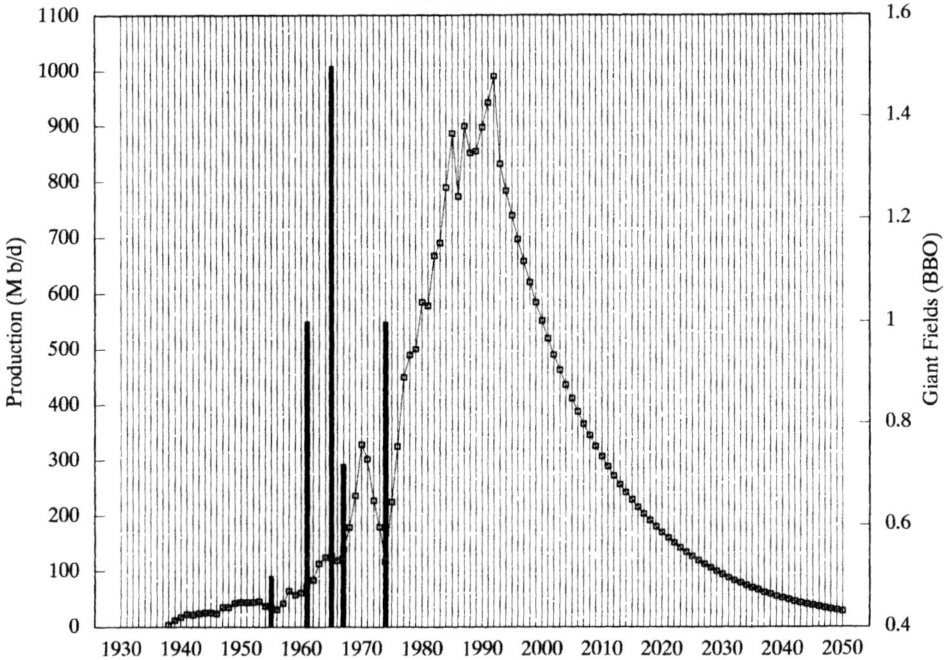

FRANCE

ULTIMATE 1 BBO

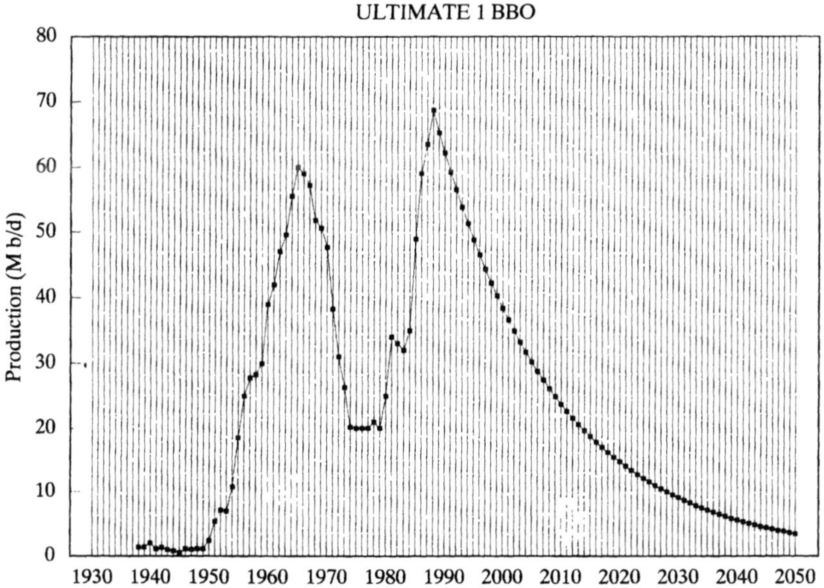

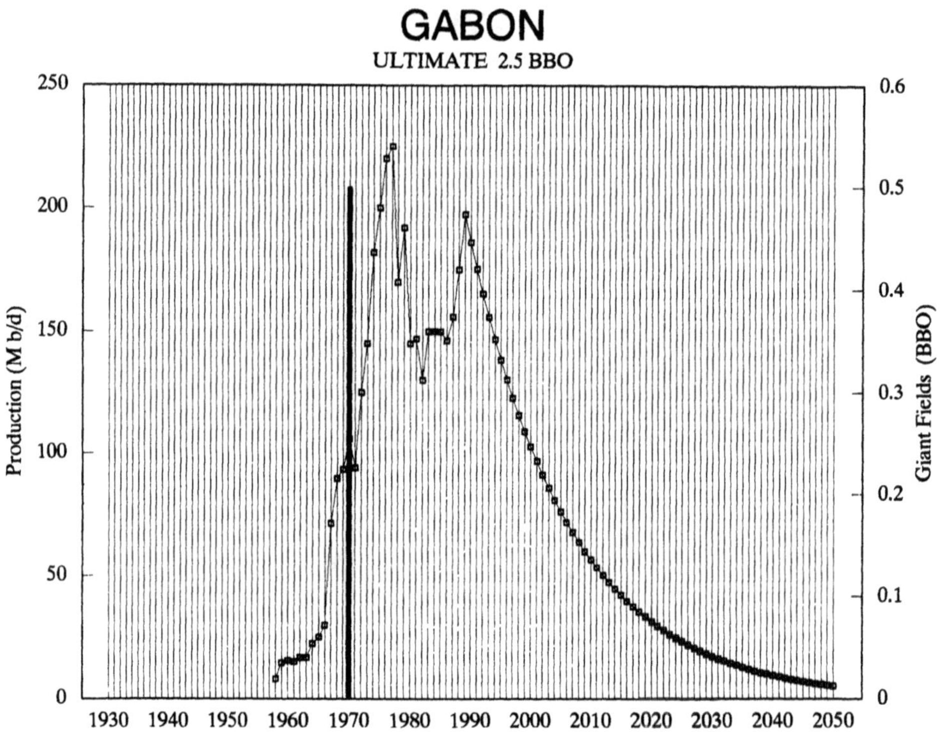
GABON
ULTIMATE 2.5 BBO
Production (M b/d)
Giant Fields (BBO)
250
200
150
100
50
0
0.6
0.5
0.4
0.3
0.2
0.1
0
1930 1940 1950 1960 1970 1980 1990 2000 2010 2020 2030 2040 2050

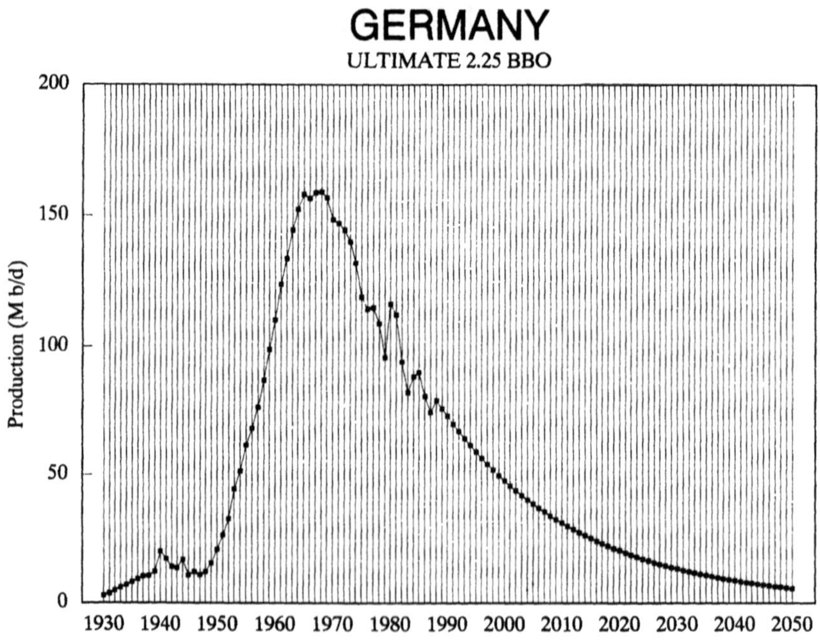
GERMANY
ULTIMATE 2.25 BBO
Production (M b/d)
200
150
100
50
0
1930 1940 1950 1960 1970 1980 1990 2000 2010 2020 2030 2040 2050

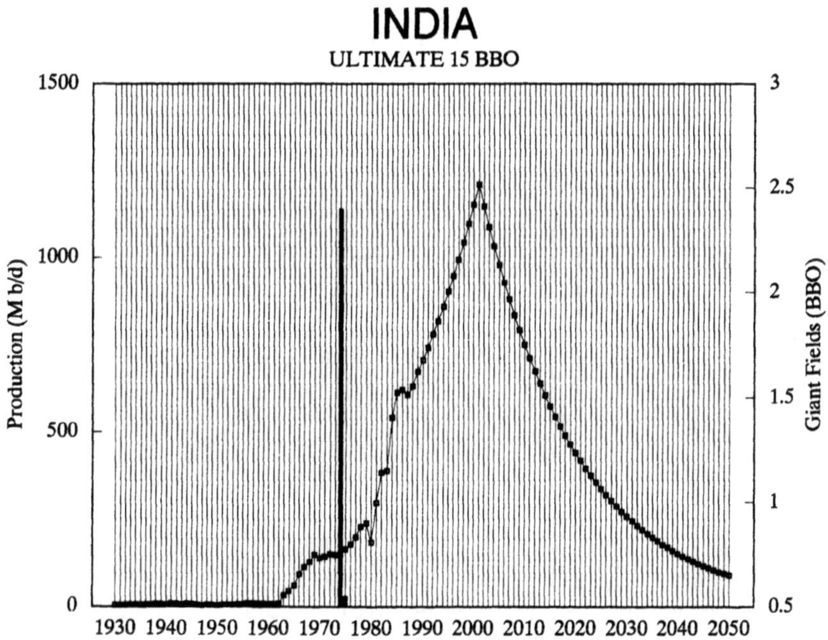
INDIA
ULTIMATE 15 BBO
Production (M b/d)
1500
1000
500
0
Giant Fields (BBO)
3
2.5
2
1.5
1
0.5
1930 1940 1950 1960 1970 1980 1990 2000 2010 2020 2030 2040 2050

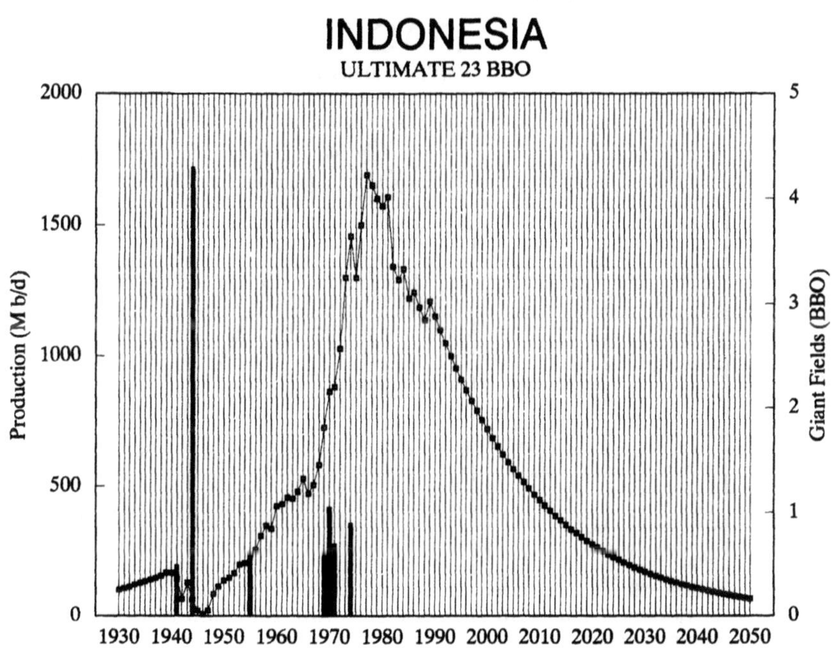
INDONESIA
ULTIMATE 23 BBO
Production (M b/d)
2000
1500
1000
500
0
Giant Fields (BBO)
5
4
3
2
1
0
1930 1940 1950 1960 1970 1980 1990 2000 2010 2020 2030 2040 2050

IRAN

ULTIMATE 115 BBO

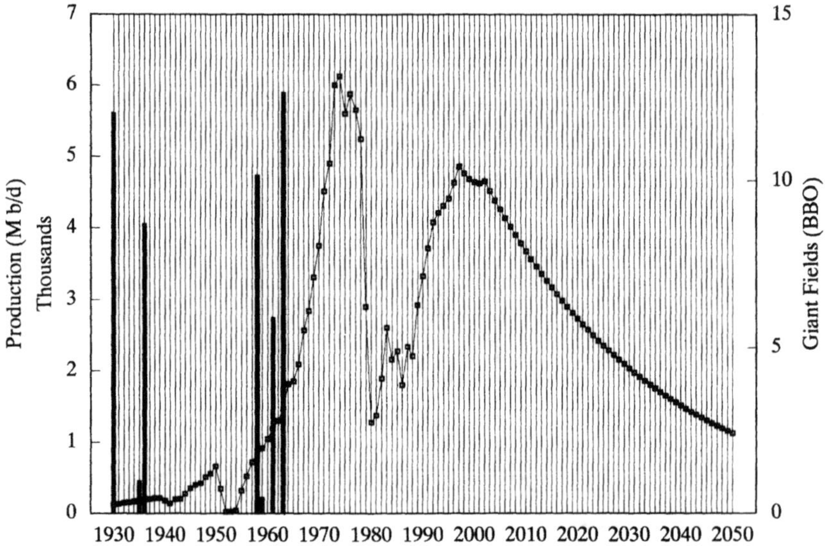

Pre – 1930 Giants included in 1930

IRAQ

ULTIMATE 100 BBO

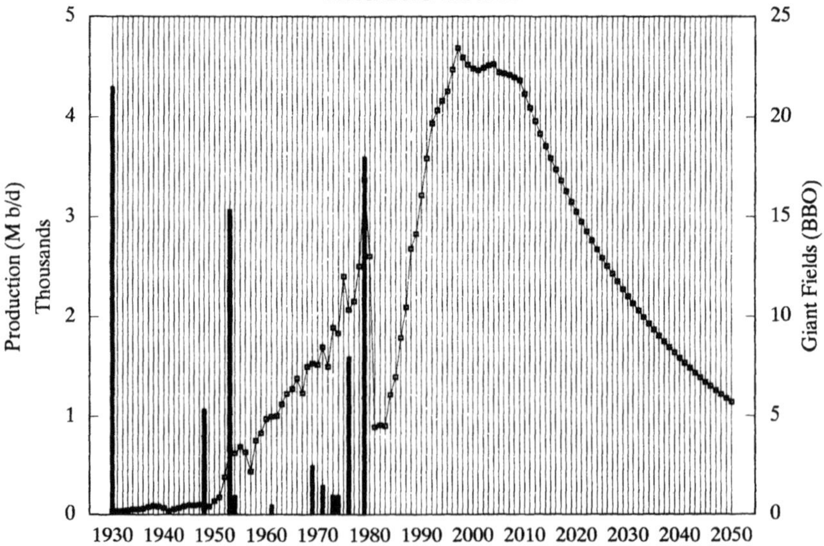

Pre – 1930 giants included in 1930

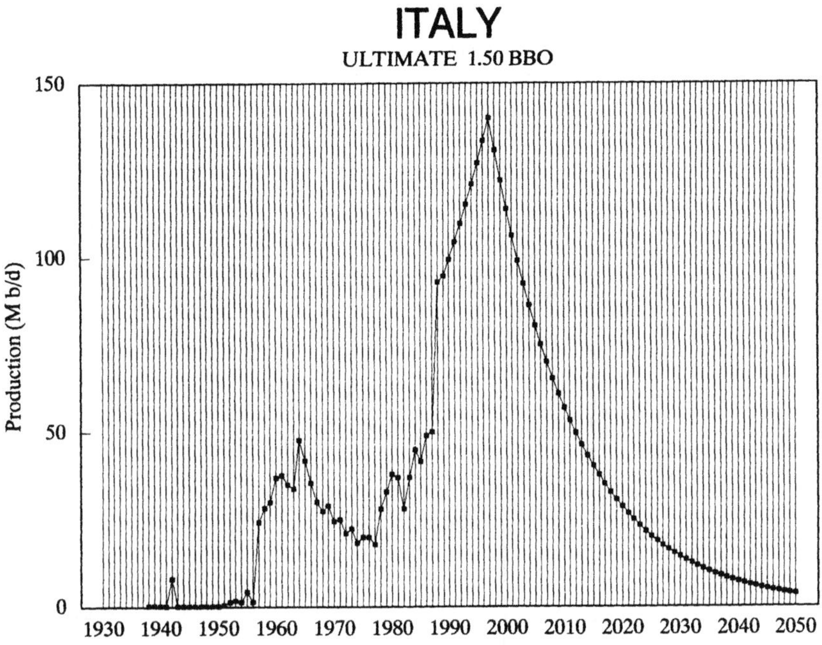
ITALY
ULTIMATE 1.50 BBO
Production (M b/d)
150
100
50
0
1930 1940 1950 1960 1970 1980 1990 2000 2010 2020 2030 2040 2050

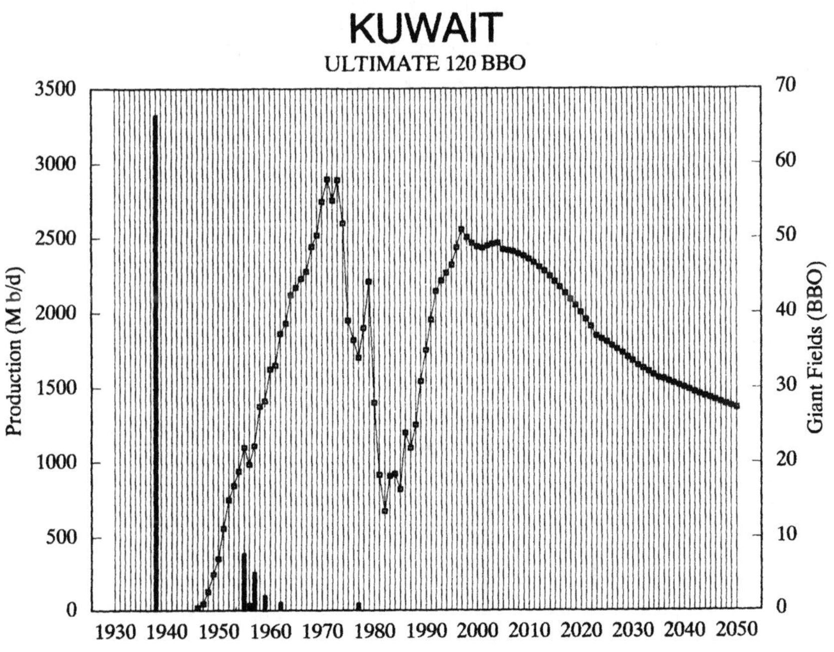
KUWAIT
ULTIMATE 120 BBO
Production (M b/d)
Giant Fields (BBO)
3500
3000
2500
2000
1500
1000
500
0
70
60
50
40
30
20
10
0
1930 1940 1950 1960 1970 1980 1990 2000 2010 2020 2030 2040 2050

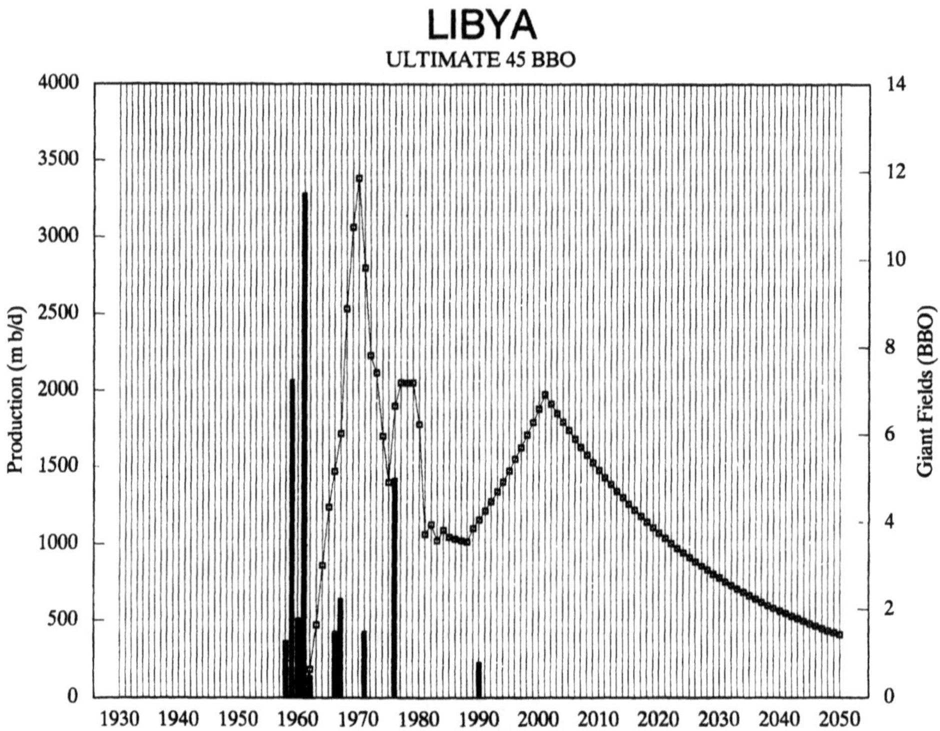

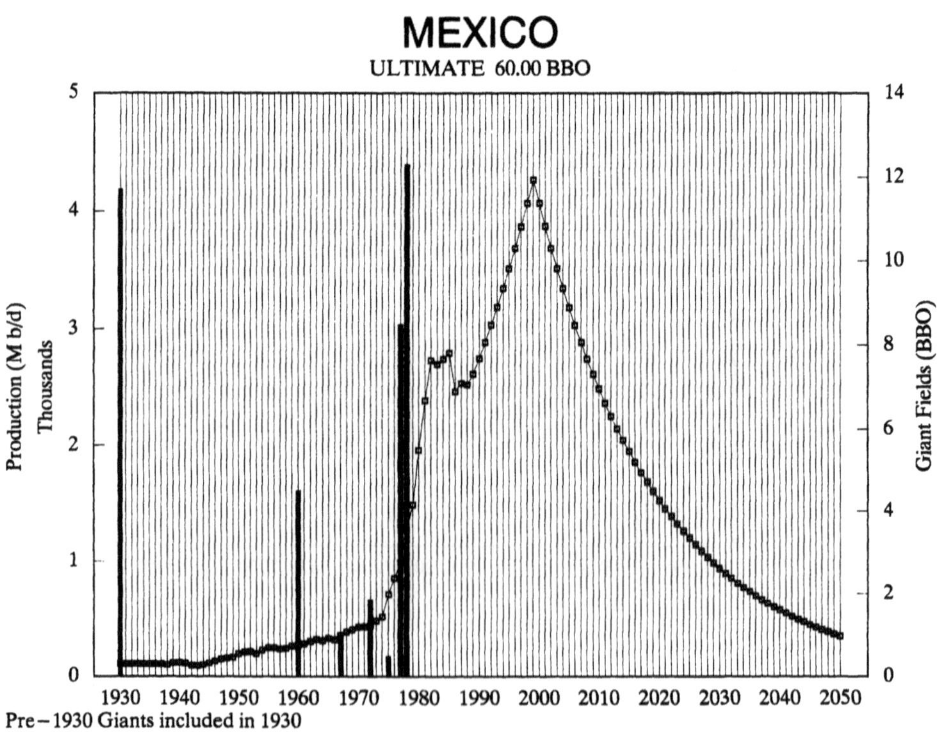

Pre–1930 Giants included in 1930

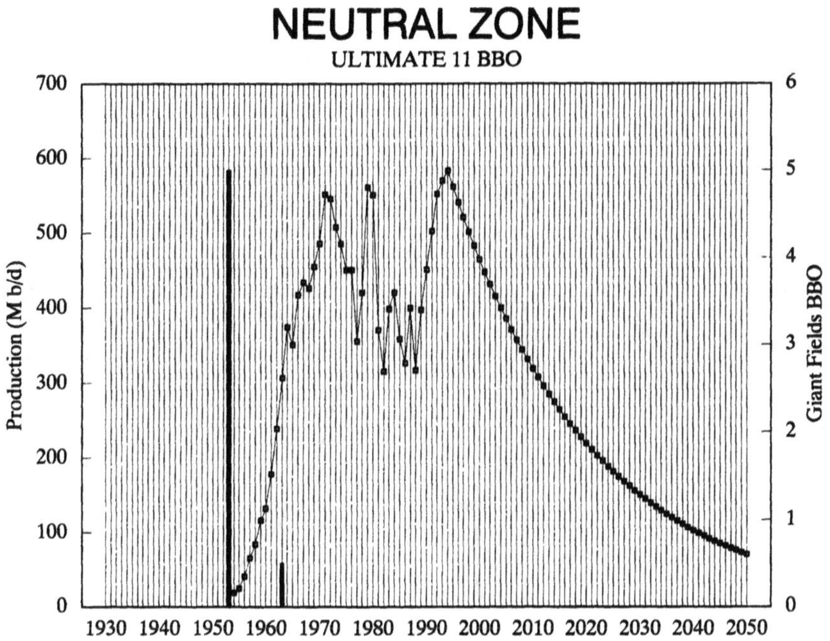
NEUTRAL ZONE
ULTIMATE 11 BBO
Production (M b/d)
Giant Fields BBO
700
600
500
400
300
200
100
0
6
5
4
3
2
1
0
1930 1940 1950 1960 1970 1980 1990 2000 2010 2020 2030 2040 2050

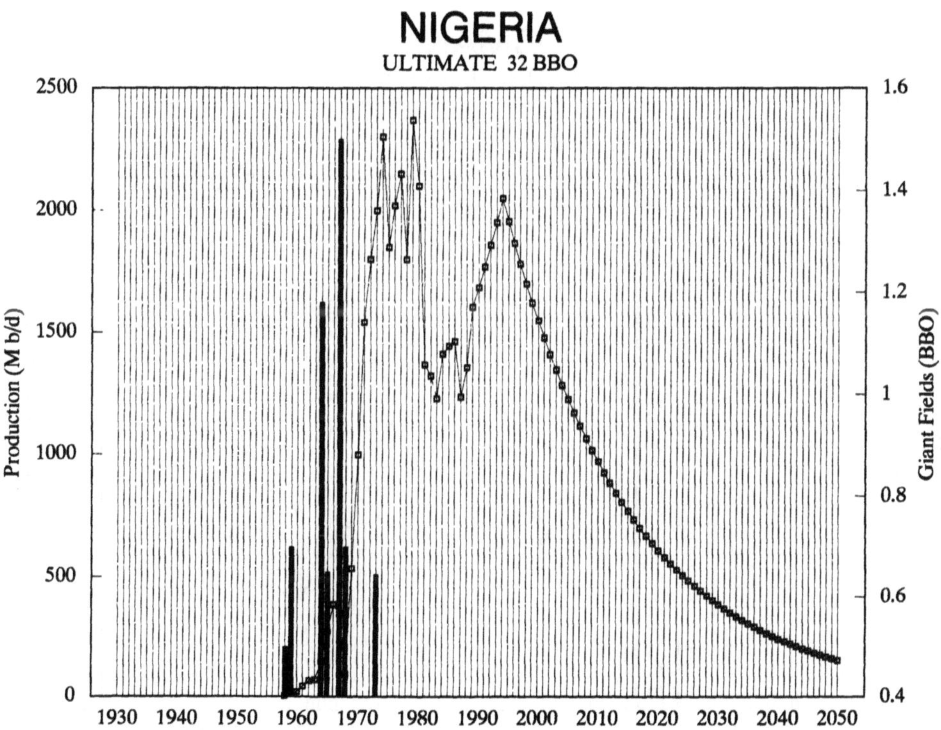
NIGERIA
ULTIMATE 32 BBO
Production (M b/d)
Giant Fields (BBO)
2500
2000
1500
1000
500
0
1.6
1.4
1.2
1
0.8
0.6
0.4
1930 1940 1950 1960 1970 1980 1990 2000 2010 2020 2030 2040 2050

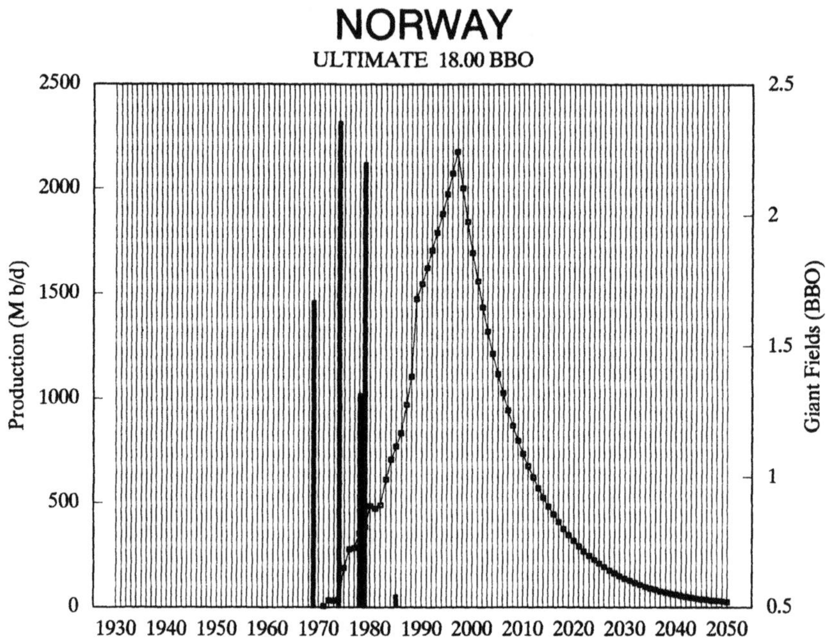
NORWAY
ULTIMATE 18.00 BBO
2500
2000
1500
1000
500
0
Production (M b/d)
2.5
2
1.5
1
0.5
Giant Fields (BBO)
1930 1940 1950 1960 1970 1980 1990 2000 2010 2020 2030 2040 2050

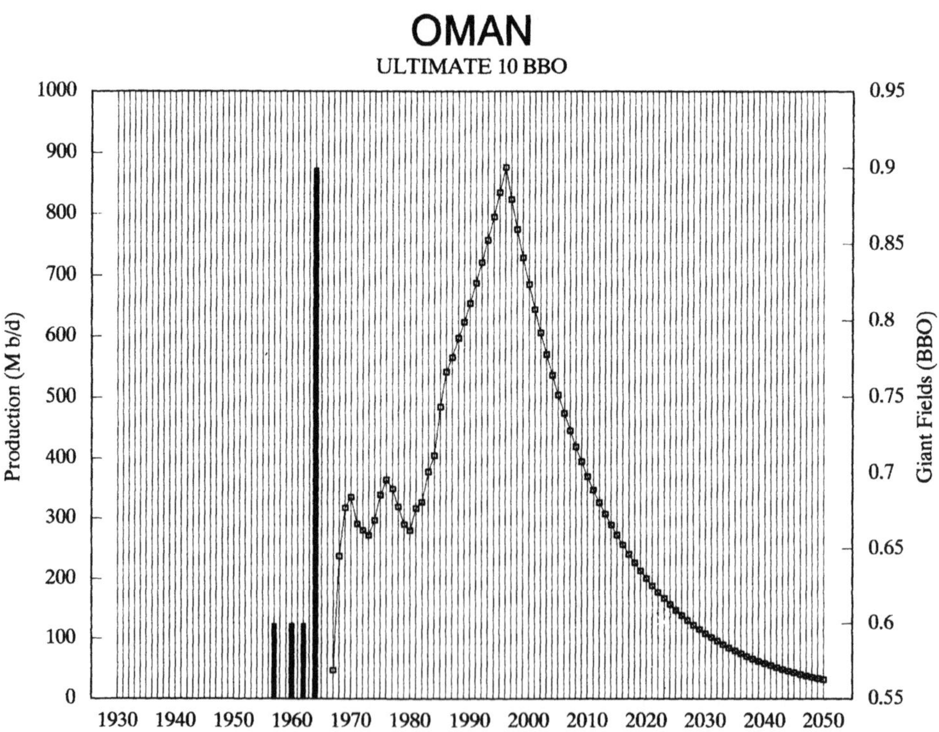
OMAN
ULTIMATE 10 BBO
1000
900
800
700
600
500
400
300
200
100
0
Production (M b/d)
0.95
0.9
0.85
0.8
0.75
0.7
0.65
0.6
0.55
Giant Fields (BBO)
1930 1940 1950 1960 1970 1980 1990 2000 2010 2020 2030 2040 2050

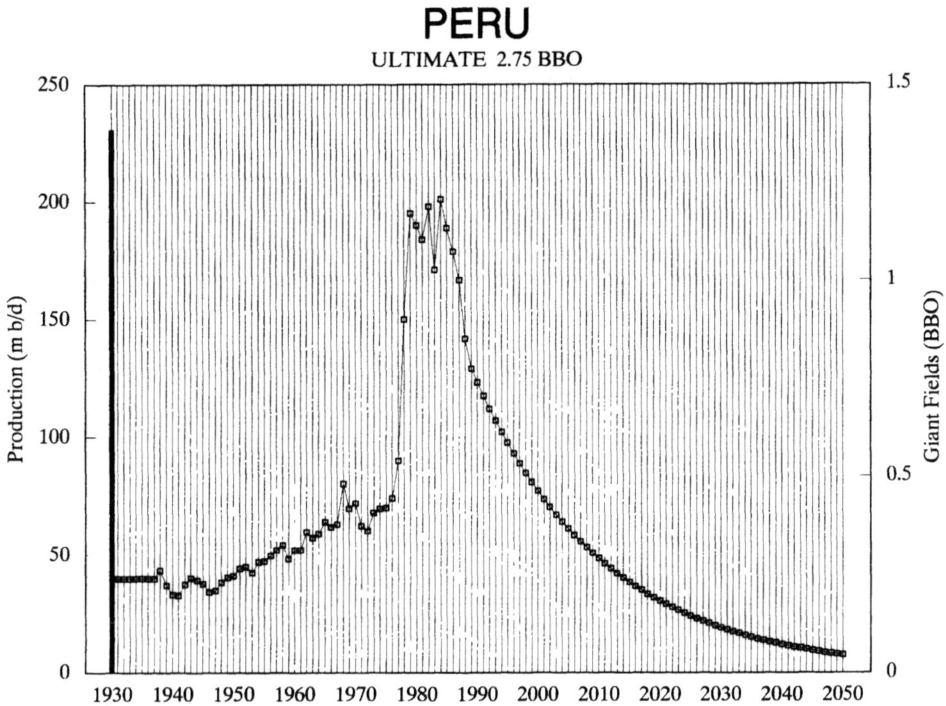

Pre – 1930 Giants included in 1930

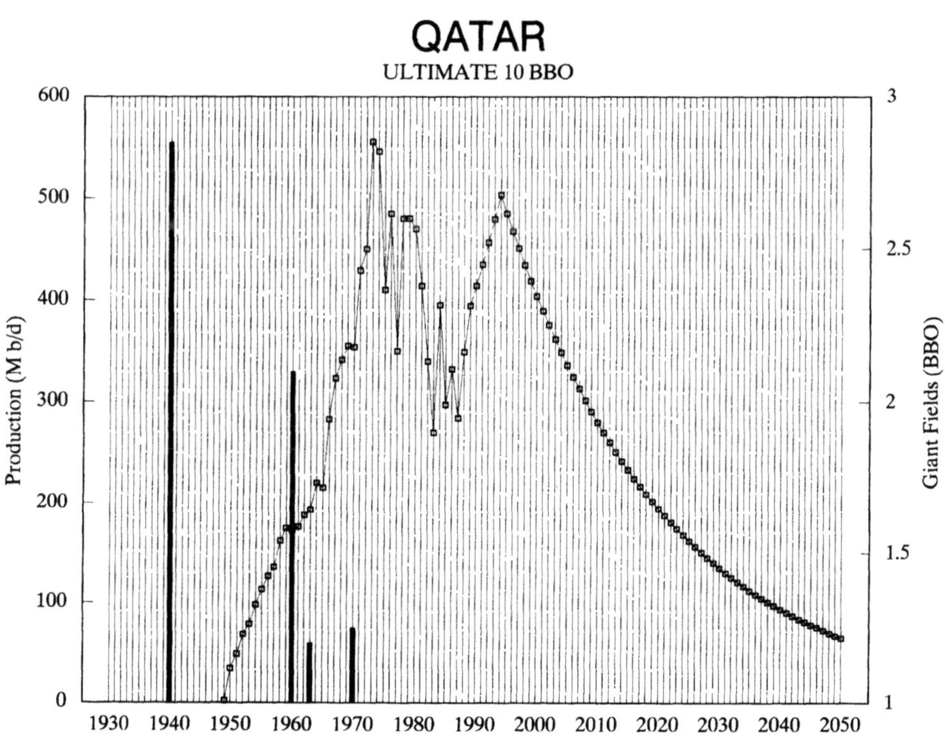

SAUDI ARABIA

ULTIMATE 275 BBO

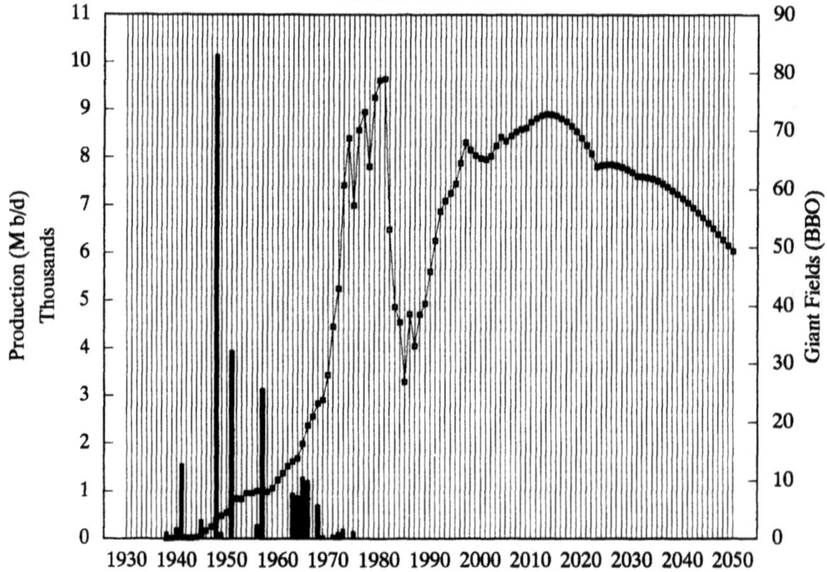

SHARJAH

ULTIMATE 2 BBO

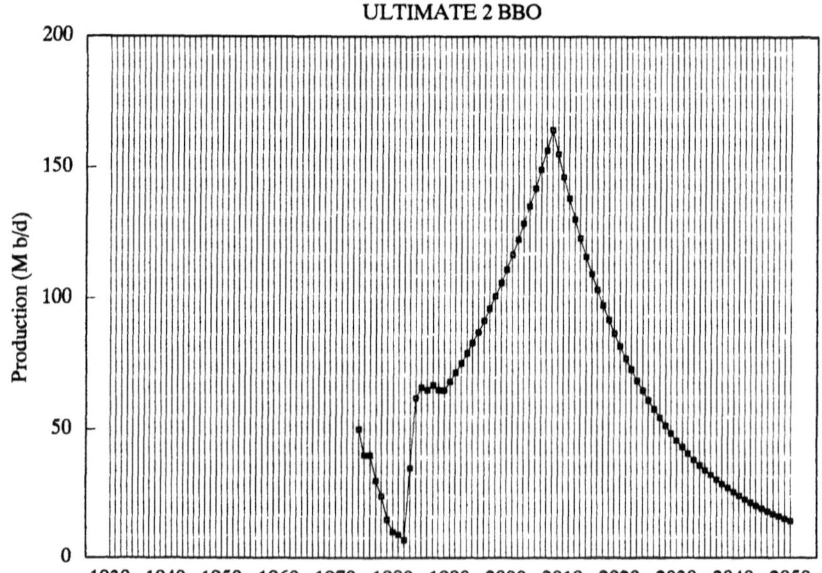

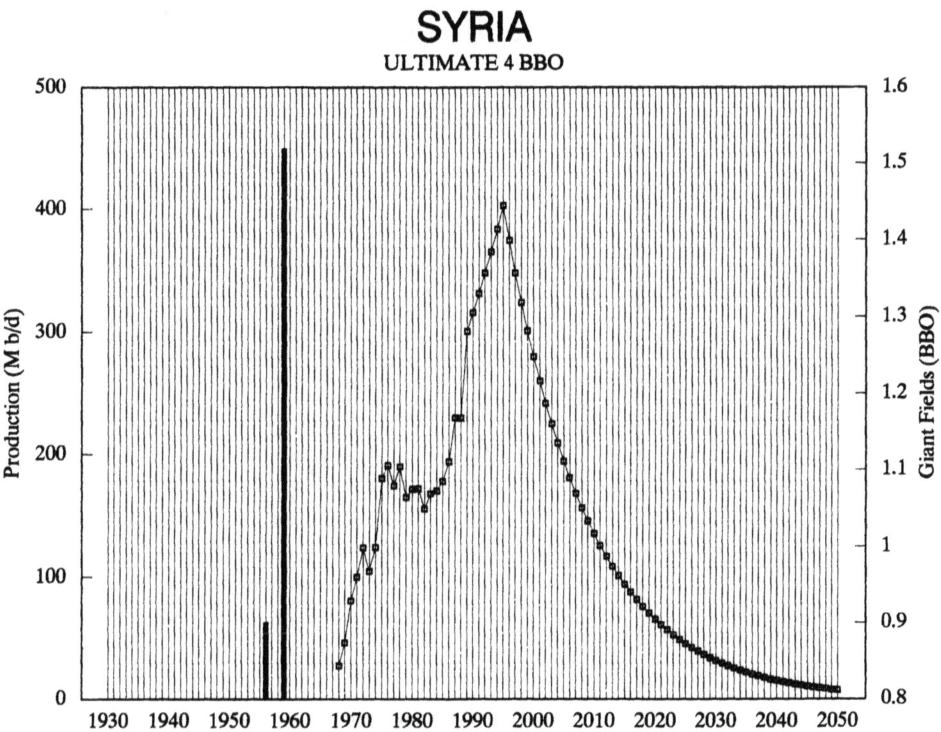

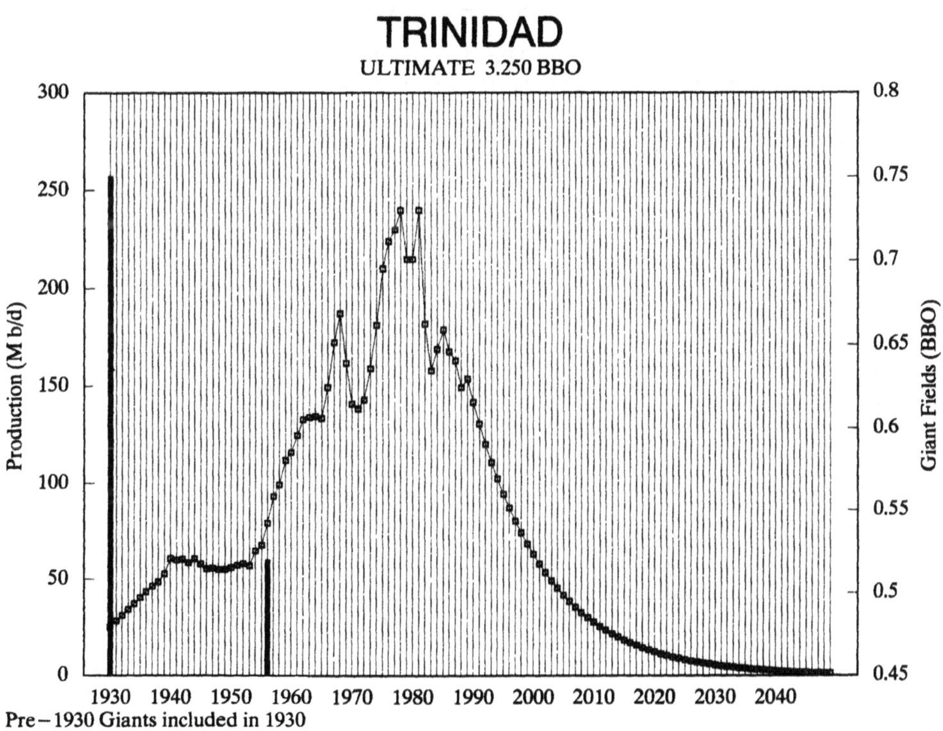

Pre–1930 Giants included in 1930

TUNISIA

ULTIMATE 3 BBO

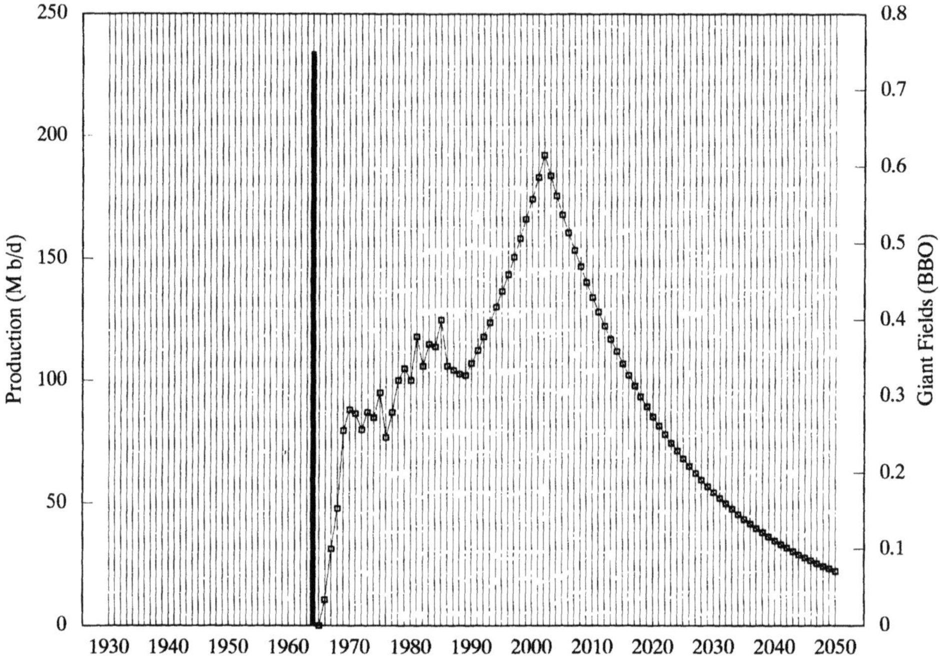

TURKEY

ULTIMATE 1.25 BBO

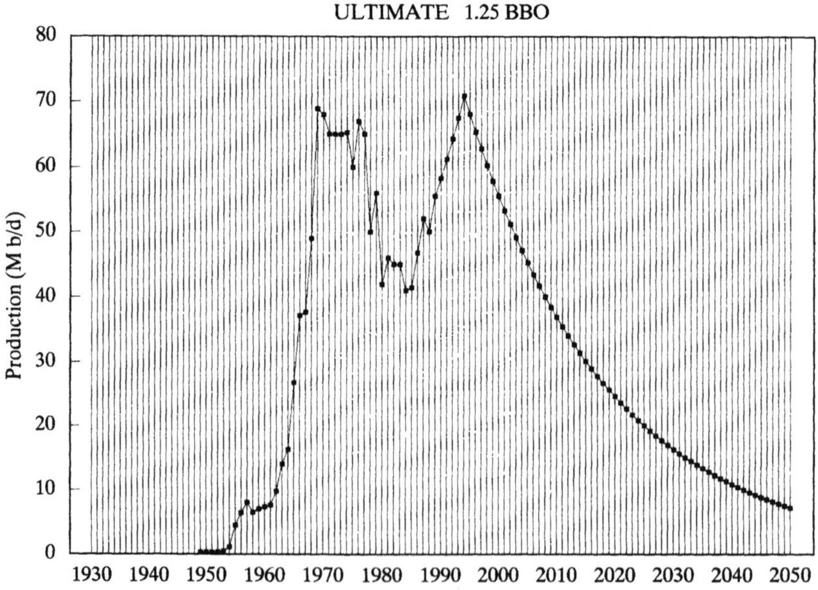

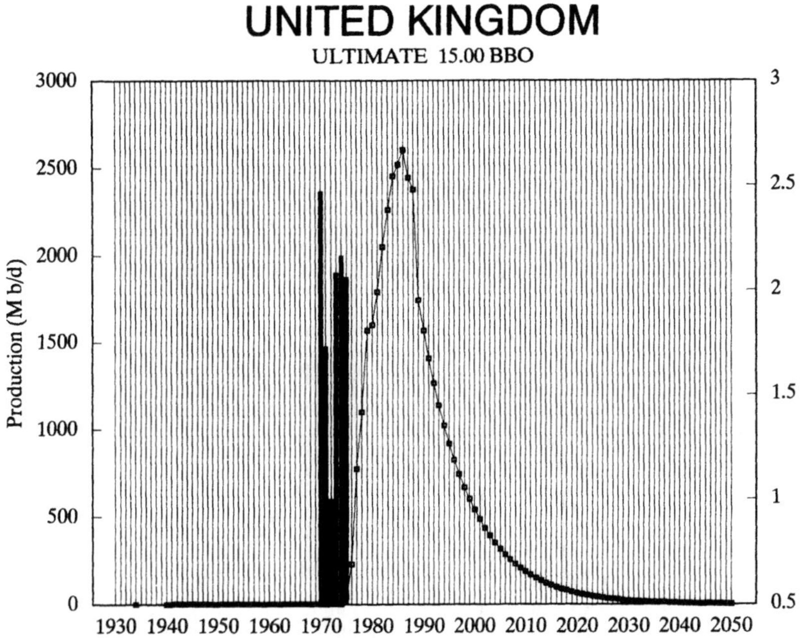

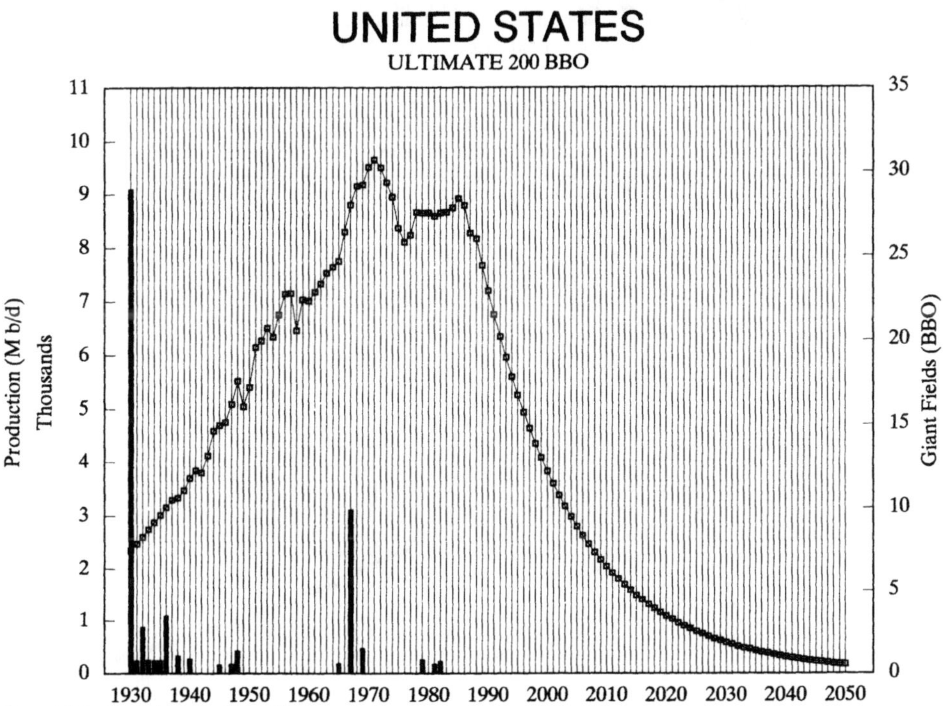

Pre−1930 giants included in 1930

U.S.S.R.

ULTIMATE 200 BBO

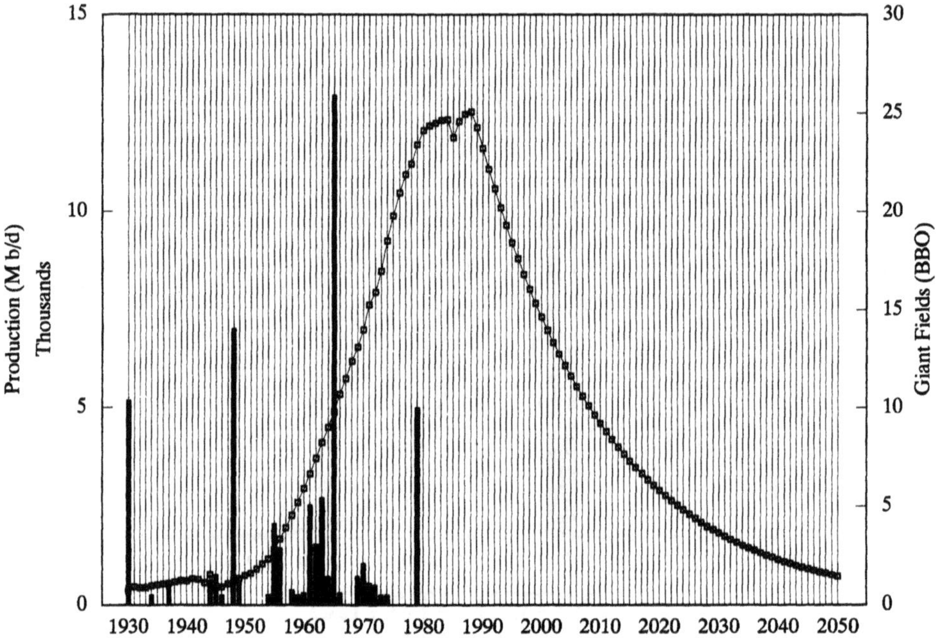

Pre–!930 Giants included in 1930

VENEZUELA

ULTIMATE 83 BBO

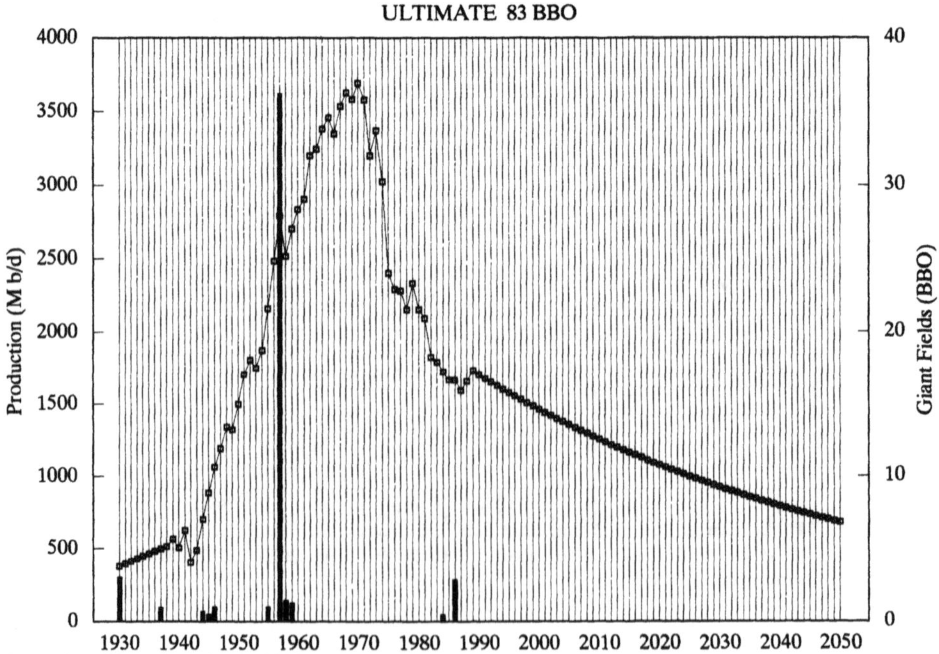

Pre–1930 Giants included in 1930

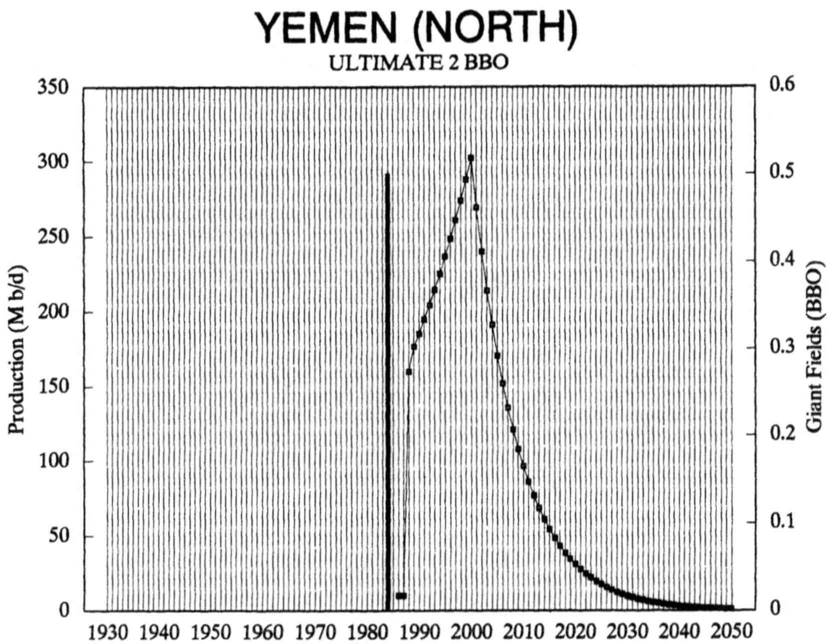
YEMEN (NORTH)
ULTIMATE 2 BBO
Production (M b/d)
Giant Fields (BBO)
350
300
250
200
150
100
50
0
0.6
0.5
0.4
0.3
0.2
0.1
0
1930 1940 1950 1960 1970 1980 1990 2000 2010 2020 2030 2040 2050

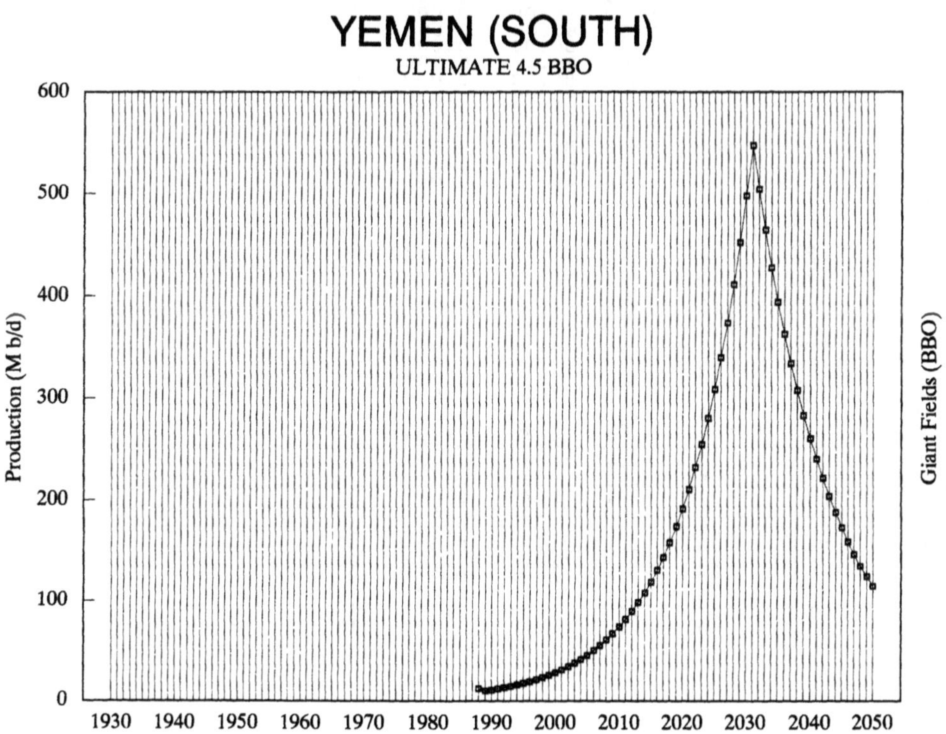
YEMEN (SOUTH)
ULTIMATE 4.5 BBO
Production (M b/d)
Giant Fields (BBO)
600
500
400
300
200
100
0
1930 1940 1950 1960 1970 1980 1990 2000 2010 2020 2030 2040 2050

CUM.PROD	9.33	BBO	GIANT FIELDS		ABU DHABI	
RESERVES	65.00	BBO	TOTAL BBO	62.2	M.EAST (MAJOR)	
DISCOVERED	74.33	BBO	% OF DISC.	84%	Scenario-2	
UNDISCOVERED	5.67	BBO	Last Giant	1973		
REMAINING	70.67	BBO			Revision Trend	
ULTIMATE	80.00	BBO			1990	n/a
Rept.Reserve	92.00		Ref.date (end)	1989	Future	DOWN

Milestones	Date	Prod Mb/d	Dep. Rate	% Produced
1st Peak	1974	1750	0.82%	4%
2nd Peak	1997	2250	1.25%	19%
Mid-Point	2034	1399	1.26%	50%
Final Decline	2035		1.26%	

Forecast "C"
A=Exponential decline
B=5% incr. to Mid-Point
C=Other
Pre-1930 Cum. 0.00
(Added to 1930 prod)

Date	Prod. M.b/d	Prod. BBO/yr	Cum. Prod.	Remain	Deplet. Rate	% Produced	Giant	Notes & Names
1930		0.000	0.00	80.00	ERR	0%		
1931		0.000	0.00	80.00	0.00%	0%		
1932		0.000	0.00	80.00	0.00%	0%		
1933		0.000	0.00	80.00	0.00%	0%		
1934		0.000	0.00	80.00	0.00%	0%		
1935		0.000	0.00	80.00	0.00%	0%		
1936		0.000	0.00	80.00	0.00%	0%		
1937		0.000	0.00	80.00	0.00%	0%		
1938		0.000	0.00	80.00	0.00%	0%		
1939		0.000	0.00	80.00	0.00%	0%		
1940		0.000	0.00	80.00	0.00%	0%		
1941		0.000	0.00	80.00	0.00%	0%		
1942		0.000	0.00	80.00	0.00%	0%		
1943		0.000	0.00	80.00	0.00%	0%		
1944		0.000	0.00	80.00	0.00%	0%		
1945		0.000	0.00	80.00	0.00%	0%		
1946		0.000	0.00	80.00	0.00%	0%		
1947		0.000	0.00	80.00	0.00%	0%		
1948		0.000	0.00	80.00	0.00%	0%		
1949		0.000	0.00	80.00	0.00%	0%		
1950		0.000	0.00	80.00	0.00%	0%		
1951		0.000	0.00	80.00	0.00%	0%		
1952		0.000	0.00	80.00	0.00%	0%		
1953		0.000	0.00	80.00	0.00%	0%		
1954		0.000	0.00	80.00	0.00%	0%	8.00	Bab
1955		0.000	0.00	80.00	0.00%	0%		
1956		0.000	0.00	80.00	0.00%	0%		
1957		0.000	0.00	80.00	0.00%	0%		
1958		0.000	0.00	80.00	0.00%	0%	5.50	Umm Shaif
1959		0.000	0.00	80.00	0.00%	0%		
1960		0.000	0.00	80.00	0.00%	0%		
1961		0.000	0.00	80.00	0.00%	0%		
1962	30	0.011	0.01	79.99	0.01%	0%	14.00	Murban, Bu Hasa
1963	49	0.018	0.03	79.97	0.02%	0%		
1964	185	0.068	0.10	79.90	0.08%	0%	21.00	Zakum
1965	285	0.104	0.20	79.80	0.13%	0%	6.50	Asab
1966	357	0.130	0.33	79.67	0.16%	0%	0.90	Shah
1967	383	0.140	0.47	79.53	0.18%	1%	1.20	Sahil
1968	511	0.187	0.66	79.34	0.23%	1%		
1969	601	0.219	0.88	79.12	0.28%	1%	3.40	Zubbaya, Mubarras
1970	641	0.234	1.11	78.89	0.30%	1%	0.90	S-A-Raazboot
1971	900	0.329	1.44	78.56	0.42%	2%		
1972	1000	0.365	1.80	78.20	0.46%	2%		
1973	1286	0.469	2.27	77.73	0.60%	3%	0.83	Umm-a-Dalkh
1974	1750	0.639	2.91	77.09	0.82%	4%		1ST PEAK
1975	1500	0.548	3.46	76.54	0.71%	4%		
1976	1590	0.580	4.04	75.96	0.76%	5%		
1977	1680	0.613	4.65	75.35	0.81%	6%		
1978	1450	0.529	5.18	74.82	0.70%	6%		
1979	1450	0.529	5.71	74.29	0.71%	7%		
1980	1380	0.504	6.21	73.79	0.68%	8%		
1981	1145	0.418	6.63	73.37	0.57%	8%		
1982	883	0.322	6.96	73.04	0.44%	9%		
1983	757	0.276	7.23	72.77	0.38%	9%		
1984	750	0.274	7.51	72.49	0.38%	9%		
1985	705	0.257	7.76	72.24	0.36%	10%		
1986	949	0.346	8.11	71.89	0.48%	10%		
1987	969	0.354	8.46	71.54	0.49%	11%		
1988	1013	0.370	8.83	71.17	0.52%	11%		% of Region

1989	1359	0.496	9.33	70.67	0.70%	12%	9.71%	
1990	1544	0.563	9.89	70.11	0.80%	12%		
1991	1720	0.628	10.52	69.48	0.90%	13%		
1992	1889	0.690	11.21	68.79	0.99%	14%		
1993	1952	0.713	11.92	68.08	1.04%	15%		
1994	1998	0.729	12.65	67.35	1.07%	16%		
1995	2044	0.746	13.40	66.60	1.11%	17%		
1996	2146	0.783	14.18	65.82	1.18%	18%		
1997	2250	0.821	15.00	65.00	1.25%	19%	2ND PEAK	
1998	2205	0.805	15.81	64.19	1.24%	20%		
1999	2171	0.792	16.60	63.40	1.23%	21%		
2000	2151	0.785	17.38	62.62	1.24%	22%		
2001	2143	0.782	18.17	61.83	1.25%	23%		
2002	2156	0.787	18.95	61.05	1.27%	24%		
2003	2167	0.791	19.74	60.26	1.30%	25%		
2004	2172	0.793	20.54	59.46	1.32%	26%		
2005	2135	0.779	21.32	58.68	1.31%	27%		
2006	2130	0.777	22.09	57.91	1.32%	28%		
2007	2121	0.774	22.87	57.13	1.34%	29%		
2008	2109	0.770	23.64	56.36	1.35%	30%		
2009	2094	0.764	24.40	55.60	1.36%	31%		
2010	2077	0.758	25.16	54.84	1.36%	31%		
2011	2056	0.750	25.91	54.09	1.37%	32%		
2012	2032	0.742	26.65	53.35	1.37%	33%		
2013	2006	0.732	27.38	52.62	1.37%	34%		
2014	1978	0.722	28.11	51.89	1.37%	35%		
2015	1947	0.711	28.82	51.18	1.37%	36%		
2016	1914	0.699	29.52	50.48	1.36%	37%		
2017	1879	0.686	30.20	49.80	1.36%	38%		
2018	1842	0.672	30.87	49.13	1.35%	39%		
2019	1804	0.659	31.53	48.47	1.34%	39%		
2020	1764	0.644	32.18	47.82	1.33%	40%		
2021	1723	0.629	32.80	47.20	1.31%	41%		
2022	1680	0.613	33.42	46.58	1.30%	42%		
2023	1626	0.593	34.01	45.99	1.27%	43%		
2024	1609	0.587	34.60	45.40	1.28%	43%		
2025	1590	0.581	35.18	44.82	1.28%	44%		
2026	1570	0.573	35.75	44.25	1.28%	45%		
2027	1549	0.565	36.32	43.68	1.28%	45%		
2028	1526	0.557	36.87	43.13	1.27%	46%		
2029	1502	0.548	37.42	42.58	1.27%	47%		
2030	1478	0.540	37.96	42.04	1.27%	47%		
2031	1453	0.530	38.49	41.51	1.26%	48%		
2032	1436	0.524	39.02	40.98	1.26%	49%	%	Add to
2033	1418	0.517	39.53	40.47	1.26%	49%		Saudi
2034	1399	0.511	40.04	39.96	1.26%	50%	50% PRODUCED	
2035	1379	0.503	40.55	39.45	1.26%	51%	1380	0.28
2036	1362	0.497	41.05	38.95	1.26%	51%	1360	-2.32
2037	1345	0.491	41.54	38.46	1.26%	52%	1339	-5.64
2038	1328	0.485	42.02	37.98	1.26%	53%	1318	-9.58
2039	1311	0.479	42.50	37.50	1.26%	53%	1297	-14.08
2040	1295	0.473	42.97	37.03	1.26%	54%	1275	-19.08
2041	1278	0.467	43.44	36.56	1.26%	54%	1254	-24.50
2042	1262	0.461	43.90	36.10	1.26%	55%	1232	-30.29
2043	1246	0.455	44.35	35.65	1.26%	55%	1210	-36.41
2044	1231	0.449	44.80	35.20	1.26%	56%	1188	-42.79
2045	1215	0.443	45.25	34.75	1.26%	57%	1166	-49.40
2046	1200	0.438	45.68	34.32	1.26%	57%	1144	-56.20
2047	1185	0.432	46.12	33.88	1.26%	58%	1121	-63.15
2048	1170	0.427	46.54	33.46	1.26%	58%	1099	-70.22
2049	1155	0.422	46.97	33.03	1.26%	59%	1078	-77.36
2050	1140	0.416	47.38	32.62	1.26%	59%	1056	-84.56

CUM.PROD	0.479 BBO	GIANT FIELDS		ALBANIA	
RESERVES	0.236 BBO	TOTAL BBO	0.0	SINO-SOVIET	
DISCOVERED	0.715 BBO	% OF DISC.	0%		
UNDISCOVERED	0.285 BBO	Last Giant	-		
REMAINING	0.521 BBO			Variance	Trend
ULTIMATE	1.000 BBO			1990	n/a
Rept.Reserve	0.236 #1	Ref. date	1989	Future	DOWN

Milestones	Date	Prod Mb/d	Dep. Rate	% Produced
1st Peak	1986	70	4.13%	41%
2nd Peak	none			
Mid-Point	1990	63	4.43%	50%
Final Decline	1990		4.43%	50%

Forecast "B"
A=Exponential decline
B=5% incr. to Mid-Point
C=Other
Pre-1930 Cum. 0.00
(Added to 1930 prod)

Date	Prod. M.b/d	Prod. BBO/yr	Cum. Prod.	Remain	Deplet. Rate	% Produced	Giant	Notes & Names
1930		0.000	0.00	1.00	ERR	0%		Data-base weak
1931		0.000	0.00	1.00	0.00%	0%		
1932		0.000	0.00	1.00	0.00%	0%		
1933	0	0.000	0.00	1.00	0.00%	0%		
1934	0	0.000	0.00	1.00	0.00%	0%		
1935	0	0.000	0.00	1.00	0.00%	0%		
1936	1	0.000	0.00	1.00	0.03%	0%		
1937	2	0.001	0.00	1.00	0.06%	0%		
1938	2	0.001	0.00	1.00	0.08%	0%		
1939	3	0.001	0.00	1.00	0.10%	0%		
1940	4	0.001	0.00	1.00	0.15%	0%		
1941	4	0.001	0.01	0.99	0.14%	1%		
1942	4	0.002	0.01	0.99	0.16%	1%		
1943	3	0.001	0.01	0.99	0.10%	1%		
1944	1	0.000	0.01	0.99	0.03%	1%		
1945	1	0.000	0.01	0.99	0.03%	1%		
1946	3	0.001	0.01	0.99	0.10%	1%		
1947	6	0.002	0.01	0.99	0.20%	1%		
1948	4	0.001	0.01	0.99	0.15%	1%		
1949	6	0.002	0.02	0.98	0.22%	2%		
1950	8	0.003	0.02	0.98	0.29%	2%		
1951	3	0.001	0.02	0.98	0.12%	2%		
1952	3	0.001	0.02	0.98	0.11%	2%		
1953	3	0.001	0.02	0.98	0.10%	2%		
1954	3	0.001	0.02	0.98	0.12%	2%		
1955	4	0.001	0.02	0.98	0.14%	2%		
1956	5	0.002	0.03	0.97	0.19%	3%		
1957	9	0.003	0.03	0.97	0.34%	3%		
1958	7	0.003	0.03	0.97	0.28%	3%		
1959	9	0.003	0.04	0.96	0.33%	4%		
1960	13	0.005	0.04	0.96	0.50%	4%		
1961	14	0.005	0.05	0.95	0.54%	5%		
1962	14	0.005	0.05	0.95	0.55%	5%		
1963	14	0.005	0.06	0.94	0.53%	6%		
1964	14	0.005	0.06	0.94	0.54%	6%		
1965	15	0.005	0.07	0.93	0.57%	7%		
1966	15	0.005	0.07	0.93	0.57%	7%		
1967	20	0.007	0.08	0.92	0.79%	8%		
1968	22	0.008	0.09	0.91	0.88%	9%		
1969	21	0.008	0.09	0.91	0.84%	9%		
1970	27	0.010	0.10	0.90	1.10%	10%		
1971	24	0.009	0.11	0.89	0.97%	11%		
1972	29	0.011	0.12	0.88	1.18%	12%		
1973	39	0.014	0.14	0.86	1.60%	14%		
1974	41	0.015	0.15	0.85	1.74%	15%		
1975	41	0.015	0.17	0.83	1.77%	17%		
1976	41	0.015	0.18	0.82	1.80%	18%		
1977	41	0.015	0.20	0.80	1.83%	20%		
1978	41	0.015	0.21	0.79	1.87%	21%		
1979	34	0.012	0.22	0.78	1.58%	22%		
1980	70	0.026	0.25	0.75	3.30%	25%		
1981	71	0.026	0.28	0.72	3.46%	28%		
1982	73	0.027	0.30	0.70	3.69%	30%		
1983	80	0.029	0.33	0.67	4.19%	33%		
1984	69	0.025	0.36	0.64	3.74%	36%		
1985	62	0.023	0.38	0.62	3.54%	38%		
1986	70	0.026	0.41	0.59	4.13%	41%		1ST PEAK
1987	70	0.026	0.43	0.57	4.30%	43%		
1988	70	0.026	0.46	0.54	4.49%	46%		

1989	60	0.022	0.48	0.52	4.05%	48%	
1990	63	0.023	0.50	0.50	4.43%	50%	50% PRODUCED
1991	60	0.022	0.52	0.48	4.43%	52%	
1992	58	0.021	0.54	0.46	4.43%	54%	
1993	55	0.020	0.57	0.43	4.43%	57%	
1994	53	0.019	0.58	0.42	4.43%	58%	
1995	50	0.018	0.60	0.40	4.43%	60%	
1996	48	0.018	0.62	0.38	4.43%	62%	
1997	46	0.017	0.64	0.36	4.43%	64%	
1998	44	0.016	0.65	0.35	4.43%	65%	
1999	42	0.015	0.67	0.33	4.43%	67%	
2000	40	0.015	0.68	0.32	4.43%	68%	
2001	38	0.014	0.70	0.30	4.43%	70%	
2002	37	0.013	0.71	0.29	4.43%	71%	
2003	35	0.013	0.72	0.28	4.43%	72%	
2004	34	0.012	0.74	0.26	4.43%	74%	
2005	32	0.012	0.75	0.25	4.43%	75%	
2006	31	0.011	0.76	0.24	4.43%	76%	
2007	29	0.011	0.77	0.23	4.43%	77%	
2008	28	0.010	0.78	0.22	4.43%	78%	
2009	27	0.010	0.79	0.21	4.43%	79%	
2010	26	0.009	0.80	0.20	4.43%	80%	
2011	24	0.009	0.81	0.19	4.43%	81%	
2012	23	0.009	0.82	0.18	4.43%	82%	
2013	22	0.008	0.82	0.18	4.43%	82%	
2014	21	0.008	0.83	0.17	4.43%	83%	
2015	20	0.007	0.84	0.16	4.43%	84%	
2016	19	0.007	0.85	0.15	4.43%	85%	
2017	19	0.007	0.85	0.15	4.43%	85%	
2018	18	0.006	0.86	0.14	4.43%	86%	
2019	17	0.006	0.87	0.13	4.43%	87%	
2020	16	0.006	0.87	0.13	4.43%	87%	
2021	16	0.006	0.88	0.12	4.43%	88%	
2022	15	0.005	0.88	0.12	4.43%	88%	
2023	14	0.005	0.89	0.11	4.43%	89%	
2024	14	0.005	0.89	0.11	4.43%	89%	
2025	13	0.005	0.90	0.10	4.43%	90%	
2026	12	0.005	0.90	0.10	4.43%	90%	
2027	12	0.004	0.91	0.09	4.43%	91%	
2028	11	0.004	0.91	0.09	4.43%	91%	
2029	11	0.004	0.91	0.09	4.43%	91%	
2030	10	0.004	0.92	0.08	4.43%	92%	
2031	10	0.004	0.92	0.08	4.43%	92%	
2032	9	0.003	0.93	0.07	4.43%	93%	
2033	9	0.003	0.93	0.07	4.43%	93%	
2034	9	0.003	0.93	0.07	4.43%	93%	
2035	8	0.003	0.94	0.06	4.43%	94%	
2036	8	0.003	0.94	0.06	4.43%	94%	
2037	8	0.003	0.94	0.06	4.43%	94%	
2038	7	0.003	0.94	0.06	4.43%	94%	
2039	7	0.003	0.95	0.05	4.43%	95%	
2040	7	0.002	0.95	0.05	4.43%	95%	
2041	6	0.002	0.95	0.05	4.43%	95%	
2042	6	0.002	0.95	0.05	4.43%	95%	
2043	6	0.002	0.95	0.05	4.43%	95%	
2044	5	0.002	0.96	0.04	4.43%	96%	
2045	5	0.002	0.96	0.04	4.43%	96%	
2046	5	0.002	0.96	0.04	4.43%	96%	
2047	5	0.002	0.96	0.04	4.43%	96%	
2048	5	0.002	0.96	0.04	4.43%	96%	
2049	4	0.002	0.97	0.03	4.43%	97%	
2050	4	0.002	0.97	0.03	4.43%	97%	

CUM.PROD	8.366 BBO	GIANT FIELDS			ALGERIA	
RESERVES	9.200 BBO	TOTAL BBO	10.9		AFRICA	
DISCOVERED	17.566 BBO	% OF DISC.	62%			
UNDISCOVERED	2.434 BBO	Last Giant	1963			
REMAINING	11.634 BBO				Variance Trend	
ULTIMATE	20.000 BBO				1990	+9%
Rept.Reserves	9.200 BBO	Ref.Date (end)		1989	Future DOWN	

Milestones	Date	Prod Mb/d	Dep Rate	% Produced	Forecast "B"
1st Peak	1978	1260	3.04%	27%	A=Exponential decline
2nd Peak	1994	891	3.10%	49%	B=5% inc. to Mid-Point
Mid-Point	1995	863	3.10%	51%	C=Other
Final Decline	1996		3.10%		Pre-1930 Cum 0.00 (Added to 1930 Prod)

Date	Prod. M.b/d	Prod. BBO/yr	Cum Prod	Remain	Dep. Rate	% Produced	Giant	Names & Notes
1930		0.000	0.00	20.00	ERR	0%		
1931		0.000	0.00	20.00	0.00%	0%		
1932		0.000	0.00	20.00	0.00%	0%		
1933		0.000	0.00	20.00	0.00%	0%		
1934		0.000	0.00	20.00	0.00%	0%		
1935		0.000	0.00	20.00	0.00%	0%		
1936		0.000	0.00	20.00	0.00%	0%		
1937		0.000	0.00	20.00	0.00%	0%		
1938		0.000	0.00	20.00	0.00%	0%		
1939		0.000	0.00	20.00	0.00%	0%		
1940		0.000	0.00	20.00	0.00%	0%		
1941		0.000	0.00	20.00	0.00%	0%		
1942		0.000	0.00	20.00	0.00%	0%		
1943		0.000	0.00	20.00	0.00%	0%		
1944		0.000	0.00	20.00	0.00%	0%		
1945		0.000	0.00	20.00	0.00%	0%		
1946		0.000	0.00	20.00	0.00%	0%		
1947		0.000	0.00	20.00	0.00%	0%		
1948		0.000	0.00	20.00	0.00%	0%		
1949		0.000	0.00	20.00	0.00%	0%		
1950		0.000	0.00	20.00	0.00%	0%		
1951		0.000	0.00	20.00	0.00%	0%		
1952		0.000	0.00	20.00	0.00%	0%		
1953	2	0.001	0.00	20.00	0.00%	0%		
1954	2	0.001	0.00	20.00	0.00%	0%		
1955	1	0.000	0.00	20.00	0.00%	0%		
1956	1	0.000	0.00	20.00	0.00%	0%	9.00	H.Massaoud
1957	0	0.000	0.00	20.00	0.00%	0%	0.90	Zarzaitine
1958	10	0.004	0.01	19.99	0.02%	0%		
1959	22	0.008	0.01	19.99	0.04%	0%		
1960	174	0.064	0.08	19.92	0.32%	0%		
1961	323	0.118	0.19	19.81	0.59%	1%		
1962	430	0.157	0.35	19.65	0.79%	2%	0.50	R.el Baguel
1963	502	0.183	0.54	19.46	0.93%	3%	0.50	Gassi Touil
1964	552	0.202	0.74	19.26	1.04%	4%		
1965	547	0.200	0.94	19.06	1.04%	5%		
1966	710	0.259	1.20	18.80	1.36%	6%		
1967	812	0.296	1.49	18.51	1.58%	7%		
1968	899	0.328	1.82	18.18	1.77%	9%		
1969	937	0.342	2.16	17.84	1.88%	11%		
1970	984	0.359	2.52	17.48	2.01%	13%		
1971	603	0.220	2.74	17.26	1.26%	14%		
1972	1061	0.387	3.13	16.87	2.24%	16%		
1973	1035	0.378	3.51	16.49	2.24%	18%		
1974	888	0.324	3.83	16.17	1.97%	19%		
1975	935	0.341	4.17	15.83	2.11%	21%		
1976	950	0.347	4.52	15.48	2.19%	23%		
1977	990	0.361	4.88	15.12	2.33%	24%		
1978	1260	0.460	5.34	14.66	3.04%	27%		1ST PEAK
1979	1240	0.453	5.79	14.21	3.09%	29%		
1980	1000	0.365	6.16	13.84	2.57%	31%		
1981	750	0.274	6.43	13.57	1.98%	32%		
1982	750	0.274	6.70	13.30	2.02%	34%		
1983	687	0.251	6.96	13.04	1.88%	35%		
1984	608	0.222	7.18	12.82	1.70%	36%		
1985	643	0.235	7.41	12.59	1.83%	37%		
1986	601	0.219	7.63	12.37	1.74%	38%		
1987	648	0.236	7.87	12.13	1.91%	39%		
1988	667	0.243	8.11	11.89	2.01%	41%		

1989	698	0.255	8.37	11.63	2.14%	42%	
1990	733	0.268	8.63	11.37	2.30%	43%	
1991	770	0.281	8.91	11.09	2.47%	45%	
1992	808	0.295	9.21	10.79	2.66%	46%	
1993	848	0.310	9.52	10.48	2.87%	48%	
1994	891	0.325	9.84	10.16	3.10%	49%	2ND PEAK
1995	863	0.315	10.16	9.84	3.10%	51%	50% PRODUCED
1996	836	0.305	10.46	9.54	3.10%	52%	
1997	810	0.296	10.76	9.24	3.10%	54%	
1998	785	0.286	11.05	8.95	3.10%	55%	
1999	760	0.278	11.32	8.68	3.10%	57%	
2000	737	0.269	11.59	8.41	3.10%	58%	
2001	714	0.261	11.85	8.15	3.10%	59%	
2002	692	0.253	12.11	7.89	3.10%	61%	
2003	670	0.245	12.35	7.65	3.10%	62%	
2004	650	0.237	12.59	7.41	3.10%	63%	
2005	630	0.230	12.82	7.18	3.10%	64%	
2006	610	0.223	13.04	6.96	3.10%	65%	
2007	591	0.216	13.26	6.74	3.10%	66%	
2008	573	0.209	13.46	6.54	3.10%	67%	
2009	555	0.203	13.67	6.33	3.10%	68%	
2010	538	0.196	13.86	6.14	3.10%	69%	
2011	521	0.190	14.05	5.95	3.10%	70%	
2012	505	0.184	14.24	5.76	3.10%	71%	
2013	489	0.179	14.42	5.58	3.10%	72%	
2014	474	0.173	14.59	5.41	3.10%	73%	
2015	459	0.168	14.76	5.24	3.10%	74%	
2016	445	0.163	14.92	5.08	3.10%	75%	
2017	431	0.157	15.08	4.92	3.10%	75%	
2018	418	0.153	15.23	4.77	3.10%	76%	
2019	405	0.148	15.38	4.62	3.10%	77%	
2020	393	0.143	15.52	4.48	3.10%	78%	
2021	380	0.139	15.66	4.34	3.10%	78%	
2022	369	0.135	15.79	4.21	3.10%	79%	
2023	357	0.130	15.93	4.07	3.10%	80%	
2024	346	0.126	16.05	3.95	3.10%	80%	
2025	335	0.122	16.17	3.83	3.10%	81%	
2026	325	0.119	16.29	3.71	3.10%	81%	
2027	315	0.115	16.41	3.59	3.10%	82%	
2028	305	0.111	16.52	3.48	3.10%	83%	
2029	296	0.108	16.63	3.37	3.10%	83%	
2030	287	0.105	16.73	3.27	3.10%	84%	
2031	278	0.101	16.83	3.17	3.10%	84%	
2032	269	0.098	16.93	3.07	3.10%	85%	
2033	261	0.095	17.03	2.97	3.10%	85%	
2034	253	0.092	17.12	2.88	3.10%	86%	
2035	245	0.089	17.21	2.79	3.10%	86%	
2036	237	0.087	17.29	2.71	3.10%	86%	
2037	230	0.084	17.38	2.62	3.10%	87%	
2038	223	0.081	17.46	2.54	3.10%	87%	
2039	216	0.079	17.54	2.46	3.10%	88%	
2040	209	0.076	17.61	2.39	3.10%	88%	
2041	203	0.074	17.69	2.31	3.10%	88%	
2042	196	0.072	17.76	2.24	3.10%	89%	
2043	190	0.069	17.83	2.17	3.10%	89%	
2044	184	0.067	17.90	2.10	3.10%	89%	
2045	179	0.065	17.96	2.04	3.10%	90%	
2046	173	0.063	18.02	1.98	3.10%	90%	
2047	168	0.061	18.09	1.91	3.10%	90%	
2048	163	0.059	18.15	1.85	3.10%	91%	
2049	158	0.057	18.20	1.80	3.10%	91%	
2050	153	0.056	18.26	1.74	3.10%	91%	

CUM.PROD	1.467 BBO	GIANT FIELDS		ANGOLA
RESERVES	2.024 BBO	TOTAL BBO	1.5	AFRICA
DISCOVERED	3.491 BBO	% OF DISC.	43%	
UNDISCOVERED	0.759 BBO	Last Giant	1967	----------------
REMAINING	2.783 BBO	-------------------		Variance Trend
ULTIMATE	4.250 BBO			1990 ZERO
Rept.Reserves	2.024 BBO	Ref.Date (end)	1989	Future DOWN

Milestones	Date	Prod Mb/d	Dep Rate	% Produced
1st Peak	1974	162	1.47%	7%
2nd Peak	1992	530	7.98%	48%
Mid-Point	1993	488	7.98%	52%
Final Decline	1993		7.98%	

Forecast "B"
A=Exponential decline
B=5% inc. to Mid-Point
C=Other
Pre-1930 Cum 0
(Added to 1930 Prod)

Date	Prod. M.b/d	Prod. BBO/yr	Cum Prod	Remain	Dep. Rate	% Produced	Giant	Names & Notes
1930		0.000	0.00	4.25	ERR	0%		
1931		0.000	0.00	4.25	0.00%	0%		
1932		0.000	0.00	4.25	0.00%	0%		
1933		0.000	0.00	4.25	0.00%	0%		
1934		0.000	0.00	4.25	0.00%	0%		
1935		0.000	0.00	4.25	0.00%	0%		
1936		0.000	0.00	4.25	0.00%	0%		
1937		0.000	0.00	4.25	0.00%	0%		
1938		0.000	0.00	4.25	0.00%	0%		
1939		0.000	0.00	4.25	0.00%	0%		
1940		0.000	0.00	4.25	0.00%	0%		
1941		0.000	0.00	4.25	0.00%	0%		
1942		0.000	0.00	4.25	0.00%	0%		
1943		0.000	0.00	4.25	0.00%	0%		
1944		0.000	0.00	4.25	0.00%	0%		
1945		0.000	0.00	4.25	0.00%	0%		
1946		0.000	0.00	4.25	0.00%	0%		
1947		0.000	0.00	4.25	0.00%	0%		
1948		0.000	0.00	4.25	0.00%	0%		
1949		0.000	0.00	4.25	0.00%	0%		
1950		0.000	0.00	4.25	0.00%	0%		
1951		0.000	0.00	4.25	0.00%	0%		
1952		0.000	0.00	4.25	0.00%	0%		
1953		0.000	0.00	4.25	0.00%	0%		
1954		0.000	0.00	4.25	0.00%	0%		
1955		0.000	0.00	4.25	0.00%	0%		
1956		0.000	0.00	4.25	0.00%	0%		
1957		0.000	0.00	4.25	0.00%	0%		
1958	1.9	0.001	0.00	4.25	0.02%	0%		
1959	0.9	0.000	0.00	4.25	0.01%	0%		
1960	1.3	0.000	0.00	4.25	0.01%	0%		
1961	1.4	0.001	0.00	4.25	0.01%	0%		
1962	7.7	0.003	0.00	4.25	0.07%	0%		
1963	14.4	0.005	0.01	4.24	0.12%	0%		
1964	23.6	0.009	0.02	4.23	0.20%	0%		
1965	15.2	0.006	0.02	4.23	0.13%	1%		
1966	12.7	0.005	0.03	4.22	0.11%	1%	0.6	Malongo
1967	13.1	0.005	0.03	4.22	0.11%	1%	0.9	Mal. W
1968	17.0	0.006	0.04	4.21	0.15%	1%		
1969	47.5	0.017	0.06	4.19	0.41%	1%		
1970	110.0	0.040	0.10	4.15	0.96%	2%		
1971	109.0	0.040	0.14	4.11	0.96%	3%		
1972	135.0	0.049	0.19	4.06	1.20%	4%		
1973	160.0	0.058	0.24	4.01	1.44%	6%		
1974	161.5	0.059	0.30	3.95	1.47%	7%		1ST PEAK
1975	120.0	0.044	0.35	3.90	1.11%	8%		
1976	47.5	0.017	0.36	3.89	0.44%	9%		
1977	195.0	0.071	0.44	3.81	1.83%	10%		
1978	130.0	0.047	0.48	3.77	1.24%	11%		
1979	143.0	0.052	0.54	3.71	1.39%	13%		
1980	150.0	0.055	0.59	3.66	1.47%	14%		
1981	140.0	0.051	0.64	3.61	1.40%	15%		
1982	122.0	0.045	0.69	3.56	1.23%	16%		
1983	174.0	0.064	0.75	3.50	1.78%	18%		
1984	207.0	0.076	0.83	3.42	2.16%	19%		
1985	228.5	0.083	0.91	3.34	2.44%	21%		
1986	280.7	0.102	1.01	3.24	3.07%	24%		
1987	341.3	0.125	1.14	3.11	3.85%	27%		
1988	449.3	0.164	1.30	2.95	5.27%	31%		

1989	457.7	0.167	1.47	2.78	5.66%	35%	
1990	480.6	0.175	1.64	2.61	6.30%	39%	
1991	504.6	0.184	1.83	2.42	7.06%	43%	
1992	529.8	0.193	2.02	2.23	7.98%	48%	2ND PEAK
1993	487.6	0.178	2.20	2.05	7.98%	52%	50% PRODUCED
1994	448.7	0.164	2.36	1.89	7.98%	56%	
1995	412.9	0.151	2.51	1.74	7.98%	59%	
1996	380.0	0.139	2.65	1.60	7.98%	62%	
1997	349.6	0.128	2.78	1.47	7.98%	65%	
1998	321.7	0.117	2.90	1.35	7.98%	68%	
1999	296.1	0.108	3.00	1.25	7.98%	71%	
2000	272.4	0.099	3.10	1.15	7.98%	73%	
2001	250.7	0.092	3.19	1.06	7.98%	75%	
2002	230.7	0.084	3.28	0.97	7.98%	77%	
2003	212.3	0.077	3.36	0.89	7.98%	79%	
2004	195.3	0.071	3.43	0.82	7.98%	81%	
2005	179.8	0.066	3.49	0.76	7.98%	82%	
2006	165.4	0.060	3.55	0.70	7.98%	84%	
2007	152.2	0.056	3.61	0.64	7.98%	85%	
2008	140.1	0.051	3.66	0.59	7.98%	86%	
2009	128.9	0.047	3.71	0.54	7.98%	87%	
2010	118.6	0.043	3.75	0.50	7.98%	88%	
2011	109.1	0.040	3.79	0.46	7.98%	89%	
2012	100.4	0.037	3.83	0.42	7.98%	90%	
2013	92.4	0.034	3.86	0.39	7.98%	91%	
2014	85.0	0.031	3.89	0.36	7.98%	92%	
2015	78.3	0.029	3.92	0.33	7.98%	92%	
2016	72.0	0.026	3.95	0.30	7.98%	93%	
2017	66.3	0.024	3.97	0.28	7.98%	93%	
2018	61.0	0.022	3.99	0.26	7.98%	94%	
2019	56.1	0.020	4.01	0.24	7.98%	94%	
2020	51.6	0.019	4.03	0.22	7.98%	95%	
2021	47.5	0.017	4.05	0.20	7.98%	95%	
2022	43.7	0.016	4.07	0.18	7.98%	96%	
2023	40.2	0.015	4.08	0.17	7.98%	96%	
2024	37.0	0.014	4.09	0.16	7.98%	96%	
2025	34.1	0.012	4.11	0.14	7.98%	97%	
2026	31.3	0.011	4.12	0.13	7.98%	97%	
2027	28.8	0.011	4.13	0.12	7.98%	97%	
2028	26.5	0.010	4.14	0.11	7.98%	97%	
2029	24.4	0.009	4.15	0.10	7.98%	98%	
2030	22.5	0.008	4.16	0.09	7.98%	98%	
2031	20.7	0.008	4.16	0.09	7.98%	98%	
2032	19.0	0.007	4.17	0.08	7.98%	98%	
2033	17.5	0.006	4.18	0.07	7.98%	98%	
2034	16.1	0.006	4.18	0.07	7.98%	98%	
2035	14.8	0.005	4.19	0.06	7.98%	99%	
2036	13.6	0.005	4.19	0.06	7.98%	99%	
2037	12.6	0.005	4.20	0.05	7.98%	99%	
2038	11.6	0.004	4.20	0.05	7.98%	99%	
2039	10.6	0.004	4.21	0.04	7.98%	99%	
2040	9.8	0.004	4.21	0.04	7.98%	99%	
2041	9.0	0.003	4.21	0.04	7.98%	99%	
2042	8.3	0.003	4.22	0.03	7.98%	99%	
2043	7.6	0.003	4.22	0.03	7.98%	99%	
2044	7.0	0.003	4.22	0.03	7.98%	99%	
2045	6.5	0.002	4.22	0.03	7.98%	99%	
2046	5.9	0.002	4.22	0.03	7.98%	99%	
2047	5.5	0.002	4.23	0.02	7.98%	99%	
2048	5.0	0.002	4.23	0.02	7.98%	100%	
2049	4.6	0.002	4.23	0.02	7.98%	100%	
2050	4.3	0.002	4.23	0.02	7.98%	100%	

CUM.PROD	3.69 BBO	GIANT FIELDS		ARGENTINA	
RESERVES	2.28 BBO	TOTAL BBO	2.00	L.AMERICA	
DISCOVERED	5.97 BBO	% OF DISC.	34%		
UNDISCOVERED	1.53 BBO	Last Giant		----------------	
REMAINING	3.81 BBO	------------------		Variance	Trend
ULTIMATE	7.50 BBO			1990	+3%
Rept.Reserves	2.28 BBO	Ref.Date(end) 1989		Future	UP

Milestones	Date	Prod Mb/d	Dep. Rate	% Produced	Forecast "A"
1st Peak	1981	497	4.60%	31%	A=Exponential decline
2nd Peak					B=5% inc. to Mid-Point
					C=Other
Mid-Point	1981	497	4.60%	31%	Pre-1930 Cum 0.16
Final Decline			4.69%		(added to 1930 Prod)

Date	Prod. M.b/d	Prod. BBO/yr	Cum Prod	Remain	Dep. Rate	% Produced	Giant	Names % Notes
1930	25	0.009	0.17	7.33	ERR	2%	2.00	Comodoro-R
1931	27	0.010	0.18	7.32	0.14%	2%		
1932	30	0.011	0.19	7.31	0.15%	3%		
1933	33	0.012	0.20	7.30	0.16%	3%		
1934	36	0.013	0.21	7.29	0.18%	3%		
1935	38	0.014	0.23	7.27	0.19%	3%		
1936	41	0.015	0.24	7.26	0.21%	3%		
1937	44	0.016	0.26	7.24	0.22%	3%		
1938	46	0.017	0.28	7.22	0.23%	4%		
1939	51	0.019	0.30	7.20	0.26%	4%		
1940	57	0.021	0.32	7.18	0.29%	4%		
1941	60	0.022	0.34	7.16	0.30%	5%		
1942	65	0.024	0.36	7.14	0.33%	5%		
1943	74	0.027	0.39	7.11	0.38%	5%		
1944	66	0.024	0.41	7.09	0.34%	6%		
1945	63	0.023	0.44	7.06	0.32%	6%		
1946	56	0.021	0.46	7.04	0.29%	6%		
1947	60	0.022	0.48	7.02	0.31%	6%		
1948	65	0.024	0.50	7.00	0.34%	7%		
1949	62	0.023	0.52	6.98	0.32%	7%		
1950	64	0.023	0.55	6.95	0.33%	7%		
1951	66	0.024	0.57	6.93	0.35%	8%		
1952	69	0.025	0.60	6.90	0.36%	8%		
1953	77	0.028	0.63	6.87	0.41%	8%		
1954	80	0.029	0.65	6.85	0.42%	9%		
1955	83	0.030	0.68	6.82	0.44%	9%		
1956	83	0.030	0.71	6.79	0.45%	10%		
1957	91	0.033	0.75	6.75	0.49%	10%		
1958	94	0.034	0.78	6.72	0.51%	10%		
1959	130	0.047	0.83	6.67	0.71%	11%		
1960	160	0.058	0.89	6.61	0.88%	12%		
1961	252	0.092	0.98	6.52	1.39%	13%		
1962	267	0.097	1.08	6.42	1.49%	14%		
1963	262	0.096	1.17	6.33	1.49%	16%		
1964	274	0.100	1.27	6.23	1.58%	17%		
1965	262	0.095	1.37	6.13	1.53%	18%		
1966	282	0.103	1.47	6.03	1.68%	20%		
1967	307	0.112	1.58	5.92	1.86%	21%		
1968	343	0.125	1.71	5.79	2.11%	23%		
1969	339	0.124	1.83	5.67	2.14%	24%		
1970	391	0.143	1.98	5.52	2.52%	26%		
1971	411	0.150	2.13	5.37	2.72%	28%		
1972	435	0.159	2.28	5.22	2.95%	30%		
1973	442	0.161	2.45	5.05	3.09%	33%		
1974	422	0.154	2.60	4.90	3.05%	35%		
1975	390	0.142	2.74	4.76	2.90%	37%		
1976	390	0.142	2.88	4.62	2.99%	38%		
1977	440	0.161	3.04	4.46	3.48%	41%		
1978	450	0.164	3.21	4.29	3.69%	43%		
1979	470	0.172	2.00	4.12	4.00%	27%		
1980	490	0.179	2.18	3.94	4.34%	29%		
1981	497	0.181	2.36	3.76	4.60%	31%		1ST PEAK
1982	483	0.176	2.54	3.58	4.69%	34%		& 50% PRODUCED
1983	481	0.176	2.71	3.41	4.90%	36%		
1984	467	0.170	2.88	3.24	5.00%	38%		
1985	448	0.163	3.05	3.07	5.05%	41%		
1986	430	0.157	3.20	2.92	5.11%	43%		
1987	419	0.153	3.36	2.76	5.25%	45%		
1988	450	0.164	3.52	2.60	5.94%	47%		

1989	459	0.168	3.69	2.43	6.45%	49%
1990	430	0.157	3.84	2.27	6.45%	51%
1991	402	0.147	3.99	2.13	6.45%	53%
1992	376	0.137	4.13	1.99	6.45%	55%
1993	352	0.128	4.26	1.86	6.45%	57%
1994	329	0.120	4.38	1.74	6.45%	58%
1995	308	0.112	4.49	1.63	6.45%	60%
1996	288	0.105	4.59	1.52	6.45%	61%
1997	269	0.098	4.69	1.43	6.45%	63%
1998	252	0.092	4.78	1.33	6.45%	64%
1999	236	0.086	4.87	1.25	6.45%	65%
2000	221	0.081	4.95	1.17	6.45%	66%
2001	206	0.075	5.03	1.09	6.45%	67%
2002	193	0.070	5.10	1.02	6.45%	68%
2003	181	0.066	5.16	0.96	6.45%	69%
2004	169	0.062	5.22	0.89	6.45%	70%
2005	158	0.058	5.28	0.84	6.45%	70%
2006	148	0.054	5.34	0.78	6.45%	71%
2007	138	0.050	5.39	0.73	6.45%	72%
2008	129	0.047	5.43	0.69	6.45%	72%
2009	121	0.044	5.48	0.64	6.45%	73%
2010	113	0.041	5.52	0.60	6.45%	74%
2011	106	0.039	5.56	0.56	6.45%	74%
2012	99	0.036	5.59	0.52	6.45%	75%
2013	93	0.034	5.63	0.49	6.45%	75%
2014	87	0.032	5.66	0.46	6.45%	75%
2015	81	0.030	5.69	0.43	6.45%	76%
2016	76	0.028	5.72	0.40	6.45%	76%
2017	71	0.026	5.74	0.38	6.45%	77%
2018	66	0.024	5.77	0.35	6.45%	77%
2019	62	0.023	5.79	0.33	6.45%	77%
2020	58	0.021	5.81	0.31	6.45%	77%
2021	54	0.020	5.83	0.29	6.45%	78%
2022	51	0.019	5.85	0.27	6.45%	78%
2023	48	0.017	5.87	0.25	6.45%	78%
2024	45	0.016	5.88	0.24	6.45%	78%
2025	42	0.015	5.90	0.22	6.45%	79%
2026	39	0.014	5.91	0.21	6.45%	79%
2027	36	0.013	5.93	0.19	6.45%	79%
2028	34	0.012	5.94	0.18	6.45%	79%
2029	32	0.012	5.95	0.17	6.45%	79%
2030	30	0.011	5.96	0.16	6.45%	79%
2031	28	0.010	5.97	0.15	6.45%	80%
2032	26	0.010	5.98	0.14	6.45%	80%
2033	24	0.009	5.99	0.13	6.45%	80%
2034	23	0.008	6.00	0.12	6.45%	80%
2035	21	0.008	6.01	0.11	6.45%	80%
2036	20	0.007	6.01	0.11	6.45%	80%
2037	19	0.007	6.02	0.10	6.45%	80%
2038	18	0.006	6.03	0.09	6.45%	80%
2039	16	0.006	6.03	0.09	6.45%	80%
2040	15	0.006	6.04	0.08	6.45%	81%
2041	14	0.005	6.04	0.08	6.45%	81%
2042	13	0.005	6.05	0.07	6.45%	81%
2043	13	0.005	6.05	0.07	6.45%	81%
2044	12	0.004	6.06	0.06	6.45%	81%
2045	11	0.004	6.06	0.06	6.45%	81%
2046	10	0.004	6.06	0.05	6.45%	81%
2047	10	0.004	6.07	0.05	6.45%	81%
2048	9	0.003	6.07	0.05	6.45%	81%
2049	8	0.003	6.07	0.04	6.45%	81%
2050	8	0.003	6.08	0.04	6.45%	81%

CUM.PROD	3.102 BBO	GIANT FIELDS		AUSTRALIA	
RESERVES	1.677 BBO	TOTAL BBO	2.6	THE EAST	
DISCOVERED	4.779 BBO	% OF DISC.	53%		
UNDISCOVERED	1.221 BBO	Last Giant	1967		
REMAINING	2.898 BBO			Variance	Trend
ULTIMATE	6.000 BBO			1990	+19%
Rept.Reserve	1.677	Ref.date (end)	1989	Future	UP

Milestones	Date	Prod Mb/d	Dep. Rate	% Produced	Forecast "A"
1st Peak	1987	566	5.93%	45%	A=Exponential decline
2nd Peak					B=5% incr. to Mid-Point
					C=Other
Mid-Point	1989	489	5.80%	52%	Pre-1930 Cum. 0.00
Final Decline	1990		5.80%		(Added to 1930 prod)

Date	Prod. M.b/d	Prod. BBO/yr	Cum. Prod.	Remain	Deplet. Rate	% Produced	Giant	Notes & Names
1930		0.000	0.00	6.00	ERR	0%		
1931		0.000	0.00	6.00	0.00%	0%		
1932		0.000	0.00	6.00	0.00%	0%		
1933		0.000	0.00	6.00	0.00%	0%		
1934		0.000	0.00	6.00	0.00%	0%		
1935		0.000	0.00	6.00	0.00%	0%		
1936		0.000	0.00	6.00	0.00%	0%		
1937		0.000	0.00	6.00	0.00%	0%		
1938		0.000	0.00	6.00	0.00%	0%		
1939		0.000	0.00	6.00	0.00%	0%		
1940		0.000	0.00	6.00	0.00%	0%		
1941		0.000	0.00	6.00	0.00%	0%		
1942		0.000	0.00	6.00	0.00%	0%		
1943		0.000	0.00	6.00	0.00%	0%		
1944		0.000	0.00	6.00	0.00%	0%		
1945		0.000	0.00	6.00	0.00%	0%		
1946		0.000	0.00	6.00	0.00%	0%		
1947		0.000	0.00	6.00	0.00%	0%		
1948		0.000	0.00	6.00	0.00%	0%		
1949		0.000	0.00	6.00	0.00%	0%		
1950		0.000	0.00	6.00	0.00%	0%		
1951		0.000	0.00	6.00	0.00%	0%		
1952		0.000	0.00	6.00	0.00%	0%		
1953		0.000	0.00	6.00	0.00%	0%		
1954		0.000	0.00	6.00	0.00%	0%		
1955		0.000	0.00	6.00	0.00%	0%		
1956		0.000	0.00	6.00	0.00%	0%		
1957		0.000	0.00	6.00	0.00%	0%		
1958		0.000	0.00	6.00	0.00%	0%		
1959		0.000	0.00	6.00	0.00%	0%		
1960		0.000	0.00	6.00	0.00%	0%		
1961		0.000	0.00	6.00	0.00%	0%		
1962		0.000	0.00	6.00	0.00%	0%		
1963		0.000	0.00	6.00	0.00%	0%		
1964	5	0.002	0.00	6.00	0.03%	0%		
1965	8	0.003	0.00	6.00	0.05%	0%	0.50	Barracouta
1966	10	0.004	0.01	5.99	0.06%	0%		
1967	20	0.007	0.02	5.98	0.12%	0%	2.05	Halibut & Kingfish
1968	45	0.016	0.03	5.97	0.27%	1%		
1969	41	0.015	0.05	5.95	0.25%	1%		
1970	170	0.062	0.11	5.89	1.04%	2%		
1971	320	0.117	0.23	5.77	1.98%	4%		
1972	308	0.112	0.34	5.66	1.95%	6%		
1973	420	0.153	0.49	5.51	2.71%	8%		
1974	371	0.135	0.63	5.37	2.46%	10%		
1975	415	0.151	0.78	5.22	2.82%	13%		
1976	430	0.157	0.94	5.06	3.01%	16%		
1977	430	0.157	1.09	4.91	3.10%	18%		
1978	430	0.157	1.25	4.75	3.20%	21%		
1979	440	0.161	1.41	4.59	3.38%	23%		
1980	380	0.139	1.55	4.45	3.02%	26%		
1981	382	0.139	1.69	4.31	3.13%	28%		
1982	353	0.129	1.82	4.18	2.99%	30%		
1983	405	0.148	1.96	4.04	3.53%	33%		
1984	481	0.176	2.14	3.86	4.35%	36%		
1985	556	0.203	2.34	3.66	5.26%	39%		
1986	472	0.172	2.52	3.48	4.71%	42%		
1987	566	0.207	2.72	3.28	5.93%	45%		1ST PEAK
1988	552	0.202	2.92	3.08	6.15%	49%		

1989	489	0.178	3.10	2.90	5.80%	52%	50% PRODUCED
1990	460	0.168	3.27	2.73	5.80%	55%	
1991	434	0.158	3.43	2.57	5.80%	57%	
1992	409	0.149	3.58	2.42	5.80%	60%	
1993	385	0.140	3.72	2.28	5.80%	62%	
1994	363	0.132	3.85	2.15	5.80%	64%	
1995	342	0.125	3.98	2.02	5.80%	66%	
1996	322	0.117	4.09	1.91	5.80%	68%	
1997	303	0.111	4.20	1.80	5.80%	70%	
1998	286	0.104	4.31	1.69	5.80%	72%	
1999	269	0.098	4.41	1.59	5.80%	73%	
2000	253	0.092	4.50	1.50	5.80%	75%	
2001	239	0.087	4.59	1.41	5.80%	76%	
2002	225	0.082	4.67	1.33	5.80%	78%	
2003	212	0.077	4.74	1.26	5.80%	79%	
2004	199	0.073	4.82	1.18	5.80%	80%	
2005	188	0.069	4.89	1.11	5.80%	81%	
2006	177	0.065	4.95	1.05	5.80%	83%	
2007	167	0.061	5.01	0.99	5.80%	84%	
2008	157	0.057	5.07	0.93	5.80%	84%	
2009	148	0.054	5.12	0.88	5.80%	85%	
2010	139	0.051	5.17	0.83	5.80%	86%	
2011	131	0.048	5.22	0.78	5.80%	87%	
2012	124	0.045	5.27	0.73	5.80%	88%	
2013	117	0.043	5.31	0.69	5.80%	88%	
2014	110	0.040	5.35	0.65	5.80%	89%	
2015	103	0.038	5.39	0.61	5.80%	90%	
2016	97	0.036	5.42	0.58	5.80%	90%	
2017	92	0.033	5.46	0.54	5.80%	91%	
2018	86	0.032	5.49	0.51	5.80%	91%	
2019	81	0.030	5.52	0.48	5.80%	92%	
2020	77	0.028	5.55	0.45	5.80%	92%	
2021	72	0.026	5.57	0.43	5.80%	93%	
2022	68	0.025	5.60	0.40	5.80%	93%	
2023	64	0.023	5.62	0.38	5.80%	94%	
2024	60	0.022	5.64	0.36	5.80%	94%	
2025	57	0.021	5.66	0.34	5.80%	94%	
2026	54	0.020	5.68	0.32	5.80%	95%	
2027	50	0.018	5.70	0.30	5.80%	95%	
2028	48	0.017	5.72	0.28	5.80%	95%	
2029	45	0.016	5.73	0.27	5.80%	96%	
2030	42	0.015	5.75	0.25	5.80%	96%	
2031	40	0.015	5.76	0.24	5.80%	96%	
2032	37	0.014	5.78	0.22	5.80%	96%	
2033	35	0.013	5.79	0.21	5.80%	97%	
2034	33	0.012	5.80	0.20	5.80%	97%	
2035	31	0.011	5.81	0.19	5.80%	97%	
2036	29	0.011	5.83	0.17	5.80%	97%	
2037	28	0.010	5.84	0.16	5.80%	97%	
2038	26	0.010	5.84	0.16	5.80%	97%	
2039	25	0.009	5.85	0.15	5.80%	98%	
2040	23	0.008	5.86	0.14	5.80%	98%	
2041	22	0.008	5.87	0.13	5.80%	98%	
2042	21	0.008	5.88	0.12	5.80%	98%	
2043	19	0.007	5.88	0.12	5.80%	98%	
2044	18	0.007	5.89	0.11	5.80%	98%	
2045	17	0.006	5.90	0.10	5.80%	98%	
2046	16	0.006	5.90	0.10	5.80%	98%	
2047	15	0.006	5.91	0.09	5.80%	98%	
2048	14	0.005	5.91	0.09	5.80%	99%	
2049	14	0.005	5.92	0.08	5.80%	99%	
2050	13	0.005	5.92	0.08	5.80%	99%	

CUM.PROD	0.678 BBO	GIANT FIELDS		AUSTRIA	
RESERVES	0.066 BBO	TOTAL BBO	0.50	W.EUROPE	
DISCOVERED	0.744 BBO	% OF DISC.	67%		
UNDISCOVERED	0.256 BBO	Last Giant	1949		
REMAINING	0.322 BBO			Variance	Trend
ULTIMATE	1.000 BBO			1990	+5.9%
Rept.Reserve	0.066 BBO	Ref.Date(end)	1989	Future	DOWN

Milestones	Date	Prod Mb/d	Dep. Rate	% Produced
1st Peak	1955	68	2.90%	17%
2nd Peak				
Mid-Point	1973	50	3.57%	51%
Final Decline			2.57%	

Forecast "A"
A=Exponential decline
B=5% incr. to Mid-Point
C=Other
Pre-1930 Cum.
(Added to 1930 Prod)

Date	Prod. M.b/d	Prod. BBO/yr	Cum Prod	Remain	Dep. Rate	% Produced	Giant	Notes & Names
1930		0.000	0.00	1.00	ERR	0%		
1931		0.000	0.00	1.00	0.00%	0%		
1932		0.000	0.00	1.00	0.00%	0%		
1933		0.000	0.00	1.00	0.00%	0%		
1934		0.000	0.00	1.00	0.00%	0%		
1935		0.000	0.00	1.00	0.00%	0%		
1936		0.000	0.00	1.00	0.00%	0%		
1937		0.000	0.00	1.00	0.00%	0%		
1938	1	0.000	0.00	1.00	0.04%	0%		
1939	3	0.001	0.00	1.00	0.12%	0%		
1940	8	0.003	0.00	1.00	0.28%	0%		
1941	12	0.004	0.01	0.99	0.43%	1%		
1942	16	0.006	0.01	0.99	0.60%	1%		
1943	21	0.007	0.02	0.98	0.76%	2%		
1944	23	0.008	0.03	0.97	0.84%	3%		
1945	8	0.003	0.03	0.97	0.32%	3%		
1946	16	0.006	0.04	0.96	0.59%	4%		
1947	16	0.006	0.04	0.96	0.60%	4%		
1948	17	0.006	0.05	0.95	0.65%	5%		
1949	20	0.007	0.06	0.94	0.77%	6%	0.50	Matzen
1950	34	0.012	0.07	0.93	1.32%	7%		
1951	42	0.015	0.09	0.91	1.65%	9%		
1952	47	0.017	0.10	0.90	1.88%	10%		
1953	50	0.018	0.12	0.88	2.04%	12%		
1954	62	0.023	0.14	0.86	2.58%	14%		
1955	68	0.025	0.17	0.83	2.90%	17%		1st PEAK
1956	65	0.024	0.19	0.81	2.85%	19%		
1957	60	0.022	0.21	0.79	2.71%	21%		
1958	57	0.021	0.24	0.76	2.63%	24%		
1959	46	0.017	0.25	0.75	2.17%	25%		
1960	46	0.017	0.27	0.73	2.26%	27%		
1961	45	0.017	0.29	0.71	2.27%	29%		
1962	43	0.016	0.30	0.70	2.22%	30%		
1963	50	0.018	0.32	0.68	2.59%	32%		
1964	53	0.019	0.34	0.66	2.83%	34%		
1965	51	0.019	0.36	0.64	2.83%	36%		
1966	54	0.020	0.38	0.62	3.06%	38%		
1967	55	0.020	0.40	0.60	3.21%	40%		
1968	47	0.017	0.41	0.59	2.85%	41%		
1969	51	0.019	0.43	0.57	3.18%	43%		
1970	55	0.020	0.45	0.55	3.52%	45%		
1971	51	0.019	0.47	0.53	3.40%	47%		
1972	47	0.017	0.49	0.51	3.22%	49%		
1973	50	0.018	0.51	0.49	3.57%	51%		50% PRODUCED
1974	47	0.017	0.52	0.48	3.48%	52%		
1975	41	0.015	0.54	0.46	3.17%	54%		
1976	37	0.014	0.55	0.45	2.93%	55%		
1977	36	0.013	0.57	0.43	2.93%	57%		
1978	35	0.013	0.58	0.42	2.94%	58%		
1979	36	0.013	0.59	0.41	3.12%	59%		
1980	29	0.011	0.60	0.40	2.59%	60%		
1981	25	0.009	0.61	0.39	2.29%	61%		
1982	25	0.009	0.62	0.38	2.35%	62%		
1983	24	0.009	0.63	0.37	2.31%	63%		
1984	23	0.008	0.64	0.36	2.26%	64%		
1985	22	0.008	0.65	0.35	2.18%	65%		
1986	22	0.008	0.65	0.35	2.28%	65%		
1987	22	0.008	0.66	0.34	2.27%	66%		
1988	23	0.008	0.67	0.33	2.50%	67%		

1989	23	0.009	0.68	0.32	2.57%	68%
1990	23	0.008	0.69	0.31	2.57%	69%
1991	22	0.008	0.69	0.31	2.57%	69%
1992	22	0.008	0.70	0.30	2.57%	70%
1993	21	0.008	0.71	0.29	2.57%	71%
1994	20	0.007	0.72	0.28	2.57%	72%
1995	20	0.007	0.72	0.28	2.57%	72%
1996	19	0.007	0.73	0.27	2.57%	73%
1997	19	0.007	0.74	0.26	2.57%	74%
1998	18	0.007	0.75	0.25	2.57%	75%
1999	18	0.007	0.75	0.25	2.57%	75%
2000	17	0.006	0.76	0.24	2.57%	76%
2001	17	0.006	0.76	0.24	2.57%	76%
2002	17	0.006	0.77	0.23	2.57%	77%
2003	16	0.006	0.78	0.22	2.57%	78%
2004	16	0.006	0.78	0.22	2.57%	78%
2005	15	0.006	0.79	0.21	2.57%	79%
2006	15	0.005	0.79	0.21	2.57%	79%
2007	15	0.005	0.80	0.20	2.57%	80%
2008	14	0.005	0.80	0.20	2.57%	80%
2009	14	0.005	0.81	0.19	2.57%	81%
2010	13	0.005	0.81	0.19	2.57%	81%
2011	13	0.005	0.82	0.18	2.57%	82%
2012	13	0.005	0.82	0.18	2.57%	82%
2013	12	0.005	0.83	0.17	2.57%	83%
2014	12	0.004	0.83	0.17	2.57%	83%
2015	12	0.004	0.84	0.16	2.57%	84%
2016	12	0.004	0.84	0.16	2.57%	84%
2017	11	0.004	0.84	0.16	2.57%	84%
2018	11	0.004	0.85	0.15	2.57%	85%
2019	11	0.004	0.85	0.15	2.57%	85%
2020	10	0.004	0.86	0.14	2.57%	86%
2021	10	0.004	0.86	0.14	2.57%	86%
2022	10	0.004	0.86	0.14	2.57%	86%
2023	10	0.004	0.87	0.13	2.57%	87%
2024	9	0.003	0.87	0.13	2.57%	87%
2025	9	0.003	0.87	0.13	2.57%	87%
2026	9	0.003	0.88	0.12	2.57%	88%
2027	9	0.003	0.88	0.12	2.57%	88%
2028	8	0.003	0.88	0.12	2.57%	88%
2029	8	0.003	0.89	0.11	2.57%	89%
2030	8	0.003	0.89	0.11	2.57%	89%
2031	8	0.003	0.89	0.11	2.57%	89%
2032	8	0.003	0.89	0.11	2.57%	89%
2033	7	0.003	0.90	0.10	2.57%	90%
2034	7	0.003	0.90	0.10	2.57%	90%
2035	7	0.003	0.90	0.10	2.57%	90%
2036	7	0.002	0.91	0.09	2.57%	91%
2037	7	0.002	0.91	0.09	2.57%	91%
2038	6	0.002	0.91	0.09	2.57%	91%
2039	6	0.002	0.91	0.09	2.57%	91%
2040	6	0.002	0.91	0.09	2.57%	91%
2041	6	0.002	0.92	0.08	2.57%	92%
2042	6	0.002	0.92	0.08	2.57%	92%
2043	6	0.002	0.92	0.08	2.57%	92%
2044	6	0.002	0.92	0.08	2.57%	92%
2045	5	0.002	0.93	0.07	2.57%	93%
2046	5	0.002	0.93	0.07	2.57%	93%
2047	5	0.002	0.93	0.07	2.57%	93%
2048	5	0.002	0.93	0.07	2.57%	93%
2049	5	0.002	0.93	0.07	2.57%	93%
2050	5	0.002	0.93	0.07	2.57%	93%

CUM.PROD	0.813 BBO	GIANT FIELDS		BAHRAIN
RESERVES	0.112 BBO	TOTAL BBO	0.9	M.EAST (MINOR)
DISCOVERED	0.924 BBO	% OF DISC.	100%	
UNDISCOVERED	0.076 BBO	Last Giant	1932	
REMAINING	0.187 BBO			Variance Trend
ULTIMATE	1.000 BBO			1990 zero
Rept.Reserve	0.112	Ref. date	1989	Future DOWN

Milestones	Date	Prod Mb/d	Dep. Rate	% Produced
1st Peak	1970	77	4.85%	45%
2nd Peak				
Mid-Point	1972	72	5.03%	50%
Final Decline	1990		7.60%	

Forecast "A"
A=Exponential decline
B=5% incr. to Mid-Point
C=Other
Pre-1930 Cum. 0.00
(Added to 1930 prod)

Date	Prod. M.b/d	Prod. BBO/yr	Cum. Prod.	Remain	Deplet. Rate	% Produced	Giant	Notes & Names
1930		0.000	0.00	1.00	ERR	0%		
1931		0.000	0.00	1.00	0.00%	0%		
1932		0.000	0.00	1.00	0.00%	0%	0.92	Bahrain
1933		0.000	0.00	1.00	0.00%	0%		
1934		0.000	0.00	1.00	0.00%	0%		
1935		0.000	0.00	1.00	0.00%	0%		
1936		0.000	0.00	1.00	0.00%	0%		
1937		0.000	0.00	1.00	0.00%	0%		
1938	23	0.008	0.01	0.99	0.83%	1%		
1939	21	0.008	0.02	0.98	0.77%	2%		
1940	19	0.007	0.02	0.98	0.72%	2%		
1941	19	0.007	0.03	0.97	0.69%	3%		
1942	17	0.006	0.04	0.96	0.64%	4%		
1943	18	0.007	0.04	0.96	0.68%	4%		
1944	18	0.007	0.05	0.95	0.70%	5%		
1945	20	0.007	0.06	0.94	0.77%	6%		
1946	23	0.008	0.06	0.94	0.89%	6%		
1947	26	0.009	0.07	0.93	1.01%	7%		
1948	30	0.011	0.09	0.91	1.18%	9%		
1949	30	0.011	0.10	0.90	1.20%	10%		
1950	30	0.011	0.11	0.89	1.22%	11%		
1951	30	0.011	0.12	0.88	1.23%	12%		
1952	30	0.011	0.13	0.87	1.25%	13%		
1953	30	0.011	0.14	0.86	1.24%	14%		
1954	30	0.011	0.15	0.85	1.28%	15%		
1955	30	0.011	0.16	0.84	1.29%	16%		
1956	30	0.011	0.17	0.83	1.31%	17%		
1957	32	0.012	0.18	0.82	1.41%	18%		
1958	41	0.015	0.20	0.80	1.82%	20%		
1959	45	0.016	0.22	0.78	2.06%	22%		
1960	45	0.016	0.23	0.77	2.10%	23%		
1961	45	0.016	0.25	0.75	2.14%	25%		
1962	45	0.016	0.27	0.73	2.19%	27%		
1963	45	0.016	0.28	0.72	2.24%	28%		
1964	49	0.018	0.30	0.70	2.49%	30%		
1965	56	0.021	0.32	0.68	2.93%	32%		
1966	61	0.022	0.34	0.66	3.28%	34%		
1967	68	0.025	0.37	0.63	3.77%	37%		
1968	75	0.027	0.39	0.61	4.33%	39%		
1969	76	0.028	0.42	0.58	4.58%	42%		
1970	77	0.028	0.45	0.55	4.85%	45%		1ST PEAK
1971	75	0.027	0.48	0.52	5.00%	48%		
1972	72	0.026	0.50	0.50	5.03%	50%		50% PRODUCED
1973	64	0.023	0.53	0.47	4.69%	53%		
1974	68	0.025	0.55	0.45	5.25%	55%		
1975	62	0.023	0.57	0.43	5.05%	57%		
1976	58	0.021	0.60	0.40	4.98%	60%		
1977	54	0.020	0.62	0.38	4.88%	62%		
1978	56	0.020	0.64	0.36	5.32%	64%		
1979	50	0.018	0.65	0.35	5.02%	65%		
1980	49	0.018	0.67	0.33	5.18%	67%		
1981	44	0.016	0.69	0.31	4.90%	69%		
1982	45	0.016	0.70	0.30	5.27%	70%		
1983	41	0.015	0.72	0.28	5.07%	72%		
1984	41	0.015	0.73	0.27	5.34%	73%		
1985	42	0.015	0.75	0.25	5.77%	75%		
1986	44	0.016	0.77	0.23	6.43%	77%		
1987	43	0.016	0.78	0.22	6.74%	78%		
1988	42	0.015	0.80	0.20	7.08%	80%		

1989	42	0.015	0.81	0.19	7.60%	81%
1990	39	0.014	0.83	0.17	7.60%	83%
1991	36	0.013	0.84	0.16	7.60%	84%
1992	33	0.012	0.85	0.15	7.60%	85%
1993	31	0.011	0.86	0.14	7.60%	86%
1994	28	0.010	0.87	0.13	7.60%	87%
1995	26	0.010	0.88	0.12	7.60%	88%
1996	24	0.009	0.89	0.11	7.60%	89%
1997	22	0.008	0.90	0.10	7.60%	90%
1998	21	0.008	0.91	0.09	7.60%	91%
1999	19	0.007	0.92	0.08	7.60%	92%
2000	18	0.006	0.92	0.08	7.60%	92%
2001	16	0.006	0.93	0.07	7.60%	93%
2002	15	0.006	0.93	0.07	7.60%	93%
2003	14	0.005	0.94	0.06	7.60%	94%
2004	13	0.005	0.94	0.06	7.60%	94%
2005	12	0.004	0.95	0.05	7.60%	95%
2006	11	0.004	0.95	0.05	7.60%	95%
2007	10	0.004	0.95	0.05	7.60%	95%
2008	9	0.003	0.96	0.04	7.60%	96%
2009	9	0.003	0.96	0.04	7.60%	96%
2010	8	0.003	0.96	0.04	7.60%	96%
2011	7	0.003	0.97	0.03	7.60%	97%
2012	7	0.003	0.97	0.03	7.60%	97%
2013	6	0.002	0.97	0.03	7.60%	97%
2014	6	0.002	0.97	0.03	7.60%	97%
2015	5	0.002	0.98	0.02	7.60%	98%
2016	5	0.002	0.98	0.02	7.60%	98%
2017	5	0.002	0.98	0.02	7.60%	98%
2018	4	0.002	0.98	0.02	7.60%	98%
2019	4	0.001	0.98	0.02	7.60%	98%
2020	4	0.001	0.98	0.02	7.60%	98%
2021	3	0.001	0.99	0.01	7.60%	99%
2022	3	0.001	0.99	0.01	7.60%	99%
2023	3	0.001	0.99	0.01	7.60%	99%
2024	3	0.001	0.99	0.01	7.60%	99%
2025	2	0.001	0.99	0.01	7.60%	99%
2026	2	0.001	0.99	0.01	7.60%	99%
2027	2	0.001	0.99	0.01	7.60%	99%
2028	2	0.001	0.99	0.01	7.60%	99%
2029	2	0.001	0.99	0.01	7.60%	99%
2030	2	0.001	0.99	0.01	7.60%	99%
2031	2	0.001	0.99	0.01	7.60%	99%
2032	1	0.001	0.99	0.01	7.60%	99%
2033	1	0.000	0.99	0.01	7.60%	99%
2034	1	0.000	0.99	0.01	7.60%	99%
2035	1	0.000	1.00	0.00	7.60%	100%
2036	1	0.000	1.00	0.00	7.60%	100%
2037	1	0.000	1.00	0.00	7.60%	100%
2038	1	0.000	1.00	0.00	7.60%	100%
2039	1	0.000	1.00	0.00	7.60%	100%
2040	1	0.000	1.00	0.00	7.60%	100%
2041	1	0.000	1.00	0.00	7.60%	100%
2042	1	0.000	1.00	0.00	7.60%	100%
2043	1	0.000	1.00	0.00	7.60%	100%
2044	1	0.000	1.00	0.00	7.60%	100%
2045	1	0.000	1.00	0.00	7.60%	100%
2046	0	0.000	1.00	0.00	7.60%	100%
2047	0	0.000	1.00	0.00	7.60%	100%
2048	0	0.000	1.00	0.00	7.60%	100%
2049	0	0.000	1.00	0.00	7.60%	100%
2050	0	0.000	1.00	0.00	7.60%	100%

CUM.PROD	0.002 BBO	GIANT FIELDS		BANGLADESH	
RESERVES	0.000 BBO	TOTAL BBO	0	THE EAST	
DISCOVERED	0.003 BBO	% OF DISC.	0%		
UNDISCOVERED	0.017 BBO	Last Giant			
REMAINING	0.018 BBO			Variance	Trend
ULTIMATE	0.020 BBO			1990	n/a
Rept.Reserve	0.000	Ref.date(end) 1989		Future	UP

Milestones	Date	Prod Mb/d	Dep. Rate	% Produced
1st Peak	2005	2	6.12%	51%
2nd Peak	None			
Mid-Point	2005	2	6.12%	51%
Final Decline	2005		6.12%	

Forecast "B"
A=Exponential decline
B=5% incr. to Mid-Point
C=Other
Pre-1930 Cum. 0
(Added to 1930 prod)

Date	Prod. M.b/d	Prod. BBO/yr	Cum. Prod.	Remain	Deplet. Rate	% Produced	Giant	Notes & Names
1930		0.000	0.00	0.02	ERR	0%		
1931		0.000	0.00	0.02	0.00%	0%		
1932		0.000	0.00	0.02	0.00%	0%		
1933		0.000	0.00	0.02	0.00%	0%		
1934		0.000	0.00	0.02	0.00%	0%		
1935		0.000	0.00	0.02	0.00%	0%		
1936		0.000	0.00	0.02	0.00%	0%		
1937		0.000	0.00	0.02	0.00%	0%		
1938		0.000	0.00	0.02	0.00%	0%		
1939		0.000	0.00	0.02	0.00%	0%		
1940		0.000	0.00	0.02	0.00%	0%		
1941		0.000	0.00	0.02	0.00%	0%		
1942		0.000	0.00	0.02	0.00%	0%		
1943		0.000	0.00	0.02	0.00%	0%		
1944		0.000	0.00	0.02	0.00%	0%		
1945		0.000	0.00	0.02	0.00%	0%		
1946		0.000	0.00	0.02	0.00%	0%		
1947		0.000	0.00	0.02	0.00%	0%		
1948		0.000	0.00	0.02	0.00%	0%		
1949		0.000	0.00	0.02	0.00%	0%		
1950		0.000	0.00	0.02	0.00%	0%		
1951		0.000	0.00	0.02	0.00%	0%		
1952		0.000	0.00	0.02	0.00%	0%		
1953		0.000	0.00	0.02	0.00%	0%		
1954		0.000	0.00	0.02	0.00%	0%		
1955		0.000	0.00	0.02	0.00%	0%		
1956		0.000	0.00	0.02	0.00%	0%		
1957		0.000	0.00	0.02	0.00%	0%		
1958		0.000	0.00	0.02	0.00%	0%		
1959		0.000	0.00	0.02	0.00%	0%		
1960		0.000	0.00	0.02	0.00%	0%		
1961	0	0.000	0.00	0.02	0.04%	0%		
1962	0	0.000	0.00	0.02	0.05%	0%		
1963	0	0.000	0.00	0.02	0.05%	0%		
1964	0	0.000	0.00	0.02	0.04%	0%		
1965	0	0.000	0.00	0.02	0.04%	0%		
1966	0	0.000	0.00	0.02	0.05%	0%		
1967	0	0.000	0.00	0.02	0.05%	0%		
1968	0	0.000	0.00	0.02	0.05%	0%		
1969	0	0.000	0.00	0.02	0.09%	0%		
1970	0	0.000	0.00	0.02	0.09%	1%		
1971	0	0.000	0.00	0.02	0.11%	1%		
1972	0	0.000	0.00	0.02	0.17%	1%		
1973	0	0.000	0.00	0.02	0.18%	1%		
1974	0	0.000	0.00	0.02	0.18%	1%		
1975	0	0.000	0.00	0.02	0.22%	1%		
1976	0	0.000	0.00	0.02	0.26%	2%		
1977	0	0.000	0.00	0.02	0.26%	2%		
1978	0	0.000	0.00	0.02	0.26%	2%		
1979	0	0.000	0.00	0.02	0.24%	2%		
1980	0	0.000	0.00	0.02	0.28%	3%		
1981	0	0.000	0.00	0.02	0.38%	3%		
1982	0	0.000	0.00	0.02	0.43%	3%		
1983	0	0.000	0.00	0.02	0.47%	4%		
1984	0	0.000	0.00	0.02	0.74%	5%		
1985	0	0.000	0.00	0.02	0.78%	5%		
1986	0	0.000	0.00	0.02	0.83%	6%		
1987	1	0.000	0.00	0.02	0.97%	7%		
1988	1	0.000	0.00	0.02	1.77%	9%		

1989	1	0.000	0.00	0.02	1.80%	10%	
1990	1	0.000	0.00	0.02	1.92%	12%	
1991	1	0.000	0.00	0.02	2.06%	14%	
1992	1	0.000	0.00	0.02	2.21%	16%	
1993	1	0.000	0.00	0.02	2.37%	18%	
1994	1	0.000	0.00	0.02	2.55%	20%	
1995	1	0.000	0.00	0.02	2.75%	22%	
1996	1	0.000	0.00	0.02	2.97%	24%	
1997	1	0.000	0.01	0.01	3.21%	27%	
1998	1	0.001	0.01	0.01	3.48%	29%	
1999	1	0.001	0.01	0.01	3.79%	32%	
2000	2	0.001	0.01	0.01	4.14%	35%	
2001	2	0.001	0.01	0.01	4.53%	38%	
2002	2	0.001	0.01	0.01	4.98%	41%	
2003	2	0.001	0.01	0.01	5.51%	44%	
2004	2	0.001	0.01	0.01	6.12%	48%	
2005	2	0.001	0.01	0.01	6.12%	51%	50% PRODUCED
2006	2	0.001	0.01	0.01	6.12%	54%	& 1ST PEAK
2007	2	0.001	0.01	0.01	6.12%	57%	
2008	1	0.001	0.01	0.01	6.12%	59%	
2009	1	0.000	0.01	0.01	6.12%	62%	
2010	1	0.000	0.01	0.01	6.12%	64%	
2011	1	0.000	0.01	0.01	6.12%	66%	
2012	1	0.000	0.01	0.01	6.12%	68%	
2013	1	0.000	0.01	0.01	6.12%	70%	
2014	1	0.000	0.01	0.01	6.12%	72%	
2015	1	0.000	0.01	0.01	6.12%	74%	
2016	1	0.000	0.02	0.00	6.12%	75%	
2017	1	0.000	0.02	0.00	6.12%	77%	
2018	1	0.000	0.02	0.00	6.12%	78%	
2019	1	0.000	0.02	0.00	6.12%	80%	
2020	1	0.000	0.02	0.00	6.12%	81%	
2021	1	0.000	0.02	0.00	6.12%	82%	
2022	1	0.000	0.02	0.00	6.12%	83%	
2023	1	0.000	0.02	0.00	6.12%	84%	
2024	1	0.000	0.02	0.00	6.12%	85%	
2025	0	0.000	0.02	0.00	6.12%	86%	
2026	0	0.000	0.02	0.00	6.12%	87%	
2027	0	0.000	0.02	0.00	6.12%	88%	
2028	0	0.000	0.02	0.00	6.12%	88%	
2029	0	0.000	0.02	0.00	6.12%	89%	
2030	0	0.000	0.02	0.00	6.12%	90%	
2031	0	0.000	0.02	0.00	6.12%	90%	
2032	0	0.000	0.02	0.00	6.12%	91%	
2033	0	0.000	0.02	0.00	6.12%	92%	
2034	0	0.000	0.02	0.00	6.12%	92%	
2035	0	0.000	0.02	0.00	6.12%	93%	
2036	0	0.000	0.02	0.00	6.12%	93%	
2037	0	0.000	0.02	0.00	6.12%	93%	
2038	0	0.000	0.02	0.00	6.12%	94%	
2039	0	0.000	0.02	0.00	6.12%	94%	
2040	0	0.000	0.02	0.00	6.12%	95%	
2041	0	0.000	0.02	0.00	6.12%	95%	
2042	0	0.000	0.02	0.00	6.12%	95%	
2043	0	0.000	0.02	0.00	6.12%	96%	
2044	0	0.000	0.02	0.00	6.12%	96%	
2045	0	0.000	0.02	0.00	6.12%	96%	
2046	0	0.000	0.02	0.00	6.12%	96%	
2047	0	0.000	0.02	0.00	6.12%	97%	
2048	0	0.000	0.02	0.00	6.12%	97%	
2049	0	0.000	0.02	0.00	6.12%	97%	
2050	0	0.000	0.02	0.00	6.12%	97%	

CUM.PROD	0.005 BBO	GIANT FIELDS		BARBADOS	
RESERVES	0.003 BBO	TOTAL BBO	0	L.AMERICA	
DISCOVERED	0.008 BBO	% OF DISC.	0%		
UNDISCOVERED	0.002 BBO	Last Giant			
REMAINING	0.005 BBO			Variance	Trend
ULTIMATE	0.010 BBO			1990	+10%
Rept.Reserves	0.003 BBO	Ref.Date (end)1989		Future	DOWN

Milestones	Date	Prod Mb/d	Dep. Rate	% Produced
1st Peak	1985	2	9.45%	34%
2nd Peak				
Mid-Point	1989	1	8.63%	54%
Final Decline	1990		8.60%	

Forecast	"A"
A=Exponential decline	
B=5% inc. to Mid-Point	
C=Other	
Pre-1930 Cum	0
(added to 1930 Prod)	

Date	Prod. M.b/d	Prod. BBO/yr	Cum Prod	Remain	Dep. Rate	% Produced	Giant	Names % Notes
1930		0.000	0.00	0.01	ERR	0%		
1931		0.000	0.00	0.01	0.0%	0%		
1932		0.000	0.00	0.01	0.0%	0%		
1933		0.000	0.00	0.01	0.0%	0%		
1934		0.000	0.00	0.01	0.0%	0%		
1935		0.000	0.00	0.01	0.0%	0%		
1936		0.000	0.00	0.01	0.0%	0%		
1937		0.000	0.00	0.01	0.0%	0%		
1938		0.000	0.00	0.01	0.0%	0%		
1939		0.000	0.00	0.01	0.0%	0%		
1940		0.000	0.00	0.01	0.0%	0%		
1941		0.000	0.00	0.01	0.0%	0%		
1942		0.000	0.00	0.01	0.0%	0%		
1943		0.000	0.00	0.01	0.0%	0%		
1944		0.000	0.00	0.01	0.0%	0%		
1945		0.000	0.00	0.01	0.0%	0%		
1946		0.000	0.00	0.01	0.0%	0%		
1947		0.000	0.00	0.01	0.0%	0%		
1948		0.000	0.00	0.01	0.0%	0%		
1949		0.000	0.00	0.01	0.0%	0%		
1950		0.000	0.00	0.01	0.0%	0%		
1951		0.000	0.00	0.01	0.0%	0%		
1952		0.000	0.00	0.01	0.0%	0%		
1953		0.000	0.00	0.01	0.0%	0%		
1954		0.000	0.00	0.01	0.0%	0%		
1955		0.000	0.00	0.01	0.0%	0%		
1956		0.000	0.00	0.01	0.0%	0%		
1957		0.000	0.00	0.01	0.0%	0%		
1958		0.000	0.00	0.01	0.0%	0%		
1959		0.000	0.00	0.01	0.0%	0%		
1960		0.000	0.00	0.01	0.0%	0%		
1961		0.000	0.00	0.01	0.0%	0%		
1962		0.000	0.00	0.01	0.0%	0%		
1963		0.000	0.00	0.01	0.0%	0%		
1964		0.000	0.00	0.01	0.0%	0%		
1965		0.000	0.00	0.01	0.0%	0%		
1966		0.000	0.00	0.01	0.0%	0%		
1967		0.000	0.00	0.01	0.0%	0%		
1968		0.000	0.00	0.01	0.0%	0%		
1969		0.000	0.00	0.01	0.0%	0%		
1970		0.000	0.00	0.01	0.0%	0%		
1971		0.000	0.00	0.01	0.0%	0%		
1972		0.000	0.00	0.01	0.0%	0%		
1973		0.000	0.00	0.01	0.0%	0%		
1974	0.30	0.000	0.00	0.01	1.1%	1%		
1975	0.30	0.000	0.00	0.01	1.1%	2%		
1976	0.40	0.000	0.00	0.01	1.5%	4%		
1977	0.30	0.000	0.00	0.01	1.1%	5%		
1978	0.70	0.000	0.00	0.01	2.7%	7%		
1979	0.80	0.000	0.00	0.01	3.1%	10%		
1980	1.00	0.000	0.00	0.01	4.1%	14%		
1981	0.70	0.000	0.00	0.01	3.0%	16%		
1982	0.60	0.000	0.00	0.01	2.6%	19%		
1983	0.50	0.000	0.00	0.01	2.2%	20%		
1984	1.70	0.001	0.00	0.01	7.8%	27%		
1985	1.90	0.001	0.00	0.01	9.5%	34%		1ST PEAK
1986	1.80	0.001	0.00	0.01	9.9%	40%		
1987	1.40	0.001	0.00	0.01	8.5%	45%		
1988	1.10	0.000	0.00	0.01	7.3%	49%		

1989	1.20	0.000	0.01	0.00	8.6%	54%	50% PRODUCED
1990	1.09	0.000	0.01	0.00	8.6%	58%	
1991	1.00	0.000	0.01	0.00	8.6%	61%	
1992	0.91	0.000	0.01	0.00	8.6%	65%	
1993	0.83	0.000	0.01	0.00	8.6%	68%	
1994	0.76	0.000	0.01	0.00	8.6%	70%	
1995	0.70	0.000	0.01	0.00	8.6%	73%	
1996	0.64	0.000	0.01	0.00	8.6%	75%	
1997	0.58	0.000	0.01	0.00	8.6%	77%	
1998	0.53	0.000	0.01	0.00	8.6%	79%	
1999	0.49	0.000	0.01	0.00	8.6%	81%	
2000	0.44	0.000	0.01	0.00	8.6%	83%	
2001	0.41	0.000	0.01	0.00	8.6%	84%	
2002	0.37	0.000	0.01	0.00	8.6%	86%	
2003	0.34	0.000	0.01	0.00	8.6%	87%	
2004	0.31	0.000	0.01	0.00	8.6%	88%	
2005	0.28	0.000	0.01	0.00	8.6%	89%	
2006	0.26	0.000	0.01	0.00	8.6%	90%	
2007	0.24	0.000	0.01	0.00	8.6%	91%	
2008	0.22	0.000	0.01	0.00	8.6%	92%	
2009	0.20	0.000	0.01	0.00	8.6%	92%	
2010	0.18	0.000	0.01	0.00	8.6%	93%	
2011	0.17	0.000	0.01	0.00	8.6%	94%	
2012	0.15	0.000	0.01	0.00	8.6%	94%	
2013	0.14	0.000	0.01	0.00	8.6%	95%	
2014	0.13	0.000	0.01	0.00	8.6%	95%	
2015	0.12	0.000	0.01	0.00	8.6%	96%	
2016	0.11	0.000	0.01	0.00	8.6%	96%	
2017	0.10	0.000	0.01	0.00	8.6%	96%	
2018	0.09	0.000	0.01	0.00	8.6%	97%	
2019	0.08	0.000	0.01	0.00	8.6%	97%	
2020	0.07	0.000	0.01	0.00	8.6%	97%	
2021	0.07	0.000	0.01	0.00	8.6%	97%	
2022	0.06	0.000	0.01	0.00	8.6%	98%	
2023	0.06	0.000	0.01	0.00	8.6%	98%	
2024	0.05	0.000	0.01	0.00	8.6%	98%	
2025	0.05	0.000	0.01	0.00	8.6%	98%	
2026	0.04	0.000	0.01	0.00	8.6%	98%	
2027	0.04	0.000	0.01	0.00	8.6%	98%	
2028	0.04	0.000	0.01	0.00	8.6%	99%	
2029	0.03	0.000	0.01	0.00	8.6%	99%	
2030	0.03	0.000	0.01	0.00	8.6%	99%	
2031	0.03	0.000	0.01	0.00	8.6%	99%	
2032	0.03	0.000	0.01	0.00	8.6%	99%	
2033	0.02	0.000	0.01	0.00	8.6%	99%	
2034	0.02	0.000	0.01	0.00	8.6%	99%	
2035	0.02	0.000	0.01	0.00	8.6%	99%	
2036	0.02	0.000	0.01	0.00	8.6%	99%	
2037	0.02	0.000	0.01	0.00	8.6%	99%	
2038	0.01	0.000	0.01	0.00	8.6%	99%	
2039	0.01	0.000	0.01	0.00	8.6%	99%	
2040	0.01	0.000	0.01	0.00	8.6%	100%	
2041	0.01	0.000	0.01	0.00	8.6%	100%	
2042	0.01	0.000	0.01	0.00	8.6%	100%	
2043	0.01	0.000	0.01	0.00	8.6%	100%	
2044	0.01	0.000	0.01	0.00	8.6%	100%	
2045	0.01	0.000	0.01	0.00	8.6%	100%	
2046	0.01	0.000	0.01	0.00	8.6%	100%	
2047	0.01	0.000	0.01	0.00	8.6%	100%	
2048	0.01	0.000	0.01	0.00	8.6%	100%	
2049	0.01	0.000	0.01	0.00	8.6%	100%	
2050	0.00	0.000	0.01	0.00	8.6%	100%	

CUM.PROD	0.017 BBO	GIANT FIELDS			BENIN
RESERVES	0.050 BBO	TOTAL BBO	0.0		AFRICA
DISCOVERED	0.067 BBO	% OF DISC.	0%		
UNDISCOVERED	0.033 BBO	Last Giant			
REMAINING	0.083 BBO				Variance Trend
ULTIMATE	0.100 BBO				1990 -18%
Rept.Reserves	0.100 BBO	Ref.Date (end)		1989	Future DOWN

Milestones	Date	Prod Mb/d	Dep Rate	% Produced
1st Peak	1987	12	4.84%	14%
2nd Peak	2002	8	5.59%	48%
Mid-Point	2003	8	5.59%	51%
Final Decline	2003		5.59%	

Forecast "B"
A=Exponential decline
B=5% inc. to Mid-Point
C=Other
Pre-1930 Cum 0
(Added to 1930 Prod)

Date	Prod. M.b/d	Prod. BBO/yr	Cum Prod	Remain	Dep. Rate	% Produced	Giant	Names & Notes
1930		0.000	0.00	0.10	ERR	0%		
1931		0.000	0.00	0.10	0.00%	0%		
1932		0.000	0.00	0.10	0.00%	0%		
1933		0.000	0.00	0.10	0.00%	0%		
1934		0.000	0.00	0.10	0.00%	0%		
1935		0.000	0.00	0.10	0.00%	0%		
1936		0.000	0.00	0.10	0.00%	0%		
1937		0.000	0.00	0.10	0.00%	0%		
1938		0.000	0.00	0.10	0.00%	0%		
1939		0.000	0.00	0.10	0.00%	0%		
1940		0.000	0.00	0.10	0.00%	0%		
1941		0.000	0.00	0.10	0.00%	0%		
1942		0.000	0.00	0.10	0.00%	0%		
1943		0.000	0.00	0.10	0.00%	0%		
1944		0.000	0.00	0.10	0.00%	0%		
1945		0.000	0.00	0.10	0.00%	0%		
1946		0.000	0.00	0.10	0.00%	0%		
1947		0.000	0.00	0.10	0.00%	0%		
1948		0.000	0.00	0.10	0.00%	0%		
1949		0.000	0.00	0.10	0.00%	0%		
1950		0.000	0.00	0.10	0.00%	0%		
1951		0.000	0.00	0.10	0.00%	0%		
1952		0.000	0.00	0.10	0.00%	0%		
1953		0.000	0.00	0.10	0.00%	0%		
1954		0.000	0.00	0.10	0.00%	0%		
1955		0.000	0.00	0.10	0.00%	0%		
1956		0.000	0.00	0.10	0.00%	0%		
1957		0.000	0.00	0.10	0.00%	0%		
1958		0.000	0.00	0.10	0.00%	0%		
1959		0.000	0.00	0.10	0.00%	0%		
1960		0.000	0.00	0.10	0.00%	0%		
1961		0.000	0.00	0.10	0.00%	0%		
1962		0.000	0.00	0.10	0.00%	0%		
1963		0.000	0.00	0.10	0.00%	0%		
1964		0.000	0.00	0.10	0.00%	0%		
1965		0.000	0.00	0.10	0.00%	0%		
1966		0.000	0.00	0.10	0.00%	0%		
1967		0.000	0.00	0.10	0.00%	0%		
1968		0.000	0.00	0.10	0.00%	0%		
1969		0.000	0.00	0.10	0.00%	0%		
1970		0.000	0.00	0.10	0.00%	0%		
1971		0.000	0.00	0.10	0.00%	0%		
1972		0.000	0.00	0.10	0.00%	0%		
1973		0.000	0.00	0.10	0.00%	0%		
1974		0.000	0.00	0.10	0.00%	0%		
1975		0.000	0.00	0.10	0.00%	0%		
1976		0.000	0.00	0.10	0.00%	0%		
1977		0.000	0.00	0.10	0.00%	0%		
1978		0.000	0.00	0.10	0.00%	0%		
1979		0.000	0.00	0.10	0.00%	0%		
1980		0.000	0.00	0.10	0.00%	0%		
1981		0.000	0.00	0.10	0.00%	0%		
1982		0.000	0.00	0.10	0.00%	0%		
1983	5.20	0.002	0.00	0.10	1.90%	2%		
1984	6.90	0.003	0.00	0.10	2.57%	4%		
1985	6.90	0.003	0.01	0.09	2.63%	7%		
1986	7.00	0.003	0.01	0.09	2.75%	9%		
1987	12.00	0.004	0.01	0.09	4.84%	14%		1ST PEAK
1988	4.60	0.002	0.02	0.08	1.95%	16%		

1989	4.50	0.002	0.02	0.08	1.94%	17%	
1990	4.73	0.002	0.02	0.08	2.08%	19%	
1991	4.96	0.002	0.02	0.08	2.23%	21%	
1992	5.21	0.002	0.02	0.08	2.40%	23%	
1993	5.47	0.002	0.02	0.08	2.58%	25%	
1994	5.74	0.002	0.03	0.07	2.78%	27%	
1995	6.03	0.002	0.03	0.07	3.00%	29%	
1996	6.33	0.002	0.03	0.07	3.25%	31%	
1997	6.65	0.002	0.03	0.07	3.53%	34%	
1998	6.98	0.003	0.04	0.06	3.84%	36%	
1999	7.33	0.003	0.04	0.06	4.19%	39%	
2000	7.70	0.003	0.04	0.06	4.60%	42%	
2001	8.08	0.003	0.04	0.06	5.06%	45%	
2002	8.49	0.003	0.05	0.05	5.59%	48%	2ND PEAK
2003	8.00	0.003	0.05	0.05	5.59%	51%	50% PRODUCED
2004	7.56	0.003	0.05	0.05	5.59%	53%	
2005	7.13	0.003	0.06	0.04	5.59%	56%	
2006	6.74	0.002	0.06	0.04	5.59%	58%	
2007	6.36	0.002	0.06	0.04	5.59%	61%	
2008	6.00	0.002	0.06	0.04	5.59%	63%	
2009	5.67	0.002	0.07	0.03	5.59%	65%	
2010	5.35	0.002	0.07	0.03	5.59%	67%	
2011	5.05	0.002	0.07	0.03	5.59%	69%	
2012	4.77	0.002	0.07	0.03	5.59%	71%	
2013	4.50	0.002	0.07	0.03	5.59%	72%	
2014	4.25	0.002	0.07	0.03	5.59%	74%	
2015	4.01	0.001	0.08	0.02	5.59%	75%	
2016	3.79	0.001	0.08	0.02	5.59%	77%	
2017	3.58	0.001	0.08	0.02	5.59%	78%	
2018	3.38	0.001	0.08	0.02	5.59%	79%	
2019	3.19	0.001	0.08	0.02	5.59%	80%	
2020	3.01	0.001	0.08	0.02	5.59%	81%	
2021	2.84	0.001	0.08	0.02	5.59%	82%	
2022	2.68	0.001	0.08	0.02	5.59%	83%	
2023	2.53	0.001	0.08	0.02	5.59%	84%	
2024	2.39	0.001	0.09	0.01	5.59%	85%	
2025	2.26	0.001	0.09	0.01	5.59%	86%	
2026	2.13	0.001	0.09	0.01	5.59%	87%	
2027	2.01	0.001	0.09	0.01	5.59%	88%	
2028	1.90	0.001	0.09	0.01	5.59%	88%	
2029	1.79	0.001	0.09	0.01	5.59%	89%	
2030	1.69	0.001	0.09	0.01	5.59%	90%	
2031	1.60	0.001	0.09	0.01	5.59%	90%	
2032	1.51	0.001	0.09	0.01	5.59%	91%	
2033	1.43	0.001	0.09	0.01	5.59%	91%	
2034	1.35	0.000	0.09	0.01	5.59%	92%	
2035	1.27	0.000	0.09	0.01	5.59%	92%	
2036	1.20	0.000	0.09	0.01	5.59%	93%	
2037	1.13	0.000	0.09	0.01	5.59%	93%	
2038	1.07	0.000	0.09	0.01	5.59%	93%	
2039	1.01	0.000	0.09	0.01	5.59%	94%	
2040	0.95	0.000	0.09	0.01	5.59%	94%	
2041	0.90	0.000	0.09	0.01	5.59%	94%	
2042	0.85	0.000	0.09	0.01	5.59%	95%	
2043	0.80	0.000	0.10	0.00	5.59%	95%	
2044	0.76	0.000	0.10	0.00	5.59%	95%	
2045	0.71	0.000	0.10	0.00	5.59%	96%	
2046	0.67	0.000	0.10	0.00	5.59%	96%	
2047	0.64	0.000	0.10	0.00	5.59%	96%	
2048	0.60	0.000	0.10	0.00	5.59%	96%	
2049	0.57	0.000	0.10	0.00	5.59%	97%	
2050	0.54	0.000	0.10	0.00	5.59%	97%	

CUM.PROD	0.297 BBO	GIANT FIELDS		BOLIVIA	
RESERVES	0.182 BBO	TOTAL BBO	0	L.AMERICA	
DISCOVERED	0.479 BBO	% OF DISC.	0%		
UNDISCOVERED	0.271 BBO	Last Giant			
REMAINING	0.453 BBO			Variance	Trend
ULTIMATE	0.750 BBO			1990	-10%
Rept.Reserves	0.182 BBO	Ref.Date (end)1989		Future	DOWN

Milestones	Date	Prod Mb/d	Dep. Rate	% Produced
1st Peak	1974	48	2.82%	20%
2nd Peak	1997	30	2.81%	50%
Mid-Point	1997	30	2.81%	50%
Final Decline	1997		2.81%	

Forecast "B"
A=Exponential decline
B=5% inc. to Mid-Point
C=Other
Pre-1930 Cum 0.005
(added to 1930 Prod)

Date	Prod. M.b/d	Prod. BBO/yr	Cum Prod	Remain	Dep. Rate	% Produced	Giant Names % Notes
1930	0.10	0.000	0.01	0.74	ERR	1%	
1931	0.10	0.000	0.01	0.74	0.00%	1%	
1932	0.10	0.000	0.01	0.74	0.00%	1%	
1933	0.10	0.000	0.01	0.74	0.00%	1%	
1934	0.20	0.000	0.01	0.74	0.01%	1%	
1935	0.30	0.000	0.01	0.74	0.01%	1%	
1936	0.40	0.000	0.01	0.74	0.02%	1%	
1937	0.50	0.000	0.01	0.74	0.02%	1%	
1938	0.60	0.000	0.01	0.74	0.03%	1%	
1939	0.60	0.000	0.01	0.74	0.03%	1%	
1940	0.80	0.000	0.01	0.74	0.04%	1%	
1941	0.60	0.000	0.01	0.74	0.03%	1%	
1942	0.80	0.000	0.01	0.74	0.04%	1%	
1943	0.90	0.000	0.01	0.74	0.04%	1%	
1944	0.90	0.000	0.01	0.74	0.04%	1%	
1945	1.00	0.000	0.01	0.74	0.05%	1%	
1946	1.00	0.000	0.01	0.74	0.05%	1%	
1947	1.30	0.000	0.01	0.74	0.06%	1%	
1948	1.30	0.000	0.01	0.74	0.06%	1%	
1949	1.90	0.001	0.01	0.74	0.09%	1%	
1950	1.60	0.001	0.01	0.74	0.08%	1%	
1951	1.40	0.001	0.01	0.74	0.07%	1%	
1952	1.50	0.001	0.01	0.74	0.07%	2%	
1953	1.38	0.001	0.01	0.74	0.07%	2%	
1954	4.50	0.002	0.01	0.74	0.22%	2%	
1955	7.40	0.003	0.02	0.73	0.37%	2%	
1956	8.70	0.003	0.02	0.73	0.43%	3%	
1957	9.80	0.004	0.02	0.73	0.49%	3%	
1958	9.10	0.003	0.03	0.72	0.46%	4%	
1959	8.70	0.003	0.03	0.72	0.44%	4%	
1960	8.30	0.003	0.03	0.72	0.42%	4%	
1961	8.70	0.003	0.04	0.71	0.44%	5%	
1962	8.20	0.003	0.04	0.71	0.42%	5%	
1963	8.80	0.003	0.04	0.71	0.45%	6%	
1964	9.20	0.003	0.05	0.70	0.47%	6%	
1965	9.20	0.003	0.05	0.70	0.48%	7%	
1966	12.50	0.005	0.05	0.70	0.65%	7%	
1967	27.30	0.010	0.06	0.69	1.43%	8%	
1968	37.00	0.014	0.08	0.67	1.97%	10%	
1969	40.00	0.015	0.09	0.66	2.17%	12%	
1970	13.00	0.005	0.10	0.65	0.72%	13%	
1971	30.00	0.011	0.11	0.64	1.67%	14%	
1972	32.00	0.012	0.12	0.63	1.82%	16%	
1973	47.00	0.017	0.14	0.61	2.72%	18%	
1974	47.50	0.017	0.15	0.60	2.82%	20%	1st PEAK
1975	42.00	0.015	0.17	0.58	2.57%	22%	
1976	41.00	0.015	0.18	0.57	2.57%	24%	
1977	35.00	0.013	0.20	0.55	2.26%	26%	
1978	30.00	0.011	0.21	0.54	1.98%	28%	
1979	30.00	0.011	0.22	0.53	2.02%	29%	
1980	30.00	0.011	0.23	0.52	2.06%	31%	
1981	24.00	0.009	0.24	0.51	1.68%	32%	
1982	24.00	0.009	0.25	0.50	1.71%	33%	
1983	22.00	0.008	0.25	0.50	1.60%	34%	
1984	20.00	0.007	0.26	0.49	1.47%	35%	
1985	20.00	0.007	0.27	0.48	1.50%	36%	
1986	18.70	0.007	0.28	0.47	1.42%	37%	
1987	18.50	0.007	0.28	0.47	1.43%	38%	
1988	18.80	0.007	0.29	0.46	1.47%	39%	

1989	20.30	0.007	0.30	0.45	1.61%	40%	
1990	21.31	0.008	0.30	0.45	1.61%	41%	
1991	22.38	0.008	0.31	0.44	1.61%	42%	
1992	23.50	0.009	0.32	0.43	1.96%	43%	
1993	24.67	0.009	0.33	0.42	2.10%	44%	
1994	25.91	0.009	0.34	0.41	2.26%	45%	
1995	27.20	0.010	0.35	0.40	2.42%	47%	
1996	28.56	0.010	0.36	0.39	2.61%	48%	
1997	29.99	0.011	0.37	0.38	2.81%	50%	50% PRODUCED
1998	29.14	0.011	0.38	0.37	2.81%	51%	& 2ND PEAK
1999	28.32	0.010	0.39	0.36	2.81%	52%	
2000	27.52	0.010	0.40	0.35	2.81%	54%	
2001	26.75	0.010	0.41	0.34	2.81%	55%	
2002	26.00	0.009	0.42	0.33	2.81%	56%	
2003	25.27	0.009	0.43	0.32	2.81%	57%	
2004	24.56	0.009	0.44	0.31	2.81%	59%	
2005	23.87	0.009	0.45	0.30	2.81%	60%	
2006	23.20	0.008	0.46	0.29	2.81%	61%	
2007	22.55	0.008	0.47	0.28	2.81%	62%	
2008	21.91	0.008	0.47	0.28	2.81%	63%	
2009	21.30	0.008	0.48	0.27	2.81%	64%	
2010	20.70	0.008	0.49	0.26	2.81%	65%	
2011	20.12	0.007	0.50	0.25	2.81%	66%	
2012	19.55	0.007	0.50	0.25	2.81%	67%	
2013	19.00	0.007	0.51	0.24	2.81%	68%	
2014	18.47	0.007	0.52	0.23	2.81%	69%	
2015	17.95	0.007	0.52	0.23	2.81%	70%	
2016	17.44	0.006	0.53	0.22	2.81%	71%	
2017	16.95	0.006	0.54	0.21	2.81%	71%	
2018	16.48	0.006	0.54	0.21	2.81%	72%	
2019	16.02	0.006	0.55	0.20	2.81%	73%	
2020	15.57	0.006	0.55	0.20	2.81%	74%	
2021	15.13	0.006	0.56	0.19	2.81%	75%	
2022	14.70	0.005	0.56	0.19	2.81%	75%	
2023	14.29	0.005	0.57	0.18	2.81%	76%	
2024	13.89	0.005	0.57	0.18	2.81%	77%	
2025	13.50	0.005	0.58	0.17	2.81%	77%	
2026	13.12	0.005	0.58	0.17	2.81%	78%	
2027	12.75	0.005	0.59	0.16	2.81%	79%	
2028	12.39	0.005	0.59	0.16	2.81%	79%	
2029	12.04	0.004	0.60	0.15	2.81%	80%	
2030	11.70	0.004	0.60	0.15	2.81%	80%	
2031	11.38	0.004	0.61	0.14	2.81%	81%	
2032	11.06	0.004	0.61	0.14	2.81%	81%	
2033	10.75	0.004	0.61	0.14	2.81%	82%	
2034	10.44	0.004	0.62	0.13	2.81%	82%	
2035	10.15	0.004	0.62	0.13	2.81%	83%	
2036	9.86	0.004	0.63	0.12	2.81%	83%	
2037	9.59	0.003	0.63	0.12	2.81%	84%	
2038	9.32	0.003	0.63	0.12	2.81%	84%	
2039	9.06	0.003	0.64	0.11	2.81%	85%	
2040	8.80	0.003	0.64	0.11	2.81%	85%	
2041	8.55	0.003	0.64	0.11	2.81%	86%	
2042	8.31	0.003	0.65	0.10	2.81%	86%	
2043	8.08	0.003	0.65	0.10	2.81%	86%	
2044	7.85	0.003	0.65	0.10	2.81%	87%	
2045	7.63	0.003	0.65	0.10	2.81%	87%	
2046	7.42	0.003	0.66	0.09	2.81%	88%	
2047	7.21	0.003	0.66	0.09	2.81%	88%	
2048	7.01	0.003	0.66	0.09	2.81%	88%	
2049	6.81	0.002	0.66	0.09	2.81%	89%	
2050	6.62	0.002	0.67	0.08	2.81%	89%	

CUM.PROD	2.640 BBO	GIANT FIELDS		BRASIL	
RESERVES	5.000 BBO	TOTAL BBO	5.99	L.AMERICA	
DISCOVERED	7.640 BBO	% OF DISC.	78%		
UNDISCOVERED	2.360 BBO	Last Giant	1986		
REMAINING	7.360 BBO			Variance	Trend
ULTIMATE	10.000 BBO			1990	+0.9%
Rept.Reserves	2.816 BBO	Ref Date(end)	1989	Future	DOWN

Milestones	Date	Prod Mb/d	Dep. Rate	% Produced
1st Peak	1998	926	6.53%	52%
2nd Peak				
Mid-Point	1998	926	6.53%	52%
Final Decline	1998		6.53%	

Forecast "B"
A=Exponential decline
B=5% inc. to Mid-Point
C=Other
Pre-1930 Cum 0
(added to 1930 Prod)

Date	Prod. M.b/d	Prod. BBO/yr	Cum Prod	Remain	Dep. Rate	% Produced	Giant	Names % Notes
1930		0.000	0.00	10.00	ERR	0%		
1931		0.000	0.00	10.00	0.00%	0%		
1932		0.000	0.00	10.00	0.00%	0%		
1933		0.000	0.00	10.00	0.00%	0%		
1934		0.000	0.00	10.00	0.00%	0%		
1935		0.000	0.00	10.00	0.00%	0%		
1936		0.000	0.00	10.00	0.00%	0%		
1937		0.000	0.00	10.00	0.00%	0%		
1938		0.000	0.00	10.00	0.00%	0%		
1939		0.000	0.00	10.00	0.00%	0%		
1940		0.000	0.00	10.00	0.00%	0%		
1941		0.000	0.00	10.00	0.00%	0%		
1942	0.10	0.000	0.00	10.00	0.00%	0%		
1943	0.10	0.000	0.00	10.00	0.00%	0%		
1944	0.20	0.000	0.00	10.00	0.00%	0%		
1945	0.20	0.000	0.00	10.00	0.00%	0%		
1946	0.20	0.000	0.00	10.00	0.00%	0%		
1947	0.30	0.000	0.00	10.00	0.00%	0%		
1948	0.40	0.000	0.00	10.00	0.00%	0%		
1949	0.30	0.000	0.00	10.00	0.00%	0%		
1950	0.80	0.000	0.00	10.00	0.00%	0%		
1951	1.90	0.001	0.00	10.00	0.01%	0%		
1952	2.00	0.001	0.00	10.00	0.01%	0%		
1953	2.39	0.001	0.00	10.00	0.01%	0%		
1954	2.60	0.001	0.00	10.00	0.01%	0%		
1955	5.20	0.002	0.01	9.99	0.02%	0%		
1956	11.00	0.004	0.01	9.99	0.04%	0%		
1957	27.40	0.010	0.02	9.98	0.10%	0%		
1958	57.20	0.021	0.04	9.96	0.21%	0%		
1959	66.30	0.024	0.07	9.93	0.24%	1%		
1960	77.00	0.028	0.09	9.91	0.28%	1%		
1961	96.00	0.035	0.13	9.87	0.35%	1%		
1962	95.00	0.035	0.16	9.84	0.35%	2%		
1963	92.00	0.034	0.20	9.80	0.34%	2%		
1964	89.00	0.032	0.23	9.77	0.33%	2%		
1965	94.40	0.034	0.26	9.74	0.35%	3%		
1966	111.00	0.041	0.30	9.70	0.42%	3%		
1967	150.00	0.055	0.36	9.64	0.56%	4%		
1968	160.30	0.059	0.42	9.58	0.61%	4%		
1969	179.10	0.065	0.48	9.52	0.68%	5%		
1970	157.60	0.058	0.54	9.46	0.60%	5%		
1971	168.00	0.061	0.60	9.40	0.65%	6%		
1972	167.00	0.061	0.66	9.34	0.65%	7%		
1973	170.00	0.062	0.72	9.28	0.66%	7%		
1974	172.20	0.063	0.79	9.21	0.68%	8%		
1975	175.00	0.064	0.85	9.15	0.69%	9%		
1976	171.00	0.062	0.91	9.09	0.68%	9%		
1977	162.00	0.059	0.97	9.03	0.65%	10%		
1978	160.00	0.058	1.03	8.97	0.65%	10%		
1979	165.00	0.060	1.09	8.91	0.67%	11%		
1980	190.00	0.069	1.16	8.84	0.78%	12%		
1981	215.00	0.078	1.24	8.76	0.89%	12%		
1982	252.00	0.092	1.33	8.67	1.05%	13%		
1983	315.00	0.115	1.45	8.55	1.33%	14%		
1984	437.00	0.160	1.61	8.39	1.86%	16%	1.1	Albaçore
1985	540.50	0.197	1.80	8.20	2.35%	18%	4.3	Marlim
1986	578.00	0.211	2.01	7.99	2.57%	20%	0.59	Miranga
1987	562.80	0.205	2.22	7.78	2.57%	22%		
1988	555.60	0.203	2.42	7.58	2.61%	24%		

1989	597.10	0.218	2.64	7.36	2.88%	26%	
1990	626.95	0.229	2.87	7.13	3.11%	29%	
1991	658.30	0.240	3.11	6.89	3.37%	31%	
1992	691.22	0.252	3.36	6.64	3.66%	34%	
1993	725.78	0.265	3.63	6.37	3.99%	36%	
1994	762.07	0.278	3.90	6.10	4.36%	39%	
1995	800.17	0.292	4.20	5.80	4.79%	42%	
1996	840.18	0.307	4.50	5.50	5.28%	45%	
1997	882.19	0.322	4.83	5.17	5.86%	48%	
1998	926.30	0.338	5.16	4.84	6.53%	52%	1st PEAK
1999	865.28	0.316	5.48	4.52	6.53%	55%	& 50% PRODUCED
2000	808.78	0.295	5.77	4.23	6.53%	58%	
2001	755.97	0.276	6.05	3.95	6.53%	61%	
2002	706.60	0.258	6.31	3.69	6.53%	63%	
2003	660.46	0.241	6.55	3.45	6.53%	65%	
2004	617.33	0.225	6.77	3.23	6.53%	68%	
2005	577.02	0.211	6.99	3.01	6.53%	70%	
2006	539.34	0.197	7.18	2.82	6.53%	72%	
2007	504.12	0.184	7.37	2.63	6.53%	74%	
2008	471.20	0.172	7.54	2.46	6.53%	75%	
2009	440.43	0.161	7.70	2.30	6.53%	77%	
2010	411.67	0.150	7.85	2.15	6.53%	78%	
2011	384.79	0.140	7.99	2.01	6.53%	80%	
2012	359.66	0.131	8.12	1.88	6.53%	81%	
2013	336.18	0.123	8.24	1.76	6.53%	82%	
2014	314.23	0.115	8.36	1.64	6.53%	84%	
2015	293.71	0.107	8.47	1.53	6.53%	85%	
2016	274.53	0.100	8.57	1.43	6.53%	86%	
2017	256.60	0.094	8.66	1.34	6.53%	87%	
2018	239.85	0.088	8.75	1.25	6.53%	87%	
2019	224.18	0.082	8.83	1.17	6.53%	88%	
2020	209.54	0.076	8.91	1.09	6.53%	89%	
2021	195.86	0.071	8.98	1.02	6.53%	90%	
2022	183.07	0.067	9.04	0.96	6.53%	90%	
2023	171.12	0.062	9.11	0.89	6.53%	91%	
2024	159.94	0.058	9.16	0.84	6.53%	92%	
2025	149.50	0.055	9.22	0.78	6.53%	92%	
2026	139.74	0.051	9.27	0.73	6.53%	93%	
2027	130.61	0.048	9.32	0.68	6.53%	93%	
2028	122.08	0.045	9.36	0.64	6.53%	94%	
2029	114.11	0.042	9.40	0.60	6.53%	94%	
2030	106.66	0.039	9.44	0.56	6.53%	94%	
2031	99.69	0.036	9.48	0.52	6.53%	95%	
2032	93.18	0.034	9.51	0.49	6.53%	95%	
2033	87.10	0.032	9.54	0.46	6.53%	95%	
2034	81.41	0.030	9.57	0.43	6.53%	96%	
2035	76.10	0.028	9.60	0.40	6.53%	96%	
2036	71.13	0.026	9.63	0.37	6.53%	96%	
2037	66.48	0.024	9.65	0.35	6.53%	97%	
2038	62.14	0.023	9.68	0.32	6.53%	97%	
2039	58.08	0.021	9.70	0.30	6.53%	97%	
2040	54.29	0.020	9.72	0.28	6.53%	97%	
2041	50.74	0.019	9.73	0.27	6.53%	97%	
2042	47.43	0.017	9.75	0.25	6.53%	98%	
2043	44.33	0.016	9.77	0.23	6.53%	98%	
2044	41.44	0.015	9.78	0.22	6.53%	98%	
2045	38.73	0.014	9.80	0.20	6.53%	98%	
2046	36.20	0.013	9.81	0.19	6.53%	98%	
2047	33.84	0.012	9.82	0.18	6.53%	98%	
2048	31.63	0.012	9.83	0.17	6.53%	98%	
2049	29.56	0.011	9.85	0.15	6.53%	98%	
2050	27.63	0.010	9.86	0.14	6.53%	99%	

Note : Date of Miranga giant uncertain
Reserves adjusted up because giant field reserves
exceed reported reserves

CUM.PROD	2.114 BBO	GIANT FIELDS		BRUNEI	
RESERVES	1.375 BBO	TOTAL BBO	2.35	THE EAST	
DISCOVERED	3.489 BBO	% OF DISC.	67%		
UNDISCOVERED	0.511 BBO	Last Giant	1963		
REMAINING	1.886 BBO			Variance	Trend
ULTIMATE	4.000 BBO			1990	+2%
Rept.Reserve	1.375	Ref.date(end)	1989	Future	DOWN

Milestones	Date	Prod Mb/d	Dep. Rate	% Produced
1st Peak	1979	255	3.05%	38%
2nd Peak				
Mid-Point	1987	134	1.93%	50%
Final Decline	1990		2.16%	

Forecast "A"
A=Exponential decline
B=5% incr. to Mid-Point
C=Other
Pre-1930 Cum. See Note
(Added to 1930 prod)

Date	Prod. M.b/d	Prod. BBO/yr	Cum. Prod.	Remain	Deplet. Rate	% Produced	Giant	Notes & Names
1930		0.000	0.00	4.00	ERR	0%	1.30	Seria(1929)
1931		0.000	0.00	4.00	0.00%	0%		
1932		0.000	0.00	4.00	0.00%	0%		Date-base lacks
1933		0.000	0.00	4.00	0.00%	0%		information on
1934		0.000	0.00	4.00	0.00%	0%		early
1935		0.000	0.00	4.00	0.00%	0%		production
1936		0.000	0.00	4.00	0.00%	0%		
1937		0.000	0.00	4.00	0.00%	0%		
1938		0.000	0.00	4.00	0.00%	0%		
1939		0.000	0.00	4.00	0.00%	0%		
1940		0.000	0.00	4.00	0.00%	0%		
1941		0.000	0.00	4.00	0.00%	0%		
1942		0.000	0.00	4.00	0.00%	0%		
1943		0.000	0.00	4.00	0.00%	0%		
1944		0.000	0.00	4.00	0.00%	0%		
1945		0.000	0.00	4.00	0.00%	0%		
1946		0.000	0.00	4.00	0.00%	0%		
1947		0.000	0.00	4.00	0.00%	0%		
1948		0.000	0.00	4.00	0.00%	0%		
1949		0.000	0.00	4.00	0.00%	0%		
1950		0.000	0.00	4.00	0.00%	0%		
1951		0.000	0.00	4.00	0.00%	0%		
1952		0.000	0.00	4.00	0.00%	0%		
1953		0.000	0.00	4.00	0.00%	0%		
1954		0.000	0.00	4.00	0.00%	0%		
1955		0.000	0.00	4.00	0.00%	0%		
1956		0.000	0.00	4.00	0.00%	0%		
1957		0.000	0.00	4.00	0.00%	0%		
1958		0.000	0.00	4.00	0.00%	0%		
1959		0.000	0.00	4.00	0.00%	0%		
1960	95	0.035	0.52	3.97	0.87%	13%		
1961	83	0.030	0.55	3.94	0.76%	14%		
1962	80	0.029	0.58	3.91	0.74%	14%		
1963	71	0.026	0.61	3.88	0.66%	15%	1.05	SW Ampa
1964	69	0.025	0.63	3.85	0.65%	16%		
1965	88	0.032	0.66	3.82	0.83%	17%		
1966	92	0.034	0.70	3.79	0.88%	17%		
1967	95	0.035	0.73	3.75	0.92%	18%		
1968	123	0.045	0.78	3.71	1.20%	19%		
1969	125	0.046	0.82	3.66	1.23%	21%		
1970	140	0.051	0.87	3.61	1.39%	22%		
1971	141	0.051	0.92	3.56	1.42%	23%		
1972	190	0.069	0.99	3.49	1.95%	25%		
1973	229	0.084	1.08	3.41	2.39%	27%		
1974	193	0.070	1.15	3.34	2.07%	29%		
1975	175	0.064	1.21	3.27	1.91%	30%		
1976	200	0.073	1.28	3.20	2.23%	32%		
1977	207	0.076	1.36	3.13	2.36%	34%		
1978	210	0.077	1.44	3.05	2.45%	36%		
1979	255	0.093	1.53	2.96	3.05%	38%		1ST PEAK
1980	230	0.084	1.61	2.87	2.84%	40%		
1981	166	0.061	1.67	2.81	2.11%	42%		
1982	155	0.057	1.73	2.75	2.01%	43%		
1983	155	0.057	1.79	2.70	2.05%	45%		
1984	160	0.058	1.85	2.64	2.16%	46%		
1985	150	0.055	1.90	2.59	2.07%	48%		
1986	170	0.062	1.96	2.52	2.40%	49%		
1987	134	0.049	2.01	2.47	1.93%	50%		50% PRODUCED
1988	138	0.050	2.06	2.42	2.04%	52%		

1989	143	0.052	2.11	2.37	2.16%	53%
1990	140	0.051	2.17	2.32	2.16%	54%
1991	137	0.050	2.22	2.27	2.16%	55%
1992	134	0.049	2.26	2.22	2.16%	57%
1993	131	0.048	2.31	2.17	2.16%	58%
1994	129	0.047	2.36	2.13	2.16%	59%
1995	126	0.046	2.41	2.08	2.16%	60%
1996	123	0.045	2.45	2.04	2.16%	61%
1997	120	0.044	2.49	1.99	2.16%	62%
1998	118	0.043	2.54	1.95	2.16%	63%
1999	115	0.042	2.58	1.91	2.16%	64%
2000	113	0.041	2.62	1.87	2.16%	66%
2001	110	0.040	2.66	1.82	2.16%	67%
2002	108	0.039	2.70	1.79	2.16%	68%
2003	106	0.039	2.74	1.75	2.16%	68%
2004	103	0.038	2.78	1.71	2.16%	69%
2005	101	0.037	2.81	1.67	2.16%	70%
2006	99	0.036	2.85	1.64	2.16%	71%
2007	97	0.035	2.88	1.60	2.16%	72%
2008	95	0.035	2.92	1.57	2.16%	73%
2009	93	0.034	2.95	1.53	2.16%	74%
2010	91	0.033	2.99	1.50	2.16%	75%
2011	89	0.032	3.02	1.47	2.16%	75%
2012	87	0.032	3.05	1.44	2.16%	76%
2013	85	0.031	3.08	1.40	2.16%	77%
2014	83	0.030	3.11	1.37	2.16%	78%
2015	81	0.030	3.14	1.34	2.16%	79%
2016	80	0.029	3.17	1.32	2.16%	79%
2017	78	0.028	3.20	1.29	2.16%	80%
2018	76	0.028	3.23	1.26	2.16%	81%
2019	74	0.027	3.25	1.23	2.16%	81%
2020	73	0.027	3.28	1.21	2.16%	82%
2021	71	0.026	3.31	1.18	2.16%	83%
2022	70	0.025	3.33	1.15	2.16%	83%
2023	68	0.025	3.36	1.13	2.16%	84%
2024	67	0.024	3.38	1.10	2.16%	85%
2025	65	0.024	3.41	1.08	2.16%	85%
2026	64	0.023	3.43	1.06	2.16%	86%
2027	63	0.023	3.45	1.03	2.16%	86%
2028	61	0.022	3.47	1.01	2.16%	87%
2029	60	0.022	3.50	0.99	2.16%	87%
2030	59	0.021	3.52	0.97	2.16%	88%
2031	57	0.021	3.54	0.95	2.16%	88%
2032	56	0.020	3.56	0.93	2.16%	89%
2033	55	0.020	3.58	0.91	2.16%	89%
2034	54	0.020	3.60	0.89	2.16%	90%
2035	53	0.019	3.62	0.87	2.16%	90%
2036	51	0.019	3.64	0.85	2.16%	91%
2037	50	0.018	3.65	0.83	2.16%	91%
2038	49	0.018	3.67	0.81	2.16%	92%
2039	48	0.018	3.69	0.80	2.16%	92%
2040	47	0.017	3.71	0.78	2.16%	93%
2041	46	0.017	3.72	0.76	2.16%	93%
2042	45	0.016	3.74	0.75	2.16%	94%
2043	44	0.016	3.76	0.73	2.16%	94%
2044	43	0.016	3.77	0.71	2.16%	94%
2045	42	0.015	3.79	0.70	2.16%	95%
2046	41	0.015	3.80	0.68	2.16%	95%
2047	40	0.015	3.82	0.67	2.16%	95%
2048	40	0.014	3.83	0.65	2.16%	96%
2049	39	0.014	3.85	0.64	2.16%	96%
2050	38	0.014	3.86	0.63	2.16%	96%

CUM.PROD	0.060 BBO	GIANT FIELDS		BULGARIA	
RESERVES	0.004 BBO	TOTAL BBO	0	SINO-SOVIET	
DISCOVERED	0.064 BBO	% OF DISC.	0%		
UNDISCOVERED	0.086 BBO	Last Giant		----------------	
REMAINING	0.090 BBO	-------------------		Revision	Trend
ULTIMATE	0.150 BBO			1990	n/a
Rept.Reserve	0.004 #1	Ref.date(end) 1989		Future	UP

Milestones	Date	Prod Mb/d	Dep. Rate	% Produced	Forecast "A"
					A=Exponential decline
1st Peak	1969	14	3.99%	21%	B=5% incr. to Mid-Point
2nd Peak	1995	8	3.23%	50%	C=Other
Mid-Point	1997	6	3.23%	53%	Pre-1930 Cum. 0
Final Decline	1997		3.23%	53%	(Added to 1930 prod)

Date	Prod. M.b/d	Prod. BBO/yr	Cum. Prod.	Remain	Deplet. Rate	% Produced	Giant	Notes & Names
1930		0.000	0.00	0.15	ERR	0%		Data-base weak
1931		0.000	0.00	0.15	0.00%	0%		
1932		0.000	0.00	0.15	0.00%	0%		
1933		0.000	0.00	0.15	0.00%	0%		
1934		0.000	0.00	0.15	0.00%	0%		
1935		0.000	0.00	0.15	0.00%	0%		
1936		0.000	0.00	0.15	0.00%	0%		
1937		0.000	0.00	0.15	0.00%	0%		
1938		0.000	0.00	0.15	0.00%	0%		
1939		0.000	0.00	0.15	0.00%	0%		
1940		0.000	0.00	0.15	0.00%	0%		
1941		0.000	0.00	0.15	0.00%	0%		
1942		0.000	0.00	0.15	0.00%	0%		
1943		0.000	0.00	0.15	0.00%	0%		
1944		0.000	0.00	0.15	0.00%	0%		
1945		0.000	0.00	0.15	0.00%	0%		
1946		0.000	0.00	0.15	0.00%	0%		
1947		0.000	0.00	0.15	0.00%	0%		
1948		0.000	0.00	0.15	0.00%	0%		
1949		0.000	0.00	0.15	0.00%	0%		
1950		0.000	0.00	0.15	0.00%	0%		
1951		0.000	0.00	0.15	0.00%	0%		
1952		0.000	0.00	0.15	0.00%	0%		
1953		0.000	0.00	0.15	0.00%	0%		
1954		0.000	0.00	0.15	0.00%	0%		
1955	3	0.001	0.00	0.15	0.73%	1%		
1956	5	0.002	0.00	0.15	1.13%	2%		
1957	6	0.002	0.00	0.15	1.41%	3%		
1958	5	0.002	0.01	0.14	1.13%	4%		
1959	4	0.001	0.01	0.14	0.97%	5%		
1960	4	0.001	0.01	0.14	1.03%	6%		
1961	4	0.001	0.01	0.14	1.06%	7%		
1962	4	0.001	0.01	0.14	1.05%	8%		
1963	4	0.001	0.01	0.14	0.93%	9%		
1964	3	0.001	0.01	0.14	0.88%	10%		
1965	5	0.002	0.02	0.13	1.24%	11%		
1966	8	0.003	0.02	0.13	2.19%	13%		
1967	10	0.004	0.02	0.13	2.77%	15%		
1968	10	0.003	0.03	0.12	2.73%	18%		
1969	14	0.005	0.03	0.12	3.99%	21%		1ST PEAK
1970	7	0.002	0.03	0.12	2.06%	23%		
1971	6	0.002	0.04	0.11	2.01%	24%		
1972	5	0.002	0.04	0.11	1.60%	25%		
1973	4	0.001	0.04	0.11	1.30%	26%		
1974	3	0.001	0.04	0.11	0.99%	27%		
1975	3	0.001	0.04	0.11	0.83%	28%		
1976	2	0.001	0.04	0.11	0.67%	28%		
1977	2	0.001	0.04	0.11	0.68%	29%		
1978	2	0.001	0.04	0.11	0.68%	29%		
1979	5	0.002	0.05	0.10	1.58%	30%		
1980	6	0.002	0.05	0.10	2.09%	32%		
1981	6	0.002	0.05	0.10	2.14%	33%		
1982	2	0.001	0.05	0.10	0.58%	34%		
1983	2	0.001	0.05	0.10	0.55%	34%		
1984	6	0.002	0.05	0.10	2.21%	35%		
1985	6	0.002	0.06	0.09	2.26%	37%		
1986	1	0.000	0.06	0.09	0.31%	37%		
1987	1	0.000	0.06	0.09	0.31%	37%		
1988	6	0.002	0.06	0.09	2.17%	39%		

1989	6	0.002	0.06	0.09	2.22%	40%	
1990	6	0.002	0.06	0.09	2.38%	41%	
1991	6	0.002	0.06	0.09	2.56%	43%	
1992	6	0.002	0.07	0.08	2.76%	44%	
1993	7	0.002	0.07	0.08	2.98%	46%	
1994	7	0.003	0.07	0.08	3.23%	48%	
1995	8	0.003	0.07	0.08	3.23%	50%	50% PRODUCED & 2ND PEAK
1996	7	0.002	0.08	0.07	3.23%	51%	
1997	6	0.002	0.08	0.07	3.23%	53%	
1998	6	0.002	0.08	0.07	3.23%	54%	
1999	6	0.002	0.08	0.07	3.23%	56%	
2000	6	0.002	0.09	0.06	3.23%	57%	
2001	6	0.002	0.09	0.06	3.23%	59%	
2002	5	0.002	0.09	0.06	3.23%	60%	
2003	5	0.002	0.09	0.06	3.23%	61%	
2004	5	0.002	0.09	0.06	3.23%	63%	
2005	5	0.002	0.10	0.05	3.23%	64%	
2006	5	0.002	0.10	0.05	3.23%	65%	
2007	5	0.002	0.10	0.05	3.23%	66%	
2008	5	0.002	0.10	0.05	3.23%	67%	
2009	4	0.002	0.10	0.05	3.23%	68%	
2010	4	0.002	0.10	0.05	3.23%	69%	
2011	4	0.001	0.11	0.04	3.23%	70%	
2012	4	0.001	0.11	0.04	3.23%	71%	
2013	4	0.001	0.11	0.04	3.23%	72%	
2014	4	0.001	0.11	0.04	3.23%	73%	
2015	4	0.001	0.11	0.04	3.23%	74%	
2016	3	0.001	0.11	0.04	3.23%	75%	
2017	3	0.001	0.11	0.04	3.23%	76%	
2018	3	0.001	0.11	0.04	3.23%	76%	
2019	3	0.001	0.12	0.03	3.23%	77%	
2020	3	0.001	0.12	0.03	3.23%	78%	
2021	3	0.001	0.12	0.03	3.23%	79%	
2022	3	0.001	0.12	0.03	3.23%	79%	
2023	3	0.001	0.12	0.03	3.23%	80%	
2024	3	0.001	0.12	0.03	3.23%	81%	
2025	3	0.001	0.12	0.03	3.23%	81%	
2026	2	0.001	0.12	0.03	3.23%	82%	
2027	2	0.001	0.12	0.03	3.23%	82%	
2028	2	0.001	0.12	0.03	3.23%	83%	
2029	2	0.001	0.13	0.02	3.23%	84%	
2030	2	0.001	0.13	0.02	3.23%	84%	
2031	2	0.001	0.13	0.02	3.23%	85%	
2032	2	0.001	0.13	0.02	3.23%	85%	
2033	2	0.001	0.13	0.02	3.23%	86%	
2034	2	0.001	0.13	0.02	3.23%	86%	
2035	2	0.001	0.13	0.02	3.23%	86%	
2036	2	0.001	0.13	0.02	3.23%	87%	
2037	2	0.001	0.13	0.02	3.23%	87%	
2038	2	0.001	0.13	0.02	3.23%	88%	
2039	2	0.001	0.13	0.02	3.23%	88%	
2040	2	0.001	0.13	0.02	3.23%	89%	
2041	2	0.001	0.13	0.02	3.23%	89%	
2042	1	0.001	0.13	0.02	3.23%	89%	
2043	1	0.001	0.13	0.02	3.23%	90%	
2044	1	0.001	0.13	0.02	3.23%	90%	
2045	1	0.000	0.14	0.01	3.23%	90%	
2046	1	0.000	0.14	0.01	3.23%	91%	
2047	1	0.000	0.14	0.01	3.23%	91%	
2048	1	0.000	0.14	0.01	3.23%	91%	
2049	1	0.000	0.14	0.01	3.23%	91%	
2050	1	0.000	0.14	0.01	3.23%	92%	

CUM.PROD	0.736 BBO	GIANT FIELDS		BURMA
RESERVES	0.051 BBO	TOTAL BBO	0	THE EAST
DISCOVERED	0.788 BBO	% OF DISC.	0%	
UNDISCOVERED	0.412 BBO	Last Giant	-	
REMAINING	0.464 BBO			Variance Trend
ULTIMATE	1.200 BBO			1990 +9%
Rept.Reserve	0.051	Ref.date(end) 1989		Future UP

Milestones	Date	Prod Mb/d	Dep. Rate	% Produced
1st Peak	1979	30	1.18%	53%
2nd Peak				
Mid-Point	1974	23	0.87%	50%
Final Decline	1990		0.53%	62%

Forecast "B"
A=Exponential decline
B=5% incr. to Mid-Point
C=Other
Pre-1930 Cum. 0.08
(Added to 1930 prod)

Date	Prod. M.b/d	Prod. BBO/yr	Cum. Prod.	Remain	Deplet. Rate	% Produced	Giant	Notes & Names
1930	10.0	0.004	0.08	1.12	ERR	7%		
1931	10.0	0.004	0.09	1.11	0.33%	7%		
1932	10.0	0.004	0.09	1.11	0.33%	8%		
1933	10.0	0.004	0.09	1.11	0.33%	8%		
1934	10.0	0.004	0.10	1.10	0.33%	8%		
1935	10.0	0.004	0.10	1.10	0.33%	8%		
1936	10.0	0.004	0.11	1.09	0.33%	9%		
1937	10.0	0.004	0.11	1.09	0.33%	9%		
1938	10.6	0.004	0.11	1.09	0.35%	9%		
1939	21.6	0.008	0.12	1.08	0.73%	10%		
1940	21.6	0.008	0.13	1.07	0.73%	11%		
1941	21.3	0.008	0.14	1.06	0.73%	11%		
1942	6.8	0.002	0.14	1.06	0.23%	12%		
1943	2.7	0.001	0.14	1.06	0.09%	12%		
1944	2.1	0.001	0.14	1.06	0.07%	12%		
1945	2.0	0.001	0.14	1.06	0.07%	12%		
1946	1.9	0.001	0.14	1.06	0.07%	12%		
1947	0.6	0.000	0.14	1.06	0.02%	12%		
1948	0.5	0.000	0.14	1.06	0.02%	12%		
1949	0.5	0.000	0.14	1.06	0.02%	12%		
1950	1.0	0.000	0.14	1.06	0.03%	12%		
1951	2.0	0.001	0.14	1.06	0.07%	12%		
1952	2.3	0.001	0.14	1.06	0.08%	12%		
1953	3.4	0.001	0.15	1.05	0.12%	12%		
1954	3.5	0.001	0.15	1.05	0.12%	12%		
1955	4.0	0.001	0.15	1.05	0.14%	12%		
1956	4.5	0.002	0.15	1.05	0.16%	13%		
1957	6.5	0.002	0.15	1.05	0.23%	13%		
1958	10.5	0.004	0.16	1.04	0.37%	13%		
1959	10.4	0.004	0.16	1.04	0.36%	13%		
1960	10.9	0.004	0.52	1.04	0.38%	43%		
1961	11.2	0.004	0.52	1.03	0.39%	44%		
1962	9.2	0.003	0.53	1.03	0.33%	44%		
1963	10.9	0.004	0.53	1.02	0.39%	44%		
1964	11.8	0.004	0.54	1.02	0.42%	45%		
1965	12.4	0.005	0.54	1.02	0.44%	45%		
1966	12.3	0.004	0.55	1.01	0.44%	45%		
1967	11.5	0.004	0.55	1.01	0.42%	46%		
1968	14.3	0.005	0.55	1.00	0.52%	46%		
1969	16.0	0.006	0.56	1.00	0.58%	47%		
1970	15.6	0.006	0.57	0.99	0.57%	47%		
1971	19.0	0.007	0.57	0.98	0.70%	48%		
1972	20.0	0.007	0.58	0.98	0.74%	48%		
1973	23.0	0.008	0.59	0.97	0.86%	49%		
1974	23.0	0.008	0.60	0.96	0.87%	50%		50% PRODUCED
1975	23.0	0.008	0.61	0.95	0.88%	50%		
1976	20.0	0.007	0.61	0.94	0.77%	51%		
1977	23.0	0.008	0.62	0.93	0.89%	52%		
1978	25.0	0.009	0.63	0.93	0.98%	53%		1ST PEAK
1979	30.0	0.011	0.64	0.91	1.18%	53%		
1980	30.0	0.011	0.65	0.90	1.20%	54%		
1981	30.0	0.011	0.66	0.89	1.21%	55%		
1982	30.0	0.011	0.67	0.88	1.23%	56%		
1983	30.0	0.011	0.68	0.87	1.24%	57%		
1984	30.0	0.011	0.70	0.86	1.26%	58%		
1985	30.0	0.011	0.71	0.85	1.27%	59%		
1986	30.0	0.011	0.72	0.84	1.29%	60%		
1987	24.0	0.009	0.73	0.83	1.05%	61%		
1988	15.0	0.005	0.73	0.82	0.66%	61%		

1989	12.0	0.004	0.74	0.82	0.53%	61%
1990	11.9	0.004	0.74	0.82	0.53%	62%
1991	11.8	0.004	0.75	0.81	0.53%	62%
1992	11.8	0.004	0.75	0.81	0.53%	62%
1993	11.7	0.004	0.75	0.80	0.53%	63%
1994	11.6	0.004	0.76	0.80	0.53%	63%
1995	11.6	0.004	0.76	0.79	0.53%	64%
1996	11.5	0.004	0.77	0.79	0.53%	64%
1997	11.5	0.004	0.77	0.79	0.53%	64%
1998	11.4	0.004	0.77	0.78	0.53%	65%
1999	11.3	0.004	0.78	0.78	0.53%	65%
2000	11.3	0.004	0.78	0.77	0.53%	65%
2001	11.2	0.004	0.79	0.77	0.53%	66%
2002	11.2	0.004	0.79	0.76	0.53%	66%
2003	11.1	0.004	0.80	0.76	0.53%	66%
2004	11.0	0.004	0.80	0.76	0.53%	67%
2005	11.0	0.004	0.80	0.75	0.53%	67%
2006	10.9	0.004	0.81	0.75	0.53%	67%
2007	10.9	0.004	0.81	0.74	0.53%	68%
2008	10.8	0.004	0.82	0.74	0.53%	68%
2009	10.8	0.004	0.82	0.74	0.53%	68%
2010	10.7	0.004	0.82	0.73	0.53%	69%
2011	10.6	0.004	0.83	0.73	0.53%	69%
2012	10.6	0.004	0.83	0.73	0.53%	69%
2013	10.5	0.004	0.83	0.72	0.53%	70%
2014	10.5	0.004	0.84	0.72	0.53%	70%
2015	10.4	0.004	0.84	0.71	0.53%	70%
2016	10.4	0.004	0.85	0.71	0.53%	70%
2017	10.3	0.004	0.85	0.71	0.53%	71%
2018	10.3	0.004	0.85	0.70	0.53%	71%
2019	10.2	0.004	0.86	0.70	0.53%	71%
2020	10.1	0.004	0.86	0.69	0.53%	72%
2021	10.1	0.004	0.86	0.69	0.53%	72%
2022	10.0	0.004	0.87	0.69	0.53%	72%
2023	10.0	0.004	0.87	0.68	0.53%	73%
2024	9.9	0.004	0.88	0.68	0.53%	73%
2025	9.9	0.004	0.88	0.68	0.53%	73%
2026	9.8	0.004	0.88	0.67	0.53%	74%
2027	9.8	0.004	0.89	0.67	0.53%	74%
2028	9.7	0.004	0.89	0.67	0.53%	74%
2029	9.7	0.004	0.89	0.66	0.53%	74%
2030	9.6	0.004	0.90	0.66	0.53%	75%
2031	9.6	0.003	0.90	0.66	0.53%	75%
2032	9.5	0.003	0.90	0.65	0.53%	75%
2033	9.5	0.003	0.91	0.65	0.53%	76%
2034	9.4	0.003	0.91	0.65	0.53%	76%
2035	9.4	0.003	0.91	0.64	0.53%	76%
2036	9.3	0.003	0.92	0.64	0.53%	76%
2037	9.3	0.003	0.92	0.63	0.53%	77%
2038	9.2	0.003	0.92	0.63	0.53%	77%
2039	9.2	0.003	0.93	0.63	0.53%	77%
2040	9.1	0.003	0.93	0.62	0.53%	78%
2041	9.1	0.003	0.93	0.62	0.53%	78%
2042	9.0	0.003	0.94	0.62	0.53%	78%
2043	9.0	0.003	0.94	0.62	0.53%	78%
2044	8.9	0.003	0.94	0.61	0.53%	79%
2045	8.9	0.003	0.95	0.61	0.53%	79%
2046	8.8	0.003	0.95	0.61	0.53%	79%
2047	8.8	0.003	0.95	0.60	0.53%	79%
2048	8.7	0.003	0.96	0.60	0.53%	80%
2049	8.7	0.003	0.96	0.60	0.53%	80%
2050	8.7	0.003	0.96	0.59	0.53%	80%

CUM.PROD	0.496 BBO	GIANT FIELDS			CAMEROON	
RESERVES	0.400 BBO	TOTAL BBO	0.0		AFRICA	
DISCOVERED	0.896 BBO	% OF DISC.	0%			
UNDISCOVERED	0.104 BBO	Last Giant				
REMAINING	0.504 BBO				Variance	Trend
ULTIMATE	1.000 BBO				1990	+7%
Rept.Reserves	0.400 BBO	Ref.Date (end)		1989	Future	DOWN

Milestones	Date	Prod Mb/d	Dep Rate	% Produced	Forecast "A"
					A=Exponential decline
1st Peak	1986	180	8.68%	31%	B=5% inc. to Mid-Point
2nd Peak					C=Other
Mid-Point	1989	173	11.12%	50%	Pre-1930 Cum 0
Final Decline	1989		11.12%		(Added to 1930 Prod)

Date	Prod. M.b/d	Prod. BBO/yr	Cum Prod	Remain	Dep. Rate	% Produced	Giant	Names & Notes
1930		0.000	0.00	1.00	ERR	0%		
1931		0.000	0.00	1.00	0.00%	0%		
1932		0.000	0.00	1.00	0.00%	0%		
1933		0.000	0.00	1.00	0.00%	0%		
1934		0.000	0.00	1.00	0.00%	0%		
1935		0.000	0.00	1.00	0.00%	0%		
1936		0.000	0.00	1.00	0.00%	0%		
1937		0.000	0.00	1.00	0.00%	0%		
1938		0.000	0.00	1.00	0.00%	0%		
1939		0.000	0.00	1.00	0.00%	0%		
1940		0.000	0.00	1.00	0.00%	0%		
1941		0.000	0.00	1.00	0.00%	0%		
1942		0.000	0.00	1.00	0.00%	0%		
1943		0.000	0.00	1.00	0.00%	0%		
1944		0.000	0.00	1.00	0.00%	0%		
1945		0.000	0.00	1.00	0.00%	0%		
1946		0.000	0.00	1.00	0.00%	0%		
1947		0.000	0.00	1.00	0.00%	0%		
1948		0.000	0.00	1.00	0.00%	0%		
1949		0.000	0.00	1.00	0.00%	0%		
1950		0.000	0.00	1.00	0.00%	0%		
1951		0.000	0.00	1.00	0.00%	0%		
1952		0.000	0.00	1.00	0.00%	0%		
1953		0.000	0.00	1.00	0.00%	0%		
1954		0.000	0.00	1.00	0.00%	0%		
1955		0.000	0.00	1.00	0.00%	0%		
1956		0.000	0.00	1.00	0.00%	0%		
1957		0.000	0.00	1.00	0.00%	0%		
1958		0.000	0.00	1.00	0.00%	0%		
1959		0.000	0.00	1.00	0.00%	0%		
1960		0.000	0.00	1.00	0.00%	0%		
1961		0.000	0.00	1.00	0.00%	0%		
1962		0.000	0.00	1.00	0.00%	0%		
1963		0.000	0.00	1.00	0.00%	0%		
1964		0.000	0.00	1.00	0.00%	0%		
1965		0.000	0.00	1.00	0.00%	0%		
1966		0.000	0.00	1.00	0.00%	0%		
1967		0.000	0.00	1.00	0.00%	0%		
1968		0.000	0.00	1.00	0.00%	0%		
1969		0.000	0.00	1.00	0.00%	0%		
1970		0.000	0.00	1.00	0.00%	0%		
1971		0.000	0.00	1.00	0.00%	0%		
1972		0.000	0.00	1.00	0.00%	0%		
1973		0.000	0.00	1.00	0.00%	0%		
1974		0.000	0.00	1.00	0.00%	0%		
1975		0.000	0.00	1.00	0.00%	0%		
1976		0.000	0.00	1.00	0.00%	0%		
1977		0.000	0.00	1.00	0.00%	0%		
1978	10	0.004	0.00	1.00	0.37%	0%		
1979	32	0.012	0.02	0.98	1.17%	2%		
1980	57	0.021	0.04	0.96	2.11%	4%		
1981	87	0.032	0.07	0.93	3.29%	7%		
1982	109	0.040	0.11	0.89	4.27%	11%		
1983	114	0.042	0.15	0.85	4.66%	15%		
1984	125	0.046	0.19	0.81	5.36%	19%		
1985	133	0.049	0.24	0.76	6.03%	24%		
1986	180	0.066	0.31	0.69	8.68%	31%		1ST PEAK
1987	170	0.062	0.37	0.63	8.98%	37%		
1988	170	0.062	0.43	0.57	9.87%	43%		

1989	173	0.063	0.50	0.50	11.12%	50%	50% PRODUCED
1990	153	0.056	0.55	0.45	11.12%	55%	
1991	136	0.050	0.60	0.40	11.12%	60%	
1992	121	0.044	0.65	0.35	11.12%	65%	
1993	108	0.039	0.69	0.31	11.12%	69%	
1994	96	0.035	0.72	0.28	11.12%	72%	
1995	85	0.031	0.75	0.25	11.12%	75%	
1996	76	0.028	0.78	0.22	11.12%	78%	
1997	67	0.025	0.80	0.20	11.12%	80%	
1998	60	0.022	0.83	0.17	11.12%	83%	
1999	53	0.019	0.85	0.15	11.12%	85%	
2000	47	0.017	0.86	0.14	11.12%	86%	
2001	42	0.015	0.88	0.12	11.12%	88%	
2002	37	0.014	0.89	0.11	11.12%	89%	
2003	33	0.012	0.90	0.10	11.12%	90%	
2004	29	0.011	0.91	0.09	11.12%	91%	
2005	26	0.010	0.92	0.08	11.12%	92%	
2006	23	0.008	0.93	0.07	11.12%	93%	
2007	21	0.008	0.94	0.06	11.12%	94%	
2008	18	0.007	0.95	0.05	11.12%	95%	
2009	16	0.006	0.95	0.05	11.12%	95%	
2010	15	0.005	0.96	0.04	11.12%	96%	
2011	13	0.005	0.96	0.04	11.12%	96%	
2012	11	0.004	0.97	0.03	11.12%	97%	
2013	10	0.004	0.97	0.03	11.12%	97%	
2014	9	0.003	0.97	0.03	11.12%	97%	
2015	8	0.003	0.98	0.02	11.12%	98%	
2016	7	0.003	0.98	0.02	11.12%	98%	
2017	6	0.002	0.98	0.02	11.12%	98%	
2018	6	0.002	0.98	0.02	11.12%	98%	
2019	5	0.002	0.99	0.01	11.12%	99%	
2020	4	0.002	0.99	0.01	11.12%	99%	
2021	4	0.001	0.99	0.01	11.12%	99%	
2022	4	0.001	0.99	0.01	11.12%	99%	
2023	3	0.001	0.99	0.01	11.12%	99%	
2024	3	0.001	0.99	0.01	11.12%	99%	
2025	2	0.001	0.99	0.01	11.12%	99%	
2026	2	0.001	0.99	0.01	11.12%	99%	
2027	2	0.001	0.99	0.01	11.12%	99%	
2028	2	0.001	0.99	0.01	11.12%	99%	
2029	2	0.001	1.00	0.00	11.12%	100%	
2030	1	0.001	1.00	0.00	11.12%	100%	
2031	1	0.000	1.00	0.00	11.12%	100%	
2032	1	0.000	1.00	0.00	11.12%	100%	
2033	1	0.000	1.00	0.00	11.12%	100%	
2034	1	0.000	1.00	0.00	11.12%	100%	
2035	1	0.000	1.00	0.00	11.12%	100%	
2036	1	0.000	1.00	0.00	11.12%	100%	
2037	1	0.000	1.00	0.00	11.12%	100%	
2038	1	0.000	1.00	0.00	11.12%	100%	
2039	0	0.000	1.00	0.00	11.12%	100%	
2040	0	0.000	1.00	0.00	11.12%	100%	
2041	0	0.000	1.00	0.00	11.12%	100%	
2042	0	0.000	1.00	0.00	11.12%	100%	
2043	0	0.000	1.00	0.00	11.12%	100%	
2044	0	0.000	1.00	0.00	11.12%	100%	
2045	0	0.000	1.00	0.00	11.12%	100%	
2046	0	0.000	1.00	0.00	11.12%	100%	
2047	0	0.000	1.00	0.00	11.12%	100%	
2048	0	0.000	1.00	0.00	11.12%	100%	
2049	0	0.000	1.00	0.00	11.12%	100%	
2050	0	0.000	1.00	0.00	11.12%	100%	

CUM.PROD	14.850 BBO	GIANT FIELDS			CANADA	
RESERVES	6.130 BBO	TOTAL BBO	6.34		N.AMERICA	
DISCOVERED	20.980 BBO	% OF DISC.	30%			
UNDISCOVERED	3.020 BBO	Last Giant	1984			
REMAINING	9.150 BBO				Variance Trend	
ULTIMATE	24.000 BBO				1990	+0.93%
Rept.Reserve	6.130	Ref.date(end)	1989		Future	UP

Milestones	Date	Prod Mb/d	Dep. Rate	% Produced	Forecast "A"
					A=Exponential decline
1st Peak	1974	1750	3.51%	27%	B=5% incr. to Mid-Point
2nd Peak	1988	1588	5.96%	62%	C=Other
Mid-Point	1984	1430	4.19%	50%	Pre-1930 Cum. 0
Final Decline	1990		5.96%	64%	(Added to 1930 prod)

Date	Prod. M.b/d	Prod. BBO/yr	Cum. Prod.	Remain	Deplet. Rate	% Produced	Giant	Notes & Names
1930	4	0.001	0.00	24.00	ERR	0%		
1931	6	0.002	0.00	24.00	0.01%	0%		
1932	7	0.003	0.01	23.99	0.01%	0%		
1933	9	0.003	0.01	23.99	0.01%	0%		
1934	11	0.004	0.01	23.99	0.02%	0%		
1935	13	0.005	0.02	23.98	0.02%	0%		
1936	15	0.006	0.02	23.98	0.02%	0%		
1937	17	0.006	0.03	23.97	0.03%	0%		
1938	19	0.007	0.04	23.96	0.03%	0%		
1939	22	0.008	0.05	23.95	0.03%	0%		
1940	24	0.009	0.05	23.95	0.04%	0%		
1941	28	0.010	0.06	23.94	0.04%	0%		
1942	28	0.010	0.07	23.93	0.04%	0%		
1943	28	0.010	0.08	23.92	0.04%	0%		
1944	28	0.010	0.09	23.91	0.04%	0%		
1945	22	0.008	0.10	23.90	0.03%	0%		
1946	21	0.008	0.11	23.89	0.03%	0%		
1947	20	0.007	0.12	23.88	0.03%	0%	0.50	Leduc
1948	33	0.012	0.13	23.87	0.05%	1%	0.81	Redwater
1949	58	0.021	0.15	23.85	0.09%	1%		
1950	78	0.029	0.18	23.82	0.12%	1%		
1951	130	0.048	0.23	23.77	0.20%	1%		
1952	170	0.062	0.29	23.71	0.26%	1%		
1953	270	0.099	0.39	23.61	0.42%	2%	1.80	Pembina
1954	255	0.093	0.48	23.52	0.39%	2%		
1955	354	0.129	0.61	23.39	0.55%	3%		
1956	518	0.189	0.80	23.20	0.81%	3%		
1957	495	0.181	0.98	23.02	0.78%	4%	1.01	Swan Hills
1958	450	0.164	1.14	22.86	0.71%	5%		
1959	508	0.185	1.33	22.67	0.81%	6%	0.54	Judy Creek
1960	532	0.194	1.52	22.48	0.86%	6%		
1961	616	0.225	1.75	22.25	1.00%	7%		
1962	668	0.244	1.99	22.01	1.10%	8%		
1963	717	0.262	2.25	21.75	1.19%	9%		
1964	763	0.278	2.53	21.47	1.28%	11%		
1965	788	0.288	2.82	21.18	1.34%	12%	0.67	Rainbow
1966	894	0.326	3.15	20.85	1.54%	13%		
1967	972	0.355	3.50	20.50	1.70%	15%		
1968	1023	0.373	3.87	20.13	1.82%	16%		
1969	1200	0.438	4.31	19.69	2.18%	18%		
1970	1278	0.466	4.78	19.22	2.37%	20%		
1971	1336	0.488	5.27	18.73	2.54%	22%		
1972	1490	0.544	5.81	18.19	2.90%	24%		
1973	1750	0.639	6.45	17.55	3.51%	27%		1st PEAK
1974	1682	0.614	7.06	16.94	3.50%	29%		
1975	1450	0.529	7.59	16.41	3.12%	32%		
1976	1300	0.475	8.07	15.93	2.89%	34%		
1977	1360	0.496	8.56	15.44	3.12%	36%		
1978	1300	0.475	9.04	14.96	3.07%	38%		
1979	1480	0.540	9.58	14.42	3.61%	40%	0.52	Hibernia
1980	1470	0.537	10.11	13.89	3.72%	42%		
1981	1287	0.470	10.58	13.42	3.38%	44%		
1982	1233	0.450	11.03	12.97	3.35%	46%		
1983	1396	0.510	11.54	12.46	3.93%	48%		
1984	1430	0.522	12.07	11.93	4.19%	50%	0.50	Amauligak
1985	1453	0.530	12.60	11.40	4.44%	52%		50% PRODUCED
1986	1475	0.538	13.13	10.87	4.72%	55%		
1987	1508	0.550	13.68	10.32	5.07%	57%		
1988	1605	0.586	14.27	9.73	5.68%	59%		

1989	1588	0.580	14.85	9.15	5.96%	62%
1990	1494	0.545	15.39	8.61	5.96%	64%
1991	1405	0.513	15.91	8.09	5.96%	66%
1992	1321	0.482	16.39	7.61	5.96%	68%
1993	1243	0.454	16.84	7.16	5.96%	70%
1994	1169	0.427	17.27	6.73	5.96%	72%
1995	1099	0.401	17.67	6.33	5.96%	74%
1996	1033	0.377	18.05	5.95	5.96%	75%
1997	972	0.355	18.40	5.60	5.96%	77%
1998	914	0.334	18.74	5.26	5.96%	78%
1999	859	0.314	19.05	4.95	5.96%	79%
2000	808	0.295	19.35	4.65	5.96%	81%
2001	760	0.277	19.62	4.38	5.96%	82%
2002	715	0.261	19.88	4.12	5.96%	83%
2003	672	0.245	20.13	3.87	5.96%	84%
2004	632	0.231	20.36	3.64	5.96%	85%
2005	594	0.217	20.58	3.42	5.96%	86%
2006	559	0.204	20.78	3.22	5.96%	87%
2007	526	0.192	20.97	3.03	5.96%	87%
2008	494	0.180	21.15	2.85	5.96%	88%
2009	465	0.170	21.32	2.68	5.96%	89%
2010	437	0.160	21.48	2.52	5.96%	90%
2011	411	0.150	21.63	2.37	5.96%	90%
2012	387	0.141	21.77	2.23	5.96%	91%
2013	364	0.133	21.91	2.09	5.96%	91%
2014	342	0.125	22.03	1.97	5.96%	92%
2015	322	0.117	22.15	1.85	5.96%	92%
2016	302	0.110	22.26	1.74	5.96%	93%
2017	284	0.104	22.36	1.64	5.96%	93%
2018	267	0.098	22.46	1.54	5.96%	94%
2019	251	0.092	22.55	1.45	5.96%	94%
2020	236	0.086	22.64	1.36	5.96%	94%
2021	222	0.081	22.72	1.28	5.96%	95%
2022	209	0.076	22.80	1.20	5.96%	95%
2023	197	0.072	22.87	1.13	5.96%	95%
2024	185	0.068	22.93	1.07	5.96%	96%
2025	174	0.063	23.00	1.00	5.96%	96%
2026	164	0.060	23.06	0.94	5.96%	96%
2027	154	0.056	23.11	0.89	5.96%	96%
2028	145	0.053	23.17	0.83	5.96%	97%
2029	136	0.050	23.22	0.78	5.96%	97%
2030	128	0.047	23.26	0.74	5.96%	97%
2031	120	0.044	23.31	0.69	5.96%	97%
2032	113	0.041	23.35	0.65	5.96%	97%
2033	106	0.039	23.39	0.61	5.96%	97%
2034	100	0.037	23.42	0.58	5.96%	98%
2035	94	0.034	23.46	0.54	5.96%	98%
2036	88	0.032	23.49	0.51	5.96%	98%
2037	83	0.030	23.52	0.48	5.96%	98%
2038	78	0.029	23.55	0.45	5.96%	98%
2039	74	0.027	23.58	0.42	5.96%	98%
2040	69	0.025	23.60	0.40	5.96%	98%
2041	65	0.024	23.63	0.37	5.96%	98%
2042	61	0.022	23.65	0.35	5.96%	99%
2043	58	0.021	23.67	0.33	5.96%	99%
2044	54	0.020	23.69	0.31	5.96%	99%
2045	51	0.019	23.71	0.29	5.96%	99%
2046	48	0.017	23.72	0.28	5.96%	99%
2047	45	0.016	23.74	0.26	5.96%	99%
2048	42	0.015	23.76	0.24	5.96%	99%
2049	40	0.015	23.77	0.23	5.96%	99%
2050	37	0.014	23.78	0.22	5.96%	99%

CUM.PROD	0.366 BBO	GIANT FIELDS			CHILE	
RESERVES	0.300 BBO	TOTAL BBO		0	L.AMERICA	
DISCOVERED	0.666 BBO	% OF DISC.		0%		
UNDISCOVERED	0.084 BBO	Last Giant			----------------	
REMAINING	0.384 BBO	-------------------			Variance	Trend
ULTIMATE	0.750 BBO				1990	-12%
Rept.Reserves	0.300 BBO	Ref. Date (end)		1989	Future	DOWN

Milestones	Date	Prod Mb/d	Dep. Rate	% Produced	Forecast "A"
					A=Exponential decline
1st Peak	1968	38	2.19%	18%	B=5% inc. to Mid-Point
2nd Peak	1982	41	3.12%	38%	C=Other
Mid-Point	1990	23	2.20%	50%	Pre-1930 Cum 0
Final Decline	1990		2.20%		(added to 1930 Prod)

Date	Prod. M.b/d	Prod. BBO/yr	Cum Prod	Remain	Dep. Rate	% Produced	Giant Names % Notes
1930		0.000	0.00	0.75	ERR	0%	
1931		0.000	0.00	0.75	0.00%	0%	
1932		0.000	0.00	0.75	0.00%	0%	
1933		0.000	0.00	0.75	0.00%	0%	
1934		0.000	0.00	0.75	0.00%	0%	
1935		0.000	0.00	0.75	0.00%	0%	
1936		0.000	0.00	0.75	0.00%	0%	
1937		0.000	0.00	0.75	0.00%	0%	
1938		0.000	0.00	0.75	0.00%	0%	
1939		0.000	0.00	0.75	0.00%	0%	
1940		0.000	0.00	0.75	0.00%	0%	
1941		0.000	0.00	0.75	0.00%	0%	
1942		0.000	0.00	0.75	0.00%	0%	
1943		0.000	0.00	0.75	0.00%	0%	
1944		0.000	0.00	0.75	0.00%	0%	
1945		0.000	0.00	0.75	0.00%	0%	
1946		0.000	0.00	0.75	0.00%	0%	
1947		0.000	0.00	0.75	0.00%	0%	
1948		0.000	0.00	0.75	0.00%	0%	
1949		0.000	0.00	0.75	0.00%	0%	
1950		0.000	0.00	0.75	0.00%	0%	
1951		0.000	0.00	0.75	0.00%	0%	
1952		0.000	0.00	0.75	0.00%	0%	
1953	3	0.001	0.00	0.75	0.15%	0%	
1954	5	0.002	0.00	0.75	0.22%	0%	
1955	7	0.003	0.01	0.74	0.35%	1%	
1956	10	0.004	0.01	0.74	0.47%	1%	
1957	12	0.004	0.01	0.74	0.58%	2%	
1958	15	0.005	0.02	0.73	0.74%	2%	
1959	18	0.006	0.03	0.72	0.88%	3%	
1960	20	0.007	0.03	0.72	1.00%	4%	
1961	25	0.009	0.04	0.71	1.28%	6%	
1962	32	0.012	0.05	0.70	1.65%	7%	
1963	36	0.013	0.07	0.68	1.89%	9%	
1964	38	0.014	0.08	0.67	2.01%	11%	
1965	34	0.012	0.09	0.66	1.86%	12%	
1966	35	0.013	0.11	0.64	1.92%	14%	
1967	35	0.013	0.12	0.63	1.98%	16%	
1968	38	0.014	0.13	0.62	2.19%	18%	1ST PEAK
1969	37	0.014	0.15	0.60	2.20%	19%	
1970	38	0.014	0.16	0.59	2.31%	21%	
1971	34	0.013	0.17	0.58	2.13%	23%	
1972	33	0.012	0.18	0.57	2.05%	25%	
1973	30	0.011	0.19	0.56	1.93%	26%	
1974	29	0.010	0.21	0.54	1.88%	27%	
1975	25	0.009	0.21	0.54	1.67%	29%	
1976	24	0.009	0.22	0.53	1.64%	30%	
1977	22	0.008	0.23	0.52	1.52%	31%	
1978	20	0.007	0.24	0.51	1.41%	32%	
1979	18	0.007	0.25	0.50	1.28%	33%	
1980	30	0.011	0.26	0.49	2.17%	34%	
1981	40	0.015	0.27	0.48	2.96%	36%	
1982	41	0.015	0.29	0.46	3.12%	38%	2ND PEAK
1983	39	0.014	0.30	0.45	3.06%	40%	
1984	38	0.014	0.31	0.44	3.08%	42%	
1985	33	0.012	0.33	0.42	2.72%	43%	
1986	34	0.013	0.34	0.41	2.96%	45%	
1987	28	0.010	0.35	0.40	2.51%	46%	
1988	25	0.009	0.36	0.39	2.25%	48%	

1989	24	0.009	0.37	0.38	2.20%	49%	
1990	23	0.008	0.37	0.38	2.20%	50%	50% PRODUCED
1991	23	0.008	0.38	0.37	2.20%	51%	
1992	22	0.008	0.39	0.36	2.20%	52%	
1993	22	0.008	0.40	0.35	2.20%	53%	
1994	21	0.008	0.41	0.34	2.20%	54%	
1995	21	0.008	0.41	0.34	2.20%	55%	
1996	20	0.007	0.42	0.33	2.20%	56%	
1997	20	0.007	0.43	0.32	2.20%	57%	
1998	19	0.007	0.44	0.31	2.20%	58%	
1999	19	0.007	0.44	0.31	2.20%	59%	
2000	19	0.007	0.45	0.30	2.20%	60%	
2001	18	0.007	0.46	0.29	2.20%	61%	
2002	18	0.006	0.46	0.29	2.20%	62%	
2003	17	0.006	0.47	0.28	2.20%	63%	
2004	17	0.006	0.48	0.27	2.20%	63%	
2005	17	0.006	0.48	0.27	2.20%	64%	
2006	16	0.006	0.49	0.26	2.20%	65%	
2007	16	0.006	0.49	0.26	2.20%	66%	
2008	16	0.006	0.50	0.25	2.20%	66%	
2009	15	0.006	0.50	0.25	2.20%	67%	
2010	15	0.005	0.51	0.24	2.20%	68%	
2011	15	0.005	0.51	0.24	2.20%	69%	
2012	14	0.005	0.52	0.23	2.20%	69%	
2013	14	0.005	0.52	0.23	2.20%	70%	
2014	14	0.005	0.53	0.22	2.20%	71%	
2015	13	0.005	0.53	0.22	2.20%	71%	
2016	13	0.005	0.54	0.21	2.20%	72%	
2017	13	0.005	0.54	0.21	2.20%	73%	
2018	12	0.005	0.55	0.20	2.20%	73%	
2019	12	0.004	0.55	0.20	2.20%	74%	
2020	12	0.004	0.56	0.19	2.20%	74%	
2021	12	0.004	0.56	0.19	2.20%	75%	
2022	11	0.004	0.57	0.18	2.20%	75%	
2023	11	0.004	0.57	0.18	2.20%	76%	
2024	11	0.004	0.57	0.18	2.20%	77%	
2025	11	0.004	0.58	0.17	2.20%	77%	
2026	10	0.004	0.58	0.17	2.20%	78%	
2027	10	0.004	0.59	0.16	2.20%	78%	
2028	10	0.004	0.59	0.16	2.20%	79%	
2029	10	0.004	0.59	0.16	2.20%	79%	
2030	10	0.003	0.60	0.15	2.20%	79%	
2031	9	0.003	0.60	0.15	2.20%	80%	
2032	9	0.003	0.60	0.15	2.20%	80%	
2033	9	0.003	0.61	0.14	2.20%	81%	
2034	9	0.003	0.61	0.14	2.20%	81%	
2035	9	0.003	0.61	0.14	2.20%	82%	
2036	8	0.003	0.62	0.13	2.20%	82%	
2037	8	0.003	0.62	0.13	2.20%	82%	
2038	8	0.003	0.62	0.13	2.20%	83%	
2039	8	0.003	0.62	0.13	2.20%	83%	
2040	8	0.003	0.63	0.12	2.20%	84%	
2041	7	0.003	0.63	0.12	2.20%	84%	
2042	7	0.003	0.63	0.12	2.20%	84%	
2043	7	0.003	0.63	0.12	2.20%	85%	
2044	7	0.003	0.64	0.11	2.20%	85%	
2045	7	0.002	0.64	0.11	2.20%	85%	
2046	7	0.002	0.64	0.11	2.20%	86%	
2047	7	0.002	0.64	0.11	2.20%	86%	
2048	6	0.002	0.65	0.10	2.20%	86%	
2049	6	0.002	0.65	0.10	2.20%	87%	
2050	6	0.002	0.65	0.10	2.20%	87%	

CUM.PROD	14.442	BBO	GIANT FIELDS			CHINA
RESERVES	24.000	BBO	TOTAL BBO	9.96		SINO-SOVIET
DISCOVERED	38.442	BBO	% OF DISC.	26%		
UNDISCOVERED	6.558	BBO	Last Giant	1959		----------------
REMAINING	30.558	BBO	-------------------			Varience Trend
ULTIMATE	45.000	BBO				1990 -5%
Rept.Reserve	24.000		Ref.date(end)	1989		Future DOWN

Milestones	Date	Prod Mb/d	Dep. Rate	% Produced	Forecast "B"
					A=Exponential decline
1st Peak	1996	3513	5.39%	51%	B=5% incr. to Mid-Point
2nd Peak	None				C=Other
Mid-Point	1996	3513	5.39%	51%	Pre-1930 Cum. 0
Final Decline	1996		5.39%	51%	(Added to 1930 prod)

Date	Prod. M.b/d	Prod. BBO/yr	Cum. Prod.	Remain	Deplet. Rate	% Produced	Giant	Notes & Names
1930		0.000	0.00	45.00	ERR	0%		
1931		0.000	0.00	45.00	0.00%	0%		
1932		0.000	0.00	45.00	0.00%	0%		
1933		0.000	0.00	45.00	0.00%	0%		
1934		0.000	0.00	45.00	0.00%	0%		
1935		0.000	0.00	45.00	0.00%	0%		
1936		0.000	0.00	45.00	0.00%	0%		
1937		0.000	0.00	45.00	0.00%	0%		
1938		0.000	0.00	45.00	0.00%	0%		
1939	0	0.000	0.00	45.00	0.00%	0%		
1940	0	0.000	0.00	45.00	0.00%	0%		
1941	0	0.000	0.00	45.00	0.00%	0%		
1942	1	0.000	0.00	45.00	0.00%	0%		
1943	1	0.000	0.00	45.00	0.00%	0%		
1944	1	0.001	0.00	45.00	0.00%	0%		
1945	1	0.000	0.00	45.00	0.00%	0%		
1946	1	0.001	0.00	45.00	0.00%	0%		
1947	1	0.000	0.00	45.00	0.00%	0%		
1948	2	0.001	0.00	45.00	0.00%	0%		
1949	2	0.001	0.00	45.00	0.00%	0%		
1950	2	0.001	0.00	45.00	0.00%	0%	0.52	Ya-erh-Hsia
1951	2	0.001	0.01	44.99	0.00%	0%		
1952	3	0.001	0.01	44.99	0.00%	0%		
1953	4	0.001	0.01	44.99	0.00%	0%		
1954	8	0.003	0.01	44.99	0.01%	0%		
1955	10	0.004	0.01	44.99	0.01%	0%	1.73	Karamay
1956	13	0.005	0.02	44.98	0.01%	0%		
1957	14	0.005	0.02	44.98	0.01%	0%		
1958	16	0.006	0.03	44.97	0.01%	0%	1.70	Leng-hu,Lung-nu
1959	42	0.015	0.05	44.95	0.03%	0%	6.00	Tach'ing
1960	62	0.023	0.52	44.93	0.05%	1%		
1961	76	0.028	0.55	44.90	0.06%	1%		
1962	87	0.032	0.58	44.87	0.07%	1%		
1963	96	0.035	0.61	44.84	0.08%	1%		
1964	103	0.038	0.65	44.80	0.08%	1%		
1965	135	0.049	0.70	44.75	0.11%	2%		
1966	205	0.075	0.78	44.68	0.17%	2%		
1967	168	0.061	0.84	44.61	0.14%	2%		
1968	207	0.076	0.91	44.54	0.17%	2%		
1969	206	0.075	0.99	44.46	0.17%	2%		
1970	400	0.146	1.13	44.32	0.33%	3%		
1971	460	0.168	1.30	44.15	0.38%	3%		
1972	592	0.216	1.52	43.93	0.49%	3%		
1973	1000	0.365	1.88	43.57	0.83%	4%		
1974	1300	0.475	2.36	43.09	1.09%	5%		
1975	1566	0.572	2.93	42.52	1.33%	7%		
1976	1769	0.646	3.58	41.88	1.52%	8%		
1977	1790	0.653	4.23	41.22	1.56%	9%		
1978	2005	0.732	4.96	40.49	1.78%	11%		
1979	2123	0.775	5.74	39.72	1.91%	13%		
1980	2122	0.775	6.51	38.94	1.95%	14%		
1981	2022	0.738	7.25	38.20	1.90%	16%		
1982	2040	0.745	7.99	37.46	1.95%	18%		
1983	2121	0.774	8.77	36.69	2.07%	19%		
1984	2292	0.837	9.60	35.85	2.28%	21%		
1985	2497	0.911	10.51	34.94	2.54%	23%		
1986	2620	0.956	11.47	33.98	2.74%	25%		
1987	2680	0.978	12.45	33.00	2.88%	28%		
1988	2690	0.982	13.43	32.02	2.98%	30%		

1989	2770	1.011	14.44	31.01	3.16%	32%	
1990	2909	1.062	15.50	29.95	3.42%	34%	
1991	3054	1.115	16.62	28.83	3.72%	37%	
1992	3207	1.170	17.79	27.66	4.06%	40%	
1993	3367	1.229	19.02	26.43	4.44%	42%	
1994	3535	1.290	20.31	25.14	4.88%	45%	
1995	3712	1.355	21.66	23.79	5.39%	48%	1ST PEAK
1996	3513	1.282	22.95	22.51	5.39%	51%	50% PRODUCED
1997	3324	1.213	24.16	21.29	5.39%	54%	
1998	3144	1.148	25.31	20.15	5.39%	56%	
1999	2975	1.086	26.39	19.06	5.39%	59%	
2000	2815	1.027	27.42	18.03	5.39%	61%	
2001	2663	0.972	28.39	17.06	5.39%	63%	
2002	2519	0.920	29.31	16.14	5.39%	65%	
2003	2384	0.870	30.18	15.27	5.39%	67%	
2004	2255	0.823	31.00	14.45	5.39%	69%	
2005	2134	0.779	31.78	13.67	5.39%	71%	
2006	2019	0.737	32.52	12.93	5.39%	72%	
2007	1910	0.697	33.22	12.24	5.39%	74%	
2008	1807	0.659	33.88	11.58	5.39%	75%	
2009	1709	0.624	34.50	10.95	5.39%	77%	
2010	1617	0.590	35.09	10.36	5.39%	78%	
2011	1530	0.558	35.65	9.80	5.39%	79%	
2012	1448	0.528	36.18	9.27	5.39%	80%	
2013	1370	0.500	36.68	8.77	5.39%	82%	
2014	1296	0.473	37.15	8.30	5.39%	83%	
2015	1226	0.447	37.60	7.85	5.39%	84%	
2016	1160	0.423	38.02	7.43	5.39%	84%	
2017	1097	0.401	38.42	7.03	5.39%	85%	
2018	1038	0.379	38.80	6.65	5.39%	86%	
2019	982	0.359	39.16	6.29	5.39%	87%	
2020	929	0.339	39.50	5.95	5.39%	88%	
2021	879	0.321	39.82	5.63	5.39%	88%	
2022	832	0.304	40.12	5.33	5.39%	89%	
2023	787	0.287	40.41	5.04	5.39%	90%	
2024	745	0.272	40.68	4.77	5.39%	90%	
2025	704	0.257	40.94	4.51	5.39%	91%	
2026	666	0.243	41.18	4.27	5.39%	92%	
2027	631	0.230	41.41	4.04	5.39%	92%	
2028	597	0.218	41.63	3.82	5.39%	93%	
2029	564	0.206	41.84	3.62	5.39%	93%	
2030	534	0.195	42.03	3.42	5.39%	93%	
2031	505	0.184	42.22	3.24	5.39%	94%	
2032	478	0.174	42.39	3.06	5.39%	94%	
2033	452	0.165	42.55	2.90	5.39%	95%	
2034	428	0.156	42.71	2.74	5.39%	95%	
2035	405	0.148	42.86	2.59	5.39%	95%	
2036	383	0.140	43.00	2.45	5.39%	96%	
2037	362	0.132	43.13	2.32	5.39%	96%	
2038	343	0.125	43.26	2.20	5.39%	96%	
2039	324	0.118	43.37	2.08	5.39%	96%	
2040	307	0.112	43.49	1.97	5.39%	97%	
2041	290	0.106	43.59	1.86	5.39%	97%	
2042	275	0.100	43.69	1.76	5.39%	97%	
2043	260	0.095	43.79	1.66	5.39%	97%	
2044	246	0.090	43.88	1.58	5.39%	98%	
2045	233	0.085	43.96	1.49	5.39%	98%	
2046	220	0.080	44.04	1.41	5.39%	98%	
2047	208	0.076	44.12	1.33	5.39%	98%	
2048	197	0.072	44.19	1.26	5.39%	98%	
2049	186	0.068	44.26	1.19	5.39%	98%	
2050	176	0.064	44.32	1.13	5.39%	98%	

```
CUM.PROD          2.993 BBO    GIANT FIELDS                    COLOMBIA
RESERVES          2.060 BBO    TOTAL BBO         1.7           L.AMERICA
DISCOVERED        5.053 BBO    % OF DISC.        34%
UNDISCOVERED      0.197 BBO    Last Giant     1983             ----------------
REMAINING         2.257 BBO    -------------------             Variance Trend
ULTIMATE          5.250 BBO                                    1990      +17%
Rept.Reserves     2.060 BBO    Ref. Date (end)        1989     Future    DOWN
-------------------------------------------------------------------------------
Milestones      Date   Prod    Dep.       %                    Forecast    "A"
                       Mb/d    Rate    Produced               A=Exponential decline
    1st Peak    1971    228    2.28%      32%                 B=5% inc. to Mid-Point
    2nd Peak    1989    405    6.15%      57%                 C=Other
    Mid-Point   1987    387    5.28%      52%                 Pre-1930 Cum      0.08
Final Decline   1990           6.15%                          (added to 1930 Prod)
===============================================================================
```

Date	Prod. M.b/d	Prod. BBO/yr	Cum Prod	Remain	Dep. Rate	% Produced	Giant	Names & Notes
1930	20.00	0.007	0.09	5.16	ERR	2%	0.7	La Cira-Infantas
1931	25.00	0.009	0.10	5.15	0.18%	2%		
1932	30.00	0.011	0.11	5.14	0.21%	2%		
1933	35.00	0.013	0.12	5.13	0.25%	2%		
1934	40.00	0.015	0.13	5.12	0.28%	3%		
1935	45.00	0.016	0.15	5.10	0.32%	3%		
1936	50.00	0.018	0.17	5.08	0.36%	3%		
1937	55.00	0.020	0.19	5.06	0.40%	4%		
1938	59.00	0.022	0.21	5.04	0.43%	4%		
1939	65.50	0.024	0.23	5.02	0.47%	4%		
1940	69.80	0.025	0.26	4.99	0.51%	5%		
1941	67.10	0.024	0.28	4.97	0.49%	5%		
1942	28.70	0.010	0.30	4.95	0.21%	6%		
1943	36.40	0.013	0.31	4.94	0.27%	6%		
1944	61.00	0.022	0.33	4.92	0.45%	6%		
1945	61.50	0.022	0.35	4.90	0.46%	7%		
1946	61.60	0.022	0.38	4.87	0.46%	7%		
1947	68.10	0.025	0.40	4.85	0.51%	8%		
1948	64.90	0.024	0.42	4.83	0.49%	8%		
1949	81.20	0.030	0.45	4.80	0.61%	9%		
1950	93.10	0.034	0.49	4.76	0.71%	9%		
1951	105.70	0.039	0.53	4.72	0.81%	10%		
1952	106.10	0.039	0.57	4.68	0.82%	11%		
1953	106.10	0.039	0.60	4.65	0.83%	12%		
1954	109.40	0.040	0.64	4.61	0.86%	12%		
1955	109.00	0.040	0.68	4.57	0.86%	13%		
1956	120.00	0.044	0.73	4.52	0.96%	14%		
1957	124.60	0.045	0.77	4.48	1.01%	15%		
1958	132.10	0.048	0.82	4.43	1.08%	16%		
1959	145.00	0.053	0.87	4.38	1.20%	17%		
1960	151.00	0.055	0.93	4.32	1.26%	18%		
1961	146.00	0.053	0.98	4.27	1.23%	19%		
1962	142.00	0.052	1.03	4.22	1.21%	20%		
1963	167.00	0.061	1.10	4.15	1.45%	21%		
1964	165.50	0.060	1.16	4.09	1.45%	22%		
1965	198.00	0.072	1.23	4.02	1.77%	23%		
1966	197.80	0.072	1.30	3.95	1.80%	25%		
1967	196.60	0.072	1.37	3.88	1.82%	26%		
1968	179.80	0.066	1.44	3.81	1.69%	27%		
1969	211.30	0.077	1.51	3.74	2.02%	29%		
1970	222.00	0.081	1.60	3.65	2.17%	30%		
1971	228.00	0.083	1.68	3.57	2.28%	32%		1ST PEAK
1972	192.00	0.070	1.75	3.50	1.96%	33%		
1973	186.00	0.068	1.82	3.43	1.94%	35%		
1974	169.70	0.062	1.88	3.37	1.80%	36%		
1975	160.00	0.058	1.94	3.31	1.73%	37%		
1976	146.00	0.053	1.99	3.26	1.61%	38%		
1977	140.00	0.051	2.04	3.21	1.57%	39%		
1978	130.00	0.047	2.09	3.16	1.48%	40%		
1979	125.00	0.046	2.13	3.12	1.44%	41%		
1980	125.00	0.046	2.18	3.07	1.46%	42%		
1981	125.00	0.046	2.23	3.02	1.49%	42%		
1982	140.00	0.051	2.28	2.97	1.69%	43%		
1983	155.00	0.057	2.33	2.92	1.90%	44%	1	Canon Limon
1984	165.00	0.060	2.39	2.86	2.07%	46%		
1985	178.00	0.065	2.46	2.79	2.27%	47%		
1986	324.00	0.118	2.58	2.67	4.24%	49%		
1987	387.00	0.141	2.72	2.53	5.28%	52%		50% PRODUCED
1988	346.00	0.126	2.84	2.41	4.99%	54%		

1989	405.00	0.148	2.99	2.26	6.15%	57%	2ND PEAK
1990	380.36	0.139	3.13	2.12	6.15%	60%	
1991	356.97	0.130	3.26	1.99	6.15%	62%	
1992	335.02	0.122	3.38	1.87	6.15%	64%	
1993	314.41	0.115	3.50	1.75	6.15%	67%	
1994	295.08	0.108	3.61	1.64	6.15%	69%	
1995	276.93	0.101	3.71	1.54	6.15%	71%	
1996	259.90	0.095	3.80	1.45	6.15%	72%	
1997	243.92	0.089	3.89	1.36	6.15%	74%	
1998	228.91	0.084	3.97	1.28	6.15%	76%	
1999	214.84	0.078	4.05	1.20	6.15%	77%	
2000	201.62	0.074	4.13	1.12	6.15%	79%	
2001	189.22	0.069	4.20	1.05	6.15%	80%	
2002	177.59	0.065	4.26	0.99	6.15%	81%	
2003	166.67	0.061	4.32	0.93	6.15%	82%	
2004	156.42	0.057	4.38	0.87	6.15%	83%	
2005	146.80	0.054	4.43	0.82	6.15%	84%	
2006	137.77	0.050	4.48	0.77	6.15%	85%	
2007	129.30	0.047	4.53	0.72	6.15%	86%	
2008	121.34	0.044	4.57	0.68	6.15%	87%	
2009	113.88	0.042	4.62	0.63	6.15%	88%	
2010	106.88	0.039	4.65	0.60	6.15%	89%	
2011	100.30	0.037	4.69	0.56	6.15%	89%	
2012	94.14	0.034	4.73	0.52	6.15%	90%	
2013	88.35	0.032	4.76	0.49	6.15%	91%	
2014	82.91	0.030	4.79	0.46	6.15%	91%	
2015	77.81	0.028	4.82	0.43	6.15%	92%	
2016	73.03	0.027	4.84	0.41	6.15%	92%	
2017	68.54	0.025	4.87	0.38	6.15%	93%	
2018	64.32	0.023	4.89	0.36	6.15%	93%	
2019	60.37	0.022	4.91	0.34	6.15%	94%	
2020	56.65	0.021	4.93	0.32	6.15%	94%	
2021	53.17	0.019	4.95	0.30	6.15%	94%	
2022	49.90	0.018	4.97	0.28	6.15%	95%	
2023	46.83	0.017	4.99	0.26	6.15%	95%	
2024	43.95	0.016	5.01	0.24	6.15%	95%	
2025	41.25	0.015	5.02	0.23	6.15%	96%	
2026	38.71	0.014	5.03	0.22	6.15%	96%	
2027	36.33	0.013	5.05	0.20	6.15%	96%	
2028	34.10	0.012	5.06	0.19	6.15%	96%	
2029	32.00	0.012	5.07	0.18	6.15%	97%	
2030	30.03	0.011	5.08	0.17	6.15%	97%	
2031	28.18	0.010	5.09	0.16	6.15%	97%	
2032	26.45	0.010	5.10	0.15	6.15%	97%	
2033	24.82	0.009	5.11	0.14	6.15%	97%	
2034	23.30	0.009	5.12	0.13	6.15%	98%	
2035	21.86	0.008	5.13	0.12	6.15%	98%	
2036	20.52	0.007	5.14	0.11	6.15%	98%	
2037	19.26	0.007	5.14	0.11	6.15%	98%	
2038	18.07	0.007	5.15	0.10	6.15%	98%	
2039	16.96	0.006	5.16	0.09	6.15%	98%	
2040	15.92	0.006	5.16	0.09	6.15%	98%	
2041	14.94	0.005	5.17	0.08	6.15%	98%	
2042	14.02	0.005	5.17	0.08	6.15%	99%	
2043	13.16	0.005	5.18	0.07	6.15%	99%	
2044	12.35	0.005	5.18	0.07	6.15%	99%	
2045	11.59	0.004	5.19	0.06	6.15%	99%	
2046	10.88	0.004	5.19	0.06	6.15%	99%	
2047	10.21	0.004	5.19	0.06	6.15%	99%	
2048	9.58	0.003	5.20	0.05	6.15%	99%	
2049	8.99	0.003	5.20	0.05	6.15%	99%	
2050	8.44	0.003	5.20	0.05	6.15%	99%	

CUM.PROD	0.503 BBO	GIANT FIELDS		CONGO	
RESERVES	0.830 BBO	TOTAL BBO	1.0	AFRICA	
DISCOVERED	1.333 BBO	% OF DISC.	75%		
UNDISCOVERED	0.667 BBO	Last Giant	1970		
REMAINING	1.497 BBO			Variance	Trend
ULTIMATE	2.000 BBO			1990	+2%
Rept.Reserves	0.830 BBO	Ref.Date (end)	1989	Future	UP

Milestones	Date	Prod Mb/d	Dep Rate	% Produced
1st Peak	1996	211	6.97%	49%
2nd Peak				
Mid-Point	1997	196	6.97%	52%
Final Decline	1997		6.97%	

Forecast "B"
A=Exponential decline
B=5% inc. to Mid-Point
C=Other
Pre-1930 Cum 0
(Added to 1930 Prod)

Date	Prod. M.b/d	Prod. BBO/yr	Cum Prod	Remain	Dep. Rate	% Produced	Giant	Names & Notes
1930		0.000	0.00	2.00	ERR	0%		
1931		0.000	0.00	2.00	0.00%	0%		
1932		0.000	0.00	2.00	0.00%	0%		
1933		0.000	0.00	2.00	0.00%	0%		
1934		0.000	0.00	2.00	0.00%	0%		
1935		0.000	0.00	2.00	0.00%	0%		
1936		0.000	0.00	2.00	0.00%	0%		
1937		0.000	0.00	2.00	0.00%	0%		
1938		0.000	0.00	2.00	0.00%	0%		
1939		0.000	0.00	2.00	0.00%	0%		
1940		0.000	0.00	2.00	0.00%	0%		
1941		0.000	0.00	2.00	0.00%	0%		
1942		0.000	0.00	2.00	0.00%	0%		
1943		0.000	0.00	2.00	0.00%	0%		
1944		0.000	0.00	2.00	0.00%	0%		
1945		0.000	0.00	2.00	0.00%	0%		
1946		0.000	0.00	2.00	0.00%	0%		
1947		0.000	0.00	2.00	0.00%	0%		
1948		0.000	0.00	2.00	0.00%	0%		
1949		0.000	0.00	2.00	0.00%	0%		
1950		0.000	0.00	2.00	0.00%	0%		
1951		0.000	0.00	2.00	0.00%	0%		
1952		0.000	0.00	2.00	0.00%	0%		
1953		0.000	0.00	2.00	0.00%	0%		
1954		0.000	0.00	2.00	0.00%	0%		
1955		0.000	0.00	2.00	0.00%	0%		
1956		0.000	0.00	2.00	0.00%	0%		
1957		0.000	0.00	2.00	0.00%	0%		
1958		0.000	0.00	2.00	0.00%	0%		
1959		0.000	0.00	2.00	0.00%	0%		
1960	1.0	0.000	0.00	2.00	0.02%	0%		
1961	2.0	0.001	0.00	2.00	0.04%	0%		
1962	2.5	0.001	0.00	2.00	0.05%	0%		
1963	2.4	0.001	0.00	2.00	0.04%	0%		
1964	1.7	0.001	0.00	2.00	0.03%	0%		
1965	1.5	0.001	0.00	2.00	0.03%	0%		
1966	1.3	0.000	0.00	2.00	0.02%	0%		
1967	1.2	0.000	0.00	2.00	0.02%	0%		
1968	2.0	0.001	0.01	1.99	0.04%	0%		
1969	0.8	0.000	0.01	1.99	0.01%	0%	0.5	Emeraude
1970	0.5	0.000	0.01	1.99	0.01%	0%	0.5	Loango
1971	0.4	0.000	0.01	1.99	0.01%	0%		
1972	7.5	0.003	0.01	1.99	0.14%	0%		
1973	39.0	0.014	0.02	1.98	0.71%	1%		
1974	51.2	0.019	0.04	1.96	0.95%	2%		
1975	38.0	0.014	0.06	1.94	0.71%	3%		
1976	38.0	0.014	0.07	1.93	0.71%	3%		
1977	35.0	0.013	0.08	1.92	0.66%	4%		
1978	28.0	0.010	0.09	1.91	0.53%	5%		
1979	57.0	0.021	0.11	1.89	1.09%	6%		
1980	56.0	0.020	0.13	1.87	1.08%	7%		
1981	79.0	0.029	0.16	1.84	1.55%	8%		
1982	87.0	0.032	0.19	1.81	1.73%	10%		
1983	95.0	0.035	0.23	1.77	1.92%	11%		
1984	118.0	0.043	0.27	1.73	2.43%	14%		
1985	115.1	0.042	0.31	1.69	2.43%	16%		
1986	115.0	0.042	0.36	1.64	2.49%	18%		
1987	118.0	0.043	0.40	1.60	2.62%	20%		
1988	134.0	0.049	0.45	1.55	3.06%	22%		

1989	150.0	0.055	0.50	1.50	3.53%	25%	
1990	157.5	0.057	0.56	1.44	3.84%	28%	
1991	165.4	0.060	0.62	1.38	4.19%	31%	
1992	173.6	0.063	0.68	1.32	4.60%	34%	
1993	182.3	0.067	0.75	1.25	5.06%	38%	
1994	191.4	0.070	0.82	1.18	5.59%	41%	
1995	201.0	0.073	0.89	1.11	6.22%	45%	
1996	211.1	0.077	0.97	1.03	6.97%	49%	1ST PEAK
1997	196.5	0.072	1.04	0.96	6.97%	52%	50% PRODUCED
1998	182.8	0.067	1.11	0.89	6.97%	55%	
1999	170.0	0.062	1.17	0.83	6.97%	59%	
2000	158.2	0.058	1.23	0.77	6.97%	61%	
2001	147.2	0.054	1.28	0.72	6.97%	64%	
2002	136.9	0.050	1.33	0.67	6.97%	67%	
2003	127.4	0.046	1.38	0.62	6.97%	69%	
2004	118.5	0.043	1.42	0.58	6.97%	71%	
2005	110.2	0.040	1.46	0.54	6.97%	73%	
2006	102.5	0.037	1.50	0.50	6.97%	75%	
2007	95.4	0.035	1.54	0.46	6.97%	77%	
2008	88.8	0.032	1.57	0.43	6.97%	78%	
2009	82.6	0.030	1.60	0.40	6.97%	80%	
2010	76.8	0.028	1.63	0.37	6.97%	81%	
2011	71.5	0.026	1.65	0.35	6.97%	83%	
2012	66.5	0.024	1.68	0.32	6.97%	84%	
2013	61.8	0.023	1.70	0.30	6.97%	85%	
2014	57.5	0.021	1.72	0.28	6.97%	86%	
2015	53.5	0.020	1.74	0.26	6.97%	87%	
2016	49.8	0.018	1.76	0.24	6.97%	88%	
2017	46.3	0.017	1.77	0.23	6.97%	89%	
2018	43.1	0.016	1.79	0.21	6.97%	90%	
2019	40.1	0.015	1.80	0.20	6.97%	90%	
2020	37.3	0.014	1.82	0.18	6.97%	91%	
2021	34.7	0.013	1.83	0.17	6.97%	92%	
2022	32.3	0.012	1.84	0.16	6.97%	92%	
2023	30.0	0.011	1.85	0.15	6.97%	93%	
2024	27.9	0.010	1.86	0.14	6.97%	93%	
2025	26.0	0.009	1.87	0.13	6.97%	94%	
2026	24.2	0.009	1.88	0.12	6.97%	94%	
2027	22.5	0.008	1.89	0.11	6.97%	95%	
2028	20.9	0.008	1.90	0.10	6.97%	95%	
2029	19.5	0.007	1.91	0.09	6.97%	95%	
2030	18.1	0.007	1.91	0.09	6.97%	96%	
2031	16.8	0.006	1.92	0.08	6.97%	96%	
2032	15.7	0.006	1.92	0.08	6.97%	96%	
2033	14.6	0.005	1.93	0.07	6.97%	96%	
2034	13.6	0.005	1.93	0.07	6.97%	97%	
2035	12.6	0.005	1.94	0.06	6.97%	97%	
2036	11.7	0.004	1.94	0.06	6.97%	97%	
2037	10.9	0.004	1.95	0.05	6.97%	97%	
2038	10.2	0.004	1.95	0.05	6.97%	98%	
2039	9.5	0.003	1.95	0.05	6.97%	98%	
2040	8.8	0.003	1.96	0.04	6.97%	98%	
2041	8.2	0.003	1.96	0.04	6.97%	98%	
2042	7.6	0.003	1.96	0.04	6.97%	98%	
2043	7.1	0.003	1.97	0.03	6.97%	98%	
2044	6.6	0.002	1.97	0.03	6.97%	98%	
2045	6.1	0.002	1.97	0.03	6.97%	99%	
2046	5.7	0.002	1.97	0.03	6.97%	99%	
2047	5.3	0.002	1.97	0.03	6.97%	99%	
2048	4.9	0.002	1.98	0.02	6.97%	99%	
2049	4.6	0.002	1.98	0.02	6.97%	99%	
2050	4.3	0.002	1.98	0.02	6.97%	99%	

CUM.PROD	0.075 BBO	GIANT FIELDS			CUBA	
RESERVES	0.050 BBO	TOTAL BBO	0		L.AMERICA	
DISCOVERED	0.125 BBO	% OF DISC.	0%			
UNDISCOVERED	0.025 BBO	Last Giant				
REMAINING	0.075 BBO				Variance	Trend
ULTIMATE	0.150 BBO				1990	-13%
Rept.Reserves	0.100 BBO	Ref. Date (end)		1989	Future	DOWN

Milestones	Date	Prod Mb/d	Dep. Rate	% Produced
1st Peak	1989	19	8.51%	50%
2nd Peak				
Mid-Point	1989	19	8.51%	50%
Final Decline	1990		8.51%	

Forecast "A"
A=Exponential decline
B=5% inc. to Mid-Point
C=Other
Pre-1930 Cum 0
(added to 1930 Prod)

Date	Prod. M.b/d	Prod. BBO/yr	Cum Prod	Remain	Dep. Rate	% Produced	Giant Names & Notes
1930	0	0.000	0.00	0.15	ERR	0%	
1931	0	0.000	0.00	0.15	0.01%	0%	
1932	0	0.000	0.00	0.15	0.01%	0%	
1933	0	0.000	0.00	0.15	0.01%	0%	
1934	0	0.000	0.00	0.15	0.02%	0%	
1935	0	0.000	0.00	0.15	0.02%	0%	
1936	0	0.000	0.00	0.15	0.05%	0%	
1937	0	0.000	0.00	0.15	0.02%	0%	
1938	0	0.000	0.00	0.15	0.05%	0%	
1939	0	0.000	0.00	0.15	0.07%	0%	
1940	0	0.000	0.00	0.15	0.10%	0%	
1941	0	0.000	0.00	0.15	0.10%	0%	
1942	0	0.000	0.00	0.15	0.10%	1%	
1943	0	0.000	0.00	0.15	0.07%	1%	
1944	0	0.000	0.00	0.15	0.07%	1%	
1945	0	0.000	0.00	0.15	0.10%	1%	
1946	1	0.000	0.00	0.15	0.17%	1%	
1947	1	0.000	0.00	0.15	0.20%	1%	
1948	0	0.000	0.00	0.15	0.10%	1%	
1949	1	0.000	0.00	0.15	0.15%	1%	
1950	0	0.000	0.00	0.15	0.10%	2%	
1951	0	0.000	0.00	0.15	0.10%	2%	
1952	0	0.000	0.00	0.15	0.02%	2%	
1953	0	0.000	0.00	0.15	0.01%	2%	
1954	0	0.000	0.00	0.15	0.01%	2%	
1955	1	0.000	0.00	0.15	0.25%	2%	
1956	2	0.001	0.00	0.15	0.37%	2%	
1957	1	0.000	0.00	0.15	0.27%	3%	
1958	1	0.000	0.00	0.15	0.22%	3%	
1959	1	0.000	0.00	0.15	0.13%	3%	
1960	0	0.000	0.00	0.15	0.08%	3%	
1961	0	0.000	0.00	0.15	0.05%	3%	
1962	0	0.000	0.00	0.15	0.05%	3%	
1963	1	0.000	0.00	0.15	0.15%	3%	
1964	1	0.000	0.01	0.14	0.18%	3%	
1965	1	0.000	0.01	0.14	0.18%	4%	
1966	1	0.000	0.01	0.14	0.18%	4%	
1967	2	0.001	0.01	0.14	0.53%	4%	
1968	3	0.001	0.01	0.14	0.74%	5%	
1969	2	0.001	0.01	0.14	0.59%	5%	
1970	2	0.001	0.01	0.14	0.57%	6%	
1971	2	0.001	0.01	0.14	0.57%	7%	
1972	2	0.001	0.01	0.14	0.55%	7%	
1973	3	0.001	0.01	0.14	0.73%	8%	
1974	3	0.001	0.01	0.14	0.90%	9%	
1975	5	0.002	0.01	0.14	1.22%	10%	
1976	5	0.002	0.02	0.13	1.27%	11%	
1977	5	0.002	0.02	0.13	1.42%	12%	
1978	6	0.002	0.02	0.13	1.61%	14%	
1979	6	0.002	0.02	0.13	1.63%	15%	
1980	6	0.002	0.02	0.13	1.57%	16%	
1981	5	0.002	0.03	0.12	1.51%	18%	
1982	11	0.004	0.03	0.12	3.22%	20%	
1983	15	0.005	0.04	0.11	4.57%	24%	
1984	16	0.006	0.04	0.11	4.95%	28%	
1985	17	0.006	0.05	0.10	5.85%	32%	
1986	19	0.007	0.05	0.10	6.75%	36%	
1987	19	0.007	0.06	0.09	7.27%	41%	
1988	19	0.007	0.07	0.08	7.84%	46%	

1989	19	0.007	0.08	0.07	8.51%	50%	1ST PEAK
1990	17	0.006	0.08	0.07	8.51%	55%	& 50% PRODUCED
1991	16	0.006	0.09	0.06	8.51%	58%	
1992	15	0.005	0.09	0.06	8.51%	62%	
1993	13	0.005	0.10	0.05	8.51%	65%	
1994	12	0.004	0.10	0.05	8.51%	68%	
1995	11	0.004	0.11	0.04	8.51%	71%	
1996	10	0.004	0.11	0.04	8.51%	73%	
1997	9	0.003	0.11	0.04	8.51%	76%	
1998	9	0.003	0.12	0.03	8.51%	78%	
1999	8	0.003	0.12	0.03	8.51%	80%	
2000	7	0.003	0.12	0.03	8.51%	81%	
2001	7	0.002	0.12	0.03	8.51%	83%	
2002	6	0.002	0.13	0.02	8.51%	84%	
2003	5	0.002	0.13	0.02	8.51%	86%	
2004	5	0.002	0.13	0.02	8.51%	87%	
2005	5	0.002	0.13	0.02	8.51%	88%	
2006	4	0.002	0.13	0.02	8.51%	89%	
2007	4	0.001	0.13	0.02	8.51%	90%	
2008	4	0.001	0.14	0.01	8.51%	91%	
2009	3	0.001	0.14	0.01	8.51%	92%	
2010	3	0.001	0.14	0.01	8.51%	92%	
2011	3	0.001	0.14	0.01	8.51%	93%	
2012	2	0.001	0.14	0.01	8.51%	94%	
2013	2	0.001	0.14	0.01	8.51%	94%	
2014	2	0.001	0.14	0.01	8.51%	95%	
2015	2	0.001	0.14	0.01	8.51%	95%	
2016	2	0.001	0.14	0.01	8.51%	95%	
2017	2	0.001	0.14	0.01	8.51%	96%	
2018	1	0.001	0.14	0.01	8.51%	96%	
2019	1	0.000	0.14	0.01	8.51%	97%	
2020	1	0.000	0.15	0.00	8.51%	97%	
2021	1	0.000	0.15	0.00	8.51%	97%	
2022	1	0.000	0.15	0.00	8.51%	97%	
2023	1	0.000	0.15	0.00	8.51%	98%	
2024	1	0.000	0.15	0.00	8.51%	98%	
2025	1	0.000	0.15	0.00	8.51%	98%	
2026	1	0.000	0.15	0.00	8.51%	98%	
2027	1	0.000	0.15	0.00	8.51%	98%	
2028	1	0.000	0.15	0.00	8.51%	98%	
2029	1	0.000	0.15	0.00	8.51%	99%	
2030	0	0.000	0.15	0.00	8.51%	99%	
2031	0	0.000	0.15	0.00	8.51%	99%	
2032	0	0.000	0.15	0.00	8.51%	99%	
2033	0	0.000	0.15	0.00	8.51%	99%	
2034	0	0.000	0.15	0.00	8.51%	99%	
2035	0	0.000	0.15	0.00	8.51%	99%	
2036	0	0.000	0.15	0.00	8.51%	99%	
2037	0	0.000	0.15	0.00	8.51%	99%	
2038	0	0.000	0.15	0.00	8.51%	99%	
2039	0	0.000	0.15	0.00	8.51%	99%	
2040	0	0.000	0.15	0.00	8.51%	99%	
2041	0	0.000	0.15	0.00	8.51%	100%	
2042	0	0.000	0.15	0.00	8.51%	100%	
2043	0	0.000	0.15	0.00	8.51%	100%	
2044	0	0.000	0.15	0.00	8.51%	100%	
2045	0	0.000	0.15	0.00	8.51%	100%	
2046	0	0.000	0.15	0.00	8.51%	100%	
2047	0	0.000	0.15	0.00	8.51%	100%	
2048	0	0.000	0.15	0.00	8.51%	100%	
2049	0	0.000	0.15	0.00	8.51%	100%	
2050	0	0.000	0.15	0.00	8.51%	100%	

Reserves
Very uncertain and approximate estimate

CUM.PROD	0.042	BBO	GIANT FIELDS		CZECKOSLOVAKIA	
RESERVES	0.043	BBO	TOTAL BBO	0	SINO-SOVIET	
DISCOVERED	0.085	BBO	% OF DISC.	0%		
UNDISCOVERED	0.015	BBO	Last Giant	-		
REMAINING	0.058	BBO			Variance Trend	
ULTIMATE	0.100	BBO			1990	n/a
Rept.Reserve	0.040	#1	Ref.date(end) 1989		Future	UP

Milestones	Date	Prod Mb/d	Dep. Rate	% Produced	Forecast "B"
1st Peak	1970	4	1.86%	25%	A=Exponential decline
2nd Peak	None				B=5% incr. to Mid-Point
					C=Other
Mid-Point	1997	3	1.95%	50%	Pre-1930 Cum. 0
Final Decline	1998		1.95%	51%	(Added to 1930 prod)

Date	Prod. M.b/d	Prod. BBO/yr	Cum. Prod.	Remain	Deplet. Rate	% Produced	Giant	Notes & Names
1930	0	0.000	0.00	0.10	ERR	0%		Data-base weak
1931	0	0.000	0.00	0.10	0.15%	0%		
1932	0	0.000	0.00	0.10	0.11%	0%		
1933	0	0.000	0.00	0.10	0.11%	1%		
1934	1	0.000	0.00	0.10	0.18%	1%		
1935	0	0.000	0.00	0.10	0.15%	1%		
1936	0	0.000	0.00	0.10	0.11%	1%		
1937	0	0.000	0.00	0.10	0.11%	1%		
1938	0	0.000	0.00	0.10	0.15%	1%		
1939	0	0.000	0.00	0.10	0.11%	1%		
1940	0	0.000	0.00	0.10	0.15%	1%		
1941	1	0.000	0.00	0.10	0.19%	2%		
1942	1	0.000	0.00	0.10	0.26%	2%		
1943	1	0.000	0.00	0.10	0.19%	2%		
1944	1	0.000	0.00	0.10	0.19%	2%		
1945	0	0.000	0.00	0.10	0.07%	2%		
1946	1	0.000	0.00	0.10	0.19%	3%		
1947	1	0.000	0.00	0.10	0.22%	3%		
1948	1	0.000	0.00	0.10	0.23%	3%		
1949	1	0.000	0.00	0.10	0.30%	3%		
1950	1	0.000	0.00	0.10	0.30%	4%		
1951	2	0.001	0.00	0.10	0.64%	4%		
1952	2	0.001	0.00	0.10	0.76%	5%		
1953	4	0.001	0.01	0.09	1.38%	6%		
1954	3	0.001	0.01	0.09	1.17%	7%		
1955	3	0.001	0.01	0.09	1.02%	8%		
1956	2	0.001	0.01	0.09	0.80%	9%		
1957	2	0.001	0.01	0.09	0.80%	10%		
1958	2	0.001	0.01	0.09	0.81%	10%		
1959	2	0.001	0.01	0.09	0.94%	11%		
1960	3	0.001	0.01	0.09	1.03%	12%		
1961	3	0.001	0.01	0.09	1.21%	13%		
1962	3	0.001	0.01	0.09	1.39%	14%		
1963	3	0.001	0.02	0.08	1.41%	16%		
1964	4	0.001	0.02	0.08	1.56%	17%		
1965	4	0.001	0.02	0.08	1.67%	18%		
1966	4	0.001	0.02	0.08	1.70%	20%		
1967	4	0.001	0.02	0.08	1.59%	21%		
1968	4	0.001	0.02	0.08	1.62%	22%		
1969	4	0.001	0.02	0.08	1.74%	24%		
1970	4	0.001	0.03	0.07	1.86%	25%		1ST PEAK
1971	4	0.001	0.03	0.07	1.80%	26%		
1972	4	0.001	0.03	0.07	1.79%	28%		
1973	3	0.001	0.03	0.07	1.67%	29%		
1974	3	0.001	0.03	0.07	1.54%	30%		
1975	3	0.001	0.03	0.07	1.46%	31%		
1976	3	0.001	0.03	0.07	1.38%	32%		
1977	2	0.001	0.03	0.07	1.07%	33%		
1978	2	0.001	0.03	0.07	1.09%	33%		
1979	2	0.001	0.03	0.07	1.32%	34%		
1980	2	0.001	0.04	0.06	1.06%	35%		
1981	2	0.001	0.04	0.06	0.96%	36%		
1982	2	0.001	0.04	0.06	1.02%	36%		
1983	2	0.001	0.04	0.06	1.03%	37%		
1984	2	0.001	0.04	0.06	0.98%	38%		
1985	2	0.001	0.04	0.06	1.35%	38%		
1986	3	0.001	0.04	0.06	1.66%	39%		
1987	3	0.001	0.04	0.06	1.75%	41%		
1988	3	0.001	0.04	0.06	1.72%	42%		

1989	3	0.001	0.04	0.06	1.62%	42%
1990	3	0.001	0.04	0.06	1.66%	43%
1991	3	0.001	0.04	0.06	1.69%	44%
1992	3	0.001	0.05	0.05	1.73%	45%
1993	3	0.001	0.05	0.05	1.77%	46%
1994	3	0.001	0.05	0.05	1.81%	47%
1995	3	0.001	0.05	0.05	1.86%	48%
1996	3	0.001	0.05	0.05	1.90%	49%
1997	3	0.001	0.05	0.05	1.95%	50%
1998	3	0.001	0.05	0.05	1.95%	51%
1999	3	0.001	0.05	0.05	1.95%	52%
2000	3	0.001	0.05	0.05	1.95%	53%
2001	3	0.001	0.05	0.05	1.95%	54%
2002	2	0.001	0.05	0.05	1.95%	55%
2003	2	0.001	0.06	0.04	1.95%	56%
2004	2	0.001	0.06	0.04	1.95%	57%
2005	2	0.001	0.06	0.04	1.95%	58%
2006	2	0.001	0.06	0.04	1.95%	58%
2007	2	0.001	0.06	0.04	1.95%	59%
2008	2	0.001	0.06	0.04	1.95%	60%
2009	2	0.001	0.06	0.04	1.95%	61%
2010	2	0.001	0.06	0.04	1.95%	61%
2011	2	0.001	0.06	0.04	1.95%	62%
2012	2	0.001	0.06	0.04	1.95%	63%
2013	2	0.001	0.06	0.04	1.95%	64%
2014	2	0.001	0.06	0.04	1.95%	64%
2015	2	0.001	0.07	0.03	1.95%	65%
2016	2	0.001	0.07	0.03	1.95%	66%
2017	2	0.001	0.07	0.03	1.95%	66%
2018	2	0.001	0.07	0.03	1.95%	67%
2019	2	0.001	0.07	0.03	1.95%	68%
2020	2	0.001	0.07	0.03	1.95%	68%
2021	2	0.001	0.07	0.03	1.95%	69%
2022	2	0.001	0.07	0.03	1.95%	70%
2023	2	0.001	0.07	0.03	1.95%	70%
2024	2	0.001	0.07	0.03	1.95%	71%
2025	2	0.001	0.07	0.03	1.95%	71%
2026	2	0.001	0.07	0.03	1.95%	72%
2027	2	0.001	0.07	0.03	1.95%	72%
2028	1	0.001	0.07	0.03	1.95%	73%
2029	1	0.001	0.07	0.03	1.95%	74%
2030	1	0.001	0.07	0.03	1.95%	74%
2031	1	0.001	0.07	0.03	1.95%	75%
2032	1	0.000	0.08	0.02	1.95%	75%
2033	1	0.000	0.08	0.02	1.95%	76%
2034	1	0.000	0.08	0.02	1.95%	76%
2035	1	0.000	0.08	0.02	1.95%	76%
2036	1	0.000	0.08	0.02	1.95%	77%
2037	1	0.000	0.08	0.02	1.95%	77%
2038	1	0.000	0.08	0.02	1.95%	78%
2039	1	0.000	0.08	0.02	1.95%	78%
2040	1	0.000	0.08	0.02	1.95%	79%
2041	1	0.000	0.08	0.02	1.95%	79%
2042	1	0.000	0.08	0.02	1.95%	79%
2043	1	0.000	0.08	0.02	1.95%	80%
2044	1	0.000	0.08	0.02	1.95%	80%
2045	1	0.000	0.08	0.02	1.95%	81%
2046	1	0.000	0.08	0.02	1.95%	81%
2047	1	0.000	0.08	0.02	1.95%	81%
2048	1	0.000	0.08	0.02	1.95%	82%
2049	1	0.000	0.08	0.02	1.95%	82%
2050	1	0.000	0.08	0.02	1.95%	82%

CUM.PROD	0.230 BBO	GIANT FIELDS		DENMARK
RESERVES	0.849 BBO	TOTAL BBO	0	W.EUROPE
DISCOVERED	1.079 BBO	% OF DISC.	0%	
UNDISCOVERED	0.171 BBO	Last Giant	-	
REMAINING	1.020 BBO			Variance Trend
ULTIMATE	1.250 BBO			1990 +1.3%
Rept.Reserve	0.849 BBO	Ref.Date(end) 1989		Future UP

Milestones	Date	Prod Mb/d	Dep. Rate	% Produced
1st Peak	1996	156	7.81%	46%
2nd Peak				
Mid-Point	1997	144	7.80%	50%
Final Decline	1997		7.80%	

Forecast "B"
A=Exponential decline
B=5% incr. to Mid-Point
C=Other
Pre-1930 Cum.
(Added to 1930 Prod)

Date	Prod. M.b/d	Prod. BBO/yr	Cum Prod	Remain	Dep. Rate	% Produced	Giant Names & Notes
1930		0.000	0.00	1.25	ERR	0%	
1931		0.000	0.00	1.25	0.0%	0%	
1932		0.000	0.00	1.25	0.0%	0%	
1933		0.000	0.00	1.25	0.0%	0%	
1934		0.000	0.00	1.25	0.0%	0%	
1935		0.000	0.00	1.25	0.0%	0%	
1936		0.000	0.00	1.25	0.0%	0%	
1937		0.000	0.00	1.25	0.0%	0%	
1938		0.000	0.00	1.25	0.0%	0%	
1939		0.000	0.00	1.25	0.0%	0%	
1940		0.000	0.00	1.25	0.0%	0%	
1941		0.000	0.00	1.25	0.0%	0%	
1942		0.000	0.00	1.25	0.0%	0%	
1943		0.000	0.00	1.25	0.0%	0%	
1944		0.000	0.00	1.25	0.0%	0%	
1945		0.000	0.00	1.25	0.0%	0%	
1946		0.000	0.00	1.25	0.0%	0%	
1947		0.000	0.00	1.25	0.0%	0%	
1948		0.000	0.00	1.25	0.0%	0%	
1949		0.000	0.00	1.25	0.0%	0%	
1950		0.000	0.00	1.25	0.0%	0%	
1951		0.000	0.00	1.25	0.0%	0%	
1952		0.000	0.00	1.25	0.0%	0%	
1953		0.000	0.00	1.25	0.0%	0%	
1954		0.000	0.00	1.25	0.0%	0%	
1955		0.000	0.00	1.25	0.0%	0%	
1956		0.000	0.00	1.25	0.0%	0%	
1957		0.000	0.00	1.25	0.0%	0%	
1958		0.000	0.00	1.25	0.0%	0%	
1959		0.000	0.00	1.25	0.0%	0%	
1960		0.000	0.00	1.25	0.0%	0%	
1961		0.000	0.00	1.25	0.0%	0%	
1962		0.000	0.00	1.25	0.0%	0%	
1963		0.000	0.00	1.25	0.0%	0%	
1964		0.000	0.00	1.25	0.0%	0%	
1965		0.000	0.00	1.25	0.0%	0%	
1966		0.000	0.00	1.25	0.0%	0%	
1967		0.000	0.00	1.25	0.0%	0%	
1968		0.000	0.00	1.25	0.0%	0%	
1969		0.000	0.00	1.25	0.0%	0%	
1970		0.000	0.00	1.25	0.0%	0%	
1971		0.000	0.00	1.25	0.0%	0%	
1972	3.00	0.001	0.00	1.25	0.1%	0%	
1973	3.10	0.001	0.00	1.25	0.1%	0%	
1974	4.00	0.001	0.00	1.25	0.1%	0%	
1975	3.00	0.001	0.00	1.25	0.1%	0%	
1976	3.00	0.001	0.01	1.24	0.1%	0%	
1977	9.00	0.003	0.01	1.24	0.3%	1%	
1978	10.00	0.004	0.01	1.24	0.3%	1%	
1979	12.00	0.004	0.02	1.23	0.4%	1%	
1980	6.00	0.002	0.02	1.23	0.2%	2%	
1981	16.00	0.006	0.03	1.22	0.5%	2%	
1982	38.00	0.014	0.04	1.21	1.1%	3%	
1983	46.00	0.017	0.06	1.19	1.4%	4%	
1984	45.00	0.016	0.07	1.18	1.4%	6%	
1985	58.00	0.021	0.09	1.16	1.8%	7%	
1986	74.70	0.027	0.12	1.13	2.4%	10%	
1987	94.00	0.034	0.16	1.09	3.0%	12%	
1988	95.50	0.035	0.19	1.06	3.2%	15%	

1989	111.00	0.041	0.23	1.02	3.8%	18%	
1990	116.55	0.043	0.27	0.98	4.2%	22%	
1991	122.38	0.045	0.32	0.93	4.6%	25%	
1992	128.50	0.047	0.36	0.89	5.0%	29%	
1993	134.92	0.049	0.41	0.84	5.6%	33%	
1994	141.67	0.052	0.47	0.78	6.2%	37%	
1995	148.75	0.054	0.52	0.73	6.9%	42%	
1996	156.19	0.057	0.58	0.67	7.8%	46%	1st PEAK
1997	143.86	0.053	0.63	0.62	7.8%	50%	50% PRODUCED
1998	132.64	0.048	0.68	0.57	7.8%	54%	
1999	122.30	0.045	0.72	0.53	7.8%	58%	
2000	112.76	0.041	0.76	0.49	7.8%	61%	
2001	103.96	0.038	0.80	0.45	7.8%	64%	
2002	95.85	0.035	0.84	0.41	7.8%	67%	
2003	88.38	0.032	0.87	0.38	7.8%	69%	
2004	81.48	0.030	0.90	0.35	7.8%	72%	
2005	75.13	0.027	0.93	0.32	7.8%	74%	
2006	69.27	0.025	0.95	0.30	7.8%	76%	
2007	63.86	0.023	0.97	0.28	7.8%	78%	
2008	58.88	0.021	1.00	0.25	7.8%	80%	
2009	54.29	0.020	1.02	0.23	7.8%	81%	
2010	50.06	0.018	1.03	0.22	7.8%	83%	
2011	46.15	0.017	1.05	0.20	7.8%	84%	
2012	42.55	0.016	1.07	0.18	7.8%	85%	
2013	39.23	0.014	1.08	0.17	7.8%	86%	
2014	36.17	0.013	1.09	0.16	7.8%	88%	
2015	33.35	0.012	1.11	0.14	7.8%	88%	
2016	30.75	0.011	1.12	0.13	7.8%	89%	
2017	28.35	0.010	1.13	0.12	7.8%	90%	
2018	26.14	0.010	1.14	0.11	7.8%	91%	
2019	24.10	0.009	1.15	0.10	7.8%	92%	
2020	22.22	0.008	1.15	0.10	7.8%	92%	
2021	20.49	0.007	1.16	0.09	7.8%	93%	
2022	18.89	0.007	1.17	0.08	7.8%	93%	
2023	17.42	0.006	1.17	0.08	7.8%	94%	
2024	16.06	0.006	1.18	0.07	7.8%	94%	
2025	14.81	0.005	1.19	0.06	7.8%	95%	
2026	13.65	0.005	1.19	0.06	7.8%	95%	
2027	12.59	0.005	1.20	0.05	7.8%	96%	
2028	11.60	0.004	1.20	0.05	7.8%	96%	
2029	10.70	0.004	1.20	0.05	7.8%	96%	
2030	9.86	0.004	1.21	0.04	7.8%	97%	
2031	9.09	0.003	1.21	0.04	7.8%	97%	
2032	8.39	0.003	1.21	0.04	7.8%	97%	
2033	7.73	0.003	1.22	0.03	7.8%	97%	
2034	7.13	0.003	1.22	0.03	7.8%	98%	
2035	6.57	0.002	1.22	0.03	7.8%	98%	
2036	6.06	0.002	1.22	0.03	7.8%	98%	
2037	5.59	0.002	1.23	0.02	7.8%	98%	
2038	5.15	0.002	1.23	0.02	7.8%	98%	
2039	4.75	0.002	1.23	0.02	7.8%	98%	
2040	4.38	0.002	1.23	0.02	7.8%	98%	
2041	4.04	0.001	1.23	0.02	7.8%	99%	
2042	3.72	0.001	1.23	0.02	7.8%	99%	
2043	3.43	0.001	1.24	0.01	7.8%	99%	
2044	3.16	0.001	1.24	0.01	7.8%	99%	
2045	2.92	0.001	1.24	0.01	7.8%	99%	
2046	2.69	0.001	1.24	0.01	7.8%	99%	
2047	2.48	0.001	1.24	0.01	7.8%	99%	
2048	2.29	0.001	1.24	0.01	7.8%	99%	
2049	2.11	0.001	1.24	0.01	7.8%	99%	
2050	1.94	0.001	1.24	0.01	7.8%	99%	

	2.177 BBO	GIANT FIELDS		DUBAI
RESERVES	3.000 BBO	TOTAL BBO	2.6	M.EAST (MINOR)
DISCOVERED	5.177 BBO	% OF DISC.	50%	
UNDISCOVERED	0.323 BBO	Last Giant	1970	----------------
REMAINING	3.323 BBO	------------------		Variance Trend
ULTIMATE	5.500 BBO			1990 +13%
Rept.Reserve	4.000 #1	Ref.date(end)	1989	Future DOWN

Milestones	Date	Prod Mb/d	Dep. Rate	% Produced
1st Peak	1992	475	5.78%	49%
2nd Peak				
Mid-Point	1993	448	5.78%	52%
Final Decline	1994		5.78%	54%

Forecast "B"
A=Exponential decline
B=5% incr. to Mid-Point
C=Other
Pre-1930 Cum. 0
(Added to 1930 prod)

Date	Prod. M.b/d	Prod. BBO/yr	Cum. Prod.	Remain	Deplet. Rate	% Produced	Giant	Notes & Names
1930		0.000	0.00	5.50	ERR	0%		
1931		0.000	0.00	5.50	0.00%	0%		
1932		0.000	0.00	5.50	0.00%	0%		
1933		0.000	0.00	5.50	0.00%	0%		
1934		0.000	0.00	5.50	0.00%	0%		
1935		0.000	0.00	5.50	0.00%	0%		
1936		0.000	0.00	5.50	0.00%	0%		
1937		0.000	0.00	5.50	0.00%	0%		
1938		0.000	0.00	5.50	0.00%	0%		
1939		0.000	0.00	5.50	0.00%	0%		
1940		0.000	0.00	5.50	0.00%	0%		
1941		0.000	0.00	5.50	0.00%	0%		
1942		0.000	0.00	5.50	0.00%	0%		
1943		0.000	0.00	5.50	0.00%	0%		
1944		0.000	0.00	5.50	0.00%	0%		
1945		0.000	0.00	5.50	0.00%	0%		
1946		0.000	0.00	5.50	0.00%	0%		
1947		0.000	0.00	5.50	0.00%	0%		
1948		0.000	0.00	5.50	0.00%	0%		
1949		0.000	0.00	5.50	0.00%	0%		
1950		0.000	0.00	5.50	0.00%	0%		
1951		0.000	0.00	5.50	0.00%	0%		
1952		0.000	0.00	5.50	0.00%	0%		
1953		0.000	0.00	5.50	0.00%	0%		
1954		0.000	0.00	5.50	0.00%	0%		
1955		0.000	0.00	5.50	0.00%	0%		
1956		0.000	0.00	5.50	0.00%	0%		
1957		0.000	0.00	5.50	0.00%	0%		
1958		0.000	0.00	5.50	0.00%	0%		
1959		0.000	0.00	5.50	0.00%	0%		
1960		0.000	0.00	5.50	0.00%	0%		
1961		0.000	0.00	5.50	0.00%	0%		
1962		0.000	0.00	5.50	0.00%	0%		
1963		0.000	0.00	5.50	0.00%	0%		
1964		0.000	0.00	5.50	0.00%	0%		
1965		0.000	0.00	5.50	0.00%	0%		
1966		0.000	0.00	5.50	0.00%	0%	1.50	Fateh
1967		0.000	0.00	5.50	0.00%	0%	1.10	Fateh SW
1968		0.000	0.00	5.50	0.00%	0%		
1969		0.000	0.00	5.50	0.00%	0%		
1970	78	0.028	0.03	5.47	0.52%	1%		
1971	126	0.046	0.07	5.43	0.84%	1%		
1972	130	0.047	0.12	5.38	0.87%	2%		
1973	223	0.081	0.20	5.30	1.51%	4%		
1974	232	0.085	0.29	5.21	1.60%	5%		
1975	260	0.095	0.38	5.12	1.82%	7%		
1976	315	0.115	0.50	5.00	2.25%	9%		
1977	320	0.117	0.61	4.89	2.33%	11%		
1978	360	0.131	0.75	4.75	2.69%	14%		
1979	360	0.131	0.88	4.62	2.76%	16%		
1980	350	0.128	1.01	4.49	2.76%	18%		
1981	358	0.131	1.14	4.36	2.91%	21%		
1982	357	0.130	1.27	4.23	2.99%	23%		
1983	327	0.119	1.39	4.11	2.82%	25%		
1984	324	0.118	1.50	4.00	2.87%	27%		
1985	348	0.127	1.63	3.87	3.18%	30%		
1986	350	0.128	1.76	3.74	3.30%	32%		
1987	381	0.139	1.90	3.60	3.72%	35%		
1988	355	0.130	2.03	3.47	3.60%	37%		

1989	411	0.150	2.18	3.32	4.32%	40%	
1990	431	0.157	2.33	3.17	4.74%	42%	
1991	453	0.165	2.50	3.00	5.22%	45%	
1992	475	0.173	2.67	2.83	5.78%	49%	1ST PEAK
1993	448	0.163	2.84	2.66	5.78%	52%	50% PRODUCED
1994	422	0.154	2.99	2.51	5.78%	54%	
1995	397	0.145	3.14	2.36	5.78%	57%	
1996	374	0.137	3.27	2.23	5.78%	59%	
1997	353	0.129	3.40	2.10	5.78%	62%	
1998	332	0.121	3.52	1.98	5.78%	64%	
1999	313	0.114	3.64	1.86	5.78%	66%	
2000	295	0.108	3.74	1.76	5.78%	68%	
2001	278	0.101	3.85	1.65	5.78%	70%	
2002	262	0.096	3.94	1.56	5.78%	72%	
2003	247	0.090	4.03	1.47	5.78%	73%	
2004	233	0.085	4.12	1.38	5.78%	75%	
2005	219	0.080	4.20	1.30	5.78%	76%	
2006	206	0.075	4.27	1.23	5.78%	78%	
2007	194	0.071	4.34	1.16	5.78%	79%	
2008	183	0.067	4.41	1.09	5.78%	80%	
2009	173	0.063	4.47	1.03	5.78%	81%	
2010	163	0.059	4.53	0.97	5.78%	82%	
2011	153	0.056	4.59	0.91	5.78%	83%	
2012	144	0.053	4.64	0.86	5.78%	84%	
2013	136	0.050	4.69	0.81	5.78%	85%	
2014	128	0.047	4.74	0.76	5.78%	86%	
2015	121	0.044	4.78	0.72	5.78%	87%	
2016	114	0.042	4.82	0.68	5.78%	88%	
2017	107	0.039	4.86	0.64	5.78%	88%	
2018	101	0.037	4.90	0.60	5.78%	89%	
2019	95	0.035	4.93	0.57	5.78%	90%	
2020	90	0.033	4.97	0.53	5.78%	90%	
2021	85	0.031	5.00	0.50	5.78%	91%	
2022	80	0.029	5.03	0.47	5.78%	91%	
2023	75	0.027	5.05	0.45	5.78%	92%	
2024	71	0.026	5.08	0.42	5.78%	92%	
2025	67	0.024	5.10	0.40	5.78%	93%	
2026	63	0.023	5.13	0.37	5.78%	93%	
2027	59	0.022	5.15	0.35	5.78%	94%	
2028	56	0.020	5.17	0.33	5.78%	94%	
2029	52	0.019	5.19	0.31	5.78%	94%	
2030	49	0.018	5.21	0.29	5.78%	95%	
2031	47	0.017	5.22	0.28	5.78%	95%	
2032	44	0.016	5.24	0.26	5.78%	95%	
2033	41	0.015	5.25	0.25	5.78%	96%	
2034	39	0.014	5.27	0.23	5.78%	96%	
2035	37	0.013	5.28	0.22	5.78%	96%	
2036	35	0.013	5.29	0.21	5.78%	96%	
2037	33	0.012	5.31	0.19	5.78%	96%	
2038	31	0.011	5.32	0.18	5.78%	97%	
2039	29	0.011	5.33	0.17	5.78%	97%	
2040	27	0.010	5.34	0.16	5.78%	97%	
2041	26	0.009	5.35	0.15	5.78%	97%	
2042	24	0.009	5.36	0.14	5.78%	97%	
2043	23	0.008	5.36	0.14	5.78%	98%	
2044	21	0.008	5.37	0.13	5.78%	98%	
2045	20	0.007	5.38	0.12	5.78%	98%	
2046	19	0.007	5.39	0.11	5.78%	98%	
2047	18	0.007	5.39	0.11	5.78%	98%	
2048	17	0.006	5.40	0.10	5.78%	98%	
2049	16	0.006	5.41	0.09	5.78%	98%	
2050	15	0.005	5.41	0.09	5.78%	98%	

Note
#1 Reported reserve adjusted downwards
Implausable increase from 1.4 to 4 BBO in 1987

CUM.PROD	1.558 BBO	GIANT FIELDS			ECUADOR
RESERVES	1.514 BBO	TOTAL BBO	2.46		L.AMERICA
DISCOVERED	3.072 BBO	% OF DISC.	80%		
UNDISCOVERED	0.428 BBO	Last Giant	1972		
REMAINING	1.942 BBO				Variance Trend
ULTIMATE	3.500 BBO				1990 zero
Rept.Reserves	1.514 BBO	Ref. Date (end)		1989	Future DOWN

Milestones	Date	Prod Mb/d	Dep. Rate	% Produced
1st Peak	1991	289	5.74%	51%
2nd Peak				
Mid-Point	1991	289	5.74%	51%
Final Decline	1992		5.74%	

Forecast "B"
A=Exponential decline
B=5% inc. to Mid-Point
C=Other
Pre-1930 Cum 0.04
(added to 1930 Prod)

Date	Prod. M.b/d	Prod. BBO/yr	Cum Prod	Remain	Dep. Rate	% Produced	Giant	Names & Notes
1930	5	0.002	0.04	3.46	ERR	1%		
1931	5	0.002	0.04	3.46	0.05%	1%		
1932	5	0.002	0.05	3.45	0.05%	1%		
1933	5	0.002	0.05	3.45	0.05%	1%		
1934	5	0.002	0.05	3.45	0.05%	1%		
1935	5	0.002	0.05	3.45	0.05%	1%		
1936	5	0.002	0.05	3.45	0.05%	2%		
1937	5	0.002	0.05	3.45	0.05%	2%		
1938	6	0.002	0.06	3.44	0.06%	2%		
1939	6	0.002	0.06	3.44	0.07%	2%		
1940	6	0.002	0.06	3.44	0.07%	2%		
1941	4	0.002	0.06	3.44	0.05%	2%		
1942	6	0.002	0.07	3.43	0.07%	2%		
1943	6	0.002	0.07	3.43	0.07%	2%		
1944	8	0.003	0.07	3.43	0.09%	2%		
1945	7	0.003	0.07	3.43	0.08%	2%		
1946	6	0.002	0.08	3.42	0.07%	2%		
1947	7	0.002	0.08	3.42	0.07%	2%		
1948	7	0.003	0.08	3.42	0.07%	2%		
1949	7	0.003	0.08	3.42	0.08%	2%		
1950	7	0.003	0.09	3.41	0.08%	2%		
1951	8	0.003	0.09	3.41	0.08%	3%		
1952	8	0.003	0.09	3.41	0.08%	3%		
1953	8	0.003	0.09	3.41	0.09%	3%		
1954	9	0.003	0.10	3.40	0.09%	3%		
1955	10	0.004	0.10	3.40	0.10%	3%		
1956	9	0.003	0.10	3.40	0.10%	3%		
1957	8	0.003	0.11	3.39	0.09%	3%		
1958	9	0.003	0.11	3.39	0.10%	3%		
1959	8	0.003	0.11	3.39	0.08%	3%		
1960	8	0.003	0.12	3.38	0.08%	3%		
1961	8	0.003	0.12	3.38	0.09%	3%		
1962	8	0.003	0.12	3.38	0.08%	3%		
1963	7	0.003	0.12	3.38	0.07%	4%		
1964	8	0.003	0.13	3.37	0.08%	4%		
1965	8	0.003	0.13	3.37	0.09%	4%		
1966	9	0.003	0.13	3.37	0.09%	4%		
1967	10	0.003	0.14	3.36	0.10%	4%		
1968	7	0.003	0.14	3.36	0.08%	4%		
1969	5	0.002	0.14	3.36	0.05%	4%		
1970	4	0.001	0.14	3.36	0.04%	4%	0.5	Auca
1971	4	0.001	0.14	3.36	0.04%	4%	0.75	Sacha
1972	59	0.022	0.17	3.33	0.65%	5%	1.212	Shushufindi
1973	197	0.072	0.24	3.26	2.16%	7%		
1974	232	0.085	0.32	3.18	2.60%	9%		
1975	165	0.060	0.38	3.12	1.90%	11%		
1976	185	0.068	0.45	3.05	2.17%	13%		
1977	180	0.066	0.52	2.98	2.15%	15%		
1978	200	0.073	0.59	2.91	2.45%	17%		
1979	220	0.080	0.67	2.83	2.76%	19%		
1980	230	0.084	0.75	2.75	2.97%	22%		
1981	204	0.074	0.83	2.67	2.71%	24%		
1982	215	0.078	0.91	2.59	2.94%	26%		
1983	236	0.086	0.99	2.51	3.32%	28%		
1984	254	0.093	1.08	2.42	3.70%	31%		
1985	273	0.100	1.18	2.32	4.13%	34%		
1986	270	0.099	1.28	2.22	4.26%	37%		
1987	157	0.057	1.34	2.16	2.59%	38%		
1988	310	0.113	1.45	2.05	5.24%	42%		

1989	286	0.104	1.56	1.94	5.10%	45%	
1990	287	0.105	1.66	1.84	5.40%	48%	
1991	289	0.105	1.77	1.73	5.74%	51%	1ST PEAK
1992	272	0.099	1.87	1.63	5.74%	53%	& 50% PRODUCED
1993	257	0.094	1.96	1.54	5.74%	56%	
1994	242	0.088	2.05	1.45	5.74%	59%	
1995	228	0.083	2.13	1.37	5.74%	61%	
1996	215	0.078	2.21	1.29	5.74%	63%	
1997	203	0.074	2.29	1.21	5.74%	65%	
1998	191	0.070	2.36	1.14	5.74%	67%	
1999	180	0.066	2.42	1.08	5.74%	69%	
2000	170	0.062	2.48	1.02	5.74%	71%	
2001	160	0.058	2.54	0.96	5.74%	73%	
2002	151	0.055	2.60	0.90	5.74%	74%	
2003	142	0.052	2.65	0.85	5.74%	76%	
2004	134	0.049	2.70	0.80	5.74%	77%	
2005	126	0.046	2.74	0.76	5.74%	78%	
2006	119	0.043	2.79	0.71	5.74%	80%	
2007	112	0.041	2.83	0.67	5.74%	81%	
2008	106	0.039	2.87	0.63	5.74%	82%	
2009	100	0.036	2.90	0.60	5.74%	83%	
2010	94	0.034	2.94	0.56	5.74%	84%	
2011	89	0.032	2.97	0.53	5.74%	85%	
2012	83	0.030	3.00	0.50	5.74%	86%	
2013	79	0.029	3.03	0.47	5.74%	87%	
2014	74	0.027	3.06	0.44	5.74%	87%	
2015	70	0.026	3.08	0.42	5.74%	88%	
2016	66	0.024	3.10	0.40	5.74%	89%	
2017	62	0.023	3.13	0.37	5.74%	89%	
2018	59	0.021	3.15	0.35	5.74%	90%	
2019	55	0.020	3.17	0.33	5.74%	91%	
2020	52	0.019	3.19	0.31	5.74%	91%	
2021	49	0.018	3.21	0.29	5.74%	92%	
2022	46	0.017	3.22	0.28	5.74%	92%	
2023	44	0.016	3.24	0.26	5.74%	93%	
2024	41	0.015	3.25	0.25	5.74%	93%	
2025	39	0.014	3.27	0.23	5.74%	93%	
2026	36	0.013	3.28	0.22	5.74%	94%	
2027	34	0.013	3.29	0.21	5.74%	94%	
2028	32	0.012	3.31	0.19	5.74%	94%	
2029	31	0.011	3.32	0.18	5.74%	95%	
2030	29	0.011	3.33	0.17	5.74%	95%	
2031	27	0.010	3.34	0.16	5.74%	95%	
2032	26	0.009	3.35	0.15	5.74%	96%	
2033	24	0.009	3.36	0.14	5.74%	96%	
2034	23	0.008	3.36	0.14	5.74%	96%	
2035	21	0.008	3.37	0.13	5.74%	96%	
2036	20	0.007	3.38	0.12	5.74%	97%	
2037	19	0.007	3.39	0.11	5.74%	97%	
2038	18	0.007	3.39	0.11	5.74%	97%	
2039	17	0.006	3.40	0.10	5.74%	97%	
2040	16	0.006	3.40	0.10	5.74%	97%	
2041	15	0.005	3.41	0.09	5.74%	97%	
2042	14	0.005	3.42	0.08	5.74%	98%	
2043	13	0.005	3.42	0.08	5.74%	98%	
2044	13	0.005	3.42	0.08	5.74%	98%	
2045	12	0.004	3.43	0.07	5.74%	98%	
2046	11	0.004	3.43	0.07	5.74%	98%	
2047	11	0.004	3.44	0.06	5.74%	98%	
2048	10	0.004	3.44	0.06	5.74%	98%	
2049	9	0.003	3.44	0.06	5.74%	98%	
2050	9	0.003	3.45	0.05	5.74%	98%	

CUM.PROD	4.645 BBO	GIANT FIELDS		EGYPT	
RESERVES	4.500 BBO	TOTAL BBO	4.7	AFRICA	
DISCOVERED	9.145 BBO	% OF DISC.	52%		
UNDISCOVERED	1.855 BBO	Last Giant	1974		
REMAINING	6.355 BBO			Variance Trend	
ULTIMATE	11.000 BBO			1990	-3%
Rept.Reserves	4.500 BBO	Ref.Date (end)	1989	Future	UP

Milestones		Prod Mb/d	Dep Rate	% Produced	Forecast "B"
1st Peak	1992	989	5.70%	52%	A=Exponential decline
2nd Peak					B=5% inc. to Mid-Point
					C=Other
Mid-Point	1992	989	5.70%	52%	Pre-1930 Cum 0
Final Decline	1992		5.70%		(Added to 1930 Prod)

Date	Prod. M.b/d	Prod. BBO/yr	Cum Prod	Remain	Dep. Rate	% Produced	Giant	Names & Notes
1930		0.000	0.00	11.00	ERR	0%		
1931		0.000	0.00	11.00	0.00%	0%		
1932		0.000	0.00	11.00	0.00%	0%		
1933		0.000	0.00	11.00	0.00%	0%		
1934		0.000	0.00	11.00	0.00%	0%		
1935		0.000	0.00	11.00	0.00%	0%		
1936		0.000	0.00	11.00	0.00%	0%		
1937		0.000	0.00	11.00	0.00%	0%		
1938	4	0.002	0.00	11.00	0.01%	0%		
1939	13	0.005	0.01	10.99	0.04%	0%		
1940	18	0.006	0.01	10.99	0.06%	0%		
1941	23	0.009	0.02	10.98	0.08%	0%		
1942	23	0.008	0.03	10.97	0.08%	0%		
1943	25	0.009	0.04	10.96	0.08%	0%		
1944	26	0.009	0.05	10.95	0.09%	0%		
1945	26	0.009	0.06	10.94	0.09%	1%		
1946	25	0.009	0.07	10.93	0.08%	1%		
1947	36	0.013	0.08	10.92	0.12%	1%		
1948	36	0.013	0.09	10.91	0.12%	1%		
1949	43	0.016	0.11	10.89	0.14%	1%		
1950	45	0.016	0.12	10.88	0.15%	1%		
1951	45	0.016	0.14	10.86	0.15%	1%		
1952	45	0.016	0.16	10.84	0.15%	1%		
1953	46	0.017	0.17	10.83	0.15%	2%		
1954	38	0.014	0.19	10.81	0.13%	2%		
1955	35	0.013	0.20	10.80	0.12%	2%	0.5	Belayim
1956	32	0.012	0.21	10.79	0.11%	2%		
1957	43	0.016	0.23	10.77	0.14%	2%		
1958	65	0.024	0.25	10.75	0.22%	2%		
1959	58	0.021	0.27	10.73	0.20%	2%		
1960	62	0.023	0.30	10.70	0.21%	3%		
1961	74	0.027	0.32	10.68	0.25%	3%	1	Belayim M
1962	84	0.031	0.35	10.65	0.29%	3%		
1963	113	0.041	0.39	10.61	0.39%	4%		
1964	125	0.046	0.44	10.56	0.43%	4%		
1965	126	0.046	0.49	10.51	0.44%	4%	1.5	El Morgan
1966	120	0.044	0.53	10.47	0.42%	5%		
1967	140	0.051	0.58	10.42	0.49%	5%	0.72	July
1968	180	0.066	0.65	10.35	0.63%	6%		
1969	236	0.086	0.73	10.27	0.83%	7%		
1970	329	0.120	0.85	10.15	1.17%	8%		
1971	302	0.110	0.96	10.04	1.09%	9%		
1972	227	0.083	1.05	9.95	0.83%	10%		
1973	180	0.066	1.11	9.89	0.66%	10%		
1974	118	0.043	1.15	9.85	0.44%	10%	1	Ramadan
1975	225	0.082	1.24	9.76	0.83%	11%		
1976	325	0.119	1.36	9.64	1.21%	12%		
1977	450	0.164	1.52	9.48	1.70%	14%		
1978	490	0.179	1.70	9.30	1.89%	15%		
1979	500	0.183	1.88	9.12	1.96%	17%		
1980	585	0.214	2.09	8.91	2.34%	19%		
1981	578	0.211	2.31	8.69	2.37%	21%		
1982	667	0.243	2.55	8.45	2.80%	23%		
1983	690	0.252	2.80	8.20	2.98%	25%		
1984	790	0.288	3.09	7.91	3.52%	28%		
1985	886	0.323	3.41	7.59	4.09%	31%		
1986	774	0.282	3.69	7.31	3.72%	34%		
1987	899	0.328	4.02	6.98	4.49%	37%		
1988	851	0.311	4.33	6.67	4.45%	39%		

1989	854	0.312	4.65	6.35	4.68%	42%	
1990	897	0.327	4.97	6.03	5.15%	45%	
1991	942	0.344	5.32	5.68	5.70%	48%	
1992	989	0.361	5.68	5.32	5.70%	52%	50% PRODUCED
1993	831	0.303	5.98	5.02	5.70%	54%	& 1ST PEAK
1994	784	0.286	6.27	4.73	5.70%	57%	
1995	739	0.270	6.54	4.46	5.70%	59%	
1996	697	0.254	6.79	4.21	5.70%	62%	
1997	657	0.240	7.03	3.97	5.70%	64%	
1998	620	0.226	7.26	3.74	5.70%	66%	
1999	584	0.213	7.47	3.53	5.70%	68%	
2000	551	0.201	7.67	3.33	5.70%	70%	
2001	520	0.190	7.86	3.14	5.70%	71%	
2002	490	0.179	8.04	2.96	5.70%	73%	
2003	462	0.169	8.21	2.79	5.70%	75%	
2004	436	0.159	8.37	2.63	5.70%	76%	
2005	411	0.150	8.52	2.48	5.70%	77%	
2006	388	0.141	8.66	2.34	5.70%	79%	
2007	365	0.133	8.79	2.21	5.70%	80%	
2008	345	0.126	8.92	2.08	5.70%	81%	
2009	325	0.119	9.04	1.96	5.70%	82%	
2010	306	0.112	9.15	1.85	5.70%	83%	
2011	289	0.105	9.25	1.75	5.70%	84%	
2012	273	0.099	9.35	1.65	5.70%	85%	
2013	257	0.094	9.45	1.55	5.70%	86%	
2014	242	0.088	9.54	1.46	5.70%	87%	
2015	229	0.083	9.62	1.38	5.70%	87%	
2016	216	0.079	9.70	1.30	5.70%	88%	
2017	203	0.074	9.77	1.23	5.70%	89%	
2018	192	0.070	9.84	1.16	5.70%	89%	
2019	181	0.066	9.91	1.09	5.70%	90%	
2020	170	0.062	9.97	1.03	5.70%	91%	
2021	161	0.059	10.03	0.97	5.70%	91%	
2022	152	0.055	10.08	0.92	5.70%	92%	
2023	143	0.052	10.14	0.86	5.70%	92%	
2024	135	0.049	10.19	0.81	5.70%	93%	
2025	127	0.046	10.23	0.77	5.70%	93%	
2026	120	0.044	10.28	0.72	5.70%	93%	
2027	113	0.041	10.32	0.68	5.70%	94%	
2028	107	0.039	10.36	0.64	5.70%	94%	
2029	100	0.037	10.39	0.61	5.70%	94%	
2030	95	0.035	10.43	0.57	5.70%	95%	
2031	89	0.033	10.46	0.54	5.70%	95%	
2032	84	0.031	10.49	0.51	5.70%	95%	
2033	79	0.029	10.52	0.48	5.70%	96%	
2034	75	0.027	10.55	0.45	5.70%	96%	
2035	71	0.026	10.57	0.43	5.70%	96%	
2036	67	0.024	10.60	0.40	5.70%	96%	
2037	63	0.023	10.62	0.38	5.70%	97%	
2038	59	0.022	10.64	0.36	5.70%	97%	
2039	56	0.020	10.66	0.34	5.70%	97%	
2040	53	0.019	10.68	0.32	5.70%	97%	
2041	50	0.018	10.70	0.30	5.70%	97%	
2042	47	0.017	10.72	0.28	5.70%	97%	
2043	44	0.016	10.73	0.27	5.70%	98%	
2044	42	0.015	10.75	0.25	5.70%	98%	
2045	39	0.014	10.76	0.24	5.70%	98%	
2046	37	0.014	10.78	0.22	5.70%	98%	
2047	35	0.013	10.79	0.21	5.70%	98%	
2048	33	0.012	10.80	0.20	5.70%	98%	
2049	31	0.011	10.81	0.19	5.70%	98%	
2050	29	0.011	10.82	0.18	5.70%	98%	

CUM.PROD	0.517 BBO	GIANT FIELDS		FRANCE	
RESERVES	0.190 BBO	TOTAL BBO	0	W.EUROPE	
DISCOVERED	0.707 BBO	% OF DISC.	0%		
UNDISCOVERED	0.293 BBO	Last Giant	-	---------------	
REMAINING	0.483 BBO	------------------		Variance	Trend
ULTIMATE	1.000 BBO			1990	zero
Rept.Reserve	0.190 BBO	Ref.Date(end) 1989		Future	DOWN

Milestones	Date	Prod Mb/d	Dep. Rate	% Produced	Forecast "A"
					A=Exponential decline
1st Peak	1965	60	2.57%	17%	B=5% incr. to Mid-Point
2nd Peak	1988	69	4.72%	49%	C=Other
Mid-Point	1989	65	4.70%	52%	Pre-1930 Cum. 0
Final Decline	1989		4.70%		(Added to 1930 Prod)

Date	Prod. M.b/d	Prod. BBO/yr	Cum Prod	Remain	Dep. Rate	% Produced	Giant	Names & Notes
1930		0.000	0.00	1.00	ERR	0%		
1931		0.000	0.00	1.00	0.00%	0%		
1932		0.000	0.00	1.00	0.00%	0%		
1933		0.000	0.00	1.00	0.00%	0%		
1934		0.000	0.00	1.00	0.00%	0%		
1935		0.000	0.00	1.00	0.00%	0%		
1936		0.000	0.00	1.00	0.00%	0%		
1937		0.000	0.00	1.00	0.00%	0%		
1938	1	0.001	0.00	1.00	0.05%	0%		
1939	1	0.001	0.00	1.00	0.05%	0%		
1940	2	0.001	0.00	1.00	0.07%	0%		
1941	1	0.000	0.00	1.00	0.04%	0%		
1942	1	0.000	0.00	1.00	0.05%	0%		
1943	1	0.000	0.00	1.00	0.04%	0%		
1944	1	0.000	0.00	1.00	0.03%	0%		
1945	1	0.000	0.00	1.00	0.02%	0%		
1946	1	0.000	0.00	1.00	0.04%	0%		
1947	1	0.000	0.00	1.00	0.04%	0%		
1948	1	0.000	0.00	1.00	0.04%	0%		
1949	1	0.000	0.01	0.99	0.04%	1%		
1950	2	0.001	0.01	0.99	0.09%	1%		
1951	5	0.002	0.01	0.99	0.20%	1%		
1952	7	0.003	0.01	0.99	0.26%	1%		
1953	7	0.003	0.01	0.99	0.26%	1%		
1954	11	0.004	0.02	0.98	0.40%	2%		
1955	19	0.007	0.02	0.98	0.69%	2%		
1956	25	0.009	0.03	0.97	0.93%	3%		
1957	28	0.010	0.04	0.96	1.05%	4%		
1958	28	0.010	0.05	0.95	1.08%	5%		
1959	30	0.011	0.06	0.94	1.15%	6%		
1960	39	0.014	0.08	0.92	1.52%	8%		
1961	42	0.015	0.09	0.91	1.66%	9%		
1962	47	0.017	0.11	0.89	1.89%	11%		
1963	50	0.018	0.13	0.87	2.04%	13%		
1964	56	0.020	0.15	0.85	2.33%	15%		
1965	60	0.022	0.17	0.83	2.57%	17%		1st PEAK
1966	59	0.022	0.19	0.81	2.60%	19%		
1967	57	0.021	0.21	0.79	2.59%	21%		
1968	52	0.019	0.23	0.77	2.40%	23%		
1969	51	0.018	0.25	0.75	2.41%	25%		
1970	48	0.017	0.27	0.73	2.32%	27%		
1971	38	0.014	0.28	0.72	1.91%	28%		
1972	31	0.011	0.29	0.71	1.58%	29%		
1973	26	0.010	0.30	0.70	1.36%	30%		
1974	20	0.007	0.31	0.69	1.06%	31%		
1975	20	0.007	0.32	0.68	1.06%	32%		
1976	20	0.007	0.33	0.67	1.07%	33%		
1977	20	0.007	0.33	0.67	1.08%	33%		
1978	21	0.008	0.34	0.66	1.15%	34%		
1979	20	0.007	0.35	0.65	1.11%	35%		
1980	25	0.009	0.36	0.64	1.40%	36%		
1981	34	0.012	0.37	0.63	1.93%	37%		
1982	33	0.012	0.38	0.62	1.91%	38%		
1983	32	0.012	0.39	0.61	1.89%	39%		
1984	35	0.013	0.41	0.59	2.10%	41%		
1985	49	0.018	0.42	0.58	3.01%	42%		
1986	59	0.022	0.45	0.55	3.74%	45%		
1987	64	0.023	0.47	0.53	4.18%	47%		
1988	69	0.025	0.49	0.51	4.72%	49%		2nd PEAK

1989	65	0.024	0.52	0.48	4.70%	52%	50% PRODUCED
1990	62	0.023	0.54	0.46	4.70%	54%	
1991	59	0.022	0.56	0.44	4.70%	56%	
1992	56	0.021	0.58	0.42	4.70%	58%	
1993	54	0.020	0.60	0.40	4.70%	60%	
1994	51	0.019	0.62	0.38	4.70%	62%	
1995	49	0.018	0.64	0.36	4.70%	64%	
1996	47	0.017	0.66	0.34	4.70%	66%	
1997	44	0.016	0.67	0.33	4.70%	67%	
1998	42	0.015	0.69	0.31	4.70%	69%	
1999	40	0.015	0.70	0.30	4.70%	70%	
2000	38	0.014	0.72	0.28	4.70%	72%	
2001	37	0.013	0.73	0.27	4.70%	73%	
2002	35	0.013	0.74	0.26	4.70%	74%	
2003	33	0.012	0.75	0.25	4.70%	75%	
2004	32	0.012	0.77	0.23	4.70%	77%	
2005	30	0.011	0.78	0.22	4.70%	78%	
2006	29	0.011	0.79	0.21	4.70%	79%	
2007	27	0.010	0.80	0.20	4.70%	80%	
2008	26	0.010	0.81	0.19	4.70%	81%	
2009	25	0.009	0.82	0.18	4.70%	82%	
2010	24	0.009	0.82	0.18	4.70%	82%	
2011	23	0.008	0.83	0.17	4.70%	83%	
2012	22	0.008	0.84	0.16	4.70%	84%	
2013	21	0.008	0.85	0.15	4.70%	85%	
2014	20	0.007	0.86	0.14	4.70%	86%	
2015	19	0.007	0.86	0.14	4.70%	86%	
2016	18	0.006	0.87	0.13	4.70%	87%	
2017	17	0.006	0.87	0.13	4.70%	87%	
2018	16	0.006	0.88	0.12	4.70%	88%	
2019	15	0.006	0.89	0.11	4.70%	89%	
2020	15	0.005	0.89	0.11	4.70%	89%	
2021	14	0.005	0.90	0.10	4.70%	90%	
2022	13	0.005	0.90	0.10	4.70%	90%	
2023	13	0.005	0.91	0.09	4.70%	91%	
2024	12	0.004	0.91	0.09	4.70%	91%	
2025	12	0.004	0.91	0.09	4.70%	91%	
2026	11	0.004	0.92	0.08	4.70%	92%	
2027	10	0.004	0.92	0.08	4.70%	92%	
2028	10	0.004	0.93	0.07	4.70%	93%	
2029	10	0.003	0.93	0.07	4.70%	93%	
2030	9	0.003	0.93	0.07	4.70%	93%	
2031	9	0.003	0.94	0.06	4.70%	94%	
2032	8	0.003	0.94	0.06	4.70%	94%	
2033	8	0.003	0.94	0.06	4.70%	94%	
2034	7	0.003	0.94	0.06	4.70%	94%	
2035	7	0.003	0.95	0.05	4.70%	95%	
2036	7	0.002	0.95	0.05	4.70%	95%	
2037	6	0.002	0.95	0.05	4.70%	95%	
2038	6	0.002	0.95	0.05	4.70%	95%	
2039	6	0.002	0.96	0.04	4.70%	96%	
2040	6	0.002	0.96	0.04	4.70%	96%	
2041	5	0.002	0.96	0.04	4.70%	96%	
2042	5	0.002	0.96	0.04	4.70%	96%	
2043	5	0.002	0.96	0.04	4.70%	96%	
2044	5	0.002	0.97	0.03	4.70%	97%	
2045	4	0.002	0.97	0.03	4.70%	97%	
2046	4	0.002	0.97	0.03	4.70%	97%	
2047	4	0.001	0.97	0.03	4.70%	97%	
2048	4	0.001	0.97	0.03	4.70%	97%	
2049	4	0.001	0.97	0.03	4.70%	97%	
2050	3	0.001	0.97	0.03	4.70%	97%	

CUM.PROD	1.323 BBO	GIANT FIELDS			GABON
RESERVES	0.733 BBO	TOTAL BBO	0.5		AFRICA
DISCOVERED	2.056 BBO	% OF DISC.	24%		
UNDISCOVERED	0.444 BBO	Last Giant	1970		
REMAINING	1.177 BBO				Variance Trend
ULTIMATE	2.500 BBO				1990 +44%
Rept.Reserves	0.733 BBO	Ref.Date (end)		1989	Future UP

Milestones		Prod Mb/d	Dep Rate	% Produced
1st Peak	1977	225	4.20%	25%
2nd Peak	1989	197	5.77%	53%
Mid-Point	1988	175	4.87%	50%
Final Decline	1990		5.77%	

Forecast "A"
A=Exponential decline
B=5% inc. to Mid-Point
C=Other
Pre-1930 Cum 0
(Added to 1930 Prod)

Date	Prod. M.b/d	Prod. BBO/yr	Cum Prod	Remain	Dep. Rate	% Produced	Giant	Names & Notes
1930		0.000	0.00	2.50	ERR	0%		
1931		0.000	0.00	2.50	0.00%	0%		
1932		0.000	0.00	2.50	0.00%	0%		
1933		0.000	0.00	2.50	0.00%	0%		
1934		0.000	0.00	2.50	0.00%	0%		
1935		0.000	0.00	2.50	0.00%	0%		
1936		0.000	0.00	2.50	0.00%	0%		
1937		0.000	0.00	2.50	0.00%	0%		
1938		0.000	0.00	2.50	0.00%	0%		
1939		0.000	0.00	2.50	0.00%	0%		
1940		0.000	0.00	2.50	0.00%	0%		
1941		0.000	0.00	2.50	0.00%	0%		
1942		0.000	0.00	2.50	0.00%	0%		
1943		0.000	0.00	2.50	0.00%	0%		
1944		0.000	0.00	2.50	0.00%	0%		
1945		0.000	0.00	2.50	0.00%	0%		
1946		0.000	0.00	2.50	0.00%	0%		
1947		0.000	0.00	2.50	0.00%	0%		
1948		0.000	0.00	2.50	0.00%	0%		
1949		0.000	0.00	2.50	0.00%	0%		
1950		0.000	0.00	2.50	0.00%	0%		
1951		0.000	0.00	2.50	0.00%	0%		
1952		0.000	0.00	2.50	0.00%	0%		
1953		0.000	0.00	2.50	0.00%	0%		
1954		0.000	0.00	2.50	0.00%	0%		
1955		0.000	0.00	2.50	0.00%	0%		
1956		0.000	0.00	2.50	0.00%	0%		
1957		0.000	0.00	2.50	0.00%	0%		
1958	8	0.003	0.00	2.50	0.12%	0%		
1959	15	0.005	0.01	2.49	0.21%	0%		
1960	16	0.006	0.01	2.49	0.23%	1%		
1961	15	0.005	0.02	2.48	0.22%	1%		
1962	17	0.006	0.03	2.47	0.25%	1%		
1963	17	0.006	0.03	2.47	0.25%	1%		
1964	22	0.008	0.04	2.46	0.33%	2%		
1965	25	0.009	0.05	2.45	0.37%	2%		
1966	30	0.011	0.06	2.44	0.45%	2%		
1967	72	0.026	0.09	2.41	1.07%	3%		
1968	90	0.033	0.12	2.38	1.36%	5%		
1969	94	0.034	0.15	2.35	1.43%	6%		
1970	106	0.039	0.19	2.31	1.65%	8%	0.5	Grondin
1971	94	0.034	0.23	2.27	1.49%	9%		Date uncertain
1972	125	0.046	0.27	2.23	2.01%	11%		
1973	145	0.053	0.32	2.18	2.38%	13%		
1974	182	0.066	0.39	2.11	3.05%	16%		
1975	200	0.073	0.46	2.04	3.46%	19%		
1976	220	0.080	0.54	1.96	3.94%	22%		
1977	225	0.082	0.63	1.87	4.20%	25%		1ST PEAK
1978	170	0.062	0.69	1.81	3.31%	28%		
1979	192	0.070	0.76	1.74	3.87%	30%		
1980	145	0.053	0.81	1.69	3.04%	32%		
1981	147	0.054	0.87	1.63	3.18%	35%		
1982	130	0.047	0.91	1.59	2.90%	37%		
1983	150	0.055	0.97	1.53	3.45%	39%		
1984	150	0.055	1.02	1.48	3.57%	41%		
1985	150	0.055	1.08	1.42	3.70%	43%		
1986	146	0.053	1.13	1.37	3.75%	45%		
1987	156	0.057	1.19	1.31	4.15%	47%		
1988	175	0.064	1.25	1.25	4.87%	50%		50% PRODUCED

1989	197	0.072	1.32	1.18	5.77%	53%	2ND PEAK
1990	186	0.068	1.39	1.11	5.77%	56%	
1991	175	0.064	1.45	1.05	5.77%	58%	
1992	165	0.060	1.52	0.98	5.77%	61%	
1993	156	0.057	1.57	0.93	5.77%	63%	
1994	147	0.054	1.63	0.87	5.77%	65%	
1995	138	0.050	1.68	0.82	5.77%	67%	
1996	130	0.048	1.72	0.78	5.77%	69%	
1997	123	0.045	1.77	0.73	5.77%	71%	
1998	116	0.042	1.81	0.69	5.77%	72%	
1999	109	0.040	1.85	0.65	5.77%	74%	
2000	103	0.037	1.89	0.61	5.77%	76%	
2001	97	0.035	1.92	0.58	5.77%	77%	
2002	91	0.033	1.96	0.54	5.77%	78%	
2003	86	0.031	1.99	0.51	5.77%	80%	
2004	81	0.030	2.02	0.48	5.77%	81%	
2005	76	0.028	2.05	0.45	5.77%	82%	
2006	72	0.026	2.07	0.43	5.77%	83%	
2007	68	0.025	2.10	0.40	5.77%	84%	
2008	64	0.023	2.12	0.38	5.77%	85%	
2009	60	0.022	2.14	0.36	5.77%	86%	
2010	57	0.021	2.16	0.34	5.77%	86%	
2011	53	0.019	2.18	0.32	5.77%	87%	
2012	50	0.018	2.20	0.30	5.77%	88%	
2013	47	0.017	2.22	0.28	5.77%	89%	
2014	45	0.016	2.23	0.27	5.77%	89%	
2015	42	0.015	2.25	0.25	5.77%	90%	
2016	40	0.014	2.26	0.24	5.77%	91%	
2017	37	0.014	2.28	0.22	5.77%	91%	
2018	35	0.013	2.29	0.21	5.77%	92%	
2019	33	0.012	2.30	0.20	5.77%	92%	
2020	31	0.011	2.31	0.19	5.77%	93%	
2021	29	0.011	2.32	0.18	5.77%	93%	
2022	28	0.010	2.33	0.17	5.77%	93%	
2023	26	0.010	2.34	0.16	5.77%	94%	
2024	25	0.009	2.35	0.15	5.77%	94%	
2025	23	0.008	2.36	0.14	5.77%	94%	
2026	22	0.008	2.37	0.13	5.77%	95%	
2027	21	0.008	2.38	0.12	5.77%	95%	
2028	19	0.007	2.38	0.12	5.77%	95%	
2029	18	0.007	2.39	0.11	5.77%	96%	
2030	17	0.006	2.40	0.10	5.77%	96%	
2031	16	0.006	2.40	0.10	5.77%	96%	
2032	15	0.006	2.41	0.09	5.77%	96%	
2033	14	0.005	2.41	0.09	5.77%	97%	
2034	14	0.005	2.42	0.08	5.77%	97%	
2035	13	0.005	2.42	0.08	5.77%	97%	
2036	12	0.004	2.43	0.07	5.77%	97%	
2037	11	0.004	2.43	0.07	5.77%	97%	
2038	11	0.004	2.44	0.06	5.77%	97%	
2039	10	0.004	2.44	0.06	5.77%	98%	
2040	10	0.003	2.44	0.06	5.77%	98%	
2041	9	0.003	2.45	0.05	5.77%	98%	
2042	8	0.003	2.45	0.05	5.77%	98%	
2043	8	0.003	2.45	0.05	5.77%	98%	
2044	8	0.003	2.46	0.04	5.77%	98%	
2045	7	0.003	2.46	0.04	5.77%	98%	
2046	7	0.002	2.46	0.04	5.77%	98%	
2047	6	0.002	2.46	0.04	5.77%	99%	
2048	6	0.002	2.46	0.04	5.77%	99%	
2049	6	0.002	2.47	0.03	5.77%	99%	
2050	5	0.002	2.47	0.03	5.77%	99%	

CUM.PROD	1.643 BBO	GIANT FIELDS		GERMANY	
RESERVES	0.450 BBO	TOTAL BBO	0	W.EUROPE	
DISCOVERED	2.093 BBO	% OF DISC.	0%	(inc. E.Germany)	
UNDISCOVERED	0.157 BBO	Last Giant	-		
REMAINING	0.607 BBO			Variance	Trend
ULTIMATE	2.250 BBO			1990	+6.8%
Rept.Reserve	0.450 BBO	Ref.Date(end)	1989	Future	DOWN

Milestones	Date	Prod Mb/d	Dep. Rate	% Produced
1st Peak	1968	159	3.76%	36%
2nd Peak				
Mid-Point	1974	132	3.95%	50%
Final Decline	1989		4.15%	

Forecast "A"
A=Exponential decline
B=5% incr. to Mid-Point
C=Other
Pre-1930 Cum. 0.005
(Added to 1930 Prod)

Date	Prod. M.b/d	Prod. BBO/yr	Cum Prod	Remain	Dep. Rate	% Produced	Giant	Names & Notes
1930	2.74	0.001	0.04	2.24	ERR	2%		
1931	3.84	0.001	0.04	2.24	0.06%	2%		
1932	4.94	0.002	0.04	2.24	0.08%	2%		
1933	6.04	0.002	0.05	2.24	0.10%	2%		
1934	7.14	0.003	0.05	2.24	0.12%	2%		
1935	8.24	0.003	0.05	2.23	0.13%	2%		
1936	9.34	0.003	0.05	2.23	0.15%	2%		
1937	10.44	0.004	0.06	2.23	0.17%	3%		
1938	10.60	0.004	0.06	2.22	0.17%	3%		
1939	12.30	0.004	0.07	2.22	0.20%	3%		
1940	20.20	0.007	0.07	2.21	0.33%	3%		
1941	17.30	0.006	0.08	2.20	0.29%	4%		
1942	14.20	0.005	0.09	2.20	0.24%	4%		
1943	13.70	0.005	0.09	2.19	0.23%	4%		
1944	16.80	0.006	0.10	2.19	0.28%	4%		
1945	10.80	0.004	0.10	2.18	0.18%	4%		
1946	12.10	0.004	0.10	2.18	0.20%	5%		
1947	11.00	0.004	0.11	2.18	0.18%	5%		
1948	12.10	0.004	0.11	2.17	0.20%	5%		
1949	15.60	0.006	0.12	2.16	0.26%	5%		
1950	21.00	0.008	0.13	2.16	0.35%	6%		
1951	26.50	0.010	0.14	2.15	0.45%	6%		
1952	32.80	0.012	0.15	2.14	0.56%	7%		
1953	44.51	0.016	0.16	2.12	0.76%	7%		
1954	51.50	0.019	0.18	2.10	0.89%	8%		
1955	61.50	0.022	0.21	2.08	1.07%	9%		
1956	68.00	0.025	0.23	2.05	1.19%	10%		
1957	76.30	0.028	0.26	2.03	1.36%	11%		
1958	86.90	0.032	0.29	1.99	1.57%	13%		
1959	98.80	0.036	0.33	1.96	1.81%	15%		
1960	110.00	0.040	0.37	1.92	2.05%	16%		
1961	123.70	0.045	0.41	1.87	2.35%	18%		
1962	133.30	0.049	0.46	1.82	2.60%	20%		
1963	144.30	0.053	0.51	1.77	2.89%	23%		
1964	152.30	0.056	0.57	1.72	3.14%	25%		
1965	158.00	0.058	0.63	1.66	3.36%	28%		
1966	156.40	0.057	0.68	1.60	3.44%	30%		
1967	158.60	0.058	0.74	1.54	3.62%	33%		
1968	158.90	0.058	0.80	1.48	3.76%	36%		1st PEAK
1969	156.70	0.057	0.86	1.43	3.85%	38%		
1970	148.30	0.054	0.91	1.37	3.79%	40%		
1971	147.00	0.054	0.96	1.32	3.91%	43%		
1972	144.30	0.053	1.02	1.27	3.99%	45%		
1973	139.80	0.051	1.07	1.22	4.03%	47%		
1974	131.70	0.048	1.12	1.17	3.95%	50%		50% PRODUCED
1975	118.80	0.043	1.16	1.12	3.71%	52%		
1976	114.30	0.042	1.20	1.08	3.71%	53%		
1977	114.80	0.042	1.24	1.04	3.87%	55%		
1978	108.80	0.040	1.28	1.00	3.81%	57%		
1979	95.80	0.035	1.32	0.97	3.49%	59%		
1980	116.10	0.042	1.36	0.92	4.39%	60%		
1981	112.00	0.041	1.40	0.88	4.42%	62%		
1982	94.00	0.034	1.44	0.85	3.89%	64%		
1983	82.20	0.030	1.47	0.82	3.53%	65%		
1984	88.40	0.032	1.50	0.79	3.94%	67%		
1985	90.00	0.033	1.53	0.75	4.18%	68%		
1986	80.80	0.029	1.56	0.72	3.91%	69%		
1987	74.30	0.027	1.59	0.70	3.74%	71%		
1988	79.10	0.029	1.62	0.67	4.14%	72%		

1989	75.90	0.028	1.64	0.64	4.15%	73%
1990	72.83	0.027	1.67	0.61	4.15%	74%
1991	69.80	0.025	1.70	0.59	4.15%	75%
1992	66.91	0.024	1.72	0.56	4.15%	76%
1993	64.13	0.023	1.74	0.54	4.15%	77%
1994	61.47	0.022	1.77	0.52	4.15%	78%
1995	58.92	0.022	1.79	0.50	4.15%	79%
1996	56.47	0.021	1.81	0.48	4.15%	80%
1997	54.13	0.020	1.83	0.46	4.15%	81%
1998	51.88	0.019	1.85	0.44	4.15%	82%
1999	49.73	0.018	1.86	0.42	4.15%	83%
2000	47.67	0.017	1.88	0.40	4.15%	84%
2001	45.69	0.017	1.90	0.39	4.15%	84%
2002	43.79	0.016	1.91	0.37	4.15%	85%
2003	41.97	0.015	1.93	0.35	4.15%	86%
2004	40.23	0.015	1.94	0.34	4.15%	86%
2005	38.56	0.014	1.96	0.33	4.15%	87%
2006	36.96	0.013	1.97	0.31	4.15%	88%
2007	35.43	0.013	1.99	0.30	4.15%	88%
2008	33.96	0.012	2.00	0.29	4.15%	89%
2009	32.55	0.012	2.01	0.27	4.15%	89%
2010	31.20	0.011	2.02	0.26	4.15%	90%
2011	29.90	0.011	2.03	0.25	4.15%	90%
2012	28.66	0.010	2.04	0.24	4.15%	91%
2013	27.47	0.010	2.05	0.23	4.15%	91%
2014	26.33	0.010	2.06	0.22	4.15%	92%
2015	25.24	0.009	2.07	0.21	4.15%	92%
2016	24.19	0.009	2.08	0.20	4.15%	92%
2017	23.19	0.008	2.09	0.20	4.15%	93%
2018	22.23	0.008	2.10	0.19	4.15%	93%
2019	21.30	0.008	2.10	0.18	4.15%	94%
2020	20.42	0.007	2.11	0.17	4.15%	94%
2021	19.57	0.007	2.12	0.16	4.15%	94%
2022	18.76	0.007	2.13	0.16	4.15%	94%
2023	17.98	0.007	2.13	0.15	4.15%	95%
2024	17.24	0.006	2.14	0.15	4.15%	95%
2025	16.52	0.006	2.14	0.14	4.15%	95%
2026	15.83	0.006	2.15	0.13	4.15%	96%
2027	15.18	0.006	2.16	0.13	4.15%	96%
2028	14.55	0.005	2.16	0.12	4.15%	96%
2029	13.94	0.005	2.17	0.12	4.15%	96%
2030	13.37	0.005	2.17	0.11	4.15%	97%
2031	12.81	0.005	2.18	0.11	4.15%	97%
2032	12.28	0.004	2.18	0.10	4.15%	97%
2033	11.77	0.004	2.18	0.10	4.15%	97%
2034	11.28	0.004	2.19	0.10	4.15%	97%
2035	10.81	0.004	2.19	0.09	4.15%	97%
2036	10.36	0.004	2.20	0.09	4.15%	98%
2037	9.93	0.004	2.20	0.08	4.15%	98%
2038	9.52	0.003	2.20	0.08	4.15%	98%
2039	9.13	0.003	2.21	0.08	4.15%	98%
2040	8.75	0.003	2.21	0.07	4.15%	98%
2041	8.38	0.003	2.21	0.07	4.15%	98%
2042	8.04	0.003	2.22	0.07	4.15%	98%
2043	7.70	0.003	2.22	0.06	4.15%	99%
2044	7.38	0.003	2.22	0.06	4.15%	99%
2045	7.08	0.003	2.22	0.06	4.15%	99%
2046	6.78	0.002	2.23	0.06	4.15%	99%
2047	6.50	0.002	2.23	0.05	4.15%	99%
2048	6.23	0.002	2.23	0.05	4.15%	99%
2049	5.97	0.002	2.23	0.05	4.15%	99%
2050	5.73	0.002	2.24	0.05	4.15%	99%

```
CUM.PROD          0.005 BBO     GIANT FIELDS                    GHANA
RESERVES          0.001 BBO     TOTAL BBO         0.0           AFRICA
DISCOVERED        0.006 BBO     % OF DISC.          0%
UNDISCOVERED      0.000 BBO     Last Giant                      ----------------
REMAINING         0.001 BBO     -------------------             Variance Trend
ULTIMATE          0.006 BBO                                     1990    N/A
Rept.Reserves     0.001 BBO     Ref.Date (end)        1989      Future N/A
-------------------------------------------------------------------------------
Milestones                 Prod    Dep          %          Forecast          "A"
                   Date    Mb/d    Rate     Produced       A=Exponential decline
     1st Peak      1979       4    21.29%      21%         B=5% inc. to Mid-Point
     2nd Peak                                              C=Other
     Mid-Point     1982       2    16.78%      55%         Pre-1930 Cum        0
Final Decline              Resources exhausted             (Added to 1930 Prod)
===============================================================================
```

Date	Prod. M.b/d	Prod. BBO/yr	Cum Prod	Remain	Dep. Rate	% Produced	Giant Names & Notes
1930		0.000	0.00	0.01	ERR	0%	
1931		0.000	0.00	0.01	0.00%	0%	
1932		0.000	0.00	0.01	0.00%	0%	
1933		0.000	0.00	0.01	0.00%	0%	
1934		0.000	0.00	0.01	0.00%	0%	
1935		0.000	0.00	0.01	0.00%	0%	
1936		0.000	0.00	0.01	0.00%	0%	
1937		0.000	0.00	0.01	0.00%	0%	
1938		0.000	0.00	0.01	0.00%	0%	
1939		0.000	0.00	0.01	0.00%	0%	
1940		0.000	0.00	0.01	0.00%	0%	
1941		0.000	0.00	0.01	0.00%	0%	
1942		0.000	0.00	0.01	0.00%	0%	
1943		0.000	0.00	0.01	0.00%	0%	
1944		0.000	0.00	0.01	0.00%	0%	
1945		0.000	0.00	0.01	0.00%	0%	
1946		0.000	0.00	0.01	0.00%	0%	
1947		0.000	0.00	0.01	0.00%	0%	
1948		0.000	0.00	0.01	0.00%	0%	
1949		0.000	0.00	0.01	0.00%	0%	
1950		0.000	0.00	0.01	0.00%	0%	
1951		0.000	0.00	0.01	0.00%	0%	
1952		0.000	0.00	0.01	0.00%	0%	
1953		0.000	0.00	0.01	0.00%	0%	
1954		0.000	0.00	0.01	0.00%	0%	
1955		0.000	0.00	0.01	0.00%	0%	
1956		0.000	0.00	0.01	0.00%	0%	
1957		0.000	0.00	0.01	0.00%	0%	
1958		0.000	0.00	0.01	0.00%	0%	
1959		0.000	0.00	0.01	0.00%	0%	
1960		0.000	0.00	0.01	0.00%	0%	
1961		0.000	0.00	0.01	0.00%	0%	
1962		0.000	0.00	0.01	0.00%	0%	
1963		0.000	0.00	0.01	0.00%	0%	
1964		0.000	0.00	0.01	0.00%	0%	
1965		0.000	0.00	0.01	0.00%	0%	
1966		0.000	0.00	0.01	0.00%	0%	
1967		0.000	0.00	0.01	0.00%	0%	
1968		0.000	0.00	0.01	0.00%	0%	
1969		0.000	0.00	0.01	0.00%	0%	
1970		0.000	0.00	0.01	0.00%	0%	
1971		0.000	0.00	0.01	0.00%	0%	
1972		0.000	0.00	0.01	0.00%	0%	
1973		0.000	0.00	0.01	0.00%	0%	
1974		0.000	0.00	0.01	0.00%	0%	
1975		0.000	0.00	0.01	0.00%	0%	
1976		0.000	0.00	0.01	0.00%	0%	
1977		0.000	0.00	0.01	0.00%	0%	
1978		0.000	0.00	0.01	0.00%	0%	
1979	4	0.001	0.00	0.00	21.29%	21%	
1980	2	0.001	0.00	0.00	15.46%	33%	
1981	2	0.001	0.00	0.00	18.28%	46%	
1982	2	0.001	0.00	0.00	16.78%	55%	
1983	1	0.000	0.00	0.00	16.13%	62%	
1984	1	0.000	0.00	0.00	9.62%	66%	
1985	2	0.001	0.00	0.00	35.47%	78%	
1986	0	0.000	0.00	0.00	8.25%	80%	
1987	0	0.000	0.00	0.00	0.00%	80%	
1988	0	0.000	0.00	0.00	0.00%	80%	

1989	0	0.000	0.00	0.00	0.00%	80%
1990	0	0.000	0.00	0.00	0.00%	80%
1991	0	0.000	0.00	0.00	0.00%	80%
1992	0	0.000	0.00	0.00	0.00%	80%
1993	0	0.000	0.00	0.00	0.00%	80%
1994	0	0.000	0.00	0.00	0.00%	80%
1995	0	0.000	0.00	0.00	0.00%	80%
1996	0	0.000	0.00	0.00	0.00%	80%
1997	0	0.000	0.00	0.00	0.00%	80%
1998	0	0.000	0.00	0.00	0.00%	80%
1999	0	0.000	0.00	0.00	0.00%	80%
2000	0	0.000	0.00	0.00	0.00%	80%
2001	0	0.000	0.00	0.00	0.00%	80%
2002	0	0.000	0.00	0.00	0.00%	80%
2003	0	0.000	0.00	0.00	0.00%	80%
2004	0	0.000	0.00	0.00	0.00%	80%
2005	0	0.000	0.00	0.00	0.00%	80%
2006	0	0.000	0.00	0.00	0.00%	80%
2007	0	0.000	0.00	0.00	0.00%	80%
2008	0	0.000	0.00	0.00	0.00%	80%
2009	0	0.000	0.00	0.00	0.00%	80%
2010	0	0.000	0.00	0.00	0.00%	80%
2011	0	0.000	0.00	0.00	0.00%	80%
2012	0	0.000	0.00	0.00	0.00%	80%
2013	0	0.000	0.00	0.00	0.00%	80%
2014	0	0.000	0.00	0.00	0.00%	80%
2015	0	0.000	0.00	0.00	0.00%	80%
2016	0	0.000	0.00	0.00	0.00%	80%
2017	0	0.000	0.00	0.00	0.00%	80%
2018	0	0.000	0.00	0.00	0.00%	80%
2019	0	0.000	0.00	0.00	0.00%	80%
2020	0	0.000	0.00	0.00	0.00%	80%
2021	0	0.000	0.00	0.00	0.00%	80%
2022	0	0.000	0.00	0.00	0.00%	80%
2023	0	0.000	0.00	0.00	0.00%	80%
2024	0	0.000	0.00	0.00	0.00%	80%
2025	0	0.000	0.00	0.00	0.00%	80%
2026	0	0.000	0.00	0.00	0.00%	80%
2027	0	0.000	0.00	0.00	0.00%	80%
2028	0	0.000	0.00	0.00	0.00%	80%
2029	0	0.000	0.00	0.00	0.00%	80%
2030	0	0.000	0.00	0.00	0.00%	80%
2031	0	0.000	0.00	0.00	0.00%	80%
2032	0	0.000	0.00	0.00	0.00%	80%
2033	0	0.000	0.00	0.00	0.00%	80%
2034	0	0.000	0.00	0.00	0.00%	80%
2035	0	0.000	0.00	0.00	0.00%	80%
2036	0	0.000	0.00	0.00	0.00%	80%
2037	0	0.000	0.00	0.00	0.00%	80%
2038	0	0.000	0.00	0.00	0.00%	80%
2039	0	0.000	0.00	0.00	0.00%	80%
2040	0	0.000	0.00	0.00	0.00%	80%
2041	0	0.000	0.00	0.00	0.00%	80%
2042	0	0.000	0.00	0.00	0.00%	80%
2043	0	0.000	0.00	0.00	0.00%	80%
2044	0	0.000	0.00	0.00	0.00%	80%
2045	0	0.000	0.00	0.00	0.00%	80%
2046	0	0.000	0.00	0.00	0.00%	80%
2047	0	0.000	0.00	0.00	0.00%	80%
2048	0	0.000	0.00	0.00	0.00%	80%
2049	0	0.000	0.00	0.00	0.00%	80%
2050	0	0.000	0.00	0.00	0.00%	80%

CUM.PROD	0.072 BBO	GIANT FIELDS		GREECE	
RESERVES	0.073 BBO	TOTAL BBO	0	W.EUROPE	
DISCOVERED	0.145 BBO	% OF DISC.	0%		
UNDISCOVERED	0.005 BBO	Last Giant	-	----------------	
REMAINING	0.078 BBO	-------------------		Variance	Trend
ULTIMATE	0.150 BBO			1990	-6.3%
Rept.Reserve	0.073 BBO	Ref.Date(end)	1989	Future	DOWN

Milestones	Date	Prod Mb/d	Dep. Rate	% Produced	Forecast "A"
					A=Exponential decline
1st Peak	1984	27	7.56%	20%	B=5% incr. to Mid-Point
2nd Peak					C=Other
Mid-Point	1990	16	7.36%	52%	Pre-1930 Cum. 0
Final Decline	1989		7.36%		(Added to 1930 Prod)

Date	Prod. M.b/d	Prod. BBO/yr	Cum Prod	Remain	Dep. Rate	% Produced	Giant	Names & Notes
1930		0.000	0.00	0.15	ERR	0%		
1931		0.000	0.00	0.15	0.00%	0%		
1932		0.000	0.00	0.15	0.00%	0%		
1933		0.000	0.00	0.15	0.00%	0%		
1934		0.000	0.00	0.15	0.00%	0%		
1935		0.000	0.00	0.15	0.00%	0%		
1936		0.000	0.00	0.15	0.00%	0%		
1937		0.000	0.00	0.15	0.00%	0%		
1938		0.000	0.00	0.15	0.00%	0%		
1939		0.000	0.00	0.15	0.00%	0%		
1940		0.000	0.00	0.15	0.00%	0%		
1941		0.000	0.00	0.15	0.00%	0%		
1942		0.000	0.00	0.15	0.00%	0%		
1943		0.000	0.00	0.15	0.00%	0%		
1944		0.000	0.00	0.15	0.00%	0%		
1945		0.000	0.00	0.15	0.00%	0%		
1946		0.000	0.00	0.15	0.00%	0%		
1947		0.000	0.00	0.15	0.00%	0%		
1948		0.000	0.00	0.15	0.00%	0%		
1949		0.000	0.00	0.15	0.00%	0%		
1950		0.000	0.00	0.15	0.00%	0%		
1951		0.000	0.00	0.15	0.00%	0%		
1952		0.000	0.00	0.15	0.00%	0%		
1953		0.000	0.00	0.15	0.00%	0%		
1954		0.000	0.00	0.15	0.00%	0%		
1955		0.000	0.00	0.15	0.00%	0%		
1956		0.000	0.00	0.15	0.00%	0%		
1957		0.000	0.00	0.15	0.00%	0%		
1958		0.000	0.00	0.15	0.00%	0%		
1959		0.000	0.00	0.15	0.00%	0%		
1960		0.000	0.00	0.15	0.00%	0%		
1961		0.000	0.00	0.15	0.00%	0%		
1962		0.000	0.00	0.15	0.00%	0%		
1963		0.000	0.00	0.15	0.00%	0%		
1964		0.000	0.00	0.15	0.00%	0%		
1965		0.000	0.00	0.15	0.00%	0%		
1966		0.000	0.00	0.15	0.00%	0%		
1967		0.000	0.00	0.15	0.00%	0%		
1968		0.000	0.00	0.15	0.00%	0%		
1969		0.000	0.00	0.15	0.00%	0%		
1970		0.000	0.00	0.15	0.00%	0%		
1971		0.000	0.00	0.15	0.00%	0%		
1972		0.000	0.00	0.15	0.00%	0%		
1973		0.000	0.00	0.15	0.00%	0%		
1974		0.000	0.00	0.15	0.00%	0%		
1975		0.000	0.00	0.15	0.00%	0%		
1976		0.000	0.00	0.15	0.00%	0%		
1977		0.000	0.00	0.15	0.00%	0%		
1978		0.000	0.00	0.15	0.00%	0%		
1979		0.000	0.00	0.15	0.00%	0%		
1980		0.000	0.00	0.15	0.00%	0%		
1981	8	0.003	0.00	0.15	1.95%	2%		
1982	21	0.008	0.01	0.14	5.21%	7%		
1983	25	0.009	0.02	0.13	6.55%	13%		
1984	27	0.010	0.03	0.12	7.56%	20%		1st PEAK
1985	26	0.010	0.04	0.11	8.00%	26%		
1986	26	0.010	0.05	0.10	8.70%	33%		
1987	24	0.009	0.06	0.09	8.80%	38%		
1988	22	0.008	0.07	0.08	8.62%	44%		

1989	17	0.006	0.07	0.08	7.36%	48%	50% PRODUCED
1990	16	0.006	0.08	0.07	7.36%	52%	
1991	15	0.005	0.08	0.07	7.36%	55%	
1992	14	0.005	0.09	0.06	7.36%	59%	
1993	13	0.005	0.09	0.06	7.36%	62%	
1994	12	0.004	0.10	0.05	7.36%	64%	
1995	11	0.004	0.10	0.05	7.36%	67%	
1996	10	0.004	0.10	0.05	7.36%	70%	
1997	9	0.003	0.11	0.04	7.36%	72%	
1998	9	0.003	0.11	0.04	7.36%	74%	
1999	8	0.003	0.11	0.04	7.36%	76%	
2000	7	0.003	0.12	0.03	7.36%	78%	
2001	7	0.002	0.12	0.03	7.36%	79%	
2002	6	0.002	0.12	0.03	7.36%	81%	
2003	6	0.002	0.12	0.03	7.36%	82%	
2004	5	0.002	0.13	0.02	7.36%	83%	
2005	5	0.002	0.13	0.02	7.36%	85%	
2006	5	0.002	0.13	0.02	7.36%	86%	
2007	4	0.002	0.13	0.02	7.36%	87%	
2008	4	0.001	0.13	0.02	7.36%	88%	
2009	4	0.001	0.13	0.02	7.36%	89%	
2010	3	0.001	0.13	0.02	7.36%	90%	
2011	3	0.001	0.14	0.01	7.36%	90%	
2012	3	0.001	0.14	0.01	7.36%	91%	
2013	3	0.001	0.14	0.01	7.36%	92%	
2014	3	0.001	0.14	0.01	7.36%	92%	
2015	2	0.001	0.14	0.01	7.36%	93%	
2016	2	0.001	0.14	0.01	7.36%	93%	
2017	2	0.001	0.14	0.01	7.36%	94%	
2018	2	0.001	0.14	0.01	7.36%	94%	
2019	2	0.001	0.14	0.01	7.36%	95%	
2020	2	0.001	0.14	0.01	7.36%	95%	
2021	1	0.001	0.14	0.01	7.36%	95%	
2022	1	0.000	0.14	0.01	7.36%	96%	
2023	1	0.000	0.14	0.01	7.36%	96%	
2024	1	0.000	0.14	0.01	7.36%	96%	
2025	1	0.000	0.15	0.00	7.36%	97%	
2026	1	0.000	0.15	0.00	7.36%	97%	
2027	1	0.000	0.15	0.00	7.36%	97%	
2028	1	0.000	0.15	0.00	7.36%	97%	
2029	1	0.000	0.15	0.00	7.36%	98%	
2030	1	0.000	0.15	0.00	7.36%	98%	
2031	1	0.000	0.15	0.00	7.36%	98%	
2032	1	0.000	0.15	0.00	7.36%	98%	
2033	1	0.000	0.15	0.00	7.36%	98%	
2034	1	0.000	0.15	0.00	7.36%	98%	
2035	1	0.000	0.15	0.00	7.36%	98%	
2036	0	0.000	0.15	0.00	7.36%	99%	
2037	0	0.000	0.15	0.00	7.36%	99%	
2038	0	0.000	0.15	0.00	7.36%	99%	
2039	0	0.000	0.15	0.00	7.36%	99%	
2040	0	0.000	0.15	0.00	7.36%	99%	
2041	0	0.000	0.15	0.00	7.36%	99%	
2042	0	0.000	0.15	0.00	7.36%	99%	
2043	0	0.000	0.15	0.00	7.36%	99%	
2044	0	0.000	0.15	0.00	7.36%	99%	
2045	0	0.000	0.15	0.00	7.36%	99%	
2046	0	0.000	0.15	0.00	7.36%	99%	
2047	0	0.000	0.15	0.00	7.36%	99%	
2048	0	0.000	0.15	0.00	7.36%	99%	
2049	0	0.000	0.15	0.00	7.36%	99%	
2050	0	0.000	0.15	0.00	7.36%	100%	

CUM.PROD	0.018 BBO	GIANT FIELDS			GUATEMALA
RESERVES	0.042 BBO	TOTAL BBO	0		L.AMERICA
DISCOVERED	0.060 BBO	% OF DISC.	0%		
UNDISCOVERED	0.040 BBO	Last Giant			
REMAINING	0.082 BBO				Variance Trend
ULTIMATE	0.100 BBO				1990 +3%
Rept.Reserves	0.042 BBO	Ref. Date (end)		1989	Future DOWN

Milestones	Date	Prod Mb/d	Dep. Rate	% Produced
1st Peak	1983	8	2.92%	9%
2nd Peak	2005	8	5.71%	51%
Mid-Point	2005	8	5.71%	51%
Final Decline	2005		5.71%	

Forecast "B"
A=Exponential decline
B=5% inc. to Mid-Point
C=Other
Pre-1930 Cum 0
(added to 1930 Prod)

Date	Prod. M.b/d	Prod. BBO/yr	Cum Prod	Remain	Dep. Rate	% Produced	Giant	Names & Notes
1930		0.000	0.00	0.10	ERR	0%		
1931		0.000	0.00	0.10	0.00%	0%		
1932		0.000	0.00	0.10	0.00%	0%		
1933		0.000	0.00	0.10	0.00%	0%		
1934		0.000	0.00	0.10	0.00%	0%		
1935		0.000	0.00	0.10	0.00%	0%		
1936		0.000	0.00	0.10	0.00%	0%		
1937		0.000	0.00	0.10	0.00%	0%		
1938		0.000	0.00	0.10	0.00%	0%		
1939		0.000	0.00	0.10	0.00%	0%		
1940		0.000	0.00	0.10	0.00%	0%		
1941		0.000	0.00	0.10	0.00%	0%		
1942		0.000	0.00	0.10	0.00%	0%		
1943		0.000	0.00	0.10	0.00%	0%		
1944		0.000	0.00	0.10	0.00%	0%		
1945		0.000	0.00	0.10	0.00%	0%		
1946		0.000	0.00	0.10	0.00%	0%		
1947		0.000	0.00	0.10	0.00%	0%		
1948		0.000	0.00	0.10	0.00%	0%		
1949		0.000	0.00	0.10	0.00%	0%		
1950		0.000	0.00	0.10	0.00%	0%		
1951		0.000	0.00	0.10	0.00%	0%		
1952		0.000	0.00	0.10	0.00%	0%		
1953		0.000	0.00	0.10	0.00%	0%		
1954		0.000	0.00	0.10	0.00%	0%		
1955		0.000	0.00	0.10	0.00%	0%		
1956		0.000	0.00	0.10	0.00%	0%		
1957		0.000	0.00	0.10	0.00%	0%		
1958		0.000	0.00	0.10	0.00%	0%		
1959		0.000	0.00	0.10	0.00%	0%		
1960		0.000	0.00	0.10	0.00%	0%		
1961		0.000	0.00	0.10	0.00%	0%		
1962		0.000	0.00	0.10	0.00%	0%		
1963		0.000	0.00	0.10	0.00%	0%		
1964		0.000	0.00	0.10	0.00%	0%		
1965		0.000	0.00	0.10	0.00%	0%		
1966		0.000	0.00	0.10	0.00%	0%		
1967		0.000	0.00	0.10	0.00%	0%		
1968		0.000	0.00	0.10	0.00%	0%		
1969		0.000	0.00	0.10	0.00%	0%		
1970		0.000	0.00	0.10	0.00%	0%		
1971		0.000	0.00	0.10	0.00%	0%		
1972		0.000	0.00	0.10	0.00%	0%		
1973		0.000	0.00	0.10	0.00%	0%		
1974		0.000	0.00	0.10	0.00%	0%		
1975		0.000	0.00	0.10	0.00%	0%		
1976		0.000	0.00	0.10	0.00%	0%		
1977		0.000	0.00	0.10	0.00%	0%		
1978	0.70	0.000	0.00	0.10	0.26%	0%		
1979	1.60	0.001	0.00	0.10	0.59%	1%		
1980	3.00	0.001	0.00	0.10	1.10%	2%		
1981	5.00	0.002	0.00	0.10	1.86%	4%		
1982	6.40	0.002	0.01	0.09	2.43%	6%		
1983	7.50	0.003	0.01	0.09	2.92%	9%		1ST PEAK
1984	5.20	0.002	0.01	0.09	2.08%	11%		
1985	3.10	0.001	0.01	0.09	1.27%	12%		
1986	4.90	0.002	0.01	0.09	2.03%	14%		
1987	3.90	0.001	0.02	0.08	1.65%	15%		
1988	3.70	0.001	0.02	0.08	1.59%	16%		

1989	3.70	0.001	0.02	0.08	1.62%	18%	
1990	3.89	0.001	0.02	0.08	1.72%	19%	
1991	4.08	0.001	0.02	0.08	1.84%	21%	
1992	4.28	0.002	0.02	0.08	1.97%	22%	
1993	4.50	0.002	0.02	0.08	2.11%	24%	
1994	4.72	0.002	0.03	0.07	2.26%	26%	
1995	4.96	0.002	0.03	0.07	2.43%	27%	
1996	5.21	0.002	0.03	0.07	2.62%	29%	
1997	5.47	0.002	0.03	0.07	2.82%	31%	
1998	5.74	0.002	0.03	0.07	3.05%	33%	
1999	6.03	0.002	0.04	0.06	3.30%	36%	
2000	6.33	0.002	0.04	0.06	3.59%	38%	
2001	6.64	0.002	0.04	0.06	3.91%	40%	
2002	6.98	0.003	0.04	0.06	4.27%	43%	
2003	7.33	0.003	0.05	0.05	4.68%	46%	
2004	7.69	0.003	0.05	0.05	5.16%	48%	
2005	8.08	0.003	0.05	0.05	5.71%	51%	50% PRODUCED
2006	7.62	0.003	0.05	0.05	5.71%	54%	& 2ND PEAK
2007	7.18	0.003	0.06	0.04	5.71%	57%	
2008	6.77	0.002	0.06	0.04	5.71%	59%	
2009	6.38	0.002	0.06	0.04	5.71%	62%	
2010	6.02	0.002	0.06	0.04	5.71%	64%	
2011	5.68	0.002	0.07	0.03	5.71%	66%	
2012	5.35	0.002	0.07	0.03	5.71%	68%	
2013	5.05	0.002	0.07	0.03	5.71%	70%	
2014	4.76	0.002	0.07	0.03	5.71%	71%	
2015	4.49	0.002	0.07	0.03	5.71%	73%	
2016	4.23	0.002	0.07	0.03	5.71%	75%	
2017	3.99	0.001	0.08	0.02	5.71%	76%	
2018	3.76	0.001	0.08	0.02	5.71%	77%	
2019	3.55	0.001	0.08	0.02	5.71%	79%	
2020	3.34	0.001	0.08	0.02	5.71%	80%	
2021	3.15	0.001	0.08	0.02	5.71%	81%	
2022	2.97	0.001	0.08	0.02	5.71%	82%	
2023	2.80	0.001	0.08	0.02	5.71%	83%	
2024	2.64	0.001	0.08	0.02	5.71%	84%	
2025	2.49	0.001	0.08	0.02	5.71%	85%	
2026	2.35	0.001	0.09	0.01	5.71%	86%	
2027	2.22	0.001	0.09	0.01	5.71%	87%	
2028	2.09	0.001	0.09	0.01	5.71%	87%	
2029	1.97	0.001	0.09	0.01	5.71%	88%	
2030	1.86	0.001	0.09	0.01	5.71%	89%	
2031	1.75	0.001	0.09	0.01	5.71%	89%	
2032	1.65	0.001	0.09	0.01	5.71%	90%	
2033	1.56	0.001	0.09	0.01	5.71%	91%	
2034	1.47	0.001	0.09	0.01	5.71%	91%	
2035	1.38	0.001	0.09	0.01	5.71%	92%	
2036	1.31	0.000	0.09	0.01	5.71%	92%	
2037	1.23	0.000	0.09	0.01	5.71%	93%	
2038	1.16	0.000	0.09	0.01	5.71%	93%	
2039	1.09	0.000	0.09	0.01	5.71%	93%	
2040	1.03	0.000	0.09	0.01	5.71%	94%	
2041	0.97	0.000	0.09	0.01	5.71%	94%	
2042	0.92	0.000	0.09	0.01	5.71%	94%	
2043	0.86	0.000	0.09	0.01	5.71%	95%	
2044	0.82	0.000	0.10	0.00	5.71%	95%	
2045	0.77	0.000	0.10	0.00	5.71%	95%	
2046	0.72	0.000	0.10	0.00	5.71%	96%	
2047	0.68	0.000	0.10	0.00	5.71%	96%	
2048	0.64	0.000	0.10	0.00	5.71%	96%	
2049	0.61	0.000	0.10	0.00	5.71%	96%	
2050	0.57	0.000	0.10	0.00	5.71%	97%	

Note : 1989 production low due to political unrest

CUM.PROD	0.540	BBO	GIANT FIELDS		HUNGARY	
RESERVES	0.140	BBO	TOTAL BBO	0	SINO-SOVIET	
DISCOVERED	0.680	BBO	% OF DISC.	0%		
UNDISCOVERED	0.570	BBO	Last Giant	-		
REMAINING	0.710	BBO			Varience	Trend
ULTIMATE	1.250	BBO			1990	n/a
Rept.Reserve	0.130	#1	Ref.date(end)	1989	Future	DOWN

Milestones	Date	Prod Mb/d	Dep. Rate	% Produced
1st Peak	1977	46	1.86%	29%
2nd Peak	None			
Mid-Point	1995	44	2.55%	51%
Final Decline	1996		2.55%	52%

Forecast "B"
A=Exponential decline
B=5% incr. to Mid-Point
C=Other
Pre-1930 Cum. 0
(Added to 1930 prod)

Date	Prod. M.b/d	Prod. BBO/yr	Cum. Prod.	Remain	Deplet. Rate	% Produced	Giant	Notes & Names
1930		0.000	0.00	1.25	ERR	0%		Data-base weak
1931		0.000	0.00	1.25	0.00%	0%		
1932		0.000	0.00	1.25	0.00%	0%		
1933		0.000	0.00	1.25	0.00%	0%		
1934		0.000	0.00	1.25	0.00%	0%		
1935		0.000	0.00	1.25	0.00%	0%		
1936		0.000	0.00	1.25	0.00%	0%		
1937	0	0.000	0.00	1.25	0.00%	0%		
1938	1	0.000	0.00	1.25	0.02%	0%		
1939	3	0.001	0.00	1.25	0.09%	0%		
1940	5	0.002	0.00	1.25	0.15%	0%		
1941	9	0.003	0.01	1.24	0.25%	1%		
1942	14	0.005	0.01	1.24	0.41%	1%		
1943	17	0.006	0.02	1.23	0.51%	1%		
1944	17	0.006	0.02	1.23	0.51%	2%		
1945	14	0.005	0.03	1.22	0.41%	2%		
1946	14	0.005	0.03	1.22	0.42%	3%		
1947	12	0.004	0.04	1.21	0.36%	3%		
1948	10	0.004	0.04	1.21	0.30%	3%		
1949	10	0.004	0.05	1.20	0.31%	4%		
1950	10	0.004	0.05	1.20	0.31%	4%		
1951	10	0.004	0.05	1.20	0.29%	4%		
1952	10	0.004	0.06	1.19	0.29%	5%		
1953	18	0.006	0.06	1.19	0.54%	5%		
1954	25	0.009	0.07	1.18	0.78%	6%		
1955	34	0.012	0.08	1.17	1.04%	7%		
1956	25	0.009	0.09	1.16	0.79%	8%		
1957	14	0.005	0.10	1.15	0.45%	8%		
1958	17	0.006	0.11	1.14	0.55%	8%		
1959	22	0.008	0.11	1.14	0.69%	9%		
1960	25	0.009	0.12	1.13	0.82%	10%		
1961	31	0.011	0.13	1.12	0.99%	11%		
1962	34	0.013	0.15	1.10	1.12%	12%		
1963	37	0.013	0.16	1.09	1.21%	13%		
1964	38	0.014	0.17	1.08	1.26%	14%		
1965	38	0.014	0.19	1.06	1.28%	15%		
1966	36	0.013	0.20	1.05	1.22%	16%		
1967	35	0.013	0.21	1.04	1.22%	17%		
1968	38	0.014	0.23	1.02	1.33%	18%		
1969	37	0.013	0.24	1.01	1.31%	19%		
1970	41	0.015	0.25	1.00	1.46%	20%		
1971	41	0.015	0.27	0.98	1.49%	22%		
1972	41	0.015	0.28	0.97	1.54%	23%		
1973	42	0.015	0.30	0.95	1.57%	24%		
1974	42	0.015	0.32	0.93	1.60%	25%		
1975	42	0.015	0.33	0.92	1.64%	26%		
1976	45	0.016	0.35	0.90	1.78%	28%		
1977	46	0.017	0.36	0.89	1.86%	29%		
1978	43	0.016	0.38	0.87	1.77%	30%		
1979	42	0.015	0.39	0.86	1.75%	32%		
1980	41	0.015	0.41	0.84	1.73%	33%		
1981	42	0.015	0.42	0.83	1.84%	34%		
1982	41	0.015	0.44	0.81	1.79%	35%		
1983	40	0.015	0.45	0.80	1.81%	36%		
1984	42	0.015	0.47	0.78	1.93%	38%		
1985	37	0.014	0.48	0.77	1.75%	39%		
1986	40	0.015	0.50	0.75	1.91%	40%		
1987	38	0.014	0.51	0.74	1.86%	41%		
1988	39	0.014	0.53	0.72	1.92%	42%		

1989	38	0.014	0.54	0.71	1.93%	43%	
1990	40	0.015	0.55	0.70	2.06%	44%	
1991	42	0.015	0.57	0.68	2.21%	46%	
1992	44	0.016	0.59	0.66	2.37%	47%	
1993	46	0.017	0.60	0.65	2.55%	48%	
1994	49	0.018	0.62	0.63	2.55%	50%	50% PRODUCED
1995	44	0.016	0.64	0.61	2.55%	51%	& 1ST PEAK
1996	43	0.016	0.65	0.60	2.55%	52%	
1997	42	0.015	0.67	0.58	2.55%	53%	
1998	41	0.015	0.68	0.57	2.55%	55%	
1999	40	0.014	0.70	0.55	2.55%	56%	
2000	39	0.014	0.71	0.54	2.55%	57%	
2001	38	0.014	0.72	0.53	2.55%	58%	
2002	37	0.013	0.74	0.51	2.55%	59%	
2003	36	0.013	0.75	0.50	2.55%	60%	
2004	35	0.013	0.76	0.49	2.55%	61%	
2005	34	0.012	0.78	0.47	2.55%	62%	
2006	33	0.012	0.79	0.46	2.55%	63%	
2007	32	0.012	0.80	0.45	2.55%	64%	
2008	31	0.011	0.81	0.44	2.55%	65%	
2009	31	0.011	0.82	0.43	2.55%	66%	
2010	30	0.011	0.83	0.42	2.55%	67%	
2011	29	0.011	0.84	0.41	2.55%	68%	
2012	28	0.010	0.85	0.40	2.55%	68%	
2013	28	0.010	0.86	0.39	2.55%	69%	
2014	27	0.010	0.87	0.38	2.55%	70%	
2015	26	0.010	0.88	0.37	2.55%	71%	
2016	26	0.009	0.89	0.36	2.55%	71%	
2017	25	0.009	0.90	0.35	2.55%	72%	
2018	24	0.009	0.91	0.34	2.55%	73%	
2019	24	0.009	0.92	0.33	2.55%	74%	
2020	23	0.008	0.93	0.32	2.55%	74%	
2021	22	0.008	0.94	0.31	2.55%	75%	
2022	22	0.008	0.94	0.31	2.55%	76%	
2023	21	0.008	0.95	0.30	2.55%	76%	
2024	21	0.008	0.96	0.29	2.55%	77%	
2025	20	0.007	0.97	0.28	2.55%	77%	
2026	20	0.007	0.97	0.28	2.55%	78%	
2027	19	0.007	0.98	0.27	2.55%	79%	
2028	19	0.007	0.99	0.26	2.55%	79%	
2029	18	0.007	1.00	0.25	2.55%	80%	
2030	18	0.006	1.00	0.25	2.55%	80%	
2031	17	0.006	1.01	0.24	2.55%	81%	
2032	17	0.006	1.01	0.24	2.55%	81%	
2033	16	0.006	1.02	0.23	2.55%	82%	
2034	16	0.006	1.03	0.22	2.55%	82%	
2035	16	0.006	1.03	0.22	2.55%	83%	
2036	15	0.006	1.04	0.21	2.55%	83%	
2037	15	0.005	1.04	0.21	2.55%	83%	
2038	14	0.005	1.05	0.20	2.55%	84%	
2039	14	0.005	1.05	0.20	2.55%	84%	
2040	14	0.005	1.06	0.19	2.55%	85%	
2041	13	0.005	1.06	0.19	2.55%	85%	
2042	13	0.005	1.07	0.18	2.55%	85%	
2043	13	0.005	1.07	0.18	2.55%	86%	
2044	12	0.005	1.08	0.17	2.55%	86%	
2045	12	0.004	1.08	0.17	2.55%	87%	
2046	12	0.004	1.09	0.16	2.55%	87%	
2047	11	0.004	1.09	0.16	2.55%	87%	
2048	11	0.004	1.09	0.16	2.55%	88%	
2049	11	0.004	1.10	0.15	2.55%	88%	
2050	11	0.004	1.10	0.15	2.55%	88%	

CUM.PROD	3.199 BBO	GIANT FIELDS		INDIA	
RESERVES	7.516 BBO	TOTAL BBO	2.95	THE EAST	
DISCOVERED	10.716 BBO	% OF DISC.	28%		
UNDISCOVERED	4.284 BBO	Last Giant	1976		
REMAINING	11.801 BBO			Variance	Trend
ULTIMATE	15.000 BBO			1990	-4%
Rept.Reserve	7.516	Ref.date (end)	1989	Future	DOWN

Milestones	Date	Prod Mb/d	Dep. Rate	% Produced
1st Peak	2001	1209	5.17%	49%
2nd Peak				
Mid-Point	2002	1147	5.17%	52%
Final Decline	2002		5.17%	52%

Forecast "B"
A=Exponential decline
B=5% incr. to Mid-Point
C=Other
Pre-1930 Cum. 0.04
(Added to 1930 prod)

Date	Prod. M.b/d	Prod. BBO/yr	Cum. Prod.	Remain	Deplet. Rate	% Produced	Giant Notes & Names
1930	5	0.002	0.04	14.96	ERR	0%	
1931	5	0.002	0.04	14.96	0.01%	0%	
1932	5	0.002	0.05	14.95	0.01%	0%	
1933	5	0.002	0.05	14.95	0.01%	0%	
1934	5	0.002	0.05	14.95	0.01%	0%	
1935	5	0.002	0.05	14.95	0.01%	0%	
1936	5	0.002	0.05	14.95	0.01%	0%	
1937	5	0.002	0.05	14.95	0.01%	0%	
1938	7	0.002	0.06	14.94	0.02%	0%	
1939	6	0.002	0.06	14.94	0.02%	0%	
1940	6	0.002	0.06	14.94	0.02%	0%	
1941	8	0.003	0.06	14.94	0.02%	0%	
1942	8	0.003	0.07	14.93	0.02%	0%	
1943	8	0.003	0.07	14.93	0.02%	0%	
1944	8	0.003	0.07	14.93	0.02%	0%	
1945	6	0.002	0.08	14.92	0.02%	1%	
1946	6	0.002	0.08	14.92	0.01%	1%	
1947	5	0.002	0.08	14.92	0.01%	1%	
1948	5	0.002	0.08	14.92	0.01%	1%	
1949	5	0.002	0.08	14.92	0.01%	1%	
1950	5	0.002	0.09	14.91	0.01%	1%	
1951	5	0.002	0.09	14.91	0.01%	1%	
1952	6	0.002	0.09	14.91	0.01%	1%	
1953	6	0.002	0.09	14.91	0.02%	1%	
1954	7	0.003	0.09	14.91	0.02%	1%	
1955	7	0.003	0.10	14.90	0.02%	1%	
1956	9	0.003	0.10	14.90	0.02%	1%	
1957	9	0.003	0.10	14.90	0.02%	1%	
1958	9	0.003	0.11	14.89	0.02%	1%	
1959	9	0.003	0.11	14.89	0.02%	1%	
1960	8	0.003	0.52	14.89	0.02%	3%	
1961	9	0.003	0.52	14.88	0.02%	3%	
1962	10	0.004	0.53	14.88	0.02%	4%	
1963	33	0.012	0.54	14.87	0.08%	4%	
1964	44	0.016	0.56	14.85	0.11%	4%	
1965	61	0.022	0.58	14.83	0.15%	4%	
1966	93	0.034	0.61	14.80	0.23%	4%	
1967	113	0.041	0.65	14.75	0.28%	4%	
1968	129	0.047	0.70	14.71	0.32%	5%	
1969	150	0.055	0.75	14.65	0.37%	5%	
1970	139	0.051	0.81	14.60	0.35%	5%	
1971	144	0.053	0.86	14.55	0.36%	6%	
1972	151	0.055	0.91	14.49	0.38%	6%	
1973	150	0.055	0.97	14.44	0.38%	6%	
1974	149	0.054	1.02	14.39	0.38%	7%	2.40 Bombay High
1975	165	0.060	1.08	14.33	0.42%	7%	0.55 Bassein N
1976	180	0.066	1.15	14.26	0.46%	8%	
1977	200	0.073	1.22	14.19	0.51%	8%	
1978	230	0.084	1.30	14.10	0.59%	9%	
1979	240	0.088	1.39	14.02	0.62%	9%	
1980	185	0.068	1.46	13.95	0.48%	10%	
1981	298	0.109	1.57	13.84	0.78%	10%	
1982	384	0.140	1.71	13.70	1.01%	11%	
1983	390	0.142	1.85	13.56	1.04%	12%	
1984	543	0.198	2.05	13.36	1.46%	14%	
1985	614	0.224	2.27	13.13	1.68%	15%	
1986	623	0.227	2.50	12.91	1.73%	17%	
1987	609	0.222	2.72	12.68	1.72%	18%	
1988	632	0.231	2.95	12.45	1.82%	20%	

1989	673	0.246	3.20	12.21	1.97%	21%	
1990	707	0.258	3.46	11.95	2.11%	23%	
1991	742	0.271	3.73	11.68	2.27%	25%	
1992	779	0.284	4.01	11.39	2.44%	27%	
1993	818	0.299	4.31	11.10	2.62%	29%	
1994	859	0.314	4.63	10.78	2.83%	31%	
1995	902	0.329	4.95	10.45	3.05%	33%	
1996	947	0.346	5.30	10.11	3.31%	35%	
1997	995	0.363	5.66	9.74	3.59%	38%	
1998	1045	0.381	6.04	9.36	3.91%	40%	
1999	1097	0.400	6.44	8.96	4.28%	43%	
2000	1152	0.420	6.87	8.54	4.69%	46%	
2001	1209	0.441	7.31	8.10	5.17%	49%	1ST PEAK
2002	1147	0.419	7.73	7.68	5.17%	52%	50% PRODUCED
2003	1088	0.397	8.12	7.28	5.17%	54%	
2004	1032	0.377	8.50	6.91	5.17%	57%	
2005	979	0.357	8.86	6.55	5.17%	59%	
2006	928	0.339	9.20	6.21	5.17%	61%	
2007	880	0.321	9.52	5.89	5.17%	63%	
2008	834	0.305	9.82	5.59	5.17%	65%	
2009	791	0.289	10.11	5.30	5.17%	67%	
2010	750	0.274	10.38	5.02	5.17%	69%	
2011	712	0.260	10.64	4.76	5.17%	71%	
2012	675	0.246	10.89	4.52	5.17%	73%	
2013	640	0.234	11.12	4.28	5.17%	74%	
2014	607	0.221	11.34	4.06	5.17%	76%	
2015	575	0.210	11.55	3.85	5.17%	77%	
2016	546	0.199	11.75	3.65	5.17%	78%	
2017	518	0.189	11.94	3.46	5.17%	80%	
2018	491	0.179	12.12	3.29	5.17%	81%	
2019	465	0.170	12.29	3.12	5.17%	82%	
2020	441	0.161	12.45	2.95	5.17%	83%	
2021	418	0.153	12.61	2.80	5.17%	84%	
2022	397	0.145	12.75	2.66	5.17%	85%	
2023	376	0.137	12.89	2.52	5.17%	86%	
2024	357	0.130	13.02	2.39	5.17%	87%	
2025	338	0.124	13.14	2.27	5.17%	88%	
2026	321	0.117	13.26	2.15	5.17%	88%	
2027	304	0.111	13.37	2.04	5.17%	89%	
2028	289	0.105	13.48	1.93	5.17%	90%	
2029	274	0.100	13.58	1.83	5.17%	91%	
2030	260	0.095	13.67	1.74	5.17%	91%	
2031	246	0.090	13.76	1.65	5.17%	92%	
2032	233	0.085	13.84	1.56	5.17%	92%	
2033	221	0.081	13.93	1.48	5.17%	93%	
2034	210	0.077	14.00	1.41	5.17%	93%	
2035	199	0.073	14.07	1.33	5.17%	94%	
2036	189	0.069	14.14	1.26	5.17%	94%	
2037	179	0.065	14.21	1.20	5.17%	95%	
2038	170	0.062	14.27	1.14	5.17%	95%	
2039	161	0.059	14.33	1.08	5.17%	96%	
2040	153	0.056	14.39	1.02	5.17%	96%	
2041	145	0.053	14.44	0.97	5.17%	96%	
2042	137	0.050	14.49	0.92	5.17%	97%	
2043	130	0.048	14.54	0.87	5.17%	97%	
2044	123	0.045	14.58	0.83	5.17%	97%	
2045	117	0.043	14.62	0.78	5.17%	97%	
2046	111	0.041	14.66	0.74	5.17%	98%	
2047	105	0.038	14.70	0.70	5.17%	98%	
2048	100	0.036	14.74	0.67	5.17%	98%	
2049	95	0.035	14.77	0.63	5.17%	98%	
2050	90	0.033	14.81	0.60	5.17%	99%	

CUM.PROD	13.900 BBO	GIANT FIELDS		INDONESIA	
RESERVES	8.200 BBO	TOTAL BBO	8.62	THE EAST	
DISCOVERED	22.100 BBO	% OF DISC.	39%		
UNDISCOVERED	0.900 BBO	Last Giant	1974		
REMAINING	9.100 BBO			Variance	Trend
ULTIMATE	23.000 BBO			1990	+10%
Rept.Reserve	8.200	Ref.date(end)	1989	Future	DOWN

Milestones	Date	Prod Mb/d	Dep. Rate	% Produced
1st Peak	1977	1690	3.93%	34%
2nd Peak				
Mid-Point	1984	1332	3.94%	53%
Final Decline	1990		4.62%	62%

Forecast "A"
A=Exponential decline
B=5% incr. to Mid-Point
C=Other
Pre-1930 Cum. 0.76
(Added to 1930 prod)

Date	Prod. M.b/d	Prod. BBO/yr	Cum. Prod.	Remain	Deplet. Rate	% Produced	Giant	Notes & Names
1930	104	0.038	0.80	22.20	ERR	3%		
1931	111	0.040	0.84	22.16	0.18%	4%		
1932	117	0.043	0.88	22.12	0.19%	4%		
1933	124	0.045	0.93	22.07	0.20%	4%		
1934	130	0.048	0.97	22.03	0.22%	4%		
1935	137	0.050	1.02	21.98	0.23%	4%		
1936	144	0.052	1.08	21.92	0.24%	5%		
1937	150	0.055	1.13	21.87	0.25%	5%		
1938	157	0.057	1.19	21.81	0.26%	5%		
1939	171	0.062	1.25	21.75	0.29%	5%		
1940	170	0.062	1.31	21.69	0.29%	6%		
1941	147	0.054	1.37	21.63	0.25%	6%	0.50	Duri
1942	66	0.024	1.39	21.61	0.11%	6%		
1943	132	0.048	1.44	21.56	0.22%	6%		
1944	64	0.023	1.46	21.54	0.11%	6%	4.30	Minas
1945	21	0.008	1.47	21.53	0.04%	6%		
1946	6	0.002	1.47	21.53	0.01%	6%		
1947	22	0.008	1.48	21.52	0.04%	6%		
1948	87	0.032	1.51	21.49	0.15%	7%		
1949	116	0.042	1.55	21.45	0.20%	7%		
1950	141	0.051	1.61	21.39	0.24%	7%		
1951	152	0.055	1.66	21.34	0.26%	7%		
1952	168	0.061	1.72	21.28	0.29%	7%		
1953	202	0.074	1.80	21.20	0.35%	8%		
1954	208	0.076	1.87	21.13	0.36%	8%		
1955	237	0.087	1.96	21.04	0.41%	9%	0.57	Bekasap
1956	257	0.094	2.05	20.95	0.45%	9%		
1957	310	0.113	2.16	20.84	0.54%	9%		
1958	350	0.128	2.29	20.71	0.61%	10%		
1959	338	0.123	2.42	20.58	0.60%	11%		
1960	425	0.155	2.57	20.43	0.75%	11%		
1961	432	0.158	2.73	20.27	0.77%	12%		
1962	458	0.167	2.90	20.10	0.82%	13%		
1963	452	0.165	3.06	19.94	0.82%	13%		
1964	480	0.175	3.24	19.76	0.88%	14%		
1965	531	0.194	3.43	19.57	0.98%	15%		
1966	472	0.172	3.60	19.40	0.88%	16%		
1967	505	0.184	3.79	19.21	0.95%	16%		
1968	582	0.213	4.00	19.00	1.11%	17%		
1969	725	0.264	4.26	18.74	1.39%	19%	0.60	Arjuna
1970	861	0.314	4.58	18.42	1.68%	20%	1.05	Attaka Bangko
1971	880	0.321	4.90	18.10	1.74%	21%	0.70	Arun
1972	1027	0.375	5.27	17.73	2.07%	23%		
1973	1300	0.475	5.75	17.25	2.68%	25%		
1974	1457	0.532	6.28	16.72	3.08%	27%	0.90	Handil
1975	1300	0.475	6.75	16.25	2.84%	29%		
1976	1500	0.548	7.30	15.70	3.37%	32%		
1977	1690	0.617	7.92	15.08	3.93%	34%		1ST PEAK
1978	1650	0.602	8.52	14.48	3.99%	37%		
1979	1600	0.584	9.11	13.89	4.03%	40%		
1980	1570	0.573	9.68	13.32	4.12%	42%		
1981	1607	0.587	10.26	12.74	4.40%	45%		
1982	1341	0.489	10.75	12.25	3.84%	47%		
1983	1292	0.472	11.23	11.77	3.85%	49%		
1984	1332	0.486	11.71	11.29	4.13%	51%		50% PRODUCED
1985	1220	0.445	12.16	10.84	3.94%	53%		
1986	1244	0.454	12.61	10.39	4.19%	55%		
1987	1186	0.433	13.04	9.96	4.17%	57%		
1988	1138	0.415	13.46	9.54	4.17%	59%		

1989	1209	0.441	13.90	9.10	4.62%	60%
1990	1152	0.420	14.32	8.68	4.62%	62%
1991	1099	0.401	14.72	8.28	4.62%	64%
1992	1048	0.382	15.10	7.90	4.62%	66%
1993	999	0.365	15.47	7.53	4.62%	67%
1994	953	0.348	15.82	7.18	4.62%	69%
1995	909	0.332	16.15	6.85	4.62%	70%
1996	867	0.317	16.47	6.53	4.62%	72%
1997	827	0.302	16.77	6.23	4.62%	73%
1998	789	0.288	17.06	5.94	4.62%	74%
1999	752	0.275	17.33	5.67	4.62%	75%
2000	718	0.262	17.59	5.41	4.62%	76%
2001	685	0.250	17.84	5.16	4.62%	78%
2002	653	0.238	18.08	4.92	4.62%	79%
2003	623	0.227	18.31	4.69	4.62%	80%
2004	594	0.217	18.52	4.48	4.62%	81%
2005	567	0.207	18.73	4.27	4.62%	81%
2006	540	0.197	18.93	4.07	4.62%	82%
2007	515	0.188	19.12	3.88	4.62%	83%
2008	492	0.179	19.30	3.70	4.62%	84%
2009	469	0.171	19.47	3.53	4.62%	85%
2010	447	0.163	19.63	3.37	4.62%	85%
2011	427	0.156	19.79	3.21	4.62%	86%
2012	407	0.149	19.93	3.07	4.62%	87%
2013	388	0.142	20.08	2.92	4.62%	87%
2014	370	0.135	20.21	2.79	4.62%	88%
2015	353	0.129	20.34	2.66	4.62%	88%
2016	337	0.123	20.46	2.54	4.62%	89%
2017	321	0.117	20.58	2.42	4.62%	89%
2018	306	0.112	20.69	2.31	4.62%	90%
2019	292	0.107	20.80	2.20	4.62%	90%
2020	279	0.102	20.90	2.10	4.62%	91%
2021	266	0.097	21.00	2.00	4.62%	91%
2022	254	0.093	21.09	1.91	4.62%	92%
2023	242	0.088	21.18	1.82	4.62%	92%
2024	231	0.084	21.26	1.74	4.62%	92%
2025	220	0.080	21.34	1.66	4.62%	93%
2026	210	0.077	21.42	1.58	4.62%	93%
2027	200	0.073	21.49	1.51	4.62%	93%
2028	191	0.070	21.56	1.44	4.62%	94%
2029	182	0.066	21.63	1.37	4.62%	94%
2030	174	0.063	21.69	1.31	4.62%	94%
2031	166	0.060	21.75	1.25	4.62%	95%
2032	158	0.058	21.81	1.19	4.62%	95%
2033	151	0.055	21.86	1.14	4.62%	95%
2034	144	0.052	21.92	1.08	4.62%	95%
2035	137	0.050	21.97	1.03	4.62%	96%
2036	131	0.048	22.01	0.99	4.62%	96%
2037	125	0.046	22.06	0.94	4.62%	96%
2038	119	0.043	22.10	0.90	4.62%	96%
2039	113	0.041	22.15	0.85	4.62%	96%
2040	108	0.039	22.18	0.82	4.62%	96%
2041	103	0.038	22.22	0.78	4.62%	97%
2042	98	0.036	22.26	0.74	4.62%	97%
2043	94	0.034	22.29	0.71	4.62%	97%
2044	90	0.033	22.33	0.67	4.62%	97%
2045	85	0.031	22.36	0.64	4.62%	97%
2046	81	0.030	22.39	0.61	4.62%	97%
2047	78	0.028	22.41	0.59	4.62%	97%
2048	74	0.027	22.44	0.56	4.62%	98%
2049	71	0.026	22.47	0.53	4.62%	98%
2050	67	0.025	22.49	0.51	4.62%	98%

CUM PROD	37.52 BBO	GIANT FIELDS		IRAN	
RESERVES	60.00 BBO	TOTAL BBO	50.9	M.EAST (MAJOR)	
DISCOVERED	97.52 BBO	% OF DISC.	52%	Scenario-2	
UNDISCOVERED	17.48 BBO	Last Giant	1963		
REMAINING	77.48 BBO	-------	-----	Revision	Trend
ULTIMATE	115.00 BBO			1990	n/a
Rept.Reserve	92.86 #1	Ref.date(end)	1989	Future	DOWN

Milestones	Date	Prod Mb/d	Dep. Rate	% Produced
1st Peak	1974	6128	2.32%	18%
2nd Peak	1997	4859	2.65%	43%
Mid-Point	2002	4655	2.91%	51%
Final Decline	2003		2.91%	

Forecast "C"
A=Exponential decline
B=5% incr. to Mid-Point
C=Other
Pre-1930 Cum. 0.88
(Added to 1930 prod)

Date	Prod. M.b/d	Prod. BBO/yr	Cum. Prod.	Remain	Deplet. Rate	% Produced	Giant	Notes & Names
1930	121	0.044	0.92	114.08	ERR	1%	12.03	Pre-1930
1931	133	0.049	0.97	114.03	0.04%	1%		
1932	145	0.053	1.03	113.97	0.05%	1%		
1933	157	0.057	1.08	113.92	0.05%	1%		
1934	169	0.062	1.14	113.86	0.05%	1%		
1935	181	0.066	1.21	113.79	0.06%	1%	1.00	Naft-i-Safd
1936	193	0.070	1.28	113.72	0.06%	1%	8.70	Agha Jari
1937	205	0.075	1.36	113.64	0.07%	1%		
1938	215	0.078	1.43	113.57	0.07%	1%		
1939	214	0.078	1.51	113.49	0.07%	1%		
1940	181	0.066	1.58	113.42	0.06%	1%		
1941	139	0.051	1.63	113.37	0.04%	1%		
1942	198	0.072	1.70	113.30	0.06%	1%		
1943	205	0.075	1.78	113.22	0.07%	2%		
1944	280	0.102	1.88	113.12	0.09%	2%		
1945	357	0.130	2.01	112.99	0.12%	2%		
1946	402	0.147	2.16	112.84	0.13%	2%		
1947	423	0.154	2.31	112.69	0.14%	2%		
1948	519	0.189	2.50	112.50	0.17%	2%		
1949	562	0.205	2.70	112.30	0.18%	2%		
1950	665	0.243	2.95	112.05	0.22%	3%		
1951	350	0.128	3.08	111.92	0.11%	3%		
1952	26	0.009	3.08	111.92	0.01%	3%		
1953	24	0.009	3.09	111.91	0.01%	3%		
1954	43	0.016	3.11	111.89	0.01%	3%		
1955	320	0.117	3.23	111.77	0.10%	3%		
1956	530	0.193	3.42	111.58	0.17%	3%		
1957	725	0.265	3.68	111.32	0.24%	3%		
1958	808	0.295	3.98	111.02	0.27%	3%	10.16	Ahwaz
1959	920	0.336	4.31	110.69	0.30%	4%	0.50	Binak
1960	1050	0.383	4.70	110.30	0.35%	4%		
1961	1190	0.434	5.13	109.87	0.39%	4%	5.90	#2
1962	1300	0.475	5.61	109.39	0.43%	5%		
1963	1470	0.537	6.14	108.86	0.49%	5%	12.65	#3
1964	1815	0.662	6.81	108.19	0.61%	6%		
1965	1859	0.678	7.48	107.52	0.63%	7%		
1966	2094	0.764	8.25	106.75	0.71%	7%		
1967	2575	0.940	9.19	105.81	0.88%	8%		
1968	2842	1.037	10.23	104.77	0.98%	9%		
1969	3314	1.210	11.44	103.56	1.15%	10%		
1970	3753	1.370	12.81	102.19	1.32%	11%		
1971	4514	1.648	14.45	100.55	1.61%	13%		
1972	4900	1.789	16.24	98.76	1.78%	14%		
1973	6000	2.190	18.43	96.57	2.22%	16%		
1974	6128	2.237	20.67	94.33	2.32%	18%		1ST PEAK
1975	5600	2.044	22.71	92.29	2.17%	20%		
1976	5875	2.144	24.86	90.14	2.32%	22%		
1977	5650	2.062	26.92	88.08	2.29%	23%		
1978	5250	1.916	28.84	86.16	2.18%	25%		
1979	2900	1.059	29.89	85.11	1.23%	26%		
1980	1280	0.467	30.36	84.64	0.55%	26%		
1981	1375	0.502	30.86	84.14	0.59%	27%		
1982	1896	0.692	31.56	83.44	0.82%	27%		
1983	2606	0.951	32.51	82.49	1.14%	28%		
1984	2166	0.791	33.30	81.70	0.96%	29%		
1985	2279	0.832	34.13	80.87	1.02%	30%		
1986	1806	0.659	34.79	80.21	0.82%	30%		
1987	2342	0.855	35.64	79.36	1.07%	31%		
1988	2208	0.806	36.45	78.55	1.02%	32%		% of Region

1989	2934	1.071	37.52	77.48	1.36%	33%	20.96%	
1990	3333	1.217	38.74	76.26	1.57%	34%		
1991	3714	1.356	40.09	74.91	1.78%	35%		
1992	4079	1.489	41.58	73.42	1.99%	36%		
1993	4215	1.539	43.12	71.88	2.10%	37%		
1994	4313	1.574	44.69	70.31	2.19%	39%		
1995	4414	1.611	46.30	68.70	2.29%	40%		
1996	4634	1.691	48.00	67.00	2.46%	42%		
1997	4859	1.773	49.77	65.23	2.65%	43%	2ND PEAK	
1998	4761	1.738	51.51	63.49	2.66%	45%		
1999	4687	1.711	53.22	61.78	2.69%	46%		
2000	4644	1.695	54.91	60.09	2.74%	48%		Add to
2001	4626	1.689	56.60	58.40	2.81%	49%		Saudi
2002	4655	1.699	58.30	56.70	2.91%	51%	50% PRODUCED	
2003	4520	1.650	59.95	55.05	2.91%	52%	4678	158
2004	4389	1.602	61.55	53.45	2.91%	54%	4690	301
2005	4261	1.555	63.11	51.89	2.91%	55%	4609	348
2006	4137	1.510	64.62	50.38	2.91%	56%	4599	462
2007	4017	1.466	66.08	48.92	2.91%	57%	4580	564
2008	3900	1.423	67.51	47.49	2.91%	59%	4553	654
2009	3786	1.382	68.89	46.11	2.91%	60%	4522	736
2010	3676	1.342	70.23	44.77	2.91%	61%	4484	808
2011	3569	1.303	71.53	43.47	2.91%	62%	4439	870
2012	3465	1.265	72.80	42.20	2.91%	63%	4388	923
2013	3365	1.228	74.03	40.97	2.91%	64%	4332	967
2014	3267	1.192	75.22	39.78	2.91%	65%	4270	1004
2015	3172	1.158	76.38	38.62	2.91%	66%	4203	1032
2016	3079	1.124	77.50	37.50	2.91%	67%	4132	1053
2017	2990	1.091	78.59	36.41	2.91%	68%	4057	1067
2018	2903	1.059	79.65	35.35	2.91%	69%	3978	1075
2019	2818	1.029	80.68	34.32	2.91%	70%	3895	1077
2020	2736	0.999	81.68	33.32	2.91%	71%	3809	1073
2021	2657	0.970	82.65	32.35	2.91%	72%	3720	1064
2022	2579	0.941	83.59	31.41	2.91%	73%	3628	1048
2023	2504	0.914	84.50	30.50	2.91%	73%	3511	1006
2024	2431	0.887	85.39	29.61	2.91%	74%	3474	1043
2025	2361	0.862	86.25	28.75	2.91%	75%	3434	1073
2026	2292	0.837	87.09	27.91	2.91%	76%	3391	1099
2027	2225	0.812	87.90	27.10	2.91%	76%	3344	1119
2028	2160	0.789	88.69	26.31	2.91%	77%	3294	1134
2029	2098	0.766	89.46	25.54	2.91%	78%	3242	1145
2030	2037	0.743	90.20	24.80	2.91%	78%	3192	1155
2031	1977	0.722	90.92	24.08	2.91%	79%	3137	1160
2032	1920	0.701	91.62	23.38	2.91%	80%	3100	1180
2033	1864	0.680	92.30	22.70	2.91%	80%	3061	1197
2034	1810	0.661	92.96	22.04	2.91%	81%	3021	1211
2035	1757	0.641	93.60	21.40	2.91%	81%	2979	1222
2036	1706	0.623	94.23	20.77	2.91%	82%	2936	1230
2037	1656	0.605	94.83	20.17	2.91%	82%	2891	1235
2038	1608	0.587	95.42	19.58	2.91%	83%	2846	1238
2039	1561	0.570	95.99	19.01	2.91%	83%	2800	1239
2040	1516	0.553	96.54	18.46	2.91%	84%	2754	1238
2041	1472	0.537	97.08	17.92	2.91%	84%	2707	1235
2042	1429	0.522	97.60	17.40	2.91%	85%	2660	1231
2043	1387	0.506	98.11	16.89	2.91%	85%	2612	1225
2044	1347	0.492	98.60	16.40	2.91%	86%	2564	1218
2045	1308	0.477	99.07	15.93	2.91%	86%	2517	1209
2046	1270	0.463	99.54	15.46	2.91%	87%	2469	1199
2047	1233	0.450	99.99	15.01	2.91%	87%	2421	1189
2048	1197	0.437	100.43	14.57	2.91%	87%	2374	1177
2049	1162	0.424	100.85	14.15	2.91%	88%	2327	1165
2050	1128	0.412	101.26	13.74	2.91%	88%	2280	1151

Note
#1 Implausable 1987 reported increase
#2 Darood,Bibi Hakemeh,Darius-Kharg,Pazanan,Ardenshir
#3 Karanj,Marun,Mansuri

CUM.PROD	18.589 BBO	GIANT FIELDS		IRAQ	
RESERVES	60.000 BBO	TOTAL BBO	75.8	M.EAST (MAJOR)	
DISCOVERED	78.589 BBO	% OF DISC.	96%	Scenario-2	
UNDISCOVERED	21.411 BBO	Last Giant	1979		
REMAINING	81.411 BBO			Variance	Trend
ULTIMATE	100.00 BBO			1990	n/a
Rept.Reserve	100.00	Ref.date(end)	1989	Future	DOWN

Milestones	Date	Prod Mb/d	Dep. Rate	% Produced	Forecast "C"
					A=Exponential decline
1st Peak	1979	3370	1.42%	12%	B=5% incr. to Mid-Point
2nd Peak	1997	4686	2.47%	30%	C=Other
Mid-Point	2009	4362	3.22%	50%	Pre-1930 Cum. 0.1
Final Decline	2010		3.22%		(Added to 1930 prod)

Date	Prod. M.b/d	Prod. BBO/yr	Cum. Prod.	Remain	Deplet. Rate	% Produced	Giant	Notes & Names
1930	30	0.011	0.11	99.89	ERR	0%	21.50	Pre-1930
1931	35	0.013	0.12	99.88	0.01%	0%		
1932	40	0.015	0.14	99.86	0.01%	0%		
1933	45	0.016	0.15	99.85	0.02%	0%		
1934	50	0.018	0.17	99.83	0.02%	0%		
1935	55	0.020	0.19	99.81	0.02%	0%		
1936	60	0.022	0.21	99.79	0.02%	0%		
1937	75	0.027	0.24	99.76	0.03%	0%		
1938	90	0.033	0.28	99.72	0.03%	0%		
1939	84	0.033	0.31	99.69	0.03%	0%		
1940	66	0.031	0.34	99.66	0.03%	0%		
1941	35	0.024	0.36	99.64	0.02%	0%		
1942	54	0.013	0.38	99.62	0.01%	0%		
1943	68	0.020	0.40	99.60	0.02%	0%		
1944	85	0.025	0.42	99.58	0.02%	0%		
1945	96	0.031	0.45	99.55	0.03%	0%		
1946	98	0.035	0.49	99.51	0.04%	0%		
1947	99	0.036	0.52	99.48	0.04%	1%		
1948	72	0.036	0.56	99.44	0.04%	1%	5.37	Nahr Umr,Zubair
1949	86	0.026	0.58	99.42	0.03%	1%		
1950	136	0.031	0.62	99.38	0.03%	1%		
1951	176	0.050	0.67	99.33	0.05%	1%		
1952	381	0.064	0.73	99.27	0.06%	1%		
1953	669	0.139	0.87	99.13	0.14%	1%	15.38	Bai Hassan,Rumai
1954	625	0.244	1.11	98.89	0.25%	1%	1.00	Jambur
1955	690	0.228	1.34	98.66	0.23%	1%		
1956	633	0.252	1.59	98.41	0.26%	2%		
1957	440	0.231	1.82	98.18	0.23%	2%		
1958	756.2	0.161	1.98	98.02	0.16%	2%		
1959	830	0.303	2.29	97.71	0.31%	2%		
1960	975	0.356	0.52	97.36	0.36%	1%		
1961	1000	0.365	0.89	96.99	0.37%	1%	0.50	Luhais
1962	1005	0.367	1.25	96.62	0.38%	1%		
1963	1120	0.409	1.66	96.22	0.42%	2%		
1964	1227	0.448	2.11	95.77	0.47%	2%		
1965	1278	0.466	2.58	95.30	0.49%	3%		
1966	1382	0.504	3.08	94.80	0.53%	3%		
1967	1230	0.449	3.53	94.35	0.47%	4%		
1968	1497	0.546	4.07	93.80	0.58%	4%		
1969	1529	0.558	4.63	93.24	0.59%	5%	2.50	Buzurgan
1970	1518	0.554	5.19	92.69	0.59%	5%		
1971	1692	0.618	5.80	92.07	0.67%	6%	1.50	Abu Ghirab,Dehlu
1972	1500	0.548	6.35	91.53	0.59%	6%		
1973	1888	0.689	7.04	90.84	0.75%	7%	1.00	Hamrin
1974	1829	0.668	7.71	90.17	0.74%	8%	1.00	Fauqi
1975	2400	0.876	8.58	89.29	0.97%	9%		
1976	2070	0.756	9.34	88.54	0.85%	9%	8.00	Majoon
1977	2150	0.785	10.12	87.75	0.89%	10%		
1978	2500	0.913	11.04	86.84	1.04%	11%		
1979	3370	1.230	12.27	85.61	1.42%	12%	18.00	E.Baghdad
1980	2600	0.949	13.22	84.66	1.11%	13%		1ST PEAK
1981	892	0.326	13.54	84.33	0.38%	14%		
1982	914	0.334	13.88	84.00	0.40%	14%		
1983	905	0.330	14.21	83.67	0.39%	14%		
1984	1218	0.445	14.65	83.23	0.53%	15%		
1985	1397	0.510	15.16	82.72	0.61%	15%		
1986	1788	0.653	15.81	82.06	0.79%	16%		
1987	2096	0.765	16.58	81.30	0.93%	17%		
1988	2679	0.978	17.56	80.32	1.20%	18%		% of Region

1989	2830	1.033	18.59	79.29	1.29%	19%	20.22%	
1990	3214	1.173	19.76	78.11	1.48%	20%		
1991	3582	1.308	21.07	76.81	1.67%	21%		
1992	3934	1.436	22.51	75.37	1.87%	23%		
1993	4066	1.484	23.99	73.89	1.97%	24%		
1994	4160	1.518	25.51	72.37	2.06%	26%		
1995	4257	1.554	27.06	70.82	2.15%	27%		
1996	4469	1.631	28.69	69.18	2.30%	29%		
1997	4686	1.710	30.40	67.47	2.47%	30%	2ND PEAK	
1998	4591	1.676	32.08	65.80	2.48%	32%		
1999	4521	1.650	33.73	64.15	2.51%	34%		
2000	4479	1.635	35.36	62.51	2.55%	35%		
2001	4462	1.629	36.99	60.88	2.61%	37%		
2002	4489	1.639	38.63	59.25	2.69%	39%		
2003	4512	1.647	40.28	57.60	2.78%	40%		
2004	4523	1.651	41.93	55.95	2.87%	42%		
2005	4445	1.622	43.55	54.33	2.90%	44%		
2006	4436	1.619	45.17	52.71	2.98%	45%		
2007	4418	1.612	46.78	51.09	3.06%	47%		Add to
2008	4392	1.603	48.39	49.49	3.14%	48%		Saudi
2009	4362	1.592	49.98	47.90	3.22%	50%	50% PRODUCED	
2010	4226	1.542	51.52	46.36	3.22%	52%	4325	99
2011	4090	1.493	53.01	44.86	3.22%	53%	4281	192
2012	3958	1.445	54.46	43.42	3.22%	54%	4232	274
2013	3830	1.398	55.86	42.02	3.22%	56%	4178	347
2014	3707	1.353	57.21	40.67	3.22%	57%	4118	411
2015	3588	1.310	58.52	39.36	3.22%	59%	4054	466
2016	3472	1.267	59.79	38.09	3.22%	60%	3985	513
2017	3360	1.227	61.01	36.86	3.22%	61%	3912	552
2018	3252	1.187	62.20	35.68	3.22%	62%	3836	584
2019	3147	1.149	63.35	34.53	3.22%	63%	3757	609
2020	3046	1.112	64.46	33.42	3.22%	64%	3674	628
2021	2948	1.076	65.54	32.34	3.22%	66%	3588	640
2022	2853	1.041	66.58	31.30	3.22%	67%	3499	645
2023	2761	1.008	67.59	30.29	3.22%	68%	3386	625
2024	2672	0.975	68.56	29.32	3.22%	69%	3351	678
2025	2586	0.944	69.50	28.37	3.22%	70%	3312	726
2026	2503	0.914	70.42	27.46	3.22%	70%	3270	767
2027	2422	0.884	71.30	26.57	3.22%	71%	3225	803
2028	2344	0.856	72.16	25.72	3.22%	72%	3177	833
2029	2269	0.828	72.99	24.89	3.22%	73%	3127	858
2030	2196	0.801	73.79	24.09	3.22%	74%	3079	883
2031	2125	0.776	74.56	23.31	3.22%	75%	3025	900
2032	2057	0.751	75.31	22.56	3.22%	75%	2990	933
2033	1990	0.727	76.04	21.84	3.22%	76%	2952	962
2034	1926	0.703	76.74	21.13	3.22%	77%	2913	987
2035	1864	0.680	77.42	20.45	3.22%	77%	2873	1008
2036	1804	0.659	78.08	19.79	3.22%	78%	2831	1027
2037	1746	0.637	78.72	19.16	3.22%	79%	2789	1042
2038	1690	0.617	79.34	18.54	3.22%	79%	2745	1055
2039	1636	0.597	79.93	17.94	3.22%	80%	2701	1065
2040	1583	0.578	80.51	17.37	3.22%	81%	2656	1073
2041	1532	0.559	81.07	16.81	3.22%	81%	2611	1079
2042	1483	0.541	81.61	16.26	3.22%	82%	2565	1083
2043	1435	0.524	82.14	15.74	3.22%	82%	2519	1084
2044	1389	0.507	82.64	15.23	3.22%	83%	2473	1085
2045	1344	0.491	83.13	14.74	3.22%	83%	2427	1083
2046	1301	0.475	83.61	14.27	3.22%	84%	2381	1081
2047	1259	0.459	84.07	13.81	3.22%	84%	2335	1076
2048	1218	0.445	84.51	13.36	3.22%	85%	2290	1071
2049	1179	0.430	84.94	12.93	3.22%	85%	2244	1065
2050	1141	0.416	85.36	12.52	3.22%	85%	2199	1058

CUM.PROD	0.311 BBO	GIANT FIELDS		ISRAEL	
RESERVES	0.001 BBO	TOTAL BBO	0	M.EAST (MINOR)	
DISCOVERED	0.312 BBO	% OF DISC.	0%		
UNDISCOVERED	0.000 BBO	Last Giant	-		
REMAINING	0.001 BBO			Variance	Trend
ULTIMATE	0.312 BBO			1990	n/a
Rept.Reserve	0.001 #1	Ref.date(end) 1989		Future	DOWN

Milestones	Date	Prod Mb/d	Dep. Rate	% Produced
1st Peak	1971	131	22.40%	47%
2nd Peak				
Mid-Point	1972	120	26.44%	61%
Final Decline	1990		0.02%	100%

Forecast "A"
A=Exponential decline
B=5% incr. to Mid-Point
C=Other
Pre-1930 Cum. 0
(Added to 1930 prod)

Date	Prod. M.b/d	Prod. BBO/yr	Cum. Prod.	Remain	Deplet. Rate	% Produced	Giant	Notes & Names
1930		0.000	0.00	0.31	ERR	0%		
1931		0.000	0.00	0.31	0.00%	0%		
1932		0.000	0.00	0.31	0.00%	0%		
1933		0.000	0.00	0.31	0.00%	0%		
1934		0.000	0.00	0.31	0.00%	0%		
1935		0.000	0.00	0.31	0.00%	0%		
1936		0.000	0.00	0.31	0.00%	0%		
1937		0.000	0.00	0.31	0.00%	0%		
1938		0.000	0.00	0.31	0.00%	0%		
1939		0.000	0.00	0.31	0.00%	0%		
1940		0.000	0.00	0.31	0.00%	0%		
1941		0.000	0.00	0.31	0.00%	0%		
1942		0.000	0.00	0.31	0.00%	0%		
1943		0.000	0.00	0.31	0.00%	0%		
1944		0.000	0.00	0.31	0.00%	0%		
1945		0.000	0.00	0.31	0.00%	0%		
1946		0.000	0.00	0.31	0.00%	0%		
1947		0.000	0.00	0.31	0.00%	0%		
1948		0.000	0.00	0.31	0.00%	0%		
1949		0.000	0.00	0.31	0.00%	0%		
1950		0.000	0.00	0.31	0.00%	0%		
1951		0.000	0.00	0.31	0.00%	0%		
1952		0.000	0.00	0.31	0.00%	0%		
1953		0.000	0.00	0.31	0.00%	0%		
1954		0.000	0.00	0.31	0.00%	0%		
1955		0.000	0.00	0.31	0.00%	0%		
1956	1	0.000	0.00	0.31	0.07%	0%		
1957	1	0.000	0.00	0.31	0.14%	0%		
1958	2	0.001	0.00	0.31	0.23%	0%		
1959	2	0.001	0.00	0.31	0.24%	1%		
1960	3	0.001	0.00	0.31	0.32%	1%		
1961	3	0.001	0.00	0.31	0.33%	1%		
1962	3	0.001	0.01	0.31	0.31%	2%		
1963	3	0.001	0.01	0.31	0.33%	2%		
1964	4	0.001	0.01	0.30	0.48%	2%		
1965	4	0.001	0.01	0.30	0.47%	3%		
1966	4	0.001	0.01	0.30	0.49%	3%		
1967	3	0.001	0.01	0.30	0.33%	4%		
1968	93	0.034	0.05	0.27	11.29%	15%		
1969	53	0.019	0.06	0.25	7.19%	21%		
1970	93	0.034	0.10	0.21	13.72%	32%		
1971	131	0.048	0.15	0.17	22.40%	47%		1ST PEAK
1972	120	0.044	0.19	0.12	26.44%	61%		50% PRODUCED
1973	100	0.037	0.23	0.09	29.95%	73%		
1974	101	0.037	0.26	0.05	43.10%	84%		
1975	85	0.031	0.29	0.02	63.86%	94%		
1976	1	0.000	0.29	0.02	1.46%	94%		
1977	1	0.000	0.29	0.02	1.48%	95%		
1978	10	0.004	0.30	0.01	21.42%	96%		
1979	30	0.011	0.31	0.00	81.76%	99%		
1980	1	0.000	0.31	0.00	10.46%	99%		
1981	0	0.000	0.31	0.00	5.00%	99%		
1982	0	0.000	0.31	0.00	5.27%	99%		
1983	0	0.000	0.31	0.00	3.71%	99%		
1984	0	0.000	0.31	0.00	1.93%	99%		
1985	0	0.000	0.31	0.00	1.96%	99%		
1986	0	0.000	0.31	0.00	2.00%	99%		
1987	1	0.000	0.31	0.00	20.43%	100%		
1988	0	0.000	0.31	0.00	7.70%	100%		

1989	0	0.000	0.31	0.00	8.35% 100%
1990	0	0.000	0.31	0.00	0.02% 100%
1991	0	0.000	0.31	0.00	0.02% 100%
1992	0	0.000	0.31	0.00	0.02% 100%
1993	0	0.000	0.31	0.00	0.02% 100%
1994	0	0.000	0.31	0.00	0.02% 100%
1995	0	0.000	0.31	0.00	0.02% 100%
1996	0	0.000	0.31	0.00	0.02% 100%
1997	0	0.000	0.31	0.00	0.02% 100%
1998	0	0.000	0.31	0.00	0.02% 100%
1999	0	0.000	0.31	0.00	0.02% 100%
2000	0	0.000	0.31	0.00	0.02% 100%
2001	0	0.000	0.31	0.00	0.02% 100%
2002	0	0.000	0.31	0.00	0.02% 100%
2003	0	0.000	0.31	0.00	0.02% 100%
2004	0	0.000	0.31	0.00	0.02% 100%
2005	0	0.000	0.31	0.00	0.02% 100%
2006	0	0.000	0.31	0.00	0.02% 100%
2007	0	0.000	0.31	0.00	0.02% 100%
2008	0	0.000	0.31	0.00	0.02% 100%
2009	0	0.000	0.31	0.00	0.02% 100%
2010	0	0.000	0.31	0.00	0.02% 100%
2011	0	0.000	0.31	0.00	0.02% 100%
2012	0	0.000	0.31	0.00	0.02% 100%
2013	0	0.000	0.31	0.00	0.02% 100%
2014	0	0.000	0.31	0.00	0.02% 100%
2015	0	0.000	0.31	0.00	0.02% 100%
2016	0	0.000	0.31	0.00	0.02% 100%
2017	0	0.000	0.31	0.00	0.02% 100%
2018	0	0.000	0.31	0.00	0.02% 100%
2019	0	0.000	0.31	0.00	0.02% 100%
2020	0	0.000	0.31	0.00	0.02% 100%
2021	0	0.000	0.31	0.00	0.02% 100%
2022	0	0.000	0.31	0.00	0.02% 100%
2023	0	0.000	0.31	0.00	0.02% 100%
2024	0	0.000	0.31	0.00	0.02% 100%
2025	0	0.000	0.31	0.00	0.02% 100%
2026	0	0.000	0.31	0.00	0.02% 100%
2027	0	0.000	0.31	0.00	0.02% 100%
2028	0	0.000	0.31	0.00	0.02% 100%
2029	0	0.000	0.31	0.00	0.02% 100%
2030	0	0.000	0.31	0.00	0.02% 100%
2031	0	0.000	0.31	0.00	0.02% 100%
2032	0	0.000	0.31	0.00	0.02% 100%
2033	0	0.000	0.31	0.00	0.02% 100%
2034	0	0.000	0.31	0.00	0.02% 100%
2035	0	0.000	0.31	0.00	0.02% 100%
2036	0	0.000	0.31	0.00	0.02% 100%
2037	0	0.000	0.31	0.00	0.02% 100%
2038	0	0.000	0.31	0.00	0.02% 100%
2039	0	0.000	0.31	0.00	0.02% 100%
2040	0	0.000	0.31	0.00	0.02% 100%
2041	0	0.000	0.31	0.00	0.02% 100%
2042	0	0.000	0.31	0.00	0.02% 100%
2043	0	0.000	0.31	0.00	0.02% 100%
2044	0	0.000	0.31	0.00	0.02% 100%
2045	0	0.000	0.31	0.00	0.02% 100%
2046	0	0.000	0.31	0.00	0.02% 100%
2047	0	0.000	0.31	0.00	0.02% 100%
2048	0	0.000	0.31	0.00	0.02% 100%
2049	0	0.000	0.31	0.00	0.02% 100%
2050	0	0.000	0.31	0.00	0.02% 100%

Note
High production in 1970's may reflect conquest of Egyptian fields

CUM.PROD	0.439 BBO	GIANT FIELDS		ITALY	
RESERVES	0.725 BBO	TOTAL BBO	0	W.EUROPE	
DISCOVERED	1.164 BBO	% OF DISC.	0%		
UNDISCOVERED	0.336 BBO	Last Giant	-		
REMAINING	1.061 BBO			Variance	Trend
ULTIMATE	1.500 BBO			1990	-2.4%
Rept.Reserve	0.725 BBO	Ref.Date(end)	1989	Future	DOWN

Milestones	Date	Prod Mb/d	Dep. Rate	% Produced
1st Peak	1964	48	1.24%	7%
2nd Peak	1997	140	6.67%	52%
Mid-Point	1997	140	6.67%	52%
Final Decline	1997		6.67%	

Forecast "B"
A=Exponential decline
B=5% incr. to Mid-Point
C=Other
Pre-1930 Cum. 0
(Added to 1930 Prod)

Date	Prod. M.b/d	Prod. BBO/yr	Cum Prod	Remain	Dep. Rate	% Produced	Giant Names & Notes
1930		0.000	0.00	1.50	ERR	0%	
1931		0.000	0.00	1.50	0.00%	0%	
1932		0.000	0.00	1.50	0.00%	0%	
1933		0.000	0.00	1.50	0.00%	0%	
1934		0.000	0.00	1.50	0.00%	0%	
1935		0.000	0.00	1.50	0.00%	0%	
1936		0.000	0.00	1.50	0.00%	0%	
1937		0.000	0.00	1.50	0.00%	0%	
1938	0	0.000	0.00	1.50	0.01%	0%	
1939	0	0.000	0.00	1.50	0.01%	0%	
1940	0	0.000	0.00	1.50	0.00%	0%	
1941	0	0.000	0.00	1.50	0.00%	0%	
1942	8	0.003	0.00	1.50	0.19%	0%	
1943	0	0.000	0.00	1.50	0.00%	0%	
1944	0	0.000	0.00	1.50	0.00%	0%	
1945	0	0.000	0.00	1.50	0.00%	0%	
1946	0	0.000	0.00	1.50	0.00%	0%	
1947	0	0.000	0.00	1.50	0.00%	0%	
1948	0	0.000	0.00	1.50	0.00%	0%	
1949	0	0.000	0.00	1.50	0.00%	0%	
1950	0	0.000	0.00	1.50	0.00%	0%	
1951	0	0.000	0.00	1.50	0.01%	0%	
1952	1	0.001	0.00	1.50	0.03%	0%	
1953	2	0.001	0.01	1.49	0.04%	0%	
1954	1	0.001	0.01	1.49	0.03%	0%	
1955	4	0.002	0.01	1.49	0.11%	0%	
1956	1	0.001	0.01	1.49	0.03%	1%	
1957	24	0.009	0.02	1.48	0.59%	1%	
1958	28	0.010	0.03	1.47	0.70%	2%	
1959	30	0.011	0.04	1.46	0.74%	3%	
1960	37	0.014	0.05	1.45	0.92%	3%	
1961	38	0.014	0.07	1.43	0.95%	4%	
1962	35	0.013	0.08	1.42	0.89%	5%	
1963	34	0.012	0.09	1.41	0.87%	6%	
1964	48	0.017	0.11	1.39	1.24%	7%	1st PEAK
1965	42	0.015	0.12	1.38	1.10%	8%	
1966	36	0.013	0.14	1.36	0.94%	9%	
1967	30	0.011	0.15	1.35	0.80%	10%	
1968	27	0.010	0.16	1.34	0.74%	10%	
1969	29	0.011	0.17	1.33	0.79%	11%	
1970	25	0.009	0.18	1.32	0.67%	12%	
1971	25	0.009	0.19	1.31	0.69%	12%	
1972	21	0.008	0.19	1.31	0.58%	13%	
1973	22	0.008	0.20	1.30	0.62%	13%	
1974	18	0.007	0.21	1.29	0.51%	14%	
1975	20	0.007	0.22	1.28	0.57%	14%	
1976	20	0.007	0.22	1.28	0.57%	15%	
1977	18	0.007	0.23	1.27	0.51%	15%	
1978	28	0.010	0.24	1.26	0.80%	16%	
1979	33	0.012	0.25	1.25	0.96%	17%	
1980	38	0.014	0.27	1.23	1.11%	18%	
1981	37	0.014	0.28	1.22	1.09%	19%	
1982	28	0.010	0.29	1.21	0.84%	19%	
1983	37	0.014	0.30	1.20	1.12%	20%	
1984	45	0.016	0.32	1.18	1.37%	21%	
1985	42	0.015	0.33	1.17	1.29%	22%	
1986	49	0.018	0.35	1.15	1.53%	23%	
1987	50	0.018	0.37	1.13	1.59%	25%	
1988	93	0.034	0.40	1.10	3.00%	27%	

1989	95	0.035	0.44	1.06	3.15%	29%	
1990	99	0.036	0.48	1.02	3.42%	32%	
1991	104	0.038	0.51	0.99	3.72%	34%	
1992	110	0.040	0.55	0.95	4.06%	37%	
1993	115	0.042	0.60	0.90	4.44%	40%	
1994	121	0.044	0.64	0.86	4.88%	43%	
1995	127	0.046	0.69	0.81	5.38%	46%	
1996	133	0.049	0.73	0.77	5.97%	49%	
1997	140	0.051	0.79	0.71	6.67%	52%	50% PRODUCED & 2nd PEAK
1998	131	0.048	0.83	0.67	6.67%	56%	
1999	122	0.044	0.88	0.62	6.67%	59%	
2000	114	0.042	0.92	0.58	6.67%	61%	
2001	106	0.039	0.96	0.54	6.67%	64%	
2002	99	0.036	0.99	0.51	6.67%	66%	
2003	92	0.034	1.03	0.47	6.67%	69%	
2004	86	0.031	1.06	0.44	6.67%	71%	
2005	81	0.029	1.09	0.41	6.67%	73%	
2006	75	0.027	1.12	0.38	6.67%	74%	
2007	70	0.026	1.14	0.36	6.67%	76%	
2008	65	0.024	1.17	0.33	6.67%	78%	
2009	61	0.022	1.19	0.31	6.67%	79%	
2010	57	0.021	1.21	0.29	6.67%	81%	
2011	53	0.019	1.23	0.27	6.67%	82%	
2012	50	0.018	1.25	0.25	6.67%	83%	
2013	46	0.017	1.26	0.24	6.67%	84%	
2014	43	0.016	1.28	0.22	6.67%	85%	
2015	40	0.015	1.29	0.21	6.67%	86%	
2016	38	0.014	1.31	0.19	6.67%	87%	
2017	35	0.013	1.32	0.18	6.67%	88%	
2018	33	0.012	1.33	0.17	6.67%	89%	
2019	31	0.011	1.34	0.16	6.67%	90%	
2020	29	0.010	1.35	0.15	6.67%	90%	
2021	27	0.010	1.36	0.14	6.67%	91%	
2022	25	0.009	1.37	0.13	6.67%	92%	
2023	23	0.008	1.38	0.12	6.67%	92%	
2024	22	0.008	1.39	0.11	6.67%	93%	
2025	20	0.007	1.40	0.10	6.67%	93%	
2026	19	0.007	1.40	0.10	6.67%	94%	
2027	18	0.006	1.41	0.09	6.67%	94%	
2028	16	0.006	1.42	0.08	6.67%	94%	
2029	15	0.006	1.42	0.08	6.67%	95%	
2030	14	0.005	1.43	0.07	6.67%	95%	
2031	13	0.005	1.43	0.07	6.67%	95%	
2032	12	0.005	1.44	0.06	6.67%	96%	
2033	12	0.004	1.44	0.06	6.67%	96%	
2034	11	0.004	1.44	0.06	6.67%	96%	
2035	10	0.004	1.45	0.05	6.67%	97%	
2036	9	0.003	1.45	0.05	6.67%	97%	
2037	9	0.003	1.45	0.05	6.67%	97%	
2038	8	0.003	1.46	0.04	6.67%	97%	
2039	8	0.003	1.46	0.04	6.67%	97%	
2040	7	0.003	1.46	0.04	6.67%	98%	
2041	7	0.002	1.47	0.03	6.67%	98%	
2042	6	0.002	1.47	0.03	6.67%	98%	
2043	6	0.002	1.47	0.03	6.67%	98%	
2044	5	0.002	1.47	0.03	6.67%	98%	
2045	5	0.002	1.47	0.03	6.67%	98%	
2046	5	0.002	1.48	0.02	6.67%	98%	
2047	4	0.002	1.48	0.02	6.67%	98%	
2048	4	0.002	1.48	0.02	6.67%	99%	
2049	4	0.001	1.48	0.02	6.67%	99%	
2050	4	0.001	1.48	0.02	6.67%	99%	

CUM.PROD	0.051 BBO	GIANT FIELDS			IVORY COAST
RESERVES	0.049 BBO	TOTAL BBO	0.0		AFRICA
DISCOVERED	0.100 BBO	% OF DISC.	0%		
UNDISCOVERED	0.000 BBO	Last Giant			
REMAINING	0.049 BBO				Variance Trend
ULTIMATE	0.100 BBO				1990 ZERO
Rept.Reserves	0.049 BBO	Ref.Date (end)		1989	Future DOWN

Milestones		Prod Mb/d	Dep Rate	% Produced	Forecast "A"
					A=Exponential decline
1st Peak	1985	28	13.21%	33%	B=5% inc. to Mid-Point
2nd Peak					C=Other
Mid-Point	1988	13	8.72%	51%	Pre-1930 Cum 0
Final Decline	1990		1.48%		(Added to 1930 Prod)

Date	Prod. M.b/d	Prod. BBO/yr	Cum Prod	Remain	Dep. Rate	% Produced	Giant	Names & Notes
1930		0.000	0.00	0.10	ERR	0%		
1931		0.000	0.00	0.10	0.00%	0%		
1932		0.000	0.00	0.10	0.00%	0%		
1933		0.000	0.00	0.10	0.00%	0%		
1934		0.000	0.00	0.10	0.00%	0%		
1935		0.000	0.00	0.10	0.00%	0%		
1936		0.000	0.00	0.10	0.00%	0%		
1937		0.000	0.00	0.10	0.00%	0%		
1938		0.000	0.00	0.10	0.00%	0%		
1939		0.000	0.00	0.10	0.00%	0%		
1940		0.000	0.00	0.10	0.00%	0%		
1941		0.000	0.00	0.10	0.00%	0%		
1942		0.000	0.00	0.10	0.00%	0%		
1943		0.000	0.00	0.10	0.00%	0%		
1944		0.000	0.00	0.10	0.00%	0%		
1945		0.000	0.00	0.10	0.00%	0%		
1946		0.000	0.00	0.10	0.00%	0%		
1947		0.000	0.00	0.10	0.00%	0%		
1948		0.000	0.00	0.10	0.00%	0%		
1949		0.000	0.00	0.10	0.00%	0%		
1950		0.000	0.00	0.10	0.00%	0%		
1951		0.000	0.00	0.10	0.00%	0%		
1952		0.000	0.00	0.10	0.00%	0%		
1953		0.000	0.00	0.10	0.00%	0%		
1954		0.000	0.00	0.10	0.00%	0%		
1955		0.000	0.00	0.10	0.00%	0%		
1956		0.000	0.00	0.10	0.00%	0%		
1957		0.000	0.00	0.10	0.00%	0%		
1958		0.000	0.00	0.10	0.00%	0%		
1959		0.000	0.00	0.10	0.00%	0%		
1960		0.000	0.00	0.10	0.00%	0%		
1961		0.000	0.00	0.10	0.00%	0%		
1962		0.000	0.00	0.10	0.00%	0%		
1963		0.000	0.00	0.10	0.00%	0%		
1964		0.000	0.00	0.10	0.00%	0%		
1965		0.000	0.00	0.10	0.00%	0%		
1966		0.000	0.00	0.10	0.00%	0%		
1967		0.000	0.00	0.10	0.00%	0%		
1968		0.000	0.00	0.10	0.00%	0%		
1969		0.000	0.00	0.10	0.00%	0%		
1970		0.000	0.00	0.10	0.00%	0%		
1971		0.000	0.00	0.10	0.00%	0%		
1972		0.000	0.00	0.10	0.00%	0%		
1973		0.000	0.00	0.10	0.00%	0%		
1974		0.000	0.00	0.10	0.00%	0%		
1975		0.000	0.00	0.10	0.00%	0%		
1976		0.000	0.00	0.10	0.00%	0%		
1977		0.000	0.00	0.10	0.00%	0%		
1978		0.000	0.00	0.10	0.00%	0%		
1979		0.000	0.00	0.10	0.00%	0%		
1980		0.000	0.00	0.10	0.00%	0%		
1981	7	0.003	0.00	0.10	2.56%	3%		
1982	9	0.003	0.01	0.09	3.37%	6%		
1983	24	0.009	0.01	0.09	9.30%	15%		
1984	22	0.008	0.02	0.08	9.40%	23%		
1985	28	0.010	0.03	0.07	13.21%	33%		1ST PEAK
1986	20	0.007	0.04	0.06	10.87%	40%		
1987	16	0.006	0.05	0.05	9.76%	46%		
1988	13	0.005	0.05	0.05	8.72%	51%		50% PRODUCED

1989	2	0.001	0.05	0.05	1.48%	51%
1990	2	0.001	0.05	0.05	1.48%	52%
1991	2	0.001	0.05	0.05	1.48%	53%
1992	2	0.001	0.05	0.05	1.48%	54%
1993	2	0.001	0.05	0.05	1.48%	54%
1994	2	0.001	0.05	0.05	1.48%	55%
1995	2	0.001	0.06	0.04	1.48%	56%
1996	2	0.001	0.06	0.04	1.48%	56%
1997	2	0.001	0.06	0.04	1.48%	57%
1998	2	0.001	0.06	0.04	1.48%	58%
1999	2	0.001	0.06	0.04	1.48%	58%
2000	2	0.001	0.06	0.04	1.48%	59%
2001	2	0.001	0.06	0.04	1.48%	59%
2002	2	0.001	0.06	0.04	1.48%	60%
2003	2	0.001	0.06	0.04	1.48%	61%
2004	2	0.001	0.06	0.04	1.48%	61%
2005	2	0.001	0.06	0.04	1.48%	62%
2006	2	0.001	0.06	0.04	1.48%	62%
2007	2	0.001	0.06	0.04	1.48%	63%
2008	2	0.001	0.06	0.04	1.48%	63%
2009	1	0.001	0.06	0.04	1.48%	64%
2010	1	0.001	0.06	0.04	1.48%	64%
2011	1	0.001	0.07	0.03	1.48%	65%
2012	1	0.001	0.07	0.03	1.48%	66%
2013	1	0.001	0.07	0.03	1.48%	66%
2014	1	0.001	0.07	0.03	1.48%	67%
2015	1	0.000	0.07	0.03	1.48%	67%
2016	1	0.000	0.07	0.03	1.48%	68%
2017	1	0.000	0.07	0.03	1.48%	68%
2018	1	0.000	0.07	0.03	1.48%	68%
2019	1	0.000	0.07	0.03	1.48%	69%
2020	1	0.000	0.07	0.03	1.48%	69%
2021	1	0.000	0.07	0.03	1.48%	70%
2022	1	0.000	0.07	0.03	1.48%	70%
2023	1	0.000	0.07	0.03	1.48%	71%
2024	1	0.000	0.07	0.03	1.48%	71%
2025	1	0.000	0.07	0.03	1.48%	72%
2026	1	0.000	0.07	0.03	1.48%	72%
2027	1	0.000	0.07	0.03	1.48%	72%
2028	1	0.000	0.07	0.03	1.48%	73%
2029	1	0.000	0.07	0.03	1.48%	73%
2030	1	0.000	0.07	0.03	1.48%	74%
2031	1	0.000	0.07	0.03	1.48%	74%
2032	1	0.000	0.07	0.03	1.48%	74%
2033	1	0.000	0.07	0.03	1.48%	75%
2034	1	0.000	0.08	0.02	1.48%	75%
2035	1	0.000	0.08	0.02	1.48%	76%
2036	1	0.000	0.08	0.02	1.48%	76%
2037	1	0.000	0.08	0.02	1.48%	76%
2038	1	0.000	0.08	0.02	1.48%	77%
2039	1	0.000	0.08	0.02	1.48%	77%
2040	1	0.000	0.08	0.02	1.48%	77%
2041	1	0.000	0.08	0.02	1.48%	78%
2042	1	0.000	0.08	0.02	1.48%	78%
2043	1	0.000	0.08	0.02	1.48%	78%
2044	1	0.000	0.08	0.02	1.48%	79%
2045	1	0.000	0.08	0.02	1.48%	79%
2046	1	0.000	0.08	0.02	1.48%	79%
2047	1	0.000	0.08	0.02	1.48%	80%
2048	1	0.000	0.08	0.02	1.48%	80%
2049	1	0.000	0.08	0.02	1.48%	80%
2050	1	0.000	0.08	0.02	1.48%	80%

CUM.PROD	0.177 BBO	GIANT FIELDS		JAPAN	
RESERVES	0.061 BBO	TOTAL BBO	0	THE EAST	
DISCOVERED	0.238 BBO	% OF DISC.	0%		
UNDISCOVERED	0.012 BBO	Last Giant	-		
REMAINING	0.073 BBO			Variance Trend	
ULTIMATE	0.250 BBO			1990	-4%
Rept.Reserve	0.061	Ref.date(end) 1989		Future	DOWN

Milestones	Date	Prod Mb/d	Dep. Rate	% Produced
1st Peak	1970	16	3.85%	41%
2nd Peak				
Mid-Point	1975	13	3.73%	51%
Final Decline	1990		5.34%	72%

Forecast "A"
A=Exponential decline
B=5% incr. to Mid-Point
C=Other
Pre-1930 Cum. 0

Date	Prod. M.b/d	Prod. BBO/yr	Cum. Prod.	Remain	Deplet. Rate	% Produced	Giant	Notes & Names
1930		0.000	0.00	0.25	ERR	0%		
1931		0.000	0.00	0.25	0.00%	0%		
1932		0.000	0.00	0.25	0.00%	0%		
1933		0.000	0.00	0.25	0.00%	0%		
1934		0.000	0.00	0.25	0.00%	0%		
1935		0.000	0.00	0.25	0.00%	0%		
1936		0.000	0.00	0.25	0.00%	0%		
1937		0.000	0.00	0.25	0.00%	0%		
1938	7	0.003	0.00	0.25	1.01%	1%		
1939	7	0.003	0.01	0.24	1.08%	2%		
1940	7	0.003	0.01	0.24	1.07%	3%		
1941	5	0.002	0.01	0.24	0.80%	4%		
1942	5	0.002	0.01	0.24	0.68%	5%		
1943	5	0.002	0.01	0.24	0.72%	5%		
1944	4	0.002	0.01	0.24	0.68%	6%		
1945	4	0.002	0.02	0.23	0.65%	6%		
1946	4	0.001	0.02	0.23	0.58%	7%		
1947	4	0.001	0.02	0.23	0.57%	8%		
1948	3	0.001	0.02	0.23	0.49%	8%		
1949	4	0.001	0.02	0.23	0.60%	9%		
1950	6	0.002	0.02	0.23	0.89%	9%		
1951	6	0.002	0.03	0.22	1.03%	10%		
1952	6	0.002	0.03	0.22	0.96%	11%		
1953	6	0.002	0.03	0.22	0.96%	12%		
1954	6	0.002	0.03	0.22	0.95%	13%		
1955	6	0.002	0.03	0.22	1.04%	14%		
1956	6	0.002	0.04	0.21	1.02%	15%		
1957	6	0.002	0.04	0.21	1.04%	16%		
1958	7	0.002	0.04	0.21	1.18%	17%		
1959	7	0.003	0.04	0.21	1.29%	18%		
1960	10	0.003	0.05	0.20	1.68%	19%		
1961	13	0.005	0.05	0.20	2.29%	21%		
1962	15	0.005	0.06	0.19	2.71%	23%		
1963	16	0.006	0.06	0.19	3.03%	25%		
1964	12	0.004	0.07	0.18	2.41%	27%		
1965	13	0.005	0.07	0.18	2.56%	29%		
1966	16	0.006	0.08	0.17	3.19%	31%		
1967	16	0.006	0.08	0.17	3.40%	34%		
1968	16	0.006	0.09	0.16	3.50%	36%		
1969	16	0.006	0.10	0.15	3.60%	38%		
1970	16	0.006	0.10	0.15	3.85%	41%		1st PEAK
1971	16	0.006	0.11	0.14	4.03%	43%		
1972	15	0.005	0.11	0.14	3.84%	45%		
1973	15	0.005	0.12	0.13	4.00%	47%		
1974	12	0.004	0.12	0.13	3.41%	49%		
1975	13	0.005	0.13	0.12	3.73%	51%		50% PRODUCED
1976	10	0.004	0.13	0.12	2.98%	53%		
1977	10	0.004	0.13	0.12	3.08%	54%		
1978	10	0.004	0.14	0.11	3.17%	55%		
1979	10	0.004	0.14	0.11	3.28%	57%		
1980	10	0.004	0.15	0.10	3.39%	58%		
1981	7	0.003	0.15	0.10	2.46%	59%		
1982	6	0.002	0.15	0.10	2.16%	60%		
1983	7	0.003	0.15	0.10	2.54%	61%		
1984	6	0.002	0.16	0.09	2.41%	62%		
1985	10	0.004	0.16	0.09	3.86%	64%		
1986	13	0.005	0.16	0.09	5.14%	66%		
1987	12	0.004	0.17	0.08	5.13%	67%		
1988	12	0.004	0.17	0.08	5.49%	69%		

1989	11	0.004	0.18	0.07	5.34%	71%
1990	11	0.004	0.18	0.07	5.34%	72%
1991	10	0.004	0.18	0.07	5.34%	74%
1992	10	0.003	0.19	0.06	5.34%	75%
1993	9	0.003	0.19	0.06	5.34%	77%
1994	9	0.003	0.19	0.06	5.34%	78%
1995	8	0.003	0.20	0.05	5.34%	79%
1996	8	0.003	0.20	0.05	5.34%	80%
1997	7	0.003	0.20	0.05	5.34%	81%
1998	7	0.003	0.21	0.04	5.34%	82%
1999	7	0.002	0.21	0.04	5.34%	83%
2000	6	0.002	0.21	0.04	5.34%	84%
2001	6	0.002	0.21	0.04	5.34%	85%
2002	6	0.002	0.21	0.04	5.34%	86%
2003	5	0.002	0.22	0.03	5.34%	86%
2004	5	0.002	0.22	0.03	5.34%	87%
2005	5	0.002	0.22	0.03	5.34%	88%
2006	4	0.002	0.22	0.03	5.34%	88%
2007	4	0.002	0.22	0.03	5.34%	89%
2008	4	0.001	0.22	0.03	5.34%	90%
2009	4	0.001	0.23	0.02	5.34%	90%
2010	4	0.001	0.23	0.02	5.34%	91%
2011	3	0.001	0.23	0.02	5.34%	91%
2012	3	0.001	0.23	0.02	5.34%	92%
2013	3	0.001	0.23	0.02	5.34%	92%
2014	3	0.001	0.23	0.02	5.34%	93%
2015	3	0.001	0.23	0.02	5.34%	93%
2016	3	0.001	0.23	0.02	5.34%	93%
2017	2	0.001	0.23	0.02	5.34%	94%
2018	2	0.001	0.24	0.01	5.34%	94%
2019	2	0.001	0.24	0.01	5.34%	94%
2020	2	0.001	0.24	0.01	5.34%	95%
2021	2	0.001	0.24	0.01	5.34%	95%
2022	2	0.001	0.24	0.01	5.34%	95%
2023	2	0.001	0.24	0.01	5.34%	95%
2024	2	0.001	0.24	0.01	5.34%	96%
2025	2	0.001	0.24	0.01	5.34%	96%
2026	1	0.001	0.24	0.01	5.34%	96%
2027	1	0.001	0.24	0.01	5.34%	96%
2028	1	0.000	0.24	0.01	5.34%	97%
2029	1	0.000	0.24	0.01	5.34%	97%
2030	1	0.000	0.24	0.01	5.34%	97%
2031	1	0.000	0.24	0.01	5.34%	97%
2032	1	0.000	0.24	0.01	5.34%	97%
2033	1	0.000	0.24	0.01	5.34%	97%
2034	1	0.000	0.24	0.01	5.34%	98%
2035	1	0.000	0.24	0.01	5.34%	98%
2036	1	0.000	0.24	0.01	5.34%	98%
2037	1	0.000	0.24	0.01	5.34%	98%
2038	1	0.000	0.25	0.00	5.34%	98%
2039	1	0.000	0.25	0.00	5.34%	98%
2040	1	0.000	0.25	0.00	5.34%	98%
2041	1	0.000	0.25	0.00	5.34%	98%
2042	1	0.000	0.25	0.00	5.34%	98%
2043	1	0.000	0.25	0.00	5.34%	98%
2044	1	0.000	0.25	0.00	5.34%	99%
2045	1	0.000	0.25	0.00	5.34%	99%
2046	0	0.000	0.25	0.00	5.34%	99%
2047	0	0.000	0.25	0.00	5.34%	99%
2048	0	0.000	0.25	0.00	5.34%	99%
2049	0	0.000	0.25	0.00	5.34%	99%
2050	0	0.000	0.25	0.00	5.34%	99%

CUM.PROD	0.002 BBO	GIANT FIELDS		JORDAN	
RESERVES	0.005 BBO	TOTAL BBO	0	M.EAST (MINOR)	
DISCOVERED	0.007 BBO	% OF DISC.	0%		
UNDISCOVERED	0.003 BBO	Last Giant	-		
REMAINING	0.008 BBO			Variance	Trend
ULTIMATE	0.010 BBO			1990	n/a
Rept.Reserve	0.005	Ref.date(end)	1989	Future	UP

Milestones	Date	Prod Mb/d	Dep. Rate	% Produced
1st Peak	1987	4	14.60%	15%
2nd Peak				
Mid-Point	2007	1	4.54%	50%
Final Decline	2007		4.54%	19%

Forecast "B"
A=Exponential decline
B=5% incr. to Mid-Point
C=Other
Pre-1930 Cum. 0
(Added to 1930 prod)

Date	Prod. M.b/d	Prod. BBO/yr	Cum. Prod.	Remain	Deplet. Rate	% Produced	Giant	Notes & Names
1930		0.000	0.00	0.01	ERR	0%		
1931		0.000	0.00	0.01	0.00%	0%		
1932		0.000	0.00	0.01	0.00%	0%		
1933		0.000	0.00	0.01	0.00%	0%		
1934		0.000	0.00	0.01	0.00%	0%		
1935		0.000	0.00	0.01	0.00%	0%		
1936		0.000	0.00	0.01	0.00%	0%		
1937		0.000	0.00	0.01	0.00%	0%		
1938		0.000	0.00	0.01	0.00%	0%		
1939		0.000	0.00	0.01	0.00%	0%		
1940		0.000	0.00	0.01	0.00%	0%		
1941		0.000	0.00	0.01	0.00%	0%		
1942		0.000	0.00	0.01	0.00%	0%		
1943		0.000	0.00	0.01	0.00%	0%		
1944		0.000	0.00	0.01	0.00%	0%		
1945		0.000	0.00	0.01	0.00%	0%		
1946		0.000	0.00	0.01	0.00%	0%		
1947		0.000	0.00	0.01	0.00%	0%		
1948		0.000	0.00	0.01	0.00%	0%		
1949		0.000	0.00	0.01	0.00%	0%		
1950		0.000	0.00	0.01	0.00%	0%		
1951		0.000	0.00	0.01	0.00%	0%		
1952		0.000	0.00	0.01	0.00%	0%		
1953		0.000	0.00	0.01	0.00%	0%		
1954		0.000	0.00	0.01	0.00%	0%		
1955		0.000	0.00	0.01	0.00%	0%		
1956		0.000	0.00	0.01	0.00%	0%		
1957		0.000	0.00	0.01	0.00%	0%		
1958		0.000	0.00	0.01	0.00%	0%		
1959		0.000	0.00	0.01	0.00%	0%		
1960		0.000	0.00	0.01	0.00%	0%		
1961		0.000	0.00	0.01	0.00%	0%		
1962		0.000	0.00	0.01	0.00%	0%		
1963		0.000	0.00	0.01	0.00%	0%		
1964		0.000	0.00	0.01	0.00%	0%		
1965		0.000	0.00	0.01	0.00%	0%		
1966		0.000	0.00	0.01	0.00%	0%		
1967		0.000	0.00	0.01	0.00%	0%		
1968		0.000	0.00	0.01	0.00%	0%		
1969		0.000	0.00	0.01	0.00%	0%		
1970		0.000	0.00	0.01	0.00%	0%		
1971		0.000	0.00	0.01	0.00%	0%		
1972		0.000	0.00	0.01	0.00%	0%		
1973		0.000	0.00	0.01	0.00%	0%		
1974		0.000	0.00	0.01	0.00%	0%		
1975		0.000	0.00	0.01	0.00%	0%		
1976		0.000	0.00	0.01	0.00%	0%		
1977		0.000	0.00	0.01	0.00%	0%		
1978		0.000	0.00	0.01	0.00%	0%		
1979		0.000	0.00	0.01	0.00%	0%		
1980		0.000	0.00	0.01	0.00%	0%		
1981		0.000	0.00	0.01	0.00%	0%		
1982		0.000	0.00	0.01	0.00%	0%		
1983		0.000	0.00	0.01	0.00%	0%		
1984		0.000	0.00	0.01	0.00%	0%		
1985		0.000	0.00	0.01	0.00%	0%		
1986		0.000	0.00	0.01	0.00%	0%		
1987	4	0.001	0.00	0.01	14.60%	15%		1ST PEAK
1988	1	0.000	0.00	0.01	2.14%	16%		

1989	0	0.000	0.00	0.01	1.31%	18%	
1990	0	0.000	0.00	0.01	1.39%	19%	
1991	0	0.000	0.00	0.01	1.48%	20%	
1992	0	0.000	0.00	0.01	1.58%	21%	
1993	0	0.000	0.00	0.01	1.69%	22%	
1994	0	0.000	0.00	0.01	1.80%	24%	
1995	0	0.000	0.00	0.01	1.93%	25%	
1996	0	0.000	0.00	0.01	2.06%	27%	
1997	0	0.000	0.00	0.01	2.21%	28%	
1998	0	0.000	0.00	0.01	2.38%	30%	
1999	0	0.000	0.00	0.01	2.56%	32%	
2000	1	0.000	0.00	0.01	2.75%	34%	
2001	1	0.000	0.00	0.01	2.97%	36%	
2002	1	0.000	0.00	0.01	3.22%	38%	
2003	1	0.000	0.00	0.01	3.49%	40%	
2004	1	0.000	0.00	0.01	3.80%	42%	
2005	1	0.000	0.00	0.01	4.14%	45%	
2006	1	0.000	0.00	0.01	4.54%	47%	
2007	1	0.000	0.00	0.01	4.54%	50%	50% PRODUCED
2008	1	0.000	0.01	0.00	4.54%	52%	
2009	1	0.000	0.01	0.00	4.54%	54%	
2010	1	0.000	0.01	0.00	4.54%	56%	
2011	1	0.000	0.01	0.00	4.54%	58%	
2012	1	0.000	0.01	0.00	4.54%	60%	
2013	0	0.000	0.01	0.00	4.54%	62%	
2014	0	0.000	0.01	0.00	4.54%	64%	
2015	0	0.000	0.01	0.00	4.54%	65%	
2016	0	0.000	0.01	0.00	4.54%	67%	
2017	0	0.000	0.01	0.00	4.54%	68%	
2018	0	0.000	0.01	0.00	4.54%	70%	
2019	0	0.000	0.01	0.00	4.54%	71%	
2020	0	0.000	0.01	0.00	4.54%	72%	
2021	0	0.000	0.01	0.00	4.54%	74%	
2022	0	0.000	0.01	0.00	4.54%	75%	
2023	0	0.000	0.01	0.00	4.54%	76%	
2024	0	0.000	0.01	0.00	4.54%	77%	
2025	0	0.000	0.01	0.00	4.54%	78%	
2026	0	0.000	0.01	0.00	4.54%	79%	
2027	0	0.000	0.01	0.00	4.54%	80%	
2028	0	0.000	0.01	0.00	4.54%	81%	
2029	0	0.000	0.01	0.00	4.54%	82%	
2030	0	0.000	0.01	0.00	4.54%	83%	
2031	0	0.000	0.01	0.00	4.54%	83%	
2032	0	0.000	0.01	0.00	4.54%	84%	
2033	0	0.000	0.01	0.00	4.54%	85%	
2034	0	0.000	0.01	0.00	4.54%	86%	
2035	0	0.000	0.01	0.00	4.54%	86%	
2036	0	0.000	0.01	0.00	4.54%	87%	
2037	0	0.000	0.01	0.00	4.54%	88%	
2038	0	0.000	0.01	0.00	4.54%	88%	
2039	0	0.000	0.01	0.00	4.54%	89%	
2040	0	0.000	0.01	0.00	4.54%	89%	
2041	0	0.000	0.01	0.00	4.54%	90%	
2042	0	0.000	0.01	0.00	4.54%	90%	
2043	0	0.000	0.01	0.00	4.54%	91%	
2044	0	0.000	0.01	0.00	4.54%	91%	
2045	0	0.000	0.01	0.00	4.54%	91%	
2046	0	0.000	0.01	0.00	4.54%	92%	
2047	0	0.000	0.01	0.00	4.54%	92%	
2048	0	0.000	0.01	0.00	4.54%	93%	
2049	0	0.000	0.01	0.00	4.54%	93%	
2050	0	0.000	0.01	0.00	4.54%	93%	

CUM.PROD	24.025 BBO	GIANT FIELDS		KUWAIT	
RESERVES	94.252 BBO	TOTAL BBO	84.9	M.EAST (MAJOR)	
DISCOVERED	118.28 BBO	% OF DISC.	72%	Scenario-2	
UNDISCOVERED	1.723 BBO	Last Giant	1977		
REMAINING	95.975 BBO			Variance	Trend
ULTIMATE	120.00 BBO			1990	n/a
Rept.Reserve	94.252	Ref.date(end)	1989	Future	UP

Milestones	Date	Prod Mb/d	Dep. Rate	% Produced
1st Peak	1979	2210	0.51%	17%
2nd Peak	1997	2554	1.03%	25%
Mid-Point	2035	1566	0.94%	50%
Final Decline	2036		0.94%	

Forecast "C"
A=Exponential decline
B=5% incr. to Mid-Point
C=Other
Pre-1930 Cum. 0
(Added to 1930 prod)

Date	Prod. M.b/d	Prod. BBO/yr	Cum. Prod.	Remain	Deplet. Rate	% Produced	Giant	Notes & Names
1930		0.000	0.00	120.00	ERR	0%		
1931		0.000	0.00	120.00	0.00%	0%		
1932		0.000	0.00	120.00	0.00%	0%		
1933		0.000	0.00	120.00	0.00%	0%		
1934		0.000	0.00	120.00	0.00%	0%		
1935		0.000	0.00	120.00	0.00%	0%		
1936		0.000	0.00	120.00	0.00%	0%		
1937		0.000	0.00	120.00	0.00%	0%		
1938		0.000	0.00	120.00	0.00%	0%	66.50	Burgan
1939		0.000	0.00	120.00	0.00%	0%		
1940		0.000	0.00	120.00	0.00%	0%		
1941		0.000	0.00	120.00	0.00%	0%		
1942		0.000	0.00	120.00	0.00%	0%		
1943		0.000	0.00	120.00	0.00%	0%		
1944		0.000	0.00	120.00	0.00%	0%		
1945		0.006	0.01	119.99	0.00%	0%		
1946	16	0.016	0.02	119.98	0.01%	0%		
1947	45	0.046	0.07	119.93	0.04%	0%		
1948	127	0.090	0.16	119.84	0.08%	0%		
1949	247	0.130	0.29	119.71	0.11%	0%		
1950	355	0.205	0.49	119.51	0.17%	0%		
1951	561	0.274	0.77	119.23	0.23%	1%		
1952	750	0.308	1.08	118.92	0.26%	1%		
1953	845	0.344	1.42	118.58	0.29%	1%		
1954	942	0.402	1.82	118.18	0.34%	2%		
1955	1100	0.360	2.18	117.82	0.30%	2%	7.88	Raudhatain
1956	987	0.405	2.59	117.41	0.34%	2%	1.00	Bahrah
1957	1110	0.502	3.09	116.91	0.43%	3%	5.30	Sabriyah
1958	1376	0.515	3.60	116.40	0.44%	3%		
1959	1410	0.593	4.20	115.80	0.51%	3%	2.10	Minagish
1960	1625	0.602	4.80	115.20	0.52%	4%		
1961	1650	0.679	5.48	114.52	0.59%	5%		
1962	1860	0.704	6.18	113.82	0.62%	5%	1.10	Umm Gudair
1963	1930	0.774	6.96	113.04	0.68%	6%		
1964	2120	0.792	7.75	112.25	0.70%	6%		
1965	2170	0.813	8.56	111.44	0.72%	7%		
1966	2228	0.831	9.39	110.61	0.75%	8%		
1967	2277	0.891	10.28	109.72	0.81%	9%		
1968	2442	0.919	11.20	108.80	0.84%	9%		
1969	2518	1.001	12.20	107.80	0.92%	10%		
1970	2744	1.057	13.26	106.74	0.98%	11%		
1971	2895	1.004	14.26	105.74	0.94%	12%		
1972	2750	1.055	15.32	104.68	1.00%	13%		
1973	2890	0.949	16.27	103.73	0.91%	14%		
1974	2600	0.712	16.98	103.02	0.69%	14%		
1975	1950	0.664	17.64	102.36	0.64%	15%		
1976	1820	0.621	18.26	101.74	0.61%	15%		
1977	1700	0.694	18.96	101.04	0.68%	16%	1.00	Ratga
1978	1900	0.807	19.76	100.24	0.80%	16%		
1979	2210	0.511	20.28	99.72	0.51%	17%		1ST PEAK
1980	1400	0.334	20.61	99.39	0.34%	17%		
1981	916	0.246	20.86	99.14	0.25%	17%		
1982	675	0.333	21.19	98.81	0.34%	18%		
1983	912	0.338	21.53	98.47	0.34%	18%		
1984	925	0.338	21.86	98.14	0.34%	18%		
1985	823	0.301	22.17	97.83	0.31%	18%		
1986	1202	0.439	22.60	97.40	0.45%	19%		
1987	1096	0.400	23.00	97.00	0.41%	19%		
1988	1254	0.458	23.46	96.54	0.47%	20%		% of Region

1989	1543	0.563	24.02	95.98	0.58%	20%	11.02%	
1990	1752	0.640	24.66	95.34	0.67%	21%		
1991	1953	0.713	25.38	94.62	0.75%	21%		
1992	2145	0.783	26.16	93.84	0.83%	22%		
1993	2216	0.809	26.97	93.03	0.86%	22%		
1994	2268	0.828	27.80	92.20	0.89%	23%		
1995	2320	0.847	28.64	91.36	0.92%	24%		
1996	2436	0.889	29.53	90.47	0.97%	25%		
1997	2554	0.932	30.46	89.54	1.03%	25%	2ND PEAK	
1998	2503	0.913	31.38	88.62	1.02%	26%		
1999	2464	0.899	32.28	87.72	1.01%	27%		
2000	2441	0.891	33.17	86.83	1.02%	28%		
2001	2432	0.888	34.06	85.94	1.02%	28%		
2002	2447	0.893	34.95	85.05	1.04%	29%		
2003	2459	0.898	35.85	84.15	1.06%	30%		
2004	2466	0.900	36.75	83.25	1.07%	31%		
2005	2423	0.884	37.63	82.37	1.06%	31%		
2006	2418	0.882	38.51	81.49	1.07%	32%		
2007	2408	0.879	39.39	80.61	1.08%	33%		
2008	2394	0.874	40.27	79.73	1.08%	34%		
2009	2377	0.868	41.13	78.87	1.09%	34%		
2010	2357	0.860	41.99	78.01	1.09%	35%		
2011	2334	0.852	42.85	77.15	1.09%	36%		
2012	2307	0.842	43.69	76.31	1.09%	36%		
2013	2277	0.831	44.52	75.48	1.09%	37%		
2014	2245	0.819	45.34	74.66	1.09%	38%		
2015	2210	0.807	46.15	73.85	1.08%	38%		
2016	2172	0.793	46.94	73.06	1.07%	39%		
2017	2133	0.778	47.72	72.28	1.07%	40%		
2018	2091	0.763	48.48	71.52	1.06%	40%		
2019	2048	0.747	49.23	70.77	1.05%	41%		
2020	2003	0.731	49.96	70.04	1.03%	42%		
2021	1956	0.714	50.67	69.33	1.02%	42%		
2022	1907	0.696	51.37	68.63	1.00%	43%		
2023	1846	0.674	52.04	67.96	0.98%	43%		
2024	1826	0.667	52.71	67.29	0.98%	44%		
2025	1805	0.659	53.37	66.63	0.98%	44%		
2026	1782	0.651	54.02	65.98	0.98%	45%		
2027	1758	0.642	54.66	65.34	0.97%	46%		
2028	1732	0.632	55.29	64.71	0.97%	46%		
2029	1704	0.622	55.91	64.09	0.96%	47%		
2030	1678	0.612	56.53	63.47	0.96%	47%		
2031	1649	0.602	57.13	62.87	0.95%	48%		
2032	1630	0.595	57.72	62.28	0.95%	48%		
2033	1609	0.587	58.31	61.69	0.94%	49%	Add to	
2034	1588	0.580	58.89	61.11	0.94%	49%	Saudi	
2035	1566	0.572	59.46	60.54	0.94%	50%	50% PRODUCED	
2036	1559	0.569	60.03	59.97	0.94%	50%	1543	-16
2037	1544	0.564	60.60	59.40	0.94%	50%	1520	
2038	1530	0.558	61.15	58.85	0.94%	51%	1496	
2039	1515	0.553	61.71	58.29	0.94%	51%	1472	
2040	1501	0.548	62.25	57.75	0.94%	52%	1448	
2041	1487	0.543	62.80	57.20	0.94%	52%	1423	
2042	1473	0.538	63.34	56.66	0.94%	53%	1398	
2043	1459	0.533	63.87	56.13	0.94%	53%	1373	
2044	1446	0.528	64.40	55.60	0.94%	54%	1348	
2045	1432	0.523	64.92	55.08	0.94%	54%	1323	
2046	1419	0.518	65.44	54.56	0.94%	55%	1298	
2047	1405	0.513	65.95	54.05	0.94%	55%	1273	
2048	1392	0.508	66.46	53.54	0.94%	55%	1248	
2049	1379	0.503	66.96	53.04	0.94%	56%	1223	
2050	1366	0.499	67.46	52.54	0.94%	56%	1198	

CUM.PROD	16.254 BBO	GIANT FIELDS		LIBYA	
RESERVES	22.800 BBO	TOTAL BBO	33.4	AFRICA	
DISCOVERED	39.054 BBO	% OF DISC.	86%		
UNDISCOVERED	5.946 BBO	Last Giant	1976	----------------	
REMAINING	28.746 BBO	------------------		Variance	Trend
ULTIMATE	45.000 BBO			1990	+19%
Rept.Reserves	22.800 BBO	Ref.Date (end)	1989	Future	UP

Milestones		Prod Mb/d	Dep Rate	% Produced
1st Peak	1970	3385	3.03%	12%
2nd Peak	2001	1977	3.17%	51%
Mid-Point	2001	1977	3.17%	51%
Final Decline	2001		3.17%	

Forecast "B"
A=Exponential decline
B=5% inc. to Mid-Point
C=Other
Pre-1930 Cum 0
(Added to 1930 Prod)

Date	Prod. M.b/d	Prod. BBO/yr	Cum Prod	Remain	Dep. Rate	% Produced	Giant	Names & Notes
1930		0.000	0.00	45.00	ERR	0%		For names see
1931		0.000	0.00	45.00	0.00%	0%		List of Giant
1932		0.000	0.00	45.00	0.00%	0%		Fields
1933		0.000	0.00	45.00	0.00%	0%		
1934		0.000	0.00	45.00	0.00%	0%		
1935		0.000	0.00	45.00	0.00%	0%		
1936		0.000	0.00	45.00	0.00%	0%		
1937		0.000	0.00	45.00	0.00%	0%		
1938		0.000	0.00	45.00	0.00%	0%		
1939		0.000	0.00	45.00	0.00%	0%		
1940		0.000	0.00	45.00	0.00%	0%		
1941		0.000	0.00	45.00	0.00%	0%		
1942		0.000	0.00	45.00	0.00%	0%		
1943		0.000	0.00	45.00	0.00%	0%		
1944		0.000	0.00	45.00	0.00%	0%		
1945		0.000	0.00	45.00	0.00%	0%		
1946		0.000	0.00	45.00	0.00%	0%		
1947		0.000	0.00	45.00	0.00%	0%		
1948		0.000	0.00	45.00	0.00%	0%		
1949		0.000	0.00	45.00	0.00%	0%		
1950		0.000	0.00	45.00	0.00%	0%		
1951		0.000	0.00	45.00	0.00%	0%		
1952		0.000	0.00	45.00	0.00%	0%		
1953		0.000	0.00	45.00	0.00%	0%		
1954		0.000	0.00	45.00	0.00%	0%		
1955		0.000	0.00	45.00	0.00%	0%		
1956		0.000	0.00	45.00	0.00%	0%		
1957		0.000	0.00	45.00	0.00%	0%		
1958		0.000	0.00	45.00	0.00%	0%	1.3	
1959		0.000	0.00	45.00	0.00%	0%	7.25	
1960		0.000	0.00	45.00	0.00%	0%	1.8	
1961	19	0.007	0.01	44.99	0.02%	0%	11.5	
1962	184	0.067	0.07	44.93	0.15%	0%	0.5	
1963	470	0.172	0.25	44.75	0.38%	1%		
1964	859	0.313	0.56	44.44	0.70%	1%		
1965	1239	0.452	1.01	43.99	1.02%	2%		
1966	1472	0.537	1.55	43.45	1.22%	3%	1.5	
1967	1717	0.627	2.18	42.82	1.44%	5%	2.25	
1968	2535	0.925	3.10	41.90	2.16%	7%		
1969	3067	1.119	4.22	40.78	2.67%	9%		
1970	3385	1.236	5.46	39.54	3.03%	12%		1ST PEAK
1971	2800	1.022	6.48	38.52	2.58%	14%	1.5	
1972	2230	0.814	7.29	37.71	2.11%	16%		
1973	2117	0.773	8.06	36.94	2.05%	18%		
1974	1700	0.621	8.68	36.32	1.68%	19%		
1975	1400	0.511	9.20	35.80	1.41%	20%		
1976	1900	0.694	9.89	35.11	1.94%	22%	5	
1977	2050	0.748	10.64	34.36	2.13%	24%		
1978	2050	0.748	11.39	33.61	2.18%	25%		
1979	2050	0.748	12.13	32.87	2.23%	27%		
1980	1780	0.650	12.78	32.22	1.98%	28%		
1981	1063	0.388	13.17	31.83	1.20%	29%		
1982	1127	0.411	13.58	31.42	1.29%	30%		
1983	1020	0.372	13.96	31.04	1.19%	31%		
1984	1090	0.398	14.35	30.65	1.28%	32%		
1985	1045	0.382	14.73	30.27	1.24%	33%		
1986	1031	0.376	15.11	29.89	1.24%	34%		
1987	1020	0.372	15.48	29.52	1.25%	34%		
1988	1013	0.370	15.85	29.15	1.25%	35%		

1989	1101	0.402	16.25	28.75	1.38%	36%	
1990	1156	0.422	16.68	28.32	1.47%	37%	0.8
1991	1213	0.443	17.12	27.88	1.56%	38%	
1992	1274	0.465	17.58	27.42	1.67%	39%	
1993	1338	0.488	18.07	26.93	1.78%	40%	
1994	1405	0.513	18.58	26.42	1.90%	41%	
1995	1475	0.538	19.12	25.88	2.04%	42%	
1996	1549	0.565	19.69	25.31	2.18%	44%	
1997	1626	0.594	20.28	24.72	2.34%	45%	
1998	1707	0.623	20.91	24.09	2.52%	46%	
1999	1793	0.654	21.56	23.44	2.72%	48%	
2000	1882	0.687	22.25	22.75	2.93%	49%	
2001	1977	0.721	22.97	22.03	3.17%	51%	50% PRODUCED
2002	1913	0.698	23.67	21.33	3.17%	53%	& 2ND PEAK
2003	1853	0.676	24.34	20.66	3.17%	54%	
2004	1794	0.655	25.00	20.00	3.17%	56%	
2005	1737	0.634	25.63	19.37	3.17%	57%	
2006	1682	0.614	26.25	18.75	3.17%	58%	
2007	1629	0.595	26.84	18.16	3.17%	60%	
2008	1577	0.576	27.42	17.58	3.17%	61%	
2009	1527	0.557	27.97	17.03	3.17%	62%	
2010	1479	0.540	28.51	16.49	3.17%	63%	
2011	1432	0.523	29.04	15.96	3.17%	65%	
2012	1386	0.506	29.54	15.46	3.17%	66%	
2013	1343	0.490	30.03	14.97	3.17%	67%	
2014	1300	0.474	30.51	14.49	3.17%	68%	
2015	1259	0.459	30.97	14.03	3.17%	69%	
2016	1219	0.445	31.41	13.59	3.17%	70%	
2017	1180	0.431	31.84	13.16	3.17%	71%	
2018	1143	0.417	32.26	12.74	3.17%	72%	
2019	1107	0.404	32.66	12.34	3.17%	73%	
2020	1072	0.391	33.05	11.95	3.17%	73%	
2021	1038	0.379	33.43	11.57	3.17%	74%	
2022	1005	0.367	33.80	11.20	3.17%	75%	
2023	973	0.355	34.15	10.85	3.17%	76%	
2024	942	0.344	34.50	10.50	3.17%	77%	
2025	912	0.333	34.83	10.17	3.17%	77%	
2026	883	0.322	35.15	9.85	3.17%	78%	
2027	855	0.312	35.47	9.53	3.17%	79%	
2028	828	0.302	35.77	9.23	3.17%	79%	
2029	802	0.293	36.06	8.94	3.17%	80%	
2030	776	0.283	36.34	8.66	3.17%	81%	
2031	752	0.274	36.62	8.38	3.17%	81%	
2032	728	0.266	36.88	8.12	3.17%	82%	
2033	705	0.257	37.14	7.86	3.17%	83%	
2034	683	0.249	37.39	7.61	3.17%	83%	
2035	661	0.241	37.63	7.37	3.17%	84%	
2036	640	0.234	37.86	7.14	3.17%	84%	
2037	620	0.226	38.09	6.91	3.17%	85%	
2038	600	0.219	38.31	6.69	3.17%	85%	
2039	581	0.212	38.52	6.48	3.17%	86%	
2040	563	0.205	38.73	6.27	3.17%	86%	
2041	545	0.199	38.93	6.07	3.17%	87%	
2042	527	0.193	39.12	5.88	3.17%	87%	
2043	511	0.186	39.31	5.69	3.17%	87%	
2044	495	0.181	39.49	5.51	3.17%	88%	
2045	479	0.175	39.66	5.34	3.17%	88%	
2046	464	0.169	39.83	5.17	3.17%	89%	
2047	449	0.164	39.99	5.01	3.17%	89%	
2048	435	0.159	40.15	4.85	3.17%	89%	
2049	421	0.154	40.31	4.69	3.17%	90%	
2050	408	0.149	40.45	4.55	3.17%	90%	

CUM.PROD	2.073	BBO	GIANT FIELDS		MALAYSIA	
RESERVES	2.950	BBO	TOTAL BBO	1	THE EAST	
DISCOVERED	5.023	BBO	% OF DISC.	20%		
UNDISCOVERED	1.977	BBO	Last Giant	?		
REMAINING	4.927	BBO			Variance	Trend
ULTIMATE	7.000	BBO			1990	+3%
Rept.Reserve	2.950		Ref.date(end) 1989		Future	DOWN

Milestones	Date	Prod Mb/d	Dep. Rate	% Produced
1st Peak	1994	711	6.15%	46%
2nd Peak				
Mid-Point	1995	667	6.15%	50%
Final Decline	1995		6.15%	50%

Forecast "B"
A=Exponential decline
B=5% incr. to Mid-Point
C=Other
Pre-1930 Cum. See Note

Date	Prod. M.b/d	Prod. BBO/yr	Cum. Prod.	Remain	Deplet. Rate	% Produced	Giant	Notes & Names
1930		0.000	0.00	7.00	ERR	0%		
1931		0.000	0.00	7.00	0.00%	0%		
1932		0.000	0.00	7.00	0.00%	0%		
1933		0.000	0.00	7.00	0.00%	0%		
1934		0.000	0.00	7.00	0.00%	0%		
1935		0.000	0.00	7.00	0.00%	0%		
1936		0.000	0.00	7.00	0.00%	0%		
1937		0.000	0.00	7.00	0.00%	0%		
1938		0.000	0.00	7.00	0.00%	0%		
1939		0.000	0.00	7.00	0.00%	0%		
1940		0.000	0.00	7.00	0.00%	0%		
1941		0.000	0.00	7.00	0.00%	0%		
1942		0.000	0.00	7.00	0.00%	0%		
1943		0.000	0.00	7.00	0.00%	0%		
1944		0.000	0.00	7.00	0.00%	0%		
1945		0.000	0.00	7.00	0.00%	0%		
1946		0.000	0.00	7.00	0.00%	0%		
1947		0.000	0.00	7.00	0.00%	0%		
1948		0.000	0.00	7.00	0.00%	0%		
1949		0.000	0.00	7.00	0.00%	0%		
1950		0.000	0.00	7.00	0.00%	0%		
1951		0.000	0.00	7.00	0.00%	0%		
1952		0.000	0.00	7.00	0.00%	0%		
1953		0.000	0.00	7.00	0.00%	0%		
1954		0.000	0.00	7.00	0.00%	0%		
1955		0.000	0.00	7.00	0.00%	0%		
1956		0.000	0.00	7.00	0.00%	0%		
1957		0.000	0.00	7.00	0.00%	0%		
1958		0.000	0.00	7.00	0.00%	0%		
1959		0.000	0.00	7.00	0.00%	0%		
1960		0.000	0.00	7.00	0.00%	0%		
1961		0.000	0.00	7.00	0.00%	0%		
1962		0.000	0.00	7.00	0.00%	0%		
1963		0.000	0.00	7.00	0.00%	0%		
1964		0.000	0.00	7.00	0.00%	0%		
1965		0.000	0.00	7.00	0.00%	0%		
1966		0.000	0.00	7.00	0.00%	0%		
1967		0.000	0.00	7.00	0.00%	0%		
1968		0.000	0.00	7.00	0.00%	0%		
1969		0.000	0.00	7.00	0.00%	0%		
1970		0.000	0.00	7.00	0.00%	0%		
1971		0.000	0.00	7.00	0.00%	0%		
1972		0.000	0.00	7.00	0.00%	0%		
1973		0.000	0.00	7.00	0.00%	0%		
1974		0.000	0.21	7.00	0.00%	3%		
1975	100	0.037	0.25	6.96	0.52%	4%	0.50	Smarang
1976	155	0.057	0.30	6.91	0.81%	4%	0.50	Baronia
1977	190	0.069	0.37	6.84	1.00%	5%		
1978	210	0.077	0.45	6.76	1.12%	6%		
1979	270	0.099	0.55	6.66	1.46%	8%		
1980	280	0.102	0.65	6.56	1.53%	9%		
1981	259	0.095	0.75	6.47	1.44%	11%		
1982	306	0.112	0.86	6.35	1.73%	12%		
1983	370	0.135	0.99	6.22	2.13%	14%		
1984	462	0.169	1.16	6.05	2.71%	17%		
1985	432	0.158	1.32	5.89	2.61%	19%		
1986	503	0.184	1.50	5.71	3.12%	21%		
1987	467	0.171	1.67	5.54	2.99%	24%		
1988	540	0.197	1.87	5.34	3.56%	27%		

1989	557	0.203	2.07	5.14	3.81%	30%	
1990	585	0.213	2.29	4.92	4.15%	33%	
1991	614	0.224	2.51	4.70	4.55%	36%	
1992	645	0.235	2.75	4.47	5.01%	39%	
1993	677	0.247	2.99	4.22	5.53%	43%	
1994	711	0.259	3.25	3.96	6.15%	46%	1ST PEAK
1995	667	0.243	3.50	3.71	6.15%	50%	& 50 % PRODUCED
1996	626	0.228	3.72	3.49	6.15%	53%	
1997	587	0.214	3.94	3.27	6.15%	56%	
1998	551	0.201	4.14	3.07	6.15%	59%	
1999	517	0.189	4.33	2.88	6.15%	62%	
2000	486	0.177	4.51	2.70	6.15%	64%	
2001	456	0.166	4.67	2.54	6.15%	67%	
2002	428	0.156	4.83	2.38	6.15%	69%	
2003	401	0.147	4.98	2.24	6.15%	71%	
2004	377	0.138	5.11	2.10	6.15%	73%	
2005	354	0.129	5.24	1.97	6.15%	75%	
2006	332	0.121	5.36	1.85	6.15%	77%	
2007	311	0.114	5.48	1.73	6.15%	78%	
2008	292	0.107	5.58	1.63	6.15%	80%	
2009	274	0.100	5.68	1.53	6.15%	81%	
2010	257	0.094	5.78	1.43	6.15%	83%	
2011	242	0.088	5.87	1.35	6.15%	84%	
2012	227	0.083	5.95	1.26	6.15%	85%	
2013	213	0.078	6.03	1.19	6.15%	86%	
2014	200	0.073	6.10	1.11	6.15%	87%	
2015	187	0.068	6.17	1.04	6.15%	88%	
2016	176	0.064	6.23	0.98	6.15%	89%	
2017	165	0.060	6.29	0.92	6.15%	90%	
2018	155	0.057	6.35	0.86	6.15%	91%	
2019	145	0.053	6.40	0.81	6.15%	91%	
2020	136	0.050	6.45	0.76	6.15%	92%	
2021	128	0.047	6.50	0.71	6.15%	93%	
2022	120	0.044	6.54	0.67	6.15%	93%	
2023	113	0.041	6.58	0.63	6.15%	94%	
2024	106	0.039	6.62	0.59	6.15%	95%	
2025	99	0.036	6.66	0.55	6.15%	95%	
2026	93	0.034	6.69	0.52	6.15%	96%	
2027	87	0.032	6.72	0.49	6.15%	96%	
2028	82	0.030	6.75	0.46	6.15%	96%	
2029	77	0.028	6.78	0.43	6.15%	97%	
2030	72	0.026	6.81	0.40	6.15%	97%	
2031	68	0.025	6.83	0.38	6.15%	98%	
2032	64	0.023	6.86	0.35	6.15%	98%	
2033	60	0.022	6.88	0.33	6.15%	98%	
2034	56	0.020	6.90	0.31	6.15%	99%	
2035	53	0.019	6.92	0.29	6.15%	99%	
2036	49	0.018	6.94	0.28	6.15%	99%	
2037	46	0.017	6.95	0.26	6.15%	99%	
2038	44	0.016	6.97	0.24	6.15%	100%	
2039	41	0.015	6.98	0.23	6.15%	100%	
2040	38	0.014	7.00	0.21	6.15%	100%	
2041	36	0.013	7.01	0.20	6.15%	100%	
2042	34	0.012	7.02	0.19	6.15%	100%	
2043	32	0.012	7.03	0.18	6.15%	100%	
2044	30	0.011	7.05	0.17	6.15%	101%	
2045	28	0.010	7.06	0.16	6.15%	101%	
2046	26	0.010	7.07	0.15	6.15%	101%	
2047	25	0.009	7.07	0.14	6.15%	101%	
2048	23	0.008	7.08	0.13	6.15%	101%	
2049	22	0.008	7.09	0.12	6.15%	101%	
2050	20	0.007	7.10	0.11	6.15%	101%	

Notes
Data base incomplete. Cum. Prod for 1974 taken from Petroconsultants
Date of giant discovery uncertain

CUM.PROD	16.357	BBO	GIANT FIELDS			MEXICO
RESERVES	40.000	BBO	TOTAL BBO	40.3		L.AMERICA
DISCOVERED	56.357	BBO	% OF DISC.	72%		
UNDISCOVERED	3.643	BBO	Last Giant	1978		
REMAINING	43.643	BBO				Variance Trend
ULTIMATE	60.000	BBO				1990 -4%
Rept.Reserves	56.365	BBO	Ref. Date (end)		1989	Future DOWN

Milestones	Date	Prod Mb/d	Dep. Rate	% Produced	Forecast "B"
					A=Exponential decline
1st Peak	1985	2797	2.11%	21%	B=5% inc. to Mid-Point
2nd Peak	2000	4063	4.78%	51%	C=Other
Mid-Point	2000	4063	4.78%	51%	Pre-1930 Cum 1.374
Final Decline	2000		4.78%		(added to 1930 Prod)

Date	Prod. M.b/d	Prod. BBO/yr	Cum Prod	Remain	Dep. Rate	% Produced	Giant	Names & Notes
1930	112.3	0.041	1.42	58.59	ERR	2%	11.7	#1
1931	110.0	0.040	1.46	58.54	0.07%	2%		
1932	110.0	0.040	1.50	58.50	0.07%	2%		
1933	110.0	0.040	1.54	58.46	0.07%	3%		
1934	110.0	0.040	1.58	58.42	0.07%	3%		
1935	110.0	0.040	1.62	58.38	0.07%	3%		
1936	110.0	0.040	1.66	58.34	0.07%	3%		
1937	110.0	0.040	1.70	58.30	0.07%	3%		
1938	105.6	0.039	1.73	58.27	0.07%	3%		
1939	117.7	0.043	1.78	58.22	0.07%	3%		
1940	120.4	0.044	1.82	58.18	0.08%	3%		
1941	115.8	0.042	1.86	58.14	0.07%	3%		
1942	95.5	0.035	1.90	58.10	0.06%	3%		
1943	96.4	0.035	1.93	58.07	0.06%	3%		
1944	104.7	0.038	1.97	58.03	0.07%	3%		
1945	119.4	0.044	2.02	57.98	0.08%	3%		
1946	135.0	0.049	2.06	57.94	0.08%	3%		
1947	153.8	0.056	2.12	57.88	0.10%	4%		
1948	159.5	0.058	2.18	57.82	0.10%	4%		
1949	166.4	0.061	2.24	57.76	0.11%	4%		
1950	197.6	0.072	2.31	57.69	0.12%	4%		
1951	211.2	0.077	2.39	57.61	0.13%	4%		
1952	218.0	0.080	2.47	57.53	0.14%	4%		
1953	201.0	0.073	2.54	57.46	0.13%	4%		
1954	232.0	0.085	2.63	57.37	0.15%	4%		
1955	250.0	0.091	2.72	57.28	0.16%	5%		
1956	253.0	0.092	2.81	57.19	0.16%	5%		
1957	241.0	0.088	2.90	57.10	0.15%	5%		
1958	243.1	0.089	2.99	57.01	0.16%	5%		
1959	264.0	0.096	3.08	56.92	0.17%	5%		
1960	275.0	0.100	3.18	56.82	0.18%	5%	4.5	#2
1961	283.0	0.103	3.29	56.71	0.18%	5%		
1962	305.0	0.111	3.40	56.60	0.20%	6%		
1963	320.0	0.117	3.52	56.48	0.21%	6%		
1964	310.0	0.113	3.63	56.37	0.20%	6%		
1965	333.2	0.122	3.75	56.25	0.22%	6%		
1966	322.2	0.118	3.87	56.13	0.21%	6%		
1967	357.6	0.131	4.00	56.00	0.23%	7%	1	#3
1968	383.5	0.140	4.14	55.86	0.25%	7%		
1969	405.1	0.148	4.29	55.71	0.26%	7%		
1970	427.4	0.156	4.44	55.56	0.28%	7%		
1971	427.0	0.156	4.60	55.40	0.28%	8%		
1972	440.0	0.161	4.76	55.24	0.29%	8%	1.85	#4
1973	478.0	0.174	4.93	55.07	0.32%	8%		
1974	513.0	0.187	5.12	54.88	0.34%	9%		
1975	710.0	0.259	5.38	54.62	0.47%	9%	0.5	#5
1976	850.0	0.310	5.69	54.31	0.57%	9%		
1977	990.0	0.361	6.05	53.95	0.67%	10%	8.5	#6
1978	1270.0	0.464	6.51	53.49	0.86%	11%	12.285	#7
1979	1490.0	0.544	7.06	52.94	1.02%	12%		
1980	1960.0	0.715	7.77	52.23	1.35%	13%		
1981	2390.0	0.872	8.65	51.35	1.67%	14%		
1982	2734.0	0.998	9.64	50.36	1.94%	16%		
1983	2702.0	0.986	10.63	49.37	1.96%	18%		
1984	2743.0	1.001	11.63	48.37	2.03%	19%		
1985	2797.0	1.021	12.65	47.35	2.11%	21%		1ST PEAK
1986	2468.0	0.901	13.55	46.45	1.90%	23%		
1987	2537.0	0.926	14.48	45.52	1.99%	24%		
1988	2527.0	0.922	15.40	44.60	2.03%	26%		

1989	2618.0	0.956	16.36	43.64	2.14%	27%	
1990	2748.9	1.003	17.36	42.64	2.30%	29%	
1991	2886.3	1.054	18.41	41.59	2.47%	31%	
1992	3030.7	1.106	19.52	40.48	2.66%	33%	
1993	3182.2	1.162	20.68	39.32	2.87%	34%	
1994	3341.3	1.220	21.90	38.10	3.10%	37%	
1995	3508.4	1.281	23.18	36.82	3.36%	39%	
1996	3683.8	1.345	24.53	35.47	3.65%	41%	
1997	3868.0	1.412	25.94	34.06	3.98%	43%	
1998	4061.4	1.482	27.42	32.58	4.35%	46%	
1999	4264.4	1.557	28.98	31.02	4.78%	48%	
2000	4062.7	1.483	30.46	29.54	4.78%	51%	2ND PEAK
2001	3868.5	1.412	31.87	28.13	4.78%	53%	& 50% PRODUCED
2002	3683.6	1.345	33.22	26.78	4.78%	55%	
2003	3507.5	1.280	34.50	25.50	4.78%	57%	
2004	3339.9	1.219	35.72	24.28	4.78%	60%	
2005	3180.2	1.161	36.88	23.12	4.78%	61%	
2006	3028.2	1.105	37.98	22.02	4.78%	63%	
2007	2883.5	1.052	39.03	20.97	4.78%	65%	
2008	2745.6	1.002	40.04	19.96	4.78%	67%	
2009	2614.4	0.954	40.99	19.01	4.78%	68%	
2010	2489.4	0.909	41.90	18.10	4.78%	70%	
2011	2370.4	0.865	42.76	17.24	4.78%	71%	
2012	2257.1	0.824	43.59	16.41	4.78%	73%	
2013	2149.2	0.784	44.37	15.63	4.78%	74%	
2014	2046.5	0.747	45.12	14.88	4.78%	75%	
2015	1948.7	0.711	45.83	14.17	4.78%	76%	
2016	1855.5	0.677	46.51	13.49	4.78%	78%	
2017	1766.8	0.645	47.15	12.85	4.78%	79%	
2018	1682.4	0.614	47.77	12.23	4.78%	80%	
2019	1602.0	0.585	48.35	11.65	4.78%	81%	
2020	1525.4	0.557	48.91	11.09	4.78%	82%	
2021	1452.5	0.530	49.44	10.56	4.78%	82%	
2022	1383.1	0.505	49.94	10.06	4.78%	83%	
2023	1316.9	0.481	50.42	9.58	4.78%	84%	
2024	1254.0	0.458	50.88	9.12	4.78%	85%	
2025	1194.1	0.436	51.32	8.68	4.78%	86%	
2026	1137.0	0.415	51.73	8.27	4.78%	86%	
2027	1082.6	0.395	52.13	7.87	4.78%	87%	
2028	1030.9	0.376	52.50	7.50	4.78%	88%	
2029	981.6	0.358	52.86	7.14	4.78%	88%	
2030	934.7	0.341	53.20	6.80	4.78%	89%	
2031	890.0	0.325	53.53	6.47	4.78%	89%	
2032	847.5	0.309	53.84	6.16	4.78%	90%	
2033	807.0	0.295	54.13	5.87	4.78%	90%	
2034	768.4	0.280	54.41	5.59	4.78%	91%	
2035	731.7	0.267	54.68	5.32	4.78%	91%	
2036	696.7	0.254	54.93	5.07	4.78%	92%	
2037	663.4	0.242	55.18	4.82	4.78%	92%	
2038	631.7	0.231	55.41	4.59	4.78%	92%	
2039	601.5	0.220	55.63	4.37	4.78%	93%	
2040	572.7	0.209	55.84	4.16	4.78%	93%	
2041	545.3	0.199	56.03	3.97	4.78%	93%	
2042	519.3	0.190	56.22	3.78	4.78%	94%	
2043	494.5	0.180	56.40	3.60	4.78%	94%	
2044	470.8	0.172	56.58	3.42	4.78%	94%	
2045	448.3	0.164	56.74	3.26	4.78%	95%	
2046	426.9	0.156	56.90	3.10	4.78%	95%	
2047	406.5	0.148	57.04	2.96	4.78%	95%	
2048	387.1	0.141	57.19	2.81	4.78%	95%	
2049	368.6	0.135	57.32	2.68	4.78%	96%	
2050	350.9	0.128	57.45	2.55	4.78%	96%	

CUM.PROD	0.018	BBO	GIANT FIELDS			MOROCCO
RESERVES	0.003	BBO	TOTAL BBO	0.0		AFRICA
DISCOVERED	0.021	BBO	% OF DISC.	0%		
UNDISCOVERED	0.000	BBO	Last Giant			
REMAINING	0.003	BBO				Variance Trend
ULTIMATE	0.021	BBO				1990 n/a
Rept.Reserves	0.003	BBO	Ref.Date (end)		1989	Future n/a

Milestones	Date	Prod Mb/d	Dep Rate	% Produced	Forecast "A"
1st Peak	1962	3	7.31%	44%	A=Exponential decline
2nd Peak					B=5% inc. to Mid-Point
					C=Other
Mid-Point	1963	3	9.46%	50%	Pre-1930 Cum 0
Final Decline		Resources exhausted			(Added to 1930 Prod)

Date	Prod. M.b/d	Prod. BBO/yr	Cum Prod	Remain	Dep. Rate	% Produced	Giant	Names & Notes
1930		0.000	0.00	0.02	ERR	0%		
1931		0.000	0.00	0.02	0.00%	0%		
1932		0.000	0.00	0.02	0.00%	0%		
1933		0.000	0.00	0.02	0.00%	0%		
1934		0.000	0.00	0.02	0.00%	0%		
1935		0.000	0.00	0.02	0.00%	0%		
1936		0.000	0.00	0.02	0.00%	0%		
1937		0.000	0.00	0.02	0.00%	0%		
1938		0.000	0.00	0.02	0.00%	0%		
1939		0.000	0.00	0.02	0.00%	0%		
1940		0.000	0.00	0.02	0.00%	0%		
1941		0.000	0.00	0.02	0.00%	0%		
1942		0.000	0.00	0.02	0.00%	0%		
1943		0.000	0.00	0.02	0.00%	0%		
1944		0.000	0.00	0.02	0.00%	0%		
1945		0.000	0.00	0.02	0.00%	0%		
1946		0.000	0.00	0.02	0.00%	0%		
1947		0.000	0.00	0.02	0.00%	0%		
1948	0	0.000	0.00	0.02	0.35%	0%		
1949	0	0.000	0.00	0.02	0.53%	1%		
1950	1	0.000	0.00	0.02	1.41%	2%		
1951	2	0.001	0.00	0.02	2.69%	5%		
1952	3	0.001	0.00	0.02	4.79%	9%		
1953	2	0.001	0.00	0.02	4.14%	13%		
1954	2	0.001	0.00	0.02	4.85%	17%		
1955	2	0.001	0.00	0.02	4.46%	21%		
1956	2	0.001	0.01	0.02	4.44%	25%		
1957	2	0.001	0.01	0.02	3.49%	27%		
1958	2	0.001	0.01	0.01	3.61%	30%		
1959	2	0.001	0.01	0.01	4.75%	33%		
1960	2	0.001	0.01	0.01	5.25%	37%		
1961	2	0.001	0.01	0.01	5.26%	40%		
1962	3	0.001	0.01	0.01	7.31%	44%		1ST PEAK
1963	3	0.001	0.01	0.01	9.46%	50%		50% PRODUCED
1964	3	0.001	0.01	0.01	10.1%	55%		
1965	2	0.001	0.01	0.01	7.36%	58%		
1966	2	0.001	0.01	0.01	9.20%	62%		
1967	2	0.001	0.01	0.01	9.21%	65%		
1968	2	0.001	0.01	0.01	10.1%	69%		
1969	1	0.001	0.01	0.01	7.90%	71%		
1970	1	0.000	0.02	0.01	6.13%	73%		
1971	1	0.000	0.02	0.01	5.87%	75%		
1972	0	0.000	0.02	0.01	2.08%	75%		
1973	1	0.000	0.02	0.00	6.37%	77%		
1974	1	0.000	0.02	0.00	6.81%	78%		
1975	1	0.000	0.02	0.00	5.68%	80%		
1976	0	0.000	0.02	0.00	0.86%	80%		
1977	0	0.000	0.02	0.00	3.47%	81%		
1978	0	0.000	0.02	0.00	0.90%	81%		
1979	1	0.000	0.02	0.00	4.54%	82%		
1980	0	0.000	0.02	0.00	1.90%	82%		
1981	0	0.000	0.02	0.00	1.94%	82%		
1982	0	0.000	0.02	0.00	1.98%	83%		
1983	0	0.000	0.02	0.00	2.02%	83%		
1984	0	0.000	0.02	0.00	2.06%	83%		
1985	0	0.000	0.02	0.00	0.00%	83%		
1986	0	0.000	0.02	0.00	3.15%	84%		
1987	0	0.000	0.02	0.00	3.25%	84%		
1988	0	0.000	0.02	0.00	4.48%	85%		

1989	0	0.000	0.02	0.00	3.52%	86%
1990	0	0.000	0.02	0.00	3.52%	86%
1991	0	0.000	0.02	0.00	3.52%	87%
1992	0	0.000	0.02	0.00	3.52%	87%
1993	0	0.000	0.02	0.00	3.52%	88%
1994	0	0.000	0.02	0.00	3.52%	88%
1995	0	0.000	0.02	0.00	3.52%	88%
1996	0	0.000	0.02	0.00	3.52%	89%
1997	0	0.000	0.02	0.00	3.52%	89%
1998	0	0.000	0.02	0.00	3.52%	90%
1999	0	0.000	0.02	0.00	3.52%	90%
2000	0	0.000	0.02	0.00	3.52%	90%
2001	0	0.000	0.02	0.00	3.52%	91%
2002	0	0.000	0.02	0.00	3.52%	91%
2003	0	0.000	0.02	0.00	3.52%	91%
2004	0	0.000	0.02	0.00	3.52%	92%
2005	0	0.000	0.02	0.00	3.52%	92%
2006	0	0.000	0.02	0.00	3.52%	92%
2007	0	0.000	0.02	0.00	3.52%	92%
2008	0	0.000	0.02	0.00	3.52%	93%
2009	0	0.000	0.02	0.00	3.52%	93%
2010	0	0.000	0.02	0.00	3.52%	93%
2011	0	0.000	0.02	0.00	3.52%	93%
2012	0	0.000	0.02	0.00	3.52%	94%
2013	0	0.000	0.02	0.00	3.52%	94%
2014	0	0.000	0.02	0.00	3.52%	94%
2015	0	0.000	0.02	0.00	3.52%	94%
2016	0	0.000	0.02	0.00	3.52%	95%
2017	0	0.000	0.02	0.00	3.52%	95%
2018	0	0.000	0.02	0.00	3.52%	95%
2019	0	0.000	0.02	0.00	3.52%	95%
2020	0	0.000	0.02	0.00	3.52%	95%
2021	0	0.000	0.02	0.00	3.52%	95%
2022	0	0.000	0.02	0.00	3.52%	96%
2023	0	0.000	0.02	0.00	3.52%	96%
2024	0	0.000	0.02	0.00	3.52%	96%
2025	0	0.000	0.02	0.00	3.52%	96%
2026	0	0.000	0.02	0.00	3.52%	96%
2027	0	0.000	0.02	0.00	3.52%	96%
2028	0	0.000	0.02	0.00	3.52%	96%
2029	0	0.000	0.02	0.00	3.52%	97%
2030	0	0.000	0.02	0.00	3.52%	97%
2031	0	0.000	0.02	0.00	3.52%	97%
2032	0	0.000	0.02	0.00	3.52%	97%
2033	0	0.000	0.02	0.00	3.52%	97%
2034	0	0.000	0.02	0.00	3.52%	97%
2035	0	0.000	0.02	0.00	3.52%	97%
2036	0	0.000	0.02	0.00	3.52%	97%
2037	0	0.000	0.02	0.00	3.52%	97%
2038	0	0.000	0.02	0.00	3.52%	98%
2039	0	0.000	0.02	0.00	3.52%	98%
2040	0	0.000	0.02	0.00	3.52%	98%
2041	0	0.000	0.02	0.00	3.52%	98%
2042	0	0.000	0.02	0.00	3.52%	98%
2043	0	0.000	0.02	0.00	3.52%	98%
2044	0	0.000	0.02	0.00	3.52%	98%
2045	0	0.000	0.02	0.00	3.52%	98%
2046	0	0.000	0.02	0.00	3.52%	98%
2047	0	0.000	0.02	0.00	3.52%	98%
2048	0	0.000	0.02	0.00	3.52%	98%
2049	0	0.000	0.02	0.00	3.52%	98%
2050	0	0.000	0.02	0.00	3.52%	98%

CUM.PROD	0.561 BBO	GIANT FIELDS		NETHERLANDS	
RESERVES	0.170 BBO	TOTAL BBO	0	W.EUROPE	
DISCOVERED	0.731 BBO	% OF DISC.	0%		
UNDISCOVERED	0.019 BBO	Last Giant	-		
REMAINING	0.189 BBO			Variance	Trend
ULTIMATE	0.750 BBO			1990	-4%
Rept.Reserve	0.170 BBO	Ref.Date(end)	1989	Future	DOWN

Milestones	Date	Prod Mb/d	Dep. Rate	% Produced
1st Peak	1965	45	2.80%	23%
2nd Peak	1989	87	14.42%	75%
Mid-Point	1982	28	2.65%	50%
Final Decline	1990		14.40%	

Forecast "A"
A=Exponential decline
B=5% incr. to Mid-Point
C=Other
Pre-1930 Cum. 0
(Added to 1930 Prod)

Date	Prod. M.b/d	Prod. BBO/yr	Cum Prod	Remain	Dep. Rate	% Produced	Giant	Mames & Notes
1930		0.000	0.00	0.75	ERR	0%		
1931		0.000	0.00	0.75	0.00%	0%		
1932		0.000	0.00	0.75	0.00%	0%		
1933		0.000	0.00	0.75	0.00%	0%		
1934		0.000	0.00	0.75	0.00%	0%		
1935		0.000	0.00	0.75	0.00%	0%		
1936		0.000	0.00	0.75	0.00%	0%		
1937		0.000	0.00	0.75	0.00%	0%		
1938		0.000	0.00	0.75	0.00%	0%		
1939		0.000	0.00	0.75	0.00%	0%		
1940		0.000	0.00	0.75	0.00%	0%		
1941		0.000	0.00	0.75	0.00%	0%		
1942		0.000	0.00	0.75	0.00%	0%		
1943		0.000	0.00	0.75	0.00%	0%		
1944		0.000	0.00	0.75	0.00%	0%		
1945	0	0.000	0.00	0.75	0.00%	0%		
1946	1	0.000	0.00	0.75	0.06%	0%		
1947	4	0.001	0.00	0.75	0.19%	0%		
1948	9	0.003	0.01	0.74	0.46%	1%		
1949	12	0.004	0.01	0.74	0.58%	1%		
1950	13	0.005	0.01	0.74	0.66%	2%		
1951	14	0.005	0.02	0.73	0.67%	3%		
1952	14	0.005	0.02	0.73	0.68%	3%		
1953	16	0.006	0.03	0.72	0.79%	4%		
1954	18	0.006	0.04	0.71	0.90%	5%		
1955	20	0.007	0.04	0.71	1.00%	6%		
1956	21	0.007	0.05	0.70	1.06%	7%		
1957	27	0.010	0.06	0.69	1.38%	8%		
1958	30	0.011	0.07	0.68	1.57%	10%		
1959	34	0.012	0.08	0.67	1.81%	11%		
1960	36	0.013	0.10	0.65	1.97%	13%		
1961	38	0.014	0.11	0.64	2.14%	15%		
1962	41	0.015	0.13	0.62	2.34%	17%		
1963	42	0.015	0.14	0.61	2.46%	19%		
1964	43	0.016	0.16	0.59	2.57%	21%		
1965	45	0.017	0.17	0.58	2.80%	23%		1st PEAK
1966	45	0.016	0.19	0.56	2.84%	25%		
1967	44	0.016	0.21	0.54	2.84%	27%		
1968	43	0.016	0.22	0.53	2.88%	30%		
1969	39	0.014	0.24	0.51	2.68%	31%		
1970	37	0.013	0.25	0.50	2.59%	33%		
1971	33	0.012	0.26	0.49	2.38%	35%		
1972	30	0.011	0.27	0.48	2.24%	36%		
1973	29	0.010	0.28	0.47	2.18%	38%		
1974	27	0.010	0.29	0.46	2.14%	39%		
1975	26	0.009	0.30	0.45	2.07%	40%		
1976	28	0.010	0.31	0.44	2.28%	42%		
1977	30	0.011	0.32	0.43	2.50%	43%		
1978	36	0.013	0.34	0.41	3.08%	45%		
1979	25	0.009	0.35	0.40	2.21%	46%		
1980	24	0.009	0.35	0.40	2.17%	47%		
1981	27	0.010	0.36	0.39	2.49%	49%		
1982	28	0.010	0.37	0.38	2.65%	50%		50% PRODUCED
1983	42	0.015	0.39	0.36	4.08%	52%		
1984	61	0.022	0.41	0.34	6.18%	55%		
1985	69	0.025	0.44	0.31	7.45%	58%		
1986	79	0.029	0.47	0.28	9.21%	62%		
1987	91	0.033	0.50	0.25	11.7%	67%		
1988	84	0.031	0.53	0.22	12.2%	71%		

1989	87	0.032	0.56	0.19	14.4%	75%	2nd PEAK
1990	74	0.027	0.59	0.16	14.4%	78%	
1991	64	0.023	0.61	0.14	14.4%	82%	
1992	55	0.020	0.63	0.12	14.4%	84%	
1993	47	0.017	0.65	0.10	14.4%	87%	
1994	40	0.015	0.66	0.09	14.4%	88%	
1995	34	0.012	0.68	0.07	14.4%	90%	
1996	29	0.011	0.69	0.06	14.4%	92%	
1997	25	0.009	0.70	0.05	14.4%	93%	
1998	21	0.008	0.70	0.05	14.4%	94%	
1999	18	0.007	0.71	0.04	14.4%	95%	
2000	16	0.006	0.72	0.03	14.4%	95%	
2001	13	0.005	0.72	0.03	14.4%	96%	
2002	12	0.004	0.73	0.02	14.4%	97%	
2003	10	0.004	0.73	0.02	14.4%	97%	
2004	8	0.003	0.73	0.02	14.4%	98%	
2005	7	0.003	0.73	0.02	14.4%	98%	
2006	6	0.002	0.74	0.01	14.4%	98%	
2007	5	0.002	0.74	0.01	14.4%	98%	
2008	5	0.002	0.74	0.01	14.4%	99%	
2009	4	0.001	0.74	0.01	14.4%	99%	
2010	3	0.001	0.74	0.01	14.4%	99%	
2011	3	0.001	0.74	0.01	14.4%	99%	
2012	2	0.001	0.74	0.01	14.4%	99%	
2013	2	0.001	0.75	0.00	14.4%	99%	
2014		0.001	0.75	0.00	14.4%	99%	
2015		0.001	0.75	0.00	14.4%	100%	
2016		0.000	0.75	0.00	14.4%	100%	
2017		0.000	0.75	0.00	14.4%	100%	
2018		0.000	0.75	0.00	14.4%	100%	
2019		0.000	0.75	0.00	14.4%	100%	
2020		0.000	0.75	0.00	14.4%	100%	
2021		0.000	0.75	0.00	14.4%	100%	
2022		0.000	0.75	0.00	14.4%	100%	
2023		0.000	0.75	0.00	14.4%	100%	
2024		0.000	0.75	0.00	14.4%	100%	
2025		0.000	0.75	0.00	14.4%	100%	
2026		0.000	0.75	0.00	14.4%	100%	
2027		0.000	0.75	0.00	14.4%	100%	
2028		0.000	0.75	0.00	14.4%	100%	
2029		0.000	0.75	0.00	14.4%	100%	
2030		0.000	0.75	0.00	14.4%	100%	
2031		0.000	0.75	0.00	14.4%	100%	
2032		0.000	0.75	0.00	14.4%	100%	
2033		0.000	0.75	0.00	14.4%	100%	
2034		0.000	0.75	0.00	14.4%	100%	
2035		0.000	0.75	0.00	14.4%	100%	
2036		0.000	0.75	0.00	14.4%	100%	
2037		0.000	0.75	0.00	14.4%	100%	
2038		0.000	0.75	0.00	14.4%	100%	
2039		0.000	0.75	0.00	14.4%	100%	
2040		0.000	0.75	0.00	14.4%	100%	
2041		0.000	0.75	0.00	14.4%	100%	
2042		0.000	0.75	0.00	14.4%	100%	
2043		0.000	0.75	0.00	14.4%	100%	
2044		0.000	0.75	0.00	14.4%	100%	
2045		0.000	0.75	0.00	14.4%	100%	
2046		0.000	0.75	0.00	14.4%	100%	
2047		0.000	0.75	0.00	14.4%	100%	
2048		0.000	0.75	0.00	14.4%	100%	
2049		0.000	0.75	0.00	14.4%	100%	
2050		0.000	0.75	0.00	14.4%	100%	

CUM.PROD	4.496 BBO	GIANT FIELDS		NEUTRAL ZONE	
RESERVES	5.200 BBO	TOTAL BBO	5.5	M.EAST (MAJOR)	
DISCOVERED	9.696 BBO	% OF DISC.	57%	Scenario-2	
UNDISCOVERED	1.304 BBO	Last Giant	1963		
REMAINING	6.504 BBO			Variance	Trend
ULTIMATE	11.000 BBO			1990	n/a
Rept.Reserve	5.200	Ref.date(end)	1989	Future	DOWN

Milestones	Date	Prod Mb/d	Dep. Rate	% Produced
1st Peak	1971	551	2.12%	16%
2nd Peak	1994	583	3.70%	50%
50% Depl.	1994	583	3.70%	50%
Final Decline	1995		3.70%	

Forecast "C"
A=Exponential decline
B=5% incr. to Mid-Point
C=Other
Pre-1930 Cum. 0
(Added to 1930 prod)

Date	Prod. M.b/d	Prod. BBO/yr	Cum. Prod.	Remain	Deplet. Rate	% Produced	Giant	Notes & Names
1930		0.000	0.00	11.00	ERR	0%		
1931		0.000	0.00	11.00	0.00%	0%		
1932		0.000	0.00	11.00	0.00%	0%		
1933		0.000	0.00	11.00	0.00%	0%		
1934		0.000	0.00	11.00	0.00%	0%		
1935		0.000	0.00	11.00	0.00%	0%		
1936		0.000	0.00	11.00	0.00%	0%		
1937		0.000	0.00	11.00	0.00%	0%		
1938		0.000	0.00	11.00	0.00%	0%		
1939		0.000	0.00	11.00	0.00%	0%		
1940		0.000	0.00	11.00	0.00%	0%		
1941		0.000	0.00	11.00	0.00%	0%		
1942		0.000	0.00	11.00	0.00%	0%		
1943		0.000	0.00	11.00	0.00%	0%		
1944		0.000	0.00	11.00	0.00%	0%		
1945		0.000	0.00	11.00	0.00%	0%		
1946		0.000	0.00	11.00	0.00%	0%		
1947		0.000	0.00	11.00	0.00%	0%		
1948		0.000	0.00	11.00	0.00%	0%		
1949		0.000	0.00	11.00	0.00%	0%		
1950		0.000	0.00	11.00	0.00%	0%		
1951		0.000	0.00	11.00	0.00%	0%		
1952		0.000	0.00	11.00	0.00%	0%		
1953		0.000	0.00	11.00	0.00%	0%	5.00	Wafra
1954	19	0.007	0.01	10.99	0.06%	0%		
1955	25	0.009	0.02	10.98	0.08%	0%		
1956	41	0.015	0.03	10.97	0.14%	0%		
1957	65	0.024	0.05	10.95	0.22%	0%		
1958	84	0.031	0.09	10.91	0.28%	1%		
1959	116	0.042	0.13	10.87	0.39%	1%		
1960	132	0.048	0.18	10.82	0.44%	2%		
1961	178	0.065	0.24	10.76	0.60%	2%		
1962	238	0.087	0.33	10.67	0.81%	3%		
1963	306	0.112	0.44	10.56	1.05%	4%	0.50	Hout
1964	374	0.137	0.58	10.42	1.29%	5%		
1965	351	0.128	0.70	10.30	1.23%	6%		
1966	417	0.152	0.86	10.14	1.48%	8%		
1967	434	0.158	1.01	9.99	1.56%	9%		
1968	426	0.156	1.17	9.83	1.56%	11%		
1969	455	0.166	1.34	9.66	1.69%	12%		
1970	486	0.177	1.51	9.49	1.83%	14%		
1971	551	0.201	1.71	9.29	2.12%	16%		
1972	545	0.199	1.91	9.09	2.14%	17%		
1973	508	0.185	2.10	8.90	2.04%	19%		
1974	485	0.177	2.28	8.72	1.99%	21%		
1975	450	0.164	2.44	8.56	1.88%	22%		
1976	450	0.164	2.60	8.40	1.92%	24%		
1977	355	0.130	2.73	8.27	1.54%	25%		
1978	420	0.153	2.89	8.11	1.85%	26%		
1979	560	0.204	3.09	7.91	2.52%	28%		
1980	550	0.201	3.29	7.71	2.54%	30%		
1981	370	0.135	3.43	7.57	1.75%	31%		
1982	315	0.115	3.54	7.46	1.52%	32%		
1983	398	0.145	3.69	7.31	1.95%	34%		
1984	420	0.153	3.84	7.16	2.10%	35%		
1985	358	0.131	3.97	7.03	1.82%	36%		
1986	326	0.119	4.09	6.91	1.69%	37%		
1987	399	0.146	4.24	6.76	2.11%	39%		
1988	316	0.115	4.35	6.65	1.71%	40%		% of Region

							50% PRODUCED	Add to Saudi
1989	397	0.145	4.50	6.50	2.18%	41%	2.833%	
1990	450	0.164	4.66	6.34	2.53%	42%		
1991	502	0.183	4.84	6.16	2.89%	44%		
1992	551	0.201	5.04	5.96	3.27%	46%		
1993	570	0.208	5.25	5.75	3.49%	48%		
1994	583	0.213	5.47	5.53	3.70%	50%		
1995	561	0.205	5.67	5.33	3.70%	52%	597	36
1996	540	0.197	5.87	5.13	3.70%	53%	626	86
1997	520	0.190	6.06	4.94	3.70%	55%	657	136
1998	501	0.183	6.24	4.76	3.70%	57%	643	142
1999	482	0.176	6.42	4.58	3.70%	58%	634	151
2000	465	0.170	6.59	4.41	3.70%	60%	628	163
2001	447	0.163	6.75	4.25	3.70%	61%	625	178
2002	431	0.157	6.91	4.09	3.70%	63%	629	198
2003	415	0.151	7.06	3.94	3.70%	64%	632	217
2004	400	0.146	7.20	3.80	3.70%	65%	634	234
2005	385	0.140	7.34	3.66	3.70%	67%	623	238
2006	371	0.135	7.48	3.52	3.70%	68%	622	251
2007	357	0.130	7.61	3.39	3.70%	69%	619	262
2008	344	0.125	7.74	3.26	3.70%	70%	615	272
2009	331	0.121	7.86	3.14	3.70%	71%	611	280
2010	319	0.116	7.97	3.03	3.70%	72%	606	287
2011	307	0.112	8.08	2.92	3.70%	73%	600	293
2012	296	0.108	8.19	2.81	3.70%	74%	593	298
2013	285	0.104	8.30	2.70	3.70%	75%	585	301
2014	274	0.100	8.40	2.60	3.70%	76%	577	303
2015	264	0.096	8.49	2.51	3.70%	77%	568	304
2016	254	0.093	8.59	2.41	3.70%	78%	559	304
2017	245	0.089	8.67	2.33	3.70%	79%	548	304
2018	236	0.086	8.76	2.24	3.70%	80%	538	302
2019	227	0.083	8.84	2.16	3.70%	80%	527	300
2020	219	0.080	8.92	2.08	3.70%	81%	515	296
2021	211	0.077	9.00	2.00	3.70%	82%	503	292
2022	203	0.074	9.07	1.93	3.70%	82%	490	288
2023	195	0.071	9.15	1.85	3.70%	83%	474	279
2024	188	0.069	9.21	1.79	3.70%	84%	470	282
2025	181	0.066	9.28	1.72	3.70%	84%	464	283
2026	174	0.064	9.34	1.66	3.70%	85%	458	284
2027	168	0.061	9.41	1.59	3.70%	86%	452	284
2028	162	0.059	9.46	1.54	3.70%	86%	445	284
2029	156	0.057	9.52	1.48	3.70%	87%	438	283
2030	150	0.055	9.58	1.42	3.70%	87%	431	281
2031	144	0.053	9.63	1.37	3.70%	88%	424	280
2032	139	0.051	9.68	1.32	3.70%	88%	419	280
2033	134	0.049	9.73	1.27	3.70%	88%	414	280
2034	129	0.047	9.78	1.22	3.70%	89%	408	279
2035	124	0.045	9.82	1.18	3.70%	89%	403	278
2036	120	0.044	9.86	1.14	3.70%	90%	397	277
2037	115	0.042	9.91	1.09	3.70%	90%	391	276
2038	111	0.040	9.95	1.05	3.70%	90%	385	274
2039	107	0.039	9.99	1.01	3.70%	91%	379	272
2040	103	0.038	10.02	0.98	3.70%	91%	372	269
2041	99	0.036	10.06	0.94	3.70%	91%	366	267
2042	95	0.035	10.09	0.91	3.70%	92%	359	264
2043	92	0.034	10.13	0.87	3.70%	92%	353	261
2044	88	0.032	10.16	0.84	3.70%	92%	347	258
2045	85	0.031	10.19	0.81	3.70%	93%	340	255
2046	82	0.030	10.22	0.78	3.70%	93%	334	252
2047	79	0.029	10.25	0.75	3.70%	93%	327	248
2048	76	0.028	10.28	0.72	3.70%	93%	321	245
2049	73	0.027	10.30	0.70	3.70%	94%	314	241
2050	71	0.026	10.33	0.67	3.70%	94%	308	238

Note Production ceases 2021

CUM.PROD	0.097 BBO	GIANT FIELDS		NEW ZEALAND	
RESERVES	0.146 BBO	TOTAL BBO	0	THE EAST	
DISCOVERED	0.243 BBO	% OF DISC.	0%		
UNDISCOVERED	0.157 BBO	Last Giant	-		
REMAINING	0.303 BBO			Variance	Trend
ULTIMATE	0.400 BBO			Last	+34%
Rept.Reserve	0.146	Ref.date(end) 1989		Future	UP

Milestones	Date	Prod Mb/d	Dep. Rate	% Produced
1st Peak	2000	48	10.38%	62%
2nd Peak				
Mid-Point	2000	48	10.38%	62%
Final Decline	2001		6.51%	

Forecast "B"
A=Exponential decline
B=5% incr. to Mid-Point
C=Other
Pre-1930 Cum. 0

Date	Prod. M.b/d	Prod. BBO/yr	Cum. Prod.	Remain	Deplet. Rate	% Produced	Giant	Notes & Names
1930		0.000	0.00	0.40	ERR	0%		
1931		0.000	0.00	0.40	0.00%	0%		
1932		0.000	0.00	0.40	0.00%	0%		
1933		0.000	0.00	0.40	0.00%	0%		
1934		0.000	0.00	0.40	0.00%	0%		
1935		0.000	0.00	0.40	0.00%	0%		
1936		0.000	0.00	0.40	0.00%	0%		
1937		0.000	0.00	0.40	0.00%	0%		
1938		0.000	0.00	0.40	0.00%	0%		
1939		0.000	0.00	0.40	0.00%	0%		
1940		0.000	0.00	0.40	0.00%	0%		
1941		0.000	0.00	0.40	0.00%	0%		
1942		0.000	0.00	0.40	0.00%	0%		
1943		0.000	0.00	0.40	0.00%	0%		
1944		0.000	0.00	0.40	0.00%	0%		
1945		0.000	0.00	0.40	0.00%	0%		
1946		0.000	0.00	0.40	0.00%	0%		
1947		0.000	0.00	0.40	0.00%	0%		
1948		0.000	0.00	0.40	0.00%	0%		
1949		0.000	0.00	0.40	0.00%	0%		
1950		0.000	0.00	0.40	0.00%	0%		
1951		0.000	0.00	0.40	0.00%	0%		
1952		0.000	0.00	0.40	0.00%	0%		
1953		0.000	0.00	0.40	0.00%	0%		
1954		0.000	0.00	0.40	0.00%	0%		
1955		0.000	0.00	0.40	0.00%	0%		
1956		0.000	0.00	0.40	0.00%	0%		
1957		0.000	0.00	0.40	0.00%	0%		
1958		0.000	0.00	0.40	0.00%	0%		
1959		0.000	0.00	0.40	0.00%	0%		
1960		0.000	0.00	0.40	0.00%	0%		
1961		0.000	0.00	0.40	0.00%	0%		
1962		0.000	0.00	0.40	0.00%	0%		
1963		0.000	0.00	0.40	0.00%	0%		
1964		0.000	0.00	0.40	0.00%	0%		
1965		0.000	0.00	0.40	0.00%	0%		
1966		0.000	0.00	0.40	0.00%	0%		
1967		0.000	0.00	0.40	0.00%	0%		
1968		0.000	0.00	0.40	0.00%	0%		
1969		0.000	0.00	0.40	0.00%	0%		
1970	0	0.000	0.00	0.40	0.01%	0%		
1971	2	0.001	0.00	0.40	0.16%	0%		
1972	3	0.001	0.00	0.40	0.25%	0%		
1973	4	0.001	0.00	0.40	0.33%	1%		
1974	4	0.001	0.00	0.40	0.35%	1%		
1975	3	0.001	0.01	0.39	0.28%	1%		
1976	10	0.004	0.01	0.39	0.93%	2%		
1977	15	0.005	0.01	0.39	1.40%	4%		
1978	13	0.005	0.02	0.38	1.23%	5%		
1979	16	0.006	0.03	0.37	1.53%	6%		
1980	6	0.002	0.03	0.37	0.58%	7%		
1981	10	0.004	0.03	0.37	0.98%	8%		
1982	15	0.005	0.04	0.36	1.48%	9%		
1983	15	0.005	0.04	0.36	1.51%	10%		
1984	18	0.007	0.05	0.35	1.83%	12%		
1985	18	0.007	0.06	0.34	1.87%	14%		
1986	29	0.011	0.07	0.33	3.10%	16%		
1987	28	0.010	0.08	0.32	3.09%	19%		
1988	28	0.010	0.09	0.31	3.16%	22%		

1989	28	0.010	0.10	0.30	3.26%	24%	
1990	29	0.011	0.11	0.29	3.54%	27%	
1991	31	0.011	0.12	0.28	3.85%	30%	
1992	32	0.012	0.13	0.27	4.20%	33%	
1993	34	0.012	0.14	0.26	4.61%	36%	
1994	36	0.013	0.16	0.24	5.07%	39%	
1995	38	0.014	0.17	0.23	5.61%	42%	
1996	39	0.014	0.18	0.22	6.24%	46%	
1997	41	0.015	0.20	0.20	6.99%	50%	
1998	43	0.016	0.21	0.19	7.89%	54%	
1999	46	0.017	0.23	0.17	8.99%	58%	
2000	48	0.017	0.25	0.15	10.38%	62%	1st PEAK
2001	27	0.010	0.26	0.14	6.51%	65%	& 50% PRODUCED
2002	25	0.009	0.27	0.13	6.51%	67%	
2003	24	0.009	0.28	0.12	6.51%	69%	
2004	22	0.008	0.28	0.12	6.51%	71%	
2005	21	0.008	0.29	0.11	6.51%	73%	
2006	19	0.007	0.30	0.10	6.51%	75%	
2007	18	0.007	0.31	0.09	6.51%	76%	
2008	17	0.006	0.31	0.09	6.51%	78%	
2009	16	0.006	0.32	0.08	6.51%	79%	
2010	15	0.005	0.32	0.08	6.51%	81%	
2011	14	0.005	0.33	0.07	6.51%	82%	
2012	13	0.005	0.33	0.07	6.51%	83%	
2013	12	0.004	0.34	0.06	6.51%	84%	
2014	11	0.004	0.34	0.06	6.51%	85%	
2015	10	0.004	0.34	0.06	6.51%	86%	
2016	10	0.004	0.35	0.05	6.51%	87%	
2017	9	0.003	0.35	0.05	6.51%	88%	
2018	9	0.003	0.36	0.04	6.51%	89%	
2019	8	0.003	0.36	0.04	6.51%	89%	
2020	7	0.003	0.36	0.04	6.51%	90%	
2021	7	0.003	0.36	0.04	6.51%	91%	
2022	7	0.002	0.37	0.03	6.51%	91%	
2023	6	0.002	0.37	0.03	6.51%	92%	
2024	6	0.002	0.37	0.03	6.51%	92%	
2025	5	0.002	0.37	0.03	6.51%	93%	
2026	5	0.002	0.37	0.03	6.51%	93%	
2027	5	0.002	0.38	0.02	6.51%	94%	
2028	4	0.002	0.38	0.02	6.51%	94%	
2029	4	0.001	0.38	0.02	6.51%	95%	
2030	4	0.001	0.38	0.02	6.51%	95%	
2031	4	0.001	0.38	0.02	6.51%	95%	
2032	3	0.001	0.38	0.02	6.51%	96%	
2033	3	0.001	0.38	0.02	6.51%	96%	
2034	3	0.001	0.38	0.02	6.51%	96%	
2035	3	0.001	0.39	0.01	6.51%	96%	
2036	3	0.001	0.39	0.01	6.51%	97%	
2037	2	0.001	0.39	0.01	6.51%	97%	
2038	2	0.001	0.39	0.01	6.51%	97%	
2039	2	0.001	0.39	0.01	6.51%	97%	
2040	2	0.001	0.39	0.01	6.51%	97%	
2041	2	0.001	0.39	0.01	6.51%	98%	
2042	2	0.001	0.39	0.01	6.51%	98%	
2043	2	0.001	0.39	0.01	6.51%	98%	
2044	1	0.001	0.39	0.01	6.51%	98%	
2045	1	0.001	0.39	0.01	6.51%	98%	
2046	1	0.000	0.39	0.01	6.51%	98%	
2047	1	0.000	0.39	0.01	6.51%	98%	
2048	1	0.000	0.39	0.01	6.51%	99%	
2049	1	0.000	0.39	0.01	6.51%	99%	
2050	1	0.000	0.39	0.01	6.51%	99%	

CUM.PROD	12.912 BBO	GIANT FIELDS			NIGERIA	
RESERVES	16.000 BBO	TOTAL BBO	5.9		AFRICA	
DISCOVERED	28.912 BBO	% OF DISC.	20%			
UNDISCOVERED	3.088 BBO	Last Giant	1973			
REMAINING	19.088 BBO				Variance	Trend
ULTIMATE	32.000 BBO				1990	+7%
Rept.Reserves	16.000 BBO	Ref.Date (end)		1989	Future	DOWN

Milestones	Date	Prod Mb/d	Dep Rate	% Produced
1st Peak	1979	2370	3.42%	24%
2nd Peak	1994	2049	4.55%	51%
Mid-Point	1994	2049	4.55%	51%
Final Decline	1994		4.55%	

Forecast "B"
A=Exponential decline
B=5% inc. to Mid-Point
C=Other
Pre-1930 Cum 0
(Added to 1930 Prod)

Date	Prod. M.b/d	Prod. BBO/yr	Cum Prod	Remain	Dep. Rate	% Produced	Giant	Names & Notes
1930		0.000	0.00	32.00	ERR	0%		
1931		0.000	0.00	32.00	0.00%	0%		
1932		0.000	0.00	32.00	0.00%	0%		
1933		0.000	0.00	32.00	0.00%	0%		
1934		0.000	0.00	32.00	0.00%	0%		
1935		0.000	0.00	32.00	0.00%	0%		
1936		0.000	0.00	32.00	0.00%	0%		
1937		0.000	0.00	32.00	0.00%	0%		
1938		0.000	0.00	32.00	0.00%	0%		
1939		0.000	0.00	32.00	0.00%	0%		
1940		0.000	0.00	32.00	0.00%	0%		
1941		0.000	0.00	32.00	0.00%	0%		
1942		0.000	0.00	32.00	0.00%	0%		
1943		0.000	0.00	32.00	0.00%	0%		
1944		0.000	0.00	32.00	0.00%	0%		
1945		0.000	0.00	32.00	0.00%	0%		
1946		0.000	0.00	32.00	0.00%	0%		
1947		0.000	0.00	32.00	0.00%	0%		
1948		0.000	0.00	32.00	0.00%	0%		
1949		0.000	0.00	32.00	0.00%	0%		
1950		0.000	0.00	32.00	0.00%	0%		
1951		0.000	0.00	32.00	0.00%	0%		
1952		0.000	0.00	32.00	0.00%	0%		
1953		0.000	0.00	32.00	0.00%	0%		
1954		0.000	0.00	32.00	0.00%	0%		
1955		0.000	0.00	32.00	0.00%	0%		
1956		0.000	0.00	32.00	0.00%	0%		
1957		0.000	0.00	32.00	0.00%	0%		
1958	4	0.002	0.00	32.00	0.01%	0%	0.5	Bomu
1959	11	0.004	0.01	31.99	0.01%	0%	0.7	Imo R.
1960	20	0.007	0.01	31.99	0.02%	0%		
1961	44	0.016	0.03	31.97	0.05%	0%		
1962	66	0.024	0.05	31.95	0.08%	0%		
1963	72	0.026	0.08	31.92	0.08%	0%		
1964	147	0.054	0.13	31.87	0.17%	0%	1.18	Okan, Olomoro-Af
1965	270	0.099	0.23	31.77	0.31%	1%	0.65	Meren
1966	383	0.140	0.37	31.63	0.44%	1%		
1967	344	0.126	0.50	31.50	0.40%	2%	1.5	Jones C. & Kokor
1968	93	0.034	0.53	31.47	0.11%	2%	0.7	Forcados-Yokri
1969	537	0.196	0.73	31.27	0.62%	2%		
1970	1000	0.365	1.09	30.91	1.17%	3%		
1971	1543	0.563	1.66	30.34	1.82%	5%		
1972	1800	0.657	2.31	29.69	2.17%	7%		
1973	2000	0.730	3.04	28.96	2.46%	10%	0.645	Nembe Creek
1974	2300	0.840	3.88	28.12	2.90%	12%		
1975	1850	0.675	4.56	27.44	2.40%	14%		
1976	2020	0.737	5.29	26.71	2.69%	17%		
1977	2150	0.785	6.08	25.92	2.94%	19%		
1978	1800	0.657	6.74	25.26	2.53%	21%		
1979	2370	0.865	7.60	24.40	3.42%	24%		1ST PEAK
1980	2100	0.767	8.37	23.63	3.14%	26%		
1981	1369	0.500	8.87	23.13	2.11%	28%		
1982	1324	0.483	9.35	22.65	2.09%	29%		
1983	1232	0.450	9.80	22.20	1.99%	31%		
1984	1414	0.516	10.32	21.68	2.32%	32%		
1985	1446	0.528	10.84	21.16	2.43%	34%		
1986	1464	0.534	11.38	20.62	2.53%	36%		
1987	1239	0.452	11.83	20.17	2.19%	37%		
1988	1358	0.496	12.33	19.67	2.46%	39%		

1989	1605	0.586	12.91	19.09	2.98%	40%	
1990	1686	0.615	13.53	18.47	3.22%	42%	
1991	1770	0.646	14.17	17.83	3.50%	44%	
1992	1858	0.678	14.85	17.15	3.80%	46%	
1993	1951	0.712	15.56	16.44	4.15%	49%	
1994	2049	0.748	16.31	15.69	4.55%	51%	50% PRODUCED & 2ND PEAK
1995	1956	0.714	17.03	14.97	4.55%	53%	
1996	1867	0.681	17.71	14.29	4.55%	55%	
1997	1782	0.650	18.36	13.64	4.55%	57%	
1998	1701	0.621	18.98	13.02	4.55%	59%	
1999	1623	0.593	19.57	12.43	4.55%	61%	
2000	1549	0.566	20.14	11.86	4.55%	63%	
2001	1479	0.540	20.68	11.32	4.55%	65%	
2002	1412	0.515	21.19	10.81	4.55%	66%	
2003	1347	0.492	21.68	10.32	4.55%	68%	
2004	1286	0.469	22.15	9.85	4.55%	69%	
2005	1228	0.448	22.60	9.40	4.55%	71%	
2006	1172	0.428	23.03	8.97	4.55%	72%	
2007	1118	0.408	23.44	8.56	4.55%	73%	
2008	1068	0.390	23.83	8.17	4.55%	74%	
2009	1019	0.372	24.20	7.80	4.55%	76%	
2010	973	0.355	24.55	7.45	4.55%	77%	
2011	928	0.339	24.89	7.11	4.55%	78%	
2012	886	0.323	25.22	6.78	4.55%	79%	
2013	846	0.309	25.52	6.48	4.55%	80%	
2014	807	0.295	25.82	6.18	4.55%	81%	
2015	771	0.281	26.10	5.90	4.55%	82%	
2016	736	0.268	26.37	5.63	4.55%	82%	
2017	702	0.256	26.62	5.38	4.55%	83%	
2018	670	0.245	26.87	5.13	4.55%	84%	
2019	640	0.233	27.10	4.90	4.55%	85%	
2020	611	0.223	27.33	4.67	4.55%	85%	
2021	583	0.213	27.54	4.46	4.55%	86%	
2022	556	0.203	27.74	4.26	4.55%	87%	
2023	531	0.194	27.93	4.07	4.55%	87%	
2024	507	0.185	28.12	3.88	4.55%	88%	
2025	484	0.177	28.30	3.70	4.55%	88%	
2026	462	0.169	28.46	3.54	4.55%	89%	
2027	441	0.161	28.63	3.37	4.55%	89%	
2028	421	0.154	28.78	3.22	4.55%	90%	
2029	401	0.147	28.93	3.07	4.55%	90%	
2030	383	0.140	29.07	2.93	4.55%	91%	
2031	366	0.134	29.20	2.80	4.55%	91%	
2032	349	0.127	29.33	2.67	4.55%	92%	
2033	333	0.122	29.45	2.55	4.55%	92%	
2034	318	0.116	29.56	2.44	4.55%	92%	
2035	304	0.111	29.68	2.32	4.55%	93%	
2036	290	0.106	29.78	2.22	4.55%	93%	
2037	277	0.101	29.88	2.12	4.55%	93%	
2038	264	0.096	29.98	2.02	4.55%	94%	
2039	252	0.092	30.07	1.93	4.55%	94%	
2040	241	0.088	30.16	1.84	4.55%	94%	
2041	230	0.084	30.24	1.76	4.55%	95%	
2042	219	0.080	30.32	1.68	4.55%	95%	
2043	209	0.076	30.40	1.60	4.55%	95%	
2044	200	0.073	30.47	1.53	4.55%	95%	
2045	191	0.070	30.54	1.46	4.55%	95%	
2046	182	0.066	30.61	1.39	4.55%	96%	
2047	174	0.063	30.67	1.33	4.55%	96%	
2048	166	0.060	30.73	1.27	4.55%	96%	
2049	158	0.058	30.79	1.21	4.55%	96%	
2050	151	0.055	30.84	1.16	4.55%	96%	

CUM.PROD	3.481 BBO	GIANT FIELDS		NORWAY	
RESERVES	11.546 BBO	TOTAL BBO	8.1	W.EUROPE	
DISCOVERED	15.027 BBO	% OF DISC.	54%		
UNDISCOVERED	2.973 BBO	Last Giant	1985		
REMAINING	14.519 BBO			Variance	Trend
ULTIMATE	18.000 BBO			1990	+1.7%
Rept.Reserve	11.546 BBO	Ref.Date(end)	1989	Future	UP

Milestones	Date	Prod Mb/d	Dep. Rate	% Produced
1st Peak	1997	2177	8.01%	49%
2nd Peak				
Mid-Point	1998	2003	8.01%	53%
Final Decline	1998		8.01%	

Forecast "B"
A=Exponential decline
B=5% incr. to Mid-Point
C=Other
Pre-1930 Cum. 0
(Added to 1930 Prod)

Date	Prod. M.b/d	Prod. BBO/yr	Cum Prod	Remain	Dep. Rate	% Produced	Giant	Names & Notes
1930		0.000	0.00	18.00	ERR	0%		
1931		0.000	0.00	18.00	0.00%	0%		
1932		0.000	0.00	18.00	0.00%	0%		
1933		0.000	0.00	18.00	0.00%	0%		
1934		0.000	0.00	18.00	0.00%	0%		
1935		0.000	0.00	18.00	0.00%	0%		
1936		0.000	0.00	18.00	0.00%	0%		
1937		0.000	0.00	18.00	0.00%	0%		
1938		0.000	0.00	18.00	0.00%	0%		
1939		0.000	0.00	18.00	0.00%	0%		
1940		0.000	0.00	18.00	0.00%	0%		
1941		0.000	0.00	18.00	0.00%	0%		
1942		0.000	0.00	18.00	0.00%	0%		
1943		0.000	0.00	18.00	0.00%	0%		
1944		0.000	0.00	18.00	0.00%	0%		
1945		0.000	0.00	18.00	0.00%	0%		
1946		0.000	0.00	18.00	0.00%	0%		
1947		0.000	0.00	18.00	0.00%	0%		
1948		0.000	0.00	18.00	0.00%	0%		
1949		0.000	0.00	18.00	0.00%	0%		
1950		0.000	0.00	18.00	0.00%	0%		
1951		0.000	0.00	18.00	0.00%	0%		
1952		0.000	0.00	18.00	0.00%	0%		
1953		0.000	0.00	18.00	0.00%	0%		
1954		0.000	0.00	18.00	0.00%	0%		
1955		0.000	0.00	18.00	0.00%	0%		
1956		0.000	0.00	18.00	0.00%	0%		
1957		0.000	0.00	18.00	0.00%	0%		
1958		0.000	0.00	18.00	0.00%	0%		
1959		0.000	0.00	18.00	0.00%	0%		
1960		0.000	0.00	18.00	0.00%	0%		
1961		0.000	0.00	18.00	0.00%	0%		
1962		0.000	0.00	18.00	0.00%	0%		
1963		0.000	0.00	18.00	0.00%	0%		
1964		0.000	0.00	18.00	0.00%	0%		
1965		0.000	0.00	18.00	0.00%	0%		
1966		0.000	0.00	18.00	0.00%	0%		
1967		0.000	0.00	18.00	0.00%	0%		
1968		0.000	0.00	18.00	0.00%	0%		
1969		0.000	0.00	18.00	0.00%	0%	1.673	Ekofisk
1970		0.000	0.00	18.00	0.00%	0%		
1971	6.15	0.002	0.00	18.00	0.01%	0%		
1972	33.20	0.012	0.01	17.99	0.07%	0%		
1973	32.20	0.012	0.03	17.97	0.07%	0%		
1974	34.70	0.013	0.04	17.96	0.07%	0%	2.358	Statfjord (#1)
1975	189.45	0.069	0.11	17.89	0.38%	1%		
1976	279.59	0.102	0.21	17.79	0.57%	1%		
1977	286.76	0.105	0.31	17.69	0.59%	2%		
1978	355.72	0.130	0.44	17.56	0.73%	2%	1.32	Gullfaks
1979	387.30	0.141	0.59	17.41	0.81%	3%	2.199	Snorre/Oseberg
1980	486.28	0.177	0.76	17.24	1.02%	4%		
1981	473.58	0.173	0.94	17.06	1.00%	5%		
1982	491.57	0.179	1.12	16.88	1.05%	6%		
1983	614.20	0.224	1.34	16.66	1.33%	7%		
1984	708.07	0.258	1.60	16.40	1.55%	9%		
1985	771.21	0.281	1.88	16.12	1.72%	10%	0.547	Heidrun
1986	834.93	0.305	2.18	15.82	1.89%	12%		
1987	972.50	0.355	2.54	15.46	2.24%	14%		
1988	1105.26	0.403	2.94	15.06	2.61%	16%		

1989	1473.44	0.538	3.48	14.52	3.57%	19%	
1990	1547.11	0.565	4.05	13.95	3.89%	22%	
1991	1624.47	0.593	4.64	13.36	4.25%	26%	
1992	1705.69	0.623	5.26	12.74	4.66%	29%	
1993	1790.98	0.654	5.91	12.09	5.13%	33%	
1994	1880.52	0.686	6.60	11.40	5.68%	37%	
1995	1974.55	0.721	7.32	10.68	6.32%	41%	
1996	2073.28	0.757	8.08	9.92	7.09%	45%	
1997	2176.94	0.795	8.87	9.13	8.01%	49%	1st PEAK
1998	2002.93	0.731	9.60	8.40	8.01%	53%	50% PRODUCED
1999	1842.50	0.673	10.28	7.72	8.01%	57%	
2000	1694.91	0.619	10.90	7.10	8.01%	61%	
2001	1559.15	0.569	11.46	6.54	8.01%	64%	
2002	1434.26	0.524	11.99	6.01	8.01%	67%	
2003	1319.38	0.482	12.47	5.53	8.01%	69%	
2004	1213.70	0.443	12.91	5.09	8.01%	72%	
2005	1116.48	0.408	13.32	4.68	8.01%	74%	
2006	1027.05	0.375	13.69	4.31	8.01%	76%	
2007	944.78	0.345	14.04	3.96	8.01%	78%	
2008	869.11	0.317	14.36	3.64	8.01%	80%	
2009	799.49	0.292	14.65	3.35	8.01%	81%	
2010	735.45	0.268	14.92	3.08	8.01%	83%	
2011	676.54	0.247	15.16	2.84	8.01%	84%	
2012	622.35	0.227	15.39	2.61	8.01%	86%	
2013	572.50	0.209	15.60	2.40	8.01%	87%	
2014	526.64	0.192	15.79	2.21	8.01%	88%	
2015	484.46	0.177	15.97	2.03	8.01%	89%	
2016	445.65	0.163	16.13	1.87	8.01%	90%	
2017	409.96	0.150	16.28	1.72	8.01%	90%	
2018	377.12	0.138	16.42	1.58	8.01%	91%	
2019	346.91	0.127	16.55	1.45	8.01%	92%	
2020	319.12	0.116	16.66	1.34	8.01%	93%	
2021	293.56	0.107	16.77	1.23	8.01%	93%	
2022	270.05	0.099	16.87	1.13	8.01%	94%	
2023	248.42	0.091	16.96	1.04	8.01%	94%	
2024	228.52	0.083	17.04	0.96	8.01%	95%	
2025	210.21	0.077	17.12	0.88	8.01%	95%	
2026	193.38	0.071	17.19	0.81	8.01%	95%	
2027	177.89	0.065	17.25	0.75	8.01%	96%	
2028	163.64	0.060	17.31	0.69	8.01%	96%	
2029	150.53	0.055	17.37	0.63	8.01%	96%	
2030	138.47	0.051	17.42	0.58	8.01%	97%	
2031	127.38	0.046	17.47	0.53	8.01%	97%	
2032	117.18	0.043	17.51	0.49	8.01%	97%	
2033	107.79	0.039	17.55	0.45	8.01%	97%	
2034	99.16	0.036	17.58	0.42	8.01%	98%	
2035	91.22	0.033	17.62	0.38	8.01%	98%	
2036	83.91	0.031	17.65	0.35	8.01%	98%	
2037	77.19	0.028	17.68	0.32	8.01%	98%	
2038	71.01	0.026	17.70	0.30	8.01%	98%	
2039	65.32	0.024	17.73	0.27	8.01%	98%	
2040	60.09	0.022	17.75	0.25	8.01%	99%	
2041	55.27	0.020	17.77	0.23	8.01%	99%	
2042	50.85	0.019	17.79	0.21	8.01%	99%	
2043	46.77	0.017	17.80	0.20	8.01%	99%	
2044	43.03	0.016	17.82	0.18	8.01%	99%	
2045	39.58	0.014	17.83	0.17	8.01%	99%	
2046	36.41	0.013	17.85	0.15	8.01%	99%	
2047	33.49	0.012	17.86	0.14	8.01%	99%	
2048	30.81	0.011	17.87	0.13	8.01%	99%	
2049	28.34	0.010	17.88	0.12	8.01%	99%	
2050	26.07	0.010	17.89	0.11	8.01%	99%	

Notes
1) Statfjord includes UK portion

CUM.PROD	3.015 BBO	GIANT FIELDS		OMAN
RESERVES	4.250 BBO	TOTAL BBO	2.7	M.EAST (MINOR)
DISCOVERED	7.265 BBO	% OF DISC.	37%	
UNDISCOVERED	2.735 BBO	Last Giant	1964	
REMAINING	6.985 BBO			Variance Trend
ULTIMATE	10.000 BBO			1990 +0.6%
Rept.Reserve	4.250	Ref.date(end)	1989	Future DOWN

Milestones	Date	Prod Mb/d	Dep. Rate	% Produced
1st Peak	1996	876	5.96%	50%
2nd Peak				
Mid-Point	1996	876	5.96%	50%
Final Decline	1996	876	5.96%	50%

Forecast "B"
A=Exponential decline
B=5% incr. to Mid-Point
C=Other
Pre-1930 Cum. 0
(Added to 1930 prod)

Date	Prod. M.b/d	Prod. BBO/yr	Cum. Prod.	Remain	Deplet. Rate	% Produced	Giant	Notes & Names
1930		0.000	0.00	10.00	ERR	0%		
1931		0.000	0.00	10.00	0.00%	0%		
1932		0.000	0.00	10.00	0.00%	0%		
1933		0.000	0.00	10.00	0.00%	0%		
1934		0.000	0.00	10.00	0.00%	0%		
1935		0.000	0.00	10.00	0.00%	0%		
1936		0.000	0.00	10.00	0.00%	0%		
1937		0.000	0.00	10.00	0.00%	0%		
1938		0.000	0.00	10.00	0.00%	0%		
1939		0.000	0.00	10.00	0.00%	0%		
1940		0.000	0.00	10.00	0.00%	0%		
1941		0.000	0.00	10.00	0.00%	0%		
1942		0.000	0.00	10.00	0.00%	0%		
1943		0.000	0.00	10.00	0.00%	0%		
1944		0.000	0.00	10.00	0.00%	0%		
1945		0.000	0.00	10.00	0.00%	0%		
1946		0.000	0.00	10.00	0.00%	0%		
1947		0.000	0.00	10.00	0.00%	0%		
1948		0.000	0.00	10.00	0.00%	0%		
1949		0.000	0.00	10.00	0.00%	0%		
1950		0.000	0.00	10.00	0.00%	0%		
1951		0.000	0.00	10.00	0.00%	0%		
1952		0.000	0.00	10.00	0.00%	0%		
1953		0.000	0.00	10.00	0.00%	0%		
1954		0.000	0.00	10.00	0.00%	0%		
1955		0.000	0.00	10.00	0.00%	0%		
1956		0.000	0.00	10.00	0.00%	0%		
1957		0.000	0.00	10.00	0.00%	0%	0.60	Marmul
1958		0.000	0.00	10.00	0.00%	0%		
1959		0.000	0.00	10.00	0.00%	0%		
1960		0.000	0.00	10.00	0.00%	0%	0.60	Natih
1961		0.000	0.00	10.00	0.00%	0%		
1962		0.000	0.00	10.00	0.00%	0%	0.60	Yibal
1963		0.000	0.00	10.00	0.00%	0%		
1964		0.000	0.00	10.00	0.00%	0%	0.90	Fahud
1965		0.000	0.00	10.00	0.00%	0%		
1966		0.000	0.00	10.00	0.00%	0%		
1967	48	0.017	0.02	9.98	0.17%	0%		
1968	238	0.087	0.10	9.90	0.87%	1%		
1969	318	0.116	0.22	9.78	1.17%	2%		
1970	336	0.123	0.34	9.66	1.25%	3%		
1971	291	0.106	0.45	9.55	1.10%	4%		
1972	280	0.102	0.55	9.45	1.07%	6%		
1973	272	0.099	0.65	9.35	1.05%	7%		
1974	297	0.108	0.76	9.24	1.16%	8%		
1975	340	0.124	0.88	9.12	1.34%	9%		
1976	365	0.133	1.02	8.98	1.46%	10%		
1977	350	0.128	1.14	8.86	1.42%	11%		
1978	320	0.117	1.26	8.74	1.32%	13%		
1979	290	0.106	1.37	8.63	1.21%	14%		
1980	280	0.102	1.47	8.53	1.18%	15%		
1981	317	0.116	1.58	8.42	1.36%	16%		
1982	328	0.120	1.70	8.30	1.42%	17%		
1983	378	0.138	1.84	8.16	1.66%	18%		
1984	404	0.147	1.99	8.01	1.81%	20%		
1985	484	0.177	2.17	7.83	2.21%	22%		
1986	541	0.197	2.36	7.64	2.52%	24%		
1987	565	0.206	2.57	7.43	2.70%	26%		
1988	597	0.218	2.79	7.21	2.93%	28%		

1989	623	0.227	3.02	6.98	3.15%	30%	
1990	654	0.239	3.25	6.75	3.42%	33%	
1991	687	0.251	3.50	6.50	3.71%	35%	
1992	721	0.263	3.77	6.23	4.05%	38%	
1993	757	0.276	4.04	5.96	4.43%	40%	
1994	795	0.290	4.33	5.67	4.87%	43%	
1995	834	0.305	4.64	5.36	5.38%	46%	
1996	876	0.320	4.96	5.04	5.96%	50%	50% PRODUCED & 1ST PEAK
1997	823	0.300	5.26	4.74	5.96%	53%	
1998	774	0.283	5.54	4.46	5.96%	55%	
1999	728	0.266	5.81	4.19	5.96%	58%	
2000	685	0.250	6.06	3.94	5.96%	61%	
2001	644	0.235	6.29	3.71	5.96%	63%	
2002	605	0.221	6.51	3.49	5.96%	65%	
2003	569	0.208	6.72	3.28	5.96%	67%	
2004	535	0.195	6.92	3.08	5.96%	69%	
2005	504	0.184	7.10	2.90	5.96%	71%	
2006	474	0.173	7.27	2.73	5.96%	73%	
2007	445	0.163	7.44	2.56	5.96%	74%	
2008	419	0.153	7.59	2.41	5.96%	76%	
2009	394	0.144	7.73	2.27	5.96%	77%	
2010	370	0.135	7.87	2.13	5.96%	79%	
2011	348	0.127	7.99	2.01	5.96%	80%	
2012	328	0.120	8.11	1.89	5.96%	81%	
2013	308	0.112	8.23	1.77	5.96%	82%	
2014	290	0.106	8.33	1.67	5.96%	83%	
2015	272	0.099	8.43	1.57	5.96%	84%	
2016	256	0.093	8.52	1.48	5.96%	85%	
2017	241	0.088	8.61	1.39	5.96%	86%	
2018	227	0.083	8.70	1.30	5.96%	87%	
2019	213	0.078	8.77	1.23	5.96%	88%	
2020	200	0.073	8.85	1.15	5.96%	88%	
2021	188	0.069	8.92	1.08	5.96%	89%	
2022	177	0.065	8.98	1.02	5.96%	90%	
2023	167	0.061	9.04	0.96	5.96%	90%	
2024	157	0.057	9.10	0.90	5.96%	91%	
2025	147	0.054	9.15	0.85	5.96%	92%	
2026	139	0.051	9.20	0.80	5.96%	92%	
2027	130	0.048	9.25	0.75	5.96%	92%	
2028	123	0.045	9.29	0.71	5.96%	93%	
2029	115	0.042	9.34	0.66	5.96%	93%	
2030	108	0.040	9.38	0.62	5.96%	94%	
2031	102	0.037	9.41	0.59	5.96%	94%	
2032	96	0.035	9.45	0.55	5.96%	94%	
2033	90	0.033	9.48	0.52	5.96%	95%	
2034	85	0.031	9.51	0.49	5.96%	95%	
2035	80	0.029	9.54	0.46	5.96%	95%	
2036	75	0.027	9.57	0.43	5.96%	96%	
2037	70	0.026	9.59	0.41	5.96%	96%	
2038	66	0.024	9.62	0.38	5.96%	96%	
2039	62	0.023	9.64	0.36	5.96%	96%	
2040	59	0.021	9.66	0.34	5.96%	97%	
2041	55	0.020	9.68	0.32	5.96%	97%	
2042	52	0.019	9.70	0.30	5.96%	97%	
2043	49	0.018	9.72	0.28	5.96%	97%	
2044	46	0.017	9.74	0.26	5.96%	97%	
2045	43	0.016	9.75	0.25	5.96%	98%	
2046	41	0.015	9.77	0.23	5.96%	98%	
2047	38	0.014	9.78	0.22	5.96%	98%	
2048	36	0.013	9.79	0.21	5.96%	98%	
2049	34	0.012	9.81	0.19	5.96%	98%	
2050	32	0.012	9.82	0.18	5.96%	98%	

CUM.PROD	0.185 BBO	GIANT FIELDS		PAKISTAN	
RESERVES	0.119 BBO	TOTAL BBO	0	THE EAST	
DISCOVERED	0.303 BBO	% OF DISC.	0%		
UNDISCOVERED	0.297 BBO	Last Giant	-		
REMAINING	0.415 BBO			Variance	Trend
ULTIMATE	0.600 BBO			1990	+40%
Rept.Reserve	0.119	Ref.date(end)	1989	Future	UP

Milestones	Date	Prod Mb/d	Dep. Rate	% Produced	Forecast "B"
1st Peak	1995	60	6.88%	50%	A=Exponential decline
2nd Peak					B=5% incr. to Mid-Point
Mid-Point	1995	60	6.88%	50%	C=Other
Final Decline	1996		6.88%		Pre-1930 Cum. 0

Date	Prod. M.b/d	Prod. BBO/yr	Cum. Prod.	Remain	Deplet. Rate	% Produced	Giant	Notes & Names
1930		0.000	0.00	0.60	ERR	0%		
1931		0.000	0.00	0.60	0.00%	0%		
1932		0.000	0.00	0.60	0.00%	0%		
1933		0.000	0.00	0.60	0.00%	0%		
1934		0.000	0.00	0.60	0.00%	0%		
1935		0.000	0.00	0.60	0.00%	0%		
1936		0.000	0.00	0.60	0.00%	0%		
1937		0.000	0.00	0.60	0.00%	0%		
1938		0.000	0.00	0.60	0.00%	0%		
1939		0.000	0.00	0.60	0.00%	0%		
1940		0.000	0.00	0.60	0.00%	0%		
1941		0.000	0.00	0.60	0.00%	0%		
1942		0.000	0.00	0.60	0.00%	0%		
1943		0.000	0.00	0.60	0.00%	0%		
1944		0.000	0.00	0.60	0.00%	0%		
1945		0.000	0.00	0.60	0.00%	0%		
1946		0.000	0.00	0.60	0.00%	0%		
1947	1	0.000	0.00	0.60	0.06%	0%		
1948	1	0.000	0.00	0.60	0.07%	0%		
1949	2	0.001	0.00	0.60	0.12%	0%		
1950	3	0.001	0.00	0.60	0.20%	0%		
1951	3	0.001	0.00	0.60	0.20%	1%		
1952	4	0.001	0.01	0.59	0.24%	1%		
1953	4	0.002	0.01	0.59	0.27%	1%		
1954	5	0.002	0.01	0.59	0.33%	1%		
1955	6	0.002	0.01	0.59	0.35%	2%		
1956	5	0.002	0.01	0.59	0.32%	2%		
1957	6	0.002	0.01	0.59	0.36%	2%		
1958	5	0.002	0.02	0.58	0.34%	3%		
1959	6	0.002	0.02	0.58	0.37%	3%		
1960	8	0.003	0.02	0.58	0.48%	4%		
1961	8	0.003	0.02	0.58	0.48%	4%		
1962	8	0.003	0.03	0.57	0.48%	5%		
1963	8	0.003	0.03	0.57	0.54%	5%		
1964	8	0.003	0.03	0.57	0.54%	6%		
1965	10	0.004	0.04	0.56	0.62%	6%		
1966	10	0.004	0.04	0.56	0.65%	7%		
1967	10	0.004	0.04	0.56	0.65%	7%		
1968	10	0.004	0.05	0.55	0.65%	8%		
1969	10	0.004	0.05	0.55	0.65%	9%		
1970	10	0.004	0.06	0.54	0.65%	9%		
1971	10	0.004	0.06	0.54	0.67%	10%		
1972	10	0.003	0.06	0.54	0.64%	10%		
1973	8	0.003	0.07	0.53	0.54%	11%		
1974	7	0.003	0.07	0.53	0.48%	11%		
1975	6	0.002	0.07	0.53	0.41%	12%		
1976	12	0.004	0.07	0.53	0.83%	12%		
1977	10	0.004	0.08	0.52	0.69%	13%		
1978	9	0.003	0.08	0.52	0.65%	14%		
1979	11	0.004	0.09	0.51	0.77%	14%		
1980	10	0.004	0.09	0.51	0.71%	15%		
1981	10	0.004	0.09	0.51	0.71%	15%		
1982	12	0.004	0.10	0.50	0.86%	16%		
1983	13	0.005	0.10	0.50	0.94%	17%		
1984	18	0.007	0.11	0.49	1.32%	18%		
1985	34	0.013	0.12	0.48	2.55%	20%		
1986	41	0.015	0.14	0.46	3.12%	23%		
1987	42	0.015	0.15	0.45	3.27%	25%		
1988	47	0.017	0.17	0.43	3.82%	28%		

1989	45	0.016	0.18	0.42	3.80%	31%	
1990	47	0.017	0.20	0.40	4.15%	34%	
1991	50	0.018	0.22	0.38	4.55%	37%	
1992	52	0.019	0.24	0.36	5.00%	40%	
1993	55	0.020	0.26	0.34	5.53%	43%	
1994	57	0.021	0.28	0.32	6.15%	47%	
1995	60	0.022	0.30	0.30	6.88%	50%	1ST PEAK & 50% PRODUCED
1996	56	0.021	0.32	0.28	6.88%	54%	
1997	52	0.019	0.34	0.26	6.88%	57%	
1998	49	0.018	0.36	0.24	6.88%	60%	
1999	45	0.017	0.38	0.22	6.88%	63%	
2000	42	0.015	0.39	0.21	6.88%	65%	
2001	39	0.014	0.41	0.19	6.88%	68%	
2002	37	0.013	0.42	0.18	6.88%	70%	
2003	34	0.012	0.43	0.17	6.88%	72%	
2004	32	0.012	0.44	0.16	6.88%	74%	
2005	30	0.011	0.45	0.15	6.88%	76%	
2006	28	0.010	0.46	0.14	6.88%	77%	
2007	26	0.009	0.47	0.13	6.88%	79%	
2008	24	0.009	0.48	0.12	6.88%	80%	
2009	22	0.008	0.49	0.11	6.88%	82%	
2010	21	0.008	0.50	0.10	6.88%	83%	
2011	19	0.007	0.50	0.10	6.88%	84%	
2012	18	0.007	0.51	0.09	6.88%	85%	
2013	17	0.006	0.52	0.08	6.88%	86%	
2014	16	0.006	0.52	0.08	6.88%	87%	
2015	15	0.005	0.53	0.07	6.88%	88%	
2016	14	0.005	0.53	0.07	6.88%	89%	
2017	13	0.005	0.54	0.06	6.88%	90%	
2018	12	0.004	0.54	0.06	6.88%	90%	
2019	11	0.004	0.55	0.05	6.88%	91%	
2020	10	0.004	0.55	0.05	6.88%	92%	
2021	9	0.003	0.55	0.05	6.88%	92%	
2022	9	0.003	0.56	0.04	6.88%	93%	
2023	8	0.003	0.56	0.04	6.88%	93%	
2024	8	0.003	0.56	0.04	6.88%	94%	
2025	7	0.003	0.56	0.04	6.88%	94%	
2026	7	0.002	0.57	0.03	6.88%	95%	
2027	6	0.002	0.57	0.03	6.88%	95%	
2028	6	0.002	0.57	0.03	6.88%	95%	
2029	5	0.002	0.57	0.03	6.88%	96%	
2030	5	0.002	0.58	0.02	6.88%	96%	
2031	5	0.002	0.58	0.02	6.88%	96%	
2032	4	0.002	0.58	0.02	6.88%	96%	
2033	4	0.001	0.58	0.02	6.88%	97%	
2034	4	0.001	0.58	0.02	6.88%	97%	
2035	3	0.001	0.58	0.02	6.88%	97%	
2036	3	0.001	0.58	0.02	6.88%	97%	
2037	3	0.001	0.59	0.01	6.88%	98%	
2038	3	0.001	0.59	0.01	6.88%	98%	
2039	3	0.001	0.59	0.01	6.88%	98%	
2040	2	0.001	0.59	0.01	6.88%	98%	
2041	2	0.001	0.59	0.01	6.88%	98%	
2042	2	0.001	0.59	0.01	6.88%	98%	
2043	2	0.001	0.59	0.01	6.88%	98%	
2044	2	0.001	0.59	0.01	6.88%	98%	
2045	2	0.001	0.59	0.01	6.88%	99%	
2046	2	0.001	0.59	0.01	6.88%	99%	
2047	1	0.001	0.59	0.01	6.88%	99%	
2048	1	0.001	0.59	0.01	6.88%	99%	
2049	1	0.000	0.59	0.01	6.88%	99%	
2050	1	0.000	0.59	0.01	6.88%	99%	

CUM.PROD	0.000	BBO	GIANT FIELDS		PAPUA	
RESERVES	0.200	BBO	TOTAL BBO	0	THE EAST	
DISCOVERED	0.200	BBO	% OF DISC.	0%		
UNDISCOVERED	0.300	BBO	Last Giant	-		
REMAINING	0.500	BBO			Variance	Trend
ULTIMATE	0.500	BBO			1990	zero
Rept.Reserve	0.200		Ref.date(end)	1989	Future	UP

Milestones	Date	Prod Mb/d	Dep. Rate	% Produced
1st Peak	2002	85	11.10%	50%
2nd Peak				
Mid-Point	2002	85	11.10%	50%
Final Decline	2002		11.10%	50%

Forecast "C"
A=Exponential decline
B=5% incr. to Mid-Point
C=Other
Pre-1930 Cum. 0

Date	Prod. M.b/d	Prod. BBO/yr	Cum. Prod.	Remain	Deplet. Rate	% Produced	Giant	Notes & Names
1930		0.000	0.00	0.50	ERR	0%		
1931		0.000	0.00	0.50	0.00%	0%		
1932		0.000	0.00	0.50	0.00%	0%		
1933		0.000	0.00	0.50	0.00%	0%		
1934		0.000	0.00	0.50	0.00%	0%		
1935		0.000	0.00	0.50	0.00%	0%		
1936		0.000	0.00	0.50	0.00%	0%		
1937		0.000	0.00	0.50	0.00%	0%		
1938		0.000	0.00	0.50	0.00%	0%		
1939		0.000	0.00	0.50	0.00%	0%		
1940		0.000	0.00	0.50	0.00%	0%		
1941		0.000	0.00	0.50	0.00%	0%		
1942		0.000	0.00	0.50	0.00%	0%		
1943		0.000	0.00	0.50	0.00%	0%		
1944		0.000	0.00	0.50	0.00%	0%		
1945		0.000	0.00	0.50	0.00%	0%		
1946		0.000	0.00	0.50	0.00%	0%		
1947		0.000	0.00	0.50	0.00%	0%		
1948		0.000	0.00	0.50	0.00%	0%		
1949		0.000	0.00	0.50	0.00%	0%		
1950		0.000	0.00	0.50	0.00%	0%		
1951		0.000	0.00	0.50	0.00%	0%		
1952		0.000	0.00	0.50	0.00%	0%		
1953		0.000	0.00	0.50	0.00%	0%		
1954		0.000	0.00	0.50	0.00%	0%		
1955		0.000	0.00	0.50	0.00%	0%		
1956		0.000	0.00	0.50	0.00%	0%		
1957		0.000	0.00	0.50	0.00%	0%		
1958		0.000	0.00	0.50	0.00%	0%		
1959		0.000	0.00	0.50	0.00%	0%		
1960		0.000	0.00	0.50	0.00%	0%		
1961		0.000	0.00	0.50	0.00%	0%		
1962		0.000	0.00	0.50	0.00%	0%		
1963		0.000	0.00	0.50	0.00%	0%		
1964		0.000	0.00	0.50	0.00%	0%		
1965		0.000	0.00	0.50	0.00%	0%		
1966		0.000	0.00	0.50	0.00%	0%		
1967		0.000	0.00	0.50	0.00%	0%		
1968		0.000	0.00	0.50	0.00%	0%		
1969		0.000	0.00	0.50	0.00%	0%		
1970		0.000	0.00	0.50	0.00%	0%		
1971		0.000	0.00	0.50	0.00%	0%		
1972		0.000	0.00	0.50	0.00%	0%		
1973		0.000	0.00	0.50	0.00%	0%		
1974		0.000	0.00	0.50	0.00%	0%		
1975		0.000	0.00	0.50	0.00%	0%		
1976		0.000	0.00	0.50	0.00%	0%		
1977		0.000	0.00	0.50	0.00%	0%		
1978		0.000	0.00	0.50	0.00%	0%		
1979		0.000	0.00	0.50	0.00%	0%		
1980		0.000	0.00	0.50	0.00%	0%		
1981		0.000	0.00	0.50	0.00%	0%		
1982		0.000	0.00	0.50	0.00%	0%		
1983		0.000	0.00	0.50	0.00%	0%		
1984		0.000	0.00	0.50	0.00%	0%		
1985		0.000	0.00	0.50	0.00%	0%		
1986		0.000	0.00	0.50	0.00%	0%		
1987		0.000	0.00	0.50	0.00%	0%		
1988		0.000	0.00	0.50	0.00%	0%		

1989		0.000	0.00	0.50	0.00%	0%	
1990	0	0.000	0.00	0.50	0.00%	0%	
1991	0	0.000	0.00	0.50	0.00%	0%	
1992	0	0.000	0.00	0.50	0.00%	0%	
1993	55	0.020	0.02	0.48	4.00%	4%	Prod. starts
1994	58	0.021	0.04	0.46	4.38%	8%	5% increase
1995	60	0.022	0.06	0.44	4.80%	13%	begins
1996	63	0.023	0.09	0.41	5.30%	17%	
1997	67	0.024	0.11	0.39	5.87%	22%	
1998	70	0.026	0.14	0.36	6.55%	27%	
1999	73	0.027	0.16	0.34	7.36%	33%	
2000	77	0.028	0.19	0.31	8.35%	38%	
2001	81	0.030	0.22	0.28	9.56%	44%	
2002	85	0.031	0.25	0.25	11.10%	50%	50% PRODUCED
2003	76	0.028	0.28	0.22	11.10%	56%	& 1ST PEAK
2004	67	0.025	0.30	0.20	11.10%	61%	
2005	60	0.022	0.33	0.17	11.10%	65%	
2006	53	0.019	0.34	0.16	11.10%	69%	
2007	47	0.017	0.36	0.14	11.10%	72%	
2008	42	0.015	0.38	0.12	11.10%	75%	
2009	37	0.014	0.39	0.11	11.10%	78%	
2010	33	0.012	0.40	0.10	11.10%	81%	
2011	29	0.011	0.41	0.09	11.10%	83%	
2012	26	0.010	0.42	0.08	11.10%	85%	
2013	23	0.009	0.43	0.07	11.10%	86%	
2014	21	0.008	0.44	0.06	11.10%	88%	
2015	18	0.007	0.45	0.05	11.10%	89%	
2016	16	0.006	0.45	0.05	11.10%	90%	
2017	15	0.005	0.46	0.04	11.10%	91%	
2018	13	0.005	0.46	0.04	11.10%	92%	
2019	11	0.004	0.47	0.03	11.10%	93%	
2020	10	0.004	0.47	0.03	11.10%	94%	
2021	9	0.003	0.47	0.03	11.10%	95%	
2022	8	0.003	0.48	0.02	11.10%	95%	
2023	7	0.003	0.48	0.02	11.10%	96%	
2024	6	0.002	0.48	0.02	11.10%	96%	
2025	6	0.002	0.48	0.02	11.10%	97%	
2026	5	0.002	0.49	0.01	11.10%	97%	
2027	4	0.002	0.49	0.01	11.10%	97%	
2028	4	0.001	0.49	0.01	11.10%	98%	
2029	4	0.001	0.49	0.01	11.10%	98%	
2030	3	0.001	0.49	0.01	11.10%	98%	
2031	3	0.001	0.49	0.01	11.10%	98%	
2032	2	0.001	0.49	0.01	11.10%	99%	
2033	2	0.001	0.49	0.01	11.10%	99%	
2034	2	0.001	0.49	0.01	11.10%	99%	
2035	2	0.001	0.49	0.01	11.10%	99%	
2036	2	0.001	0.50	0.00	11.10%	99%	
2037	1	0.001	0.50	0.00	11.10%	99%	
2038	1	0.000	0.50	0.00	11.10%	99%	
2039	1	0.000	0.50	0.00	11.10%	99%	
2040	1	0.000	0.50	0.00	11.10%	99%	
2041	1	0.000	0.50	0.00	11.10%	99%	
2042	1	0.000	0.50	0.00	11.10%	100%	
2043	1	0.000	0.50	0.00	11.10%	100%	
2044	1	0.000	0.50	0.00	11.10%	100%	
2045	1	0.000	0.50	0.00	11.10%	100%	
2046	0	0.000	0.50	0.00	11.10%	100%	
2047	0	0.000	0.50	0.00	11.10%	100%	
2048	0	0.000	0.50	0.00	11.10%	100%	
2049	0	0.000	0.50	0.00	11.10%	100%	
2050	0	0.000	0.50	0.00	11.10%	100%	

Note : Production assumed to start in 1993 following construction of a pipeline at 55,000 b/d and rise at 5% until 50% depleted

CUM.PROD	1.764 BBO	GIANT FIELDS			PERU	
RESERVES	0.412 BBO	TOTAL BBO	1.39		L.AMERICA	
DISCOVERED	2.176 BBO	% OF DISC.	64%			
UNDISCOVERED	0.574 BBO	Last Giant	1868			
REMAINING	0.986 BBO				Variance	Trend
ULTIMATE	2.750 BBO				1990	+2%
Rept.Reserves	0.412 BBO	Ref. Date (end)		1989	Future	DOWN

Milestones	Date	Prod Mb/d	Dep. Rate	% Produced	Forecast "A"
					A=Exponential decline
1st Peak	1984	201	5.42%	53%	B=5% inc. to Mid-Point
2nd Peak					C=Other
Mid-Point	1983	171	4.41%	51%	Pre-1930 Cum 0.114
Final Decline	1990		4.56%		(added to 1930 Prod)

Date	Prod. M.b/d	Prod. BBO/yr	Cum Prod	Remain	Dep. Rate	% Produced	Giant	Names & Notes
1930	40.0	0.015	0.13	2.62	ERR	5%	1.385	La Brea
1931	40.0	0.015	0.14	2.61	0.56%	5%		
1932	40.0	0.015	0.16	2.59	0.56%	6%		
1933	40.0	0.015	0.17	2.58	0.56%	6%		
1934	40.0	0.015	0.19	2.56	0.57%	7%		
1935	40.0	0.015	0.20	2.55	0.57%	7%		
1936	40.0	0.015	0.22	2.53	0.57%	8%		
1937	40.0	0.015	0.23	2.52	0.58%	8%		
1938	43.4	0.016	0.25	2.50	0.63%	9%		
1939	37.1	0.014	0.26	2.49	0.54%	9%		
1940	33.1	0.012	0.27	2.48	0.49%	10%		
1941	32.8	0.012	0.28	2.47	0.48%	10%		
1942	37.4	0.014	0.30	2.45	0.55%	11%		
1943	40.1	0.015	0.31	2.44	0.60%	11%		
1944	39.2	0.014	0.33	2.42	0.59%	12%		
1945	37.6	0.014	0.34	2.41	0.57%	12%		
1946	34.1	0.012	0.35	2.40	0.52%	13%		
1947	34.9	0.013	0.37	2.38	0.53%	13%		
1948	38.4	0.014	0.38	2.37	0.59%	14%		
1949	40.5	0.015	0.39	2.36	0.62%	14%		
1950	41.2	0.015	0.41	2.34	0.64%	15%		
1951	44.1	0.016	0.43	2.32	0.69%	15%		
1952	44.9	0.016	0.44	2.31	0.71%	16%		
1953	42.4	0.015	0.46	2.29	0.67%	17%		
1954	47.0	0.017	0.47	2.28	0.75%	17%		
1955	47.3	0.017	0.49	2.26	0.76%	18%		
1956	49.8	0.018	0.51	2.24	0.80%	19%		
1957	52.0	0.019	0.53	2.22	0.85%	19%		
1958	54.3	0.020	0.55	2.20	0.89%	20%		
1959	48.4	0.018	0.57	2.18	0.80%	21%		
1960	52.0	0.019	0.59	2.16	0.87%	21%		
1961	52.1	0.019	0.60	2.15	0.88%	22%		
1962	59.7	0.022	0.63	2.12	1.02%	23%		
1963	57.3	0.021	0.65	2.10	0.98%	24%		
1964	59.0	0.022	0.67	2.08	1.02%	24%		
1965	64.2	0.023	0.69	2.06	1.13%	25%		
1966	62.1	0.023	0.71	2.04	1.10%	26%		
1967	63.0	0.023	0.74	2.01	1.13%	27%		
1968	80.3	0.029	0.77	1.98	1.46%	28%		
1969	69.7	0.025	0.79	1.96	1.28%	29%		
1970	72.0	0.026	0.82	1.93	1.34%	30%		
1971	62.3	0.023	0.84	1.91	1.18%	31%		
1972	60.2	0.022	0.86	1.89	1.15%	31%		
1973	68.0	0.025	0.89	1.86	1.32%	32%		
1974	69.7	0.025	0.91	1.84	1.37%	33%		
1975	70.0	0.026	0.94	1.81	1.39%	34%		
1976	74.3	0.027	0.97	1.78	1.50%	35%		
1977	90.0	0.033	1.00	1.75	1.84%	36%		
1978	150.0	0.055	1.05	1.70	3.13%	38%		
1979	195.0	0.071	1.13	1.62	4.20%	41%		
1980	190.0	0.069	1.19	1.56	4.27%	43%		
1981	184.0	0.067	1.26	1.49	4.32%	46%		
1982	198.0	0.072	1.33	1.42	4.86%	49%		
1983	171.0	0.062	1.40	1.35	4.41%	51%		50% PRODUCED
1984	201.0	0.073	1.47	1.28	5.42%	53%		1ST PEAK
1985	189.0	0.069	1.54	1.21	5.39%	56%		
1986	179.0	0.065	1.60	1.15	5.39%	58%		
1987	166.7	0.061	1.67	1.08	5.31%	61%		
1988	141.7	0.052	1.72	1.03	4.77%	62%		

1989	129.0	0.047	1.76	0.99	4.56%	64%
1990	123.2	0.045	1.81	0.94	4.56%	66%
1991	117.6	0.043	1.85	0.90	4.56%	67%
1992	112.2	0.041	1.89	0.86	4.56%	69%
1993	107.1	0.039	1.93	0.82	4.56%	70%
1994	102.2	0.037	1.97	0.78	4.56%	72%
1995	97.6	0.036	2.00	0.75	4.56%	73%
1996	93.1	0.034	2.04	0.71	4.56%	74%
1997	88.9	0.032	2.07	0.68	4.56%	75%
1998	84.8	0.031	2.10	0.65	4.56%	76%
1999	80.9	0.030	2.13	0.62	4.56%	78%
2000	77.2	0.028	2.16	0.59	4.56%	79%
2001	73.7	0.027	2.19	0.56	4.56%	80%
2002	70.4	0.026	2.21	0.54	4.56%	80%
2003	67.2	0.025	2.24	0.51	4.56%	81%
2004	64.1	0.023	2.26	0.49	4.56%	82%
2005	61.2	0.022	2.28	0.47	4.56%	83%
2006	58.4	0.021	2.30	0.45	4.56%	84%
2007	55.7	0.020	2.32	0.43	4.56%	85%
2008	53.2	0.019	2.34	0.41	4.56%	85%
2009	50.8	0.019	2.36	0.39	4.56%	86%
2010	48.4	0.018	2.38	0.37	4.56%	87%
2011	46.2	0.017	2.40	0.35	4.56%	87%
2012	44.1	0.016	2.41	0.34	4.56%	88%
2013	42.1	0.015	2.43	0.32	4.56%	88%
2014	40.2	0.015	2.44	0.31	4.56%	89%
2015	38.4	0.014	2.46	0.29	4.56%	89%
2016	36.6	0.013	2.47	0.28	4.56%	90%
2017	34.9	0.013	2.48	0.27	4.56%	90%
2018	33.3	0.012	2.50	0.25	4.56%	91%
2019	31.8	0.012	2.51	0.24	4.56%	91%
2020	30.4	0.011	2.52	0.23	4.56%	92%
2021	29.0	0.011	2.53	0.22	4.56%	92%
2022	27.7	0.010	2.54	0.21	4.56%	92%
2023	26.4	0.010	2.55	0.20	4.56%	93%
2024	25.2	0.009	2.56	0.19	4.56%	93%
2025	24.1	0.009	2.57	0.18	4.56%	93%
2026	23.0	0.008	2.57	0.18	4.56%	94%
2027	21.9	0.008	2.58	0.17	4.56%	94%
2028	20.9	0.008	2.59	0.16	4.56%	94%
2029	20.0	0.007	2.60	0.15	4.56%	94%
2030	19.0	0.007	2.60	0.15	4.56%	95%
2031	18.2	0.007	2.61	0.14	4.56%	95%
2032	17.3	0.006	2.62	0.13	4.56%	95%
2033	16.6	0.006	2.62	0.13	4.56%	95%
2034	15.8	0.006	2.63	0.12	4.56%	96%
2035	15.1	0.006	2.63	0.12	4.56%	96%
2036	14.4	0.005	2.64	0.11	4.56%	96%
2037	13.7	0.005	2.65	0.10	4.56%	96%
2038	13.1	0.005	2.65	0.10	4.56%	96%
2039	12.5	0.005	2.65	0.10	4.56%	97%
2040	11.9	0.004	2.66	0.09	4.56%	97%
2041	11.4	0.004	2.66	0.09	4.56%	97%
2042	10.9	0.004	2.67	0.08	4.56%	97%
2043	10.4	0.004	2.67	0.08	4.56%	97%
2044	9.9	0.004	2.67	0.08	4.56%	97%
2045	9.5	0.003	2.68	0.07	4.56%	97%
2046	9.0	0.003	2.68	0.07	4.56%	97%
2047	8.6	0.003	2.68	0.07	4.56%	98%
2048	8.2	0.003	2.69	0.06	4.56%	98%
2049	7.8	0.003	2.69	0.06	4.56%	98%
2050	7.5	0.003	2.69	0.06	4.56%	98%

CUM.PROD	0.036 BBO	GIANT FIELDS		PHILIPPINES		
RESERVES	0.015 BBO	TOTAL BBO	0	THE EAST		
DISCOVERED	0.051 BBO	% OF DISC.	0%			
UNDISCOVERED	0.024 BBO	Last Giant	-			
REMAINING	0.039 BBO			Variance Trend		
ULTIMATE	0.075 BBO			1990	+25%	
Rept.Reserve	0.015	Ref.date(end)	1989	Future	UP	

Milestones	Date	Prod Mb/d	Dep. Rate	% Produced
1st Peak	1983	15	8.96%	26%
2nd Peak				
Mid-Point	1990	4	3.62%	50%
Final Decline	1990		3.62%	50%

Forecast "A"
A=Exponential decline
B=5% incr. to Mid-Point
C=Other
Pre-1930 Cum. 0

Date	Prod. M.b/d	Prod. BBO/yr	Cum. Prod.	Remain	Deplet. Rate	% Produced	Giant	Notes & Names
1930		0.000	0.00	0.08	ERR	0%		
1931		0.000	0.00	0.08	0.00%	0%		
1932		0.000	0.00	0.08	0.00%	0%		
1933		0.000	0.00	0.08	0.00%	0%		
1934		0.000	0.00	0.08	0.00%	0%		
1935		0.000	0.00	0.08	0.00%	0%		
1936		0.000	0.00	0.08	0.00%	0%		
1937		0.000	0.00	0.08	0.00%	0%		
1938		0.000	0.00	0.08	0.00%	0%		
1939		0.000	0.00	0.08	0.00%	0%		
1940		0.000	0.00	0.08	0.00%	0%		
1941		0.000	0.00	0.08	0.00%	0%		
1942		0.000	0.00	0.08	0.00%	0%		
1943		0.000	0.00	0.08	0.00%	0%		
1944		0.000	0.00	0.08	0.00%	0%		
1945		0.000	0.00	0.08	0.00%	0%		
1946		0.000	0.00	0.08	0.00%	0%		
1947		0.000	0.00	0.08	0.00%	0%		
1948		0.000	0.00	0.08	0.00%	0%		
1949		0.000	0.00	0.08	0.00%	0%		
1950		0.000	0.00	0.08	0.00%	0%		
1951		0.000	0.00	0.08	0.00%	0%		
1952		0.000	0.00	0.08	0.00%	0%		
1953		0.000	0.00	0.08	0.00%	0%		
1954		0.000	0.00	0.08	0.00%	0%		
1955		0.000	0.00	0.08	0.00%	0%		
1956		0.000	0.00	0.08	0.00%	0%		
1957		0.000	0.00	0.08	0.00%	0%		
1958		0.000	0.00	0.08	0.00%	0%		
1959		0.000	0.00	0.08	0.00%	0%		
1960		0.000	0.00	0.08	0.00%	0%		
1961		0.000	0.00	0.08	0.00%	0%		
1962		0.000	0.00	0.08	0.00%	0%		
1963		0.000	0.00	0.08	0.00%	0%		
1964		0.000	0.00	0.08	0.00%	0%		
1965		0.000	0.00	0.08	0.00%	0%		
1966		0.000	0.00	0.08	0.00%	0%		
1967		0.000	0.00	0.08	0.00%	0%		
1968		0.000	0.00	0.08	0.00%	0%		
1969		0.000	0.00	0.08	0.00%	0%		
1970		0.000	0.00	0.08	0.00%	0%		
1971		0.000	0.00	0.08	0.00%	0%		
1972		0.000	0.00	0.08	0.00%	0%		
1973		0.000	0.00	0.08	0.00%	0%		
1974		0.000	0.00	0.08	0.00%	0%		
1975		0.000	0.00	0.08	0.00%	0%		
1976		0.000	0.00	0.08	0.00%	0%		
1977		0.000	0.00	0.08	0.00%	0%		
1978		0.000	0.00	0.08	0.00%	0%		
1979	15	0.005	0.01	0.07	7.30%	7%		
1980	14	0.005	0.01	0.06	7.35%	14%		
1981	2	0.001	0.01	0.06	1.13%	15%		
1982	7	0.003	0.01	0.06	4.01%	18%		
1983	15	0.005	0.02	0.06	8.96%	26%		1ST PEAK
1984	12	0.004	0.02	0.05	7.87%	32%		
1985	9	0.003	0.03	0.05	6.62%	36%		
1986	6	0.002	0.03	0.05	4.80%	39%		
1987	5	0.002	0.03	0.04	4.08%	42%		
1988	9	0.003	0.03	0.04	7.68%	46%		

1989	4	0.001	0.04	0.04	3.62%	48%	
1990	4	0.001	0.04	0.04	3.62%	50%	50% PRODUCED
1991	4	0.001	0.04	0.04	3.62%	52%	
1992	4	0.001	0.04	0.03	3.62%	54%	
1993	3	0.001	0.04	0.03	3.62%	55%	
1994	3	0.001	0.04	0.03	3.62%	57%	
1995	3	0.001	0.04	0.03	3.62%	58%	
1996	3	0.001	0.04	0.03	3.62%	60%	
1997	3	0.001	0.05	0.03	3.62%	61%	
1998	3	0.001	0.05	0.03	3.62%	63%	
1999	3	0.001	0.05	0.03	3.62%	64%	
2000	3	0.001	0.05	0.03	3.62%	65%	
2001	3	0.001	0.05	0.02	3.62%	67%	
2002	2	0.001	0.05	0.02	3.62%	68%	
2003	2	0.001	0.05	0.02	3.62%	69%	
2004	2	0.001	0.05	0.02	3.62%	70%	
2005	2	0.001	0.05	0.02	3.62%	71%	
2006	2	0.001	0.05	0.02	3.62%	72%	
2007	2	0.001	0.05	0.02	3.62%	73%	
2008	2	0.001	0.06	0.02	3.62%	74%	
2009	2	0.001	0.06	0.02	3.62%	75%	
2010	2	0.001	0.06	0.02	3.62%	76%	
2011	2	0.001	0.06	0.02	3.62%	77%	
2012	2	0.001	0.06	0.02	3.62%	78%	
2013	2	0.001	0.06	0.02	3.62%	79%	
2014	2	0.001	0.06	0.02	3.62%	79%	
2015	2	0.001	0.06	0.01	3.62%	80%	
2016	1	0.001	0.06	0.01	3.62%	81%	
2017	1	0.001	0.06	0.01	3.62%	82%	
2018	1	0.001	0.06	0.01	3.62%	82%	
2019	1	0.000	0.06	0.01	3.62%	83%	
2020	1	0.000	0.06	0.01	3.62%	83%	
2021	1	0.000	0.06	0.01	3.62%	84%	
2022	1	0.000	0.06	0.01	3.62%	85%	
2023	1	0.000	0.06	0.01	3.62%	85%	
2024	1	0.000	0.06	0.01	3.62%	86%	
2025	1	0.000	0.06	0.01	3.62%	86%	
2026	1	0.000	0.07	0.01	3.62%	87%	
2027	1	0.000	0.07	0.01	3.62%	87%	
2028	1	0.000	0.07	0.01	3.62%	88%	
2029	1	0.000	0.07	0.01	3.62%	88%	
2030	1	0.000	0.07	0.01	3.62%	89%	
2031	1	0.000	0.07	0.01	3.62%	89%	
2032	1	0.000	0.07	0.01	3.62%	89%	
2033	1	0.000	0.07	0.01	3.62%	90%	
2034	1	0.000	0.07	0.01	3.62%	90%	
2035	1	0.000	0.07	0.01	3.62%	90%	
2036	1	0.000	0.07	0.01	3.62%	91%	
2037	1	0.000	0.07	0.01	3.62%	91%	
2038	1	0.000	0.07	0.01	3.62%	91%	
2039	1	0.000	0.07	0.01	3.62%	92%	
2040	1	0.000	0.07	0.01	3.62%	92%	
2041	1	0.000	0.07	0.01	3.62%	92%	
2042	1	0.000	0.07	0.01	3.62%	93%	
2043	1	0.000	0.07	0.01	3.62%	93%	
2044	1	0.000	0.07	0.01	3.62%	93%	
2045	1	0.000	0.07	0.00	3.62%	93%	
2046	0	0.000	0.07	0.00	3.62%	94%	
2047	0	0.000	0.07	0.00	3.62%	94%	
2048	0	0.000	0.07	0.00	3.62%	94%	
2049	0	0.000	0.07	0.00	3.62%	94%	
2050	0	0.000	0.07	0.00	3.62%	95%	

CUM.PROD	0.247 BBO	GIANT FIELDS		POLAND	
RESERVES	0.085 BBO	TOTAL BBO	0	SINO-SOVIET	
DISCOVERED	0.332 BBO	% OF DISC.	0%		
UNDISCOVERED	0.168 BBO	Last Giant	-		
REMAINING	0.253 BBO			Revision Trend	
ULTIMATE	0.500 BBO			1990	n/a
Rept.Reserve	0.079 #1	Ref.date(end) 1989		Future	DOWN

Milestones	Date	Prod Mb/d	Dep. Rate	% Produced	Forecast "A"
					A=Exponential decline
1st Peak	1930	13	ERR	21%	B=5% incr. to Mid-Point
2nd Peak	None				C=Other
Mid-Point	1990	3	0.39%	50%	Pre-1930 Cum. 0.1
Final Decline	1990		0.39%	50%	(Added to 1930 prod)

Date	Prod. M.b/d	Prod. BBO/yr	Cum. Prod.	Remain	Deplet. Rate	% Produced	Giant	Notes & Names
1930	13	0.005	0.10	0.40	ERR	21%		Data-base weak
1931	13	0.005	0.11	0.39	1.18%	22%		
1932	11	0.004	0.11	0.39	1.06%	23%		
1933	11	0.004	0.12	0.38	1.06%	24%		
1934	11	0.004	0.12	0.38	1.02%	24%		
1935	10	0.004	0.13	0.37	1.00%	25%		
1936	10	0.004	0.13	0.37	1.01%	26%		
1937	10	0.004	0.13	0.37	1.00%	27%		
1938	10	0.004	0.14	0.36	1.02%	27%		
1939	11	0.004	0.14	0.36	1.08%	28%		
1940	11	0.004	0.14	0.36	1.09%	29%		
1941	5	0.002	0.15	0.35	0.50%	29%		
1942	8	0.003	0.15	0.35	0.79%	30%		
1943	10	0.004	0.15	0.35	1.00%	31%		
1944	8	0.003	0.16	0.34	0.86%	31%		
1945	2	0.001	0.16	0.34	0.22%	31%		
1946	2	0.001	0.16	0.34	0.25%	31%		
1947	3	0.001	0.16	0.34	0.28%	32%		
1948	3	0.001	0.16	0.34	0.30%	32%		
1949	3	0.001	0.16	0.34	0.33%	32%		
1950	3	0.001	0.16	0.34	0.35%	32%		
1951	4	0.001	0.16	0.34	0.44%	33%		
1952	5	0.002	0.16	0.34	0.51%	33%		
1953	4	0.001	0.17	0.33	0.41%	33%		
1954	4	0.001	0.17	0.33	0.40%	34%		
1955	4	0.001	0.17	0.33	0.41%	34%		
1956	4	0.001	0.17	0.33	0.41%	34%		
1957	4	0.001	0.17	0.33	0.41%	34%		
1958	4	0.001	0.17	0.33	0.40%	35%		
1959	4	0.001	0.17	0.33	0.39%	35%		
1960	4	0.001	0.18	0.32	0.45%	35%		
1961	4	0.001	0.18	0.32	0.46%	35%		
1962	4	0.001	0.18	0.32	0.46%	36%		
1963	4	0.002	0.18	0.32	0.49%	36%		
1964	6	0.002	0.18	0.32	0.65%	36%		
1965	7	0.003	0.18	0.32	0.79%	37%		
1966	8	0.003	0.19	0.31	0.94%	38%		
1967	9	0.003	0.19	0.31	1.06%	38%		
1968	10	0.004	0.19	0.31	1.18%	39%		
1969	11	0.004	0.20	0.30	1.34%	40%		
1970	9	0.003	0.20	0.30	1.03%	40%		
1971	9	0.003	0.21	0.29	1.04%	41%		
1972	7	0.003	0.21	0.29	0.88%	42%		
1973	8	0.003	0.21	0.29	1.00%	42%		
1974	11	0.004	0.21	0.29	1.40%	43%		
1975	11	0.004	0.22	0.28	1.43%	44%		
1976	9	0.003	0.22	0.28	1.19%	44%		
1977	9	0.003	0.23	0.27	1.18%	45%		
1978	9	0.003	0.23	0.27	1.20%	46%		
1979	8	0.003	0.23	0.27	1.09%	46%		
1980	7	0.003	0.23	0.27	0.95%	47%		
1981	6	0.002	0.24	0.26	0.88%	47%		
1982	5	0.002	0.24	0.26	0.66%	48%		
1983	4	0.001	0.24	0.26	0.56%	48%		
1984	4	0.001	0.24	0.26	0.53%	48%		
1985	4	0.001	0.24	0.26	0.56%	49%		
1986	3	0.001	0.24	0.26	0.48%	49%		
1987	4	0.001	0.25	0.25	0.51%	49%		
1988	3	0.001	0.25	0.25	0.40%	49%		

1989	3	0.001	0.25	0.25	0.37%	49%
1990	3	0.001	0.25	0.25	0.39%	50%
1991	3	0.001	0.25	0.25	0.39%	50%
1992	3	0.001	0.25	0.25	0.39%	50%
1993	3	0.001	0.25	0.25	0.39%	50%
1994	3	0.001	0.25	0.25	0.39%	50%
1995	3	0.001	0.25	0.25	0.39%	51%
1996	3	0.001	0.25	0.25	0.39%	51%
1997	3	0.001	0.25	0.25	0.39%	51%
1998	3	0.001	0.26	0.24	0.39%	51%
1999	3	0.001	0.26	0.24	0.39%	51%
2000	3	0.001	0.26	0.24	0.39%	52%
2001	3	0.001	0.26	0.24	0.39%	52%
2002	3	0.001	0.26	0.24	0.39%	52%
2003	3	0.001	0.26	0.24	0.39%	52%
2004	3	0.001	0.26	0.24	0.39%	52%
2005	3	0.001	0.26	0.24	0.39%	52%
2006	3	0.001	0.26	0.24	0.39%	53%
2007	3	0.001	0.26	0.24	0.39%	53%
2008	3	0.001	0.27	0.23	0.39%	53%
2009	3	0.001	0.27	0.23	0.39%	53%
2010	2	0.001	0.27	0.23	0.39%	53%
2011	2	0.001	0.27	0.23	0.39%	54%
2012	2	0.001	0.27	0.23	0.39%	54%
2013	2	0.001	0.27	0.23	0.39%	54%
2014	2	0.001	0.27	0.23	0.39%	54%
2015	2	0.001	0.27	0.23	0.39%	54%
2016	2	0.001	0.27	0.23	0.39%	54%
2017	2	0.001	0.27	0.23	0.39%	55%
2018	2	0.001	0.27	0.23	0.39%	55%
2019	2	0.001	0.28	0.22	0.39%	55%
2020	2	0.001	0.28	0.22	0.39%	55%
2021	2	0.001	0.28	0.22	0.39%	55%
2022	2	0.001	0.28	0.22	0.39%	56%
2023	2	0.001	0.28	0.22	0.39%	56%
2024	2	0.001	0.28	0.22	0.39%	56%
2025	2	0.001	0.28	0.22	0.39%	56%
2026	2	0.001	0.28	0.22	0.39%	56%
2027	2	0.001	0.28	0.22	0.39%	56%
2028	2	0.001	0.28	0.22	0.39%	57%
2029	2	0.001	0.28	0.22	0.39%	57%
2030	2	0.001	0.28	0.22	0.39%	57%
2031	2	0.001	0.29	0.21	0.39%	57%
2032	2	0.001	0.29	0.21	0.39%	57%
2033	2	0.001	0.29	0.21	0.39%	57%
2034	2	0.001	0.29	0.21	0.39%	58%
2035	2	0.001	0.29	0.21	0.39%	58%
2036	2	0.001	0.29	0.21	0.39%	58%
2037	2	0.001	0.29	0.21	0.39%	58%
2038	2	0.001	0.29	0.21	0.39%	58%
2039	2	0.001	0.29	0.21	0.39%	58%
2040	2	0.001	0.29	0.21	0.39%	59%
2041	2	0.001	0.29	0.21	0.39%	59%
2042	2	0.001	0.29	0.21	0.39%	59%
2043	2	0.001	0.30	0.20	0.39%	59%
2044	2	0.001	0.30	0.20	0.39%	59%
2045	2	0.001	0.30	0.20	0.39%	59%
2046	2	0.001	0.30	0.20	0.39%	60%
2047	2	0.001	0.30	0.20	0.39%	60%
2048	2	0.001	0.30	0.20	0.39%	60%
2049	2	0.001	0.30	0.20	0.39%	60%
2050	2	0.001	0.30	0.20	0.39%	60%

CUM.PROD	4.233 BBO	GIANT FIELDS		QATAR	
RESERVES	4.500 BBO	TOTAL BBO	7.4	M.EAST (MINOR)	
DISCOVERED	8.733 BBO	% OF DISC.	85%		
UNDISCOVERED	1.267 BBO	Last Giant	1970		
REMAINING	5.767 BBO			Variance Trend	
ULTIMATE	10.000 BBO			1990	-7%
Rept.Reserve	4.500	Ref.date(end)	1989	Future	DOWN

Milestones	Date	Prod Mb/d	Dep. Rate	% Produced
1st Peak	1973	556	2.45%	19%
2nd Peak	1994	504	3.59%	51%
Mid-Point	1994	504	3.59%	51%
Final Decline	1994		3.59%	51%

Forecast "B"
A=Exponential decline
B=5% incr. to Mid-Point
C=Other
Pre-1930 Cum. 0
(Added to 1930 prod)

Date	Prod. M.b/d	Prod. BBO/yr	Cum. Prod.	Remain	Deplet. Rate	% Produced	Giant	Notes & Names
1930		0.000	0.00	10.00	ERR	0%		
1931		0.000	0.00	10.00	0.00%	0%		
1932		0.000	0.00	10.00	0.00%	0%		
1933		0.000	0.00	10.00	0.00%	0%		
1934		0.000	0.00	10.00	0.00%	0%		
1935		0.000	0.00	10.00	0.00%	0%		
1936		0.000	0.00	10.00	0.00%	0%		
1937		0.000	0.00	10.00	0.00%	0%		
1938		0.000	0.00	10.00	0.00%	0%		
1939		0.000	0.00	10.00	0.00%	0%		
1940		0.000	0.00	10.00	0.00%	0%	2.85	Dukhan
1941		0.000	0.00	10.00	0.00%	0%		
1942		0.000	0.00	10.00	0.00%	0%		
1943		0.000	0.00	10.00	0.00%	0%		
1944		0.000	0.00	10.00	0.00%	0%		
1945		0.000	0.00	10.00	0.00%	0%		
1946		0.000	0.00	10.00	0.00%	0%		
1947		0.000	0.00	10.00	0.00%	0%		
1948		0.000	0.00	10.00	0.00%	0%		
1949	2	0.001	0.00	10.00	0.01%	0%		
1950	34	0.012	0.01	9.99	0.12%	0%		
1951	49	0.018	0.03	9.97	0.18%	0%		
1952	68	0.025	0.06	9.94	0.25%	1%		
1953	79	0.029	0.08	9.92	0.29%	1%		
1954	98	0.036	0.12	9.88	0.36%	1%		
1955	114	0.042	0.16	9.84	0.42%	2%		
1956	127	0.046	0.21	9.79	0.47%	2%		
1957	136	0.050	0.26	9.74	0.51%	3%		
1958	162	0.059	0.32	9.68	0.61%	3%		
1959	174	0.064	0.38	9.62	0.66%	4%		
1960	173	0.063	0.44	9.56	0.66%	4%	2.10	Idd-e-Shargi
1961	176	0.064	0.51	9.49	0.67%	5%		
1962	188	0.069	0.58	9.42	0.72%	6%		
1963	193	0.070	0.65	9.35	0.75%	6%	1.20	Maydan/Mahzan
1964	221	0.081	0.73	9.27	0.86%	7%		
1965	216	0.079	0.81	9.19	0.85%	8%		
1966	283	0.103	0.91	9.09	1.12%	9%		
1967	323	0.118	1.03	8.97	1.30%	10%		
1968	341	0.125	1.15	8.85	1.39%	12%		
1969	355	0.130	1.28	8.72	1.47%	13%		
1970	354	0.129	1.41	8.59	1.48%	14%	1.25	Bul Hanine
1971	429	0.157	1.57	8.43	1.82%	16%		
1972	450	0.164	1.73	8.27	1.95%	17%		
1973	556	0.203	1.93	8.07	2.45%	19%		1ST PEAK
1974	546	0.199	2.13	7.87	2.47%	21%		
1975	410	0.150	2.28	7.72	1.90%	23%		
1976	485	0.177	2.46	7.54	2.29%	25%		
1977	350	0.128	2.59	7.41	1.69%	26%		
1978	480	0.175	2.76	7.24	2.36%	28%		
1979	480	0.175	2.94	7.06	2.42%	29%		
1980	470	0.172	3.11	6.89	2.43%	31%		
1981	414	0.151	3.26	6.74	2.19%	33%		
1982	340	0.124	3.39	6.61	1.84%	34%		
1983	270	0.099	3.48	6.52	1.49%	35%		
1984	395	0.144	3.63	6.37	2.21%	36%		
1985	297	0.108	3.74	6.26	1.70%	37%		
1986	332	0.121	3.86	6.14	1.94%	39%		
1987	284	0.104	3.96	6.04	1.69%	40%		
1988	349	0.127	4.09	5.91	2.11%	41%		

1989	395	0.144	4.23	5.77	2.44%	42%	
1990	414	0.151	4.38	5.62	2.62%	44%	
1991	435	0.159	4.54	5.46	2.83%	45%	
1992	457	0.167	4.71	5.29	3.06%	47%	
1993	480	0.175	4.88	5.12	3.31%	49%	
1994	504	0.184	5.07	4.93	3.59%	51%	50% PRODUCED & 2ND PEAK
1995	485	0.177	5.25	4.75	3.59%	52%	
1996	468	0.171	5.42	4.58	3.59%	54%	
1997	451	0.165	5.58	4.42	3.59%	56%	
1998	435	0.159	5.74	4.26	3.59%	57%	
1999	419	0.153	5.89	4.11	3.59%	59%	
2000	404	0.147	6.04	3.96	3.59%	60%	
2001	389	0.142	6.18	3.82	3.59%	62%	
2002	376	0.137	6.32	3.68	3.59%	63%	
2003	362	0.132	6.45	3.55	3.59%	65%	
2004	349	0.127	6.58	3.42	3.59%	66%	
2005	336	0.123	6.70	3.30	3.59%	67%	
2006	324	0.118	6.82	3.18	3.59%	68%	
2007	313	0.114	6.93	3.07	3.59%	69%	
2008	302	0.110	7.04	2.96	3.59%	70%	
2009	291	0.106	7.15	2.85	3.59%	72%	
2010	280	0.102	7.25	2.75	3.59%	73%	
2011	270	0.099	7.35	2.65	3.59%	74%	
2012	261	0.095	7.45	2.55	3.59%	74%	
2013	251	0.092	7.54	2.46	3.59%	75%	
2014	242	0.088	7.63	2.37	3.59%	76%	
2015	233	0.085	7.71	2.29	3.59%	77%	
2016	225	0.082	7.79	2.21	3.59%	78%	
2017	217	0.079	7.87	2.13	3.59%	79%	
2018	209	0.076	7.95	2.05	3.59%	79%	
2019	202	0.074	8.02	1.98	3.59%	80%	
2020	194	0.071	8.09	1.91	3.59%	81%	
2021	187	0.068	8.16	1.84	3.59%	82%	
2022	181	0.066	8.23	1.77	3.59%	82%	
2023	174	0.064	8.29	1.71	3.59%	83%	
2024	168	0.061	8.35	1.65	3.59%	84%	
2025	162	0.059	8.41	1.59	3.59%	84%	
2026	156	0.057	8.47	1.53	3.59%	85%	
2027	151	0.055	8.52	1.48	3.59%	85%	
2028	145	0.053	8.58	1.42	3.59%	86%	
2029	140	0.051	8.63	1.37	3.59%	86%	
2030	135	0.049	8.68	1.32	3.59%	87%	
2031	130	0.047	8.73	1.27	3.59%	87%	
2032	125	0.046	8.77	1.23	3.59%	88%	
2033	121	0.044	8.82	1.18	3.59%	88%	
2034	117	0.043	8.86	1.14	3.59%	89%	
2035	112	0.041	8.90	1.10	3.59%	89%	
2036	108	0.040	8.94	1.06	3.59%	89%	
2037	104	0.038	8.98	1.02	3.59%	90%	
2038	101	0.037	9.01	0.99	3.59%	90%	
2039	97	0.035	9.05	0.95	3.59%	90%	
2040	94	0.034	9.08	0.92	3.59%	91%	
2041	90	0.033	9.12	0.88	3.59%	91%	
2042	87	0.032	9.15	0.85	3.59%	91%	
2043	84	0.031	9.18	0.82	3.59%	92%	
2044	81	0.030	9.21	0.79	3.59%	92%	
2045	78	0.028	9.24	0.76	3.59%	92%	
2046	75	0.027	9.26	0.74	3.59%	93%	
2047	72	0.026	9.29	0.71	3.59%	93%	
2048	70	0.025	9.32	0.68	3.59%	93%	
2049	67	0.025	9.34	0.66	3.59%	93%	
2050	65	0.024	9.36	0.64	3.59%	94%	

CUM.PROD	0.020	BBO	GIANT FIELDS		RAS AL KHAIMAH	
RESERVES	0.400	BBO	TOTAL BBO	0	M.EAST (MINOR)	
DISCOVERED	0.420	BBO	% OF DISC.	0%		
UNDISCOVERED	0.080	BBO	Last Giant	-		
REMAINING	0.480	BBO			Variance	Trend
ULTIMATE	0.500	BBO			1990	-9%
Rept.Reserve	0.400		Ref.date(end)	1989	Future	DOWN

Milestones	Date	Prod Mb/d	Dep. Rate	% Produced	Forecast "B"
1st Peak	2017	39	5.30%	49%	A=Exponential decline
2nd Peak					B=5% incr. to Mid-Point
					C=Other
Mid-Point	2018	37	5.30%	52%	Pre-1930 Cum. 0
Final Decline	2018		5.30%	52%	

Date	Prod. M.b/d	Prod. BBO/yr	Cum. Prod.	Remain	Deplet. Rate	% Produced	Giant	Notes & Names
1930		0.000	0.00	0.50	ERR	0%		
1931		0.000	0.00	0.50	0.00%	0%		
1932		0.000	0.00	0.50	0.00%	0%		
1933		0.000	0.00	0.50	0.00%	0%		
1934		0.000	0.00	0.50	0.00%	0%		
1935		0.000	0.00	0.50	0.00%	0%		
1936		0.000	0.00	0.50	0.00%	0%		
1937		0.000	0.00	0.50	0.00%	0%		
1938		0.000	0.00	0.50	0.00%	0%		
1939		0.000	0.00	0.50	0.00%	0%		
1940		0.000	0.00	0.50	0.00%	0%		
1941		0.000	0.00	0.50	0.00%	0%		
1942		0.000	0.00	0.50	0.00%	0%		
1943		0.000	0.00	0.50	0.00%	0%		
1944		0.000	0.00	0.50	0.00%	0%		
1945		0.000	0.00	0.50	0.00%	0%		
1946		0.000	0.00	0.50	0.00%	0%		
1947		0.000	0.00	0.50	0.00%	0%		
1948		0.000	0.00	0.50	0.00%	0%		
1949		0.000	0.00	0.50	0.00%	0%		
1950		0.000	0.00	0.50	0.00%	0%		
1951		0.000	0.00	0.50	0.00%	0%		
1952		0.000	0.00	0.50	0.00%	0%		
1953		0.000	0.00	0.50	0.00%	0%		
1954		0.000	0.00	0.50	0.00%	0%		
1955		0.000	0.00	0.50	0.00%	0%		
1956		0.000	0.00	0.50	0.00%	0%		
1957		0.000	0.00	0.50	0.00%	0%		
1958		0.000	0.00	0.50	0.00%	0%		
1959		0.000	0.00	0.50	0.00%	0%		
1960		0.000	0.00	0.50	0.00%	0%		
1961		0.000	0.00	0.50	0.00%	0%		
1962		0.000	0.00	0.50	0.00%	0%		
1963		0.000	0.00	0.50	0.00%	0%		
1964		0.000	0.00	0.50	0.00%	0%		
1965		0.000	0.00	0.50	0.00%	0%		
1966		0.000	0.00	0.50	0.00%	0%		
1967		0.000	0.00	0.50	0.00%	0%		
1968		0.000	0.00	0.50	0.00%	0%		
1969		0.000	0.00	0.50	0.00%	0%		
1970		0.000	0.00	0.50	0.00%	0%		
1971		0.000	0.00	0.50	0.00%	0%		
1972		0.000	0.00	0.50	0.00%	0%		
1973		0.000	0.00	0.50	0.00%	0%		
1974		0.000	0.00	0.50	0.00%	0%		
1975		0.000	0.00	0.50	0.00%	0%		
1976		0.000	0.00	0.50	0.00%	0%		
1977		0.000	0.00	0.50	0.00%	0%		
1978		0.000	0.00	0.50	0.00%	0%		
1979		0.000	0.00	0.50	0.00%	0%		
1980		0.000	0.00	0.50	0.00%	0%		
1981		0.000	0.00	0.50	0.00%	0%		
1982		0.000	0.00	0.50	0.00%	0%		
1983		0.000	0.00	0.50	0.00%	0%		
1984	6	0.002	0.00	0.50	0.42%	0%		
1985	9	0.003	0.01	0.49	0.67%	1%		
1986	11	0.004	0.01	0.49	0.81%	2%		
1987	10	0.004	0.01	0.49	0.74%	3%		
1988	10	0.004	0.02	0.48	0.75%	3%		

1989	10	0.004	0.02	0.48	0.76%	4%	
1990	11	0.004	0.02	0.48	0.80%	5%	
1991	11	0.004	0.03	0.47	0.85%	6%	
1992	12	0.004	0.03	0.47	0.90%	6%	
1993	12	0.004	0.04	0.46	0.95%	7%	
1994	13	0.005	0.04	0.46	1.01%	8%	
1995	13	0.005	0.05	0.45	1.07%	9%	
1996	14	0.005	0.05	0.45	1.13%	10%	
1997	15	0.005	0.06	0.44	1.20%	11%	
1998	16	0.006	0.06	0.44	1.28%	13%	
1999	16	0.006	0.07	0.43	1.36%	14%	
2000	17	0.006	0.07	0.43	1.45%	15%	
2001	18	0.007	0.08	0.42	1.54%	16%	
2002	19	0.007	0.09	0.41	1.64%	18%	
2003	20	0.007	0.10	0.40	1.76%	19%	
2004	21	0.008	0.10	0.40	1.88%	21%	
2005	22	0.008	0.11	0.39	2.01%	22%	
2006	23	0.008	0.12	0.38	2.15%	24%	
2007	24	0.009	0.13	0.37	2.31%	26%	
2008	25	0.009	0.14	0.36	2.48%	27%	
2009	27	0.010	0.15	0.35	2.67%	29%	
2010	28	0.010	0.16	0.34	2.88%	31%	
2011	29	0.011	0.17	0.33	3.12%	34%	
2012	31	0.011	0.18	0.32	3.38%	36%	
2013	32	0.012	0.19	0.31	3.67%	38%	
2014	34	0.012	0.20	0.30	4.00%	41%	
2015	36	0.013	0.22	0.28	4.37%	43%	
2016	37	0.014	0.23	0.27	4.80%	46%	
2017	39	0.014	0.24	0.26	5.30%	49%	1ST PEAK
2018	37	0.014	0.26	0.24	5.30%	52%	50% PRODUCED
2019	35	0.013	0.27	0.23	5.30%	54%	
2020	33	0.012	0.28	0.22	5.30%	57%	
2021	32	0.012	0.29	0.21	5.30%	59%	
2022	30	0.011	0.31	0.19	5.30%	61%	
2023	28	0.010	0.32	0.18	5.30%	63%	
2024	27	0.010	0.33	0.17	5.30%	65%	
2025	25	0.009	0.33	0.17	5.30%	67%	
2026	24	0.009	0.34	0.16	5.30%	69%	
2027	23	0.008	0.35	0.15	5.30%	70%	
2028	22	0.008	0.36	0.14	5.30%	72%	
2029	20	0.007	0.37	0.13	5.30%	73%	
2030	19	0.007	0.37	0.13	5.30%	75%	
2031	18	0.007	0.38	0.12	5.30%	76%	
2032	17	0.006	0.39	0.11	5.30%	77%	
2033	16	0.006	0.39	0.11	5.30%	79%	
2034	16	0.006	0.40	0.10	5.30%	80%	
2035	15	0.005	0.40	0.10	5.30%	81%	
2036	14	0.005	0.41	0.09	5.30%	82%	
2037	13	0.005	0.41	0.09	5.30%	83%	
2038	12	0.005	0.42	0.08	5.30%	84%	
2039	12	0.004	0.42	0.08	5.30%	85%	
2040	11	0.004	0.43	0.07	5.30%	85%	
2041	11	0.004	0.43	0.07	5.30%	86%	
2042	10	0.004	0.43	0.07	5.30%	87%	
2043	10	0.003	0.44	0.06	5.30%	88%	
2044	9	0.003	0.44	0.06	5.30%	88%	
2045	9	0.003	0.44	0.06	5.30%	89%	
2046	8	0.003	0.45	0.05	5.30%	89%	
2047	8	0.003	0.45	0.05	5.30%	90%	
2048	7	0.003	0.45	0.05	5.30%	91%	
2049	7	0.003	0.46	0.04	5.30%	91%	
2050	7	0.002	0.46	0.04	5.30%	92%	

==

CUM.PROD	5.188 BBO	GIANT FIELDS			ROMANIA	
RESERVES	0.924 BBO	TOTAL BBO	0.8		SINO-SOVIET	
DISCOVERED	6.112 BBO	% OF DISC.	13%			
UNDISCOVERED	0.888 BBO	Last Giant	1890		----------------	
REMAINING	1.812 BBO	-------------------			Variance	Trend
ULTIMATE	7.000 BBO				1990	-8%
Rept.Reserve	0.880 #1	Ref.date(end)	1989		Future	DOWN

Milestones	Date	Prod Mb/d	Dep. Rate	% Produced	Forecast "A"
					A=Exponential decline
1st Peak	1934	171	1.05%	16%	B=5% incr. to Mid-Point
2nd Peak	1976	300	3.61%	58%	C=Other
Mid-Point	1971	281	2.88%	51%	Pre-1930 Cum. 0.86
Final Decline	1990		3.50%	75%	(Added to 1930)

Date	Prod. M.b/d	Prod. BBO/yr	Cum. Prod.	Remain	Deplet. Rate	% Produced	Giant	Notes & Names
1930	117	0.043	0.90	6.10	ERR	13%	0.80	Moreni-Gura
1931	136	0.050	0.95	6.05	0.82%	14%		Ocnetei (1890)
1932	147	0.054	1.01	5.99	0.89%	14%		
1933	148	0.054	1.06	5.94	0.90%	15%		Data-base weak
1934	171	0.062	1.12	5.88	1.05%	16%		1ST PEAK
1935	168	0.061	1.18	5.82	1.04%	17%		
1936	174	0.064	1.25	5.75	1.09%	18%		
1937	144	0.052	1.30	5.70	0.91%	19%		
1938	133	0.048	1.35	5.65	0.85%	19%		
1939	125	0.046	1.39	5.61	0.81%	20%		
1940	118	0.043	1.44	5.56	0.77%	21%		
1941	111	0.041	1.48	5.52	0.73%	21%		
1942	115	0.042	1.52	5.48	0.76%	22%		
1943	107	0.039	1.56	5.44	0.71%	22%		
1944	72	0.026	1.59	5.41	0.48%	23%		
1945	95	0.035	1.62	5.38	0.64%	23%		
1946	86	0.031	1.65	5.35	0.58%	24%		
1947	78	0.029	1.68	5.32	0.53%	24%		
1948	93	0.034	1.71	5.29	0.64%	24%		
1949	92	0.034	1.75	5.25	0.64%	25%		
1950	88	0.032	1.78	5.22	0.61%	25%		
1951	85	0.031	1.81	5.19	0.59%	26%		
1952	123	0.045	1.86	5.14	0.87%	27%		
1953	185	0.067	1.92	5.08	1.31%	27%		
1954	199	0.073	2.00	5.00	1.43%	29%		
1955	216	0.079	2.07	4.93	1.57%	30%		
1956	223	0.081	2.16	4.84	1.65%	31%		
1957	228	0.083	2.24	4.76	1.72%	32%		
1958	232	0.084	2.32	4.68	1.77%	33%		
1959	229	0.084	2.41	4.59	1.79%	34%		
1960	235	0.086	2.49	4.51	1.87%	36%		
1961	237	0.086	2.58	4.42	1.92%	37%		
1962	242	0.088	2.67	4.33	2.00%	38%		
1963	250	0.091	2.76	4.24	2.10%	39%		
1964	253	0.092	2.85	4.15	2.18%	41%		
1965	257	0.094	2.95	4.05	2.26%	42%		
1966	262	0.096	3.04	3.96	2.36%	43%		
1967	268	0.098	3.14	3.86	2.47%	45%		
1968	270	0.099	3.24	3.76	2.55%	46%		
1969	269	0.098	3.34	3.66	2.61%	48%		
1970	280	0.102	3.44	3.56	2.78%	49%		
1971	281	0.102	3.54	3.46	2.88%	51%		50% PRODUCED
1972	289	0.105	3.65	3.35	3.04%	52%		
1973	292	0.107	3.75	3.25	3.18%	54%		
1974	296	0.108	3.86	3.14	3.32%	55%		
1975	298	0.109	3.97	3.03	3.46%	57%		
1976	300	0.110	4.08	2.92	3.61%	58%		2ND PEAK
1977	299	0.109	4.19	2.81	3.74%	60%		
1978	274	0.100	4.29	2.71	3.56%	61%		
1979	260	0.095	4.38	2.62	3.50%	63%		
1980	230	0.084	4.47	2.53	3.21%	64%		
1981	238	0.087	4.55	2.45	3.43%	65%		
1982	235	0.086	4.64	2.36	3.51%	66%		
1983	240	0.088	4.73	2.27	3.71%	68%		
1984	234	0.085	4.81	2.19	3.75%	69%		
1985	222	0.081	4.89	2.11	3.70%	70%		
1986	226	0.082	4.98	2.02	3.91%	71%		
1987	215	0.078	5.05	1.95	3.88%	72%		
1988	189	0.069	5.12	1.88	3.54%	73%		

1989	180	0.066	5.19	1.81	3.50%	74%
1990	174	0.063	5.25	1.75	3.50%	75%
1991	168	0.061	5.31	1.69	3.50%	76%
1992	162	0.059	5.37	1.63	3.50%	77%
1993	156	0.057	5.43	1.57	3.50%	78%
1994	151	0.055	5.48	1.52	3.50%	78%
1995	145	0.053	5.54	1.46	3.50%	79%
1996	140	0.051	5.59	1.41	3.50%	80%
1997	135	0.049	5.64	1.36	3.50%	81%
1998	131	0.048	5.69	1.31	3.50%	81%
1999	126	0.046	5.73	1.27	3.50%	82%
2000	122	0.044	5.78	1.22	3.50%	83%
2001	117	0.043	5.82	1.18	3.50%	83%
2002	113	0.041	5.86	1.14	3.50%	84%
2003	109	0.040	5.90	1.10	3.50%	84%
2004	106	0.039	5.94	1.06	3.50%	85%
2005	102	0.037	5.98	1.02	3.50%	85%
2006	98	0.036	6.01	0.99	3.50%	86%
2007	95	0.035	6.05	0.95	3.50%	86%
2008	91	0.033	6.08	0.92	3.50%	87%
2009	88	0.032	6.11	0.89	3.50%	87%
2010	85	0.031	6.14	0.86	3.50%	88%
2011	82	0.030	6.17	0.83	3.50%	88%
2012	79	0.029	6.20	0.80	3.50%	89%
2013	77	0.028	6.23	0.77	3.50%	89%
2014	74	0.027	6.26	0.74	3.50%	89%
2015	71	0.026	6.28	0.72	3.50%	90%
2016	69	0.025	6.31	0.69	3.50%	90%
2017	66	0.024	6.33	0.67	3.50%	90%
2018	64	0.023	6.36	0.64	3.50%	91%
2019	62	0.023	6.38	0.62	3.50%	91%
2020	60	0.022	6.40	0.60	3.50%	91%
2021	58	0.021	6.42	0.58	3.50%	92%
2022	56	0.020	6.44	0.56	3.50%	92%
2023	54	0.020	6.46	0.54	3.50%	92%
2024	52	0.019	6.48	0.52	3.50%	93%
2025	50	0.018	6.50	0.50	3.50%	93%
2026	48	0.018	6.52	0.48	3.50%	93%
2027	46	0.017	6.53	0.47	3.50%	93%
2028	45	0.016	6.55	0.45	3.50%	94%
2029	43	0.016	6.56	0.44	3.50%	94%
2030	42	0.015	6.58	0.42	3.50%	94%
2031	40	0.015	6.59	0.41	3.50%	94%
2032	39	0.014	6.61	0.39	3.50%	94%
2033	38	0.014	6.62	0.38	3.50%	95%
2034	36	0.013	6.64	0.36	3.50%	95%
2035	35	0.013	6.65	0.35	3.50%	95%
2036	34	0.012	6.66	0.34	3.50%	95%
2037	33	0.012	6.67	0.33	3.50%	95%
2038	31	0.011	6.68	0.32	3.50%	95%
2039	30	0.011	6.69	0.31	3.50%	96%
2040	29	0.011	6.71	0.29	3.50%	96%
2041	28	0.010	6.72	0.28	3.50%	96%
2042	27	0.010	6.73	0.27	3.50%	96%
2043	26	0.010	6.74	0.26	3.50%	96%
2044	25	0.009	6.74	0.26	3.50%	96%
2045	24	0.009	6.75	0.25	3.50%	96%
2046	24	0.009	6.76	0.24	3.50%	97%
2047	23	0.008	6.77	0.23	3.50%	97%
2048	22	0.008	6.78	0.22	3.50%	97%
2049	21	0.008	6.79	0.21	3.50%	97%
2050	20	0.007	6.79	0.21	3.50%	97%

CUM.PROD	57.64 BBO	GIANT FIELDS		SAUDI ARABIA	
RESERVES	200.00 BBO	TOTAL BBO	210	M.EAST (MAJOR)	
DISCOVERED	257.64 BBO	% OF DISC.	81%	Scenario-2	
UNDISCOVERED	17.36 BBO	Last Giant	1975		
REMAINING	217.36 BBO			Revision Trend	
ULTIMATE	275.00 BBO			1990	n/a
Rept.Reserve	254.96	Ref.date(end)	1989	Future	DOWN

Milestones	Date	Prod Mb/d	Dep. Rate	% Produced
1st Peak	1981	9642	1.50%	16%
2nd Peak	2013	8902	2.16%	46%
Mid-Point	2016	8821	2.28%	50%
Final Decline	2017		2.32%	

Forecast "C"
A=Exponential decline
B=5% incr. to Mid-Point
C=Other
Pre-1930 Cum. 0
(Added to 1930 prod)

Date	Prod. M.b/d	Prod. BBO/yr	Cum. Prod.	Remain	Deplet. Rate	% Produced	Giant	Notes & Names
1930		0.000	0.00	275.00	ERR	0%		
1931		0.000	0.00	275.00	0.00%	0%		
1932		0.000	0.00	275.00	0.00%	0%		
1933		0.000	0.00	275.00	0.00%	0%		
1934		0.000	0.00	275.00	0.00%	0%		
1935		0.000	0.00	275.00	0.00%	0%		
1936		0.000	0.00	275.00	0.00%	0%		
1937		0.000	0.00	275.00	0.00%	0%		
1938	1	0.000	0.00	275.00	0.00%	0%	1.05	Dammam
1939	11	0.004	0.00	275.00	0.00%	0%		
1940	14	0.005	0.01	274.99	0.00%	0%	1.76	Abu Hadriya
1941	12	0.004	0.01	274.99	0.00%	0%	12.80	Abqaiq
1942	12	0.005	0.02	274.98	0.00%	0%		
1943	13	0.005	0.02	274.98	0.00%	0%		
1944	21	0.008	0.03	274.97	0.00%	0%		
1945	58	0.021	0.05	274.95	0.01%	0%	3.20	Qatif
1946	164	0.060	0.11	274.89	0.02%	0%		
1947	246	0.090	0.20	274.80	0.03%	0%		
1948	390	0.142	0.34	274.66	0.05%	0%	83.00	Ghawar
1949	477	0.174	0.52	274.48	0.06%	0%	0.95	Fadhili
1950	547	0.200	0.72	274.28	0.07%	0%		
1951	762	0.278	1.00	274.00	0.10%	0%	32.30	Safaniya
1952	840	0.307	1.30	273.70	0.11%	0%		
1953	835	0.305	1.61	273.39	0.11%	1%		
1954	952	0.347	1.95	273.05	0.13%	1%		
1955	951	0.347	2.30	272.70	0.13%	1%		
1956	1000	0.365	2.67	272.33	0.13%	1%	2.30	Khursaniyah
1957	990	0.361	3.03	271.97	0.13%	1%	25.80	Khurais,Manifa
1958	991	0.362	3.39	271.61	0.13%	1%		
1959	1070	0.391	3.78	271.22	0.14%	1%		
1960	1240	0.453	4.23	270.77	0.17%	2%		
1961	1377	0.503	4.74	270.26	0.19%	2%		
1962	1523	0.556	5.29	269.71	0.21%	2%		
1963	1618	0.591	5.88	269.12	0.22%	2%	7.81	Abu Sa'Fah
1964	1696	0.619	6.50	268.50	0.23%	2%	7.30	Berri
1965	1996	0.729	7.23	267.77	0.27%	3%	10.64	Zulut
1966	2374	0.867	8.10	266.90	0.32%	3%	10.00	Fereidoon-Marjan
1967	2571	0.938	9.03	265.97	0.35%	3%		
1968	2843	1.038	10.07	264.93	0.39%	4%	5.71	Shaybah
1969	2915	1.064	11.14	263.86	0.40%	4%	0.50	Bargan
1970	3438	1.255	12.39	262.61	0.48%	5%		
1971	4456	1.626	14.02	260.98	0.62%	5%	0.63	Mazalij
1972	5255	1.918	15.94	259.06	0.73%	6%	1.03	Harmaliyah
1973	7418	2.708	18.64	256.36	1.05%	7%	1.6	Maharah,Abu Jiff
1974	8400	3.066	21.71	253.29	1.20%	8%		
1975	7000	2.555	24.26	250.74	1.01%	9%	1.17	Lawnah
1976	8570	3.128	27.39	247.61	1.25%	10%		
1977	8950	3.267	30.66	244.34	1.32%	11%		
1978	7800	2.847	33.51	241.49	1.17%	12%		
1979	9250	3.376	36.88	238.12	1.40%	13%		
1980	9620	3.511	40.39	234.61	1.47%	15%		
1981	9642	3.519	43.91	231.09	1.50%	16%		
1982	6484	2.367	46.28	228.72	1.02%	17%		
1983	4872	1.778	48.06	226.94	0.78%	17%		
1984	4545	1.659	49.72	225.28	0.73%	18%		
1985	3295	1.203	50.92	224.08	0.53%	19%		
1986	4720	1.723	52.64	222.36	0.77%	19%		
1987	4054	1.480	54.12	220.88	0.67%	20%		
1988	4708	1.719	55.84	219.16	0.78%	20%		Swing Producer

1989	4936	1.801	57.64	217.36	0.82%	21%	#1
1990	5606	2.046	59.69	215.31	0.94%	22%	
1991	6248	2.280	61.97	213.03	1.06%	23%	
1992	6862	2.505	64.47	210.53	1.18%	23%	
1993	7090	2.588	67.06	207.94	1.23%	24%	
1994	7255	2.648	69.71	205.29	1.27%	25%	
1995	7460	2.723	72.43	202.57	1.33%	26%	
1996	7880	2.876	75.31	199.69	1.42%	27%	
1997	8309	3.033	78.34	196.66	1.52%	28%	
1998	8150	2.975	81.32	193.68	1.51%	30%	
1999	8035	2.933	84.25	190.75	1.51%	31%	
2000	7974	2.911	87.16	187.84	1.53%	32%	
2001	7959	2.905	90.06	184.94	1.55%	33%	
2002	8028	2.930	92.99	182.01	1.58%	34%	
2003	8244	3.009	96.00	179.00	1.65%	35%	
2004	8424	3.075	99.08	175.92	1.72%	36%	
2005	8338	3.043	102.12	172.88	1.73%	37%	
2006	8449	3.084	105.21	169.79	1.78%	38%	
2007	8530	3.114	108.32	166.68	1.83%	39%	
2008	8584	3.133	111.45	163.55	1.88%	41%	
2009	8623	3.147	114.60	160.40	1.92%	42%	
2010	8737	3.189	117.79	157.21	1.99%	43%	
2011	8822	3.220	121.01	153.99	2.05%	44%	
2012	8876	3.240	124.25	150.75	2.10%	45%	
2013	8902	3.249	127.50	147.50	2.16%	46%	
2014	8901	3.249	130.75	144.25	2.20%	48%	
2015	8873	3.239	133.98	141.02	2.25%	49%	
2016	8821	3.220	137.20	137.80	2.28%	50%	
2017	8746	3.192	140.40	134.60	2.32%	51%	
2018	8652	3.158	143.55	131.45	2.35%	52%	
2019	8538	3.117	146.67	128.33	2.37%	53%	
2020	8405	3.068	149.74	125.26	2.39%	54%	
2021	8253	3.012	152.75	122.25	2.40%	56%	
2022	8083	2.950	155.70	119.30	2.41%	57%	
2023	7815	2.853	158.55	116.45	2.39%	58%	
2024	7846	2.864	161.42	113.58	2.46%	59%	
2025	7858	2.868	164.29	110.71	2.53%	60%	
2026	7853	2.866	167.15	107.85	2.59%	61%	
2027	7830	2.858	170.01	104.99	2.65%	62%	
2028	7792	2.844	172.86	102.14	2.71%	63%	
2029	7738	2.825	175.68	99.32	2.77%	64%	
2030	7689	2.806	178.49	96.51	2.83%	65%	
2031	7616	2.780	181.27	93.73	2.88%	66%	
2032	7607	2.777	184.04	90.96	2.96%	67%	
2033	7588	2.770	186.81	88.19	3.04%	68%	
2034	7558	2.759	189.57	85.43	3.13%	69%	
2035	7519	2.744	192.32	82.68	3.21%	70%	
2036	7453	2.720	195.04	79.96	3.29%	71%	
2037	7386	2.696	197.73	77.27	3.37%	72%	
2038	7311	2.669	200.40	74.60	3.45%	73%	
2039	7229	2.639	203.04	71.96	3.54%	74%	
2040	7140	2.606	205.65	69.35	3.62%	75%	
2041	7046	2.572	208.22	66.78	3.71%	76%	
2042	6946	2.535	210.75	64.25	3.80%	77%	
2043	6842	2.497	213.25	61.75	3.89%	78%	
2044	6734	2.458	215.71	59.29	3.98%	78%	
2045	6622	2.417	218.12	56.88	4.08%	79%	
2046	6508	2.375	220.50	54.50	4.18%	80%	
2047	6391	2.333	222.83	52.17	4.28%	81%	
2048	6272	2.289	225.12	49.88	4.39%	82%	
2049	6151	2.245	227.37	47.63	4.50%	83%	
2050	6029	2.201	229.57	45.43	4.62%	83%	

Note #1
Each M-East Six country is assumed to produce a share of the region in the same proportion as was the case in 1989 until it reaches its Mid-Point, when it declines at its then Depletion Rate, except in the case of Saudi Arabia which acts as a Swing Producer, making up the difference between what the other five produce and the region's total.

CUM.PROD	0.237 BBO	GIANT FIELDS		SHARJAH	
RESERVES	1.500 BBO	TOTAL BBO	0	M.EAST (MINOR)	
DISCOVERED	1.737 BBO	% OF DISC.	0%		
UNDISCOVERED	0.263 BBO	Last Giant	-		
REMAINING	1.763 BBO			Variance	Trend
ULTIMATE	2.000 BBO			1990	-48%
Rept.Reserve	1.500	Ref.date(end)	1989	Future	DOWN

Milestones	Date	Prod Mb/d	Dep. Rate	% Produced
1st Peak	2008	164	5.65%	50%
2nd Peak				
Mid-Point	2008	164	5.65%	50%
Final Decline	2008		5.65%	50%

Forecast "B"
A=Exponential decline
B=5% incr. to Mid-Point
C=Other
Pre-1930 Cum. 0

Date	Prod. M.b/d	Prod. BBO/yr	Cum. Prod.	Remain	Deplet. Rate	% Produced	Giant	Notes & Names
1930		0.000	0.00	2.00	ERR	0%		
1931		0.000	0.00	2.00	0.00%	0%		
1932		0.000	0.00	2.00	0.00%	0%		
1933		0.000	0.00	2.00	0.00%	0%		
1934		0.000	0.00	2.00	0.00%	0%		
1935		0.000	0.00	2.00	0.00%	0%		
1936		0.000	0.00	2.00	0.00%	0%		
1937		0.000	0.00	2.00	0.00%	0%		
1938		0.000	0.00	2.00	0.00%	0%		
1939		0.000	0.00	2.00	0.00%	0%		
1940		0.000	0.00	2.00	0.00%	0%		
1941		0.000	0.00	2.00	0.00%	0%		
1942		0.000	0.00	2.00	0.00%	0%		
1943		0.000	0.00	2.00	0.00%	0%		
1944		0.000	0.00	2.00	0.00%	0%		
1945		0.000	0.00	2.00	0.00%	0%		
1946		0.000	0.00	2.00	0.00%	0%		
1947		0.000	0.00	2.00	0.00%	0%		
1948		0.000	0.00	2.00	0.00%	0%		
1949		0.000	0.00	2.00	0.00%	0%		
1950		0.000	0.00	2.00	0.00%	0%		
1951		0.000	0.00	2.00	0.00%	0%		
1952		0.000	0.00	2.00	0.00%	0%		
1953		0.000	0.00	2.00	0.00%	0%		
1954		0.000	0.00	2.00	0.00%	0%		
1955		0.000	0.00	2.00	0.00%	0%		
1956		0.000	0.00	2.00	0.00%	0%		
1957		0.000	0.00	2.00	0.00%	0%		
1958		0.000	0.00	2.00	0.00%	0%		
1959		0.000	0.00	2.00	0.00%	0%		
1960		0.000	0.00	2.00	0.00%	0%		
1961		0.000	0.00	2.00	0.00%	0%		
1962		0.000	0.00	2.00	0.00%	0%		
1963		0.000	0.00	2.00	0.00%	0%		
1964		0.000	0.00	2.00	0.00%	0%		
1965		0.000	0.00	2.00	0.00%	0%		
1966		0.000	0.00	2.00	0.00%	0%		
1967		0.000	0.00	2.00	0.00%	0%		
1968		0.000	0.00	2.00	0.00%	0%		
1969		0.000	0.00	2.00	0.00%	0%		
1970		0.000	0.00	2.00	0.00%	0%		
1971		0.000	0.00	2.00	0.00%	0%		
1972		0.000	0.00	2.00	0.00%	0%		
1973		0.000	0.00	2.00	0.00%	0%		
1974	50	0.018	0.02	1.98	0.91%	1%		
1975	40	0.015	0.03	1.97	0.74%	2%		
1976	40	0.015	0.05	1.95	0.74%	2%		
1977	30	0.011	0.06	1.94	0.56%	3%		
1978	24	0.009	0.07	1.93	0.45%	3%		
1979	15	0.005	0.07	1.93	0.28%	4%		
1980	10	0.004	0.08	1.92	0.19%	4%		
1981	9	0.003	0.08	1.92	0.17%	4%		
1982	7	0.003	0.08	1.92	0.13%	4%		
1983	35	0.013	0.09	1.91	0.67%	5%		
1984	62	0.023	0.12	1.88	1.19%	6%		
1985	66	0.024	0.14	1.86	1.28%	7%		
1986	65	0.024	0.17	1.83	1.28%	8%		
1987	67	0.024	0.19	1.81	1.33%	9%		
1988	65	0.024	0.21	1.79	1.31%	11%		

1989	65	0.024	0.24	1.76	1.33%	12%	
1990	68	0.025	0.26	1.74	1.41%	13%	
1991	72	0.026	0.29	1.71	1.51%	14%	
1992	75	0.027	0.32	1.68	1.60%	16%	
1993	79	0.029	0.34	1.66	1.71%	17%	
1994	83	0.030	0.37	1.63	1.83%	19%	
1995	87	0.032	0.41	1.59	1.96%	20%	
1996	91	0.033	0.44	1.56	2.10%	22%	
1997	96	0.035	0.48	1.52	2.25%	24%	
1998	101	0.037	0.51	1.49	2.41%	26%	
1999	106	0.039	0.55	1.45	2.60%	28%	
2000	111	0.041	0.59	1.41	2.80%	30%	
2001	117	0.043	0.63	1.37	3.02%	32%	
2002	123	0.045	0.68	1.32	3.27%	34%	
2003	129	0.047	0.73	1.27	3.55%	36%	
2004	135	0.049	0.77	1.23	3.87%	39%	
2005	142	0.052	0.83	1.17	4.23%	41%	
2006	149	0.054	0.88	1.12	4.63%	44%	
2007	156	0.057	0.94	1.06	5.10%	47%	
2008	164	0.060	1.00	1.00	5.65%	50%	1ST PEAK & 50% PRODUCED
2009	155	0.057	1.05	0.95	5.65%	53%	
2010	146	0.053	1.11	0.89	5.65%	55%	
2011	138	0.050	1.16	0.84	5.65%	58%	
2012	130	0.048	1.21	0.79	5.65%	60%	
2013	123	0.045	1.25	0.75	5.65%	63%	
2014	116	0.042	1.29	0.71	5.65%	65%	
2015	109	0.040	1.33	0.67	5.65%	67%	
2016	103	0.038	1.37	0.63	5.65%	69%	
2017	97	0.036	1.41	0.59	5.65%	70%	
2018	92	0.034	1.44	0.56	5.65%	72%	
2019	87	0.032	1.47	0.53	5.65%	74%	
2020	82	0.030	1.50	0.50	5.65%	75%	
2021	77	0.028	1.53	0.47	5.65%	76%	
2022	73	0.027	1.56	0.44	5.65%	78%	
2023	69	0.025	1.58	0.42	5.65%	79%	
2024	65	0.024	1.60	0.40	5.65%	80%	
2025	61	0.022	1.63	0.37	5.65%	81%	
2026	58	0.021	1.65	0.35	5.65%	82%	
2027	54	0.020	1.67	0.33	5.65%	83%	
2028	51	0.019	1.69	0.31	5.65%	84%	
2029	48	0.018	1.70	0.30	5.65%	85%	
2030	46	0.017	1.72	0.28	5.65%	86%	
2031	43	0.016	1.74	0.26	5.65%	87%	
2032	41	0.015	1.75	0.25	5.65%	88%	
2033	38	0.014	1.77	0.23	5.65%	88%	
2034	36	0.013	1.78	0.22	5.65%	89%	
2035	34	0.012	1.79	0.21	5.65%	90%	
2036	32	0.012	1.80	0.20	5.65%	90%	
2037	30	0.011	1.81	0.19	5.65%	91%	
2038	29	0.010	1.82	0.18	5.65%	91%	
2039	27	0.010	1.83	0.17	5.65%	92%	
2040	26	0.009	1.84	0.16	5.65%	92%	
2041	24	0.009	1.85	0.15	5.65%	93%	
2042	23	0.008	1.86	0.14	5.65%	93%	
2043	21	0.008	1.87	0.13	5.65%	93%	
2044	20	0.007	1.88	0.12	5.65%	94%	
2045	19	0.007	1.88	0.12	5.65%	94%	
2046	18	0.007	1.89	0.11	5.65%	95%	
2047	17	0.006	1.90	0.10	5.65%	95%	
2048	16	0.006	1.90	0.10	5.65%	95%	
2049	15	0.006	1.91	0.09	5.65%	95%	
2050	14	0.005	1.91	0.09	5.65%	96%	

CUM.PROD	0.195 BBO	GIANT FIELDS		SPAIN	
RESERVES	0.025 BBO	TOTAL BBO	0	W.EUROPE	
DISCOVERED	0.220 BBO	% OF DISC.	0%		
UNDISCOVERED	0.080 BBO	Last Giant	-		
REMAINING	0.105 BBO			Variance	Trend
ULTIMATE	0.300 BBO			1990	-15%
Rept.Reserve	0.025 BBO	Ref.Date(end)	1989	Future	DOWN

Milestones	Date	Prod Mb/d	Dep. Rate	% Produced
1st Peak	1983	59	10.64%	40%
2nd Peak				
Mid-Point	1985	45	9.98%	51%
Final Decline	1990		5.90%	

Forecast "A'
A=Exponential decline
B=5% incr. to Mid-Point
C=Other
Pre-1930 Cum. 0
(Added to 1930 Prod)

Date	Prod. M.b/d	Prod. BBO/yr	Cum Prod	Remain	Dep. Rate	% Produced	Giant	Names & Notes
1930		0.000	0.00	0.30	ERR	0%		
1931		0.000	0.00	0.30	0.00%	0%		
1932		0.000	0.00	0.30	0.00%	0%		
1933		0.000	0.00	0.30	0.00%	0%		
1934		0.000	0.00	0.30	0.00%	0%		
1935		0.000	0.00	0.30	0.00%	0%		
1936		0.000	0.00	0.30	0.00%	0%		
1937		0.000	0.00	0.30	0.00%	0%		
1938		0.000	0.00	0.30	0.00%	0%		
1939		0.000	0.00	0.30	0.00%	0%		
1940		0.000	0.00	0.30	0.00%	0%		
1941		0.000	0.00	0.30	0.00%	0%		
1942		0.000	0.00	0.30	0.00%	0%		
1943		0.000	0.00	0.30	0.00%	0%		
1944		0.000	0.00	0.30	0.00%	0%		
1945		0.000	0.00	0.30	0.00%	0%		
1946		0.000	0.00	0.30	0.00%	0%		
1947		0.000	0.00	0.30	0.00%	0%		
1948		0.000	0.00	0.30	0.00%	0%		
1949		0.000	0.00	0.30	0.00%	0%		
1950		0.000	0.00	0.30	0.00%	0%		
1951		0.000	0.00	0.30	0.00%	0%		
1952		0.000	0.00	0.30	0.00%	0%		
1953		0.000	0.00	0.30	0.00%	0%		
1954		0.000	0.00	0.30	0.00%	0%		
1955		0.000	0.00	0.30	0.00%	0%		
1956		0.000	0.00	0.30	0.00%	0%		
1957		0.000	0.00	0.30	0.00%	0%		
1958		0.000	0.00	0.30	0.00%	0%		
1959		0.000	0.00	0.30	0.00%	0%		
1960		0.000	0.00	0.30	0.00%	0%		
1961		0.000	0.00	0.30	0.00%	0%		
1962		0.000	0.00	0.30	0.00%	0%		
1963		0.000	0.00	0.30	0.00%	0%		
1964		0.000	0.00	0.30	0.00%	0%		
1965		0.000	0.00	0.30	0.00%	0%		
1966		0.000	0.00	0.30	0.00%	0%		
1967		0.000	0.00	0.30	0.00%	0%		
1968		0.000	0.00	0.30	0.00%	0%		
1969		0.000	0.00	0.30	0.00%	0%		
1970		0.000	0.00	0.30	0.00%	0%		
1971		0.000	0.00	0.30	0.00%	0%		
1972		0.000	0.00	0.30	0.00%	0%		
1973	22	0.008	0.01	0.29	2.70%	3%		
1974	38	0.014	0.02	0.28	4.75%	7%		
1975	38	0.014	0.04	0.26	4.99%	12%		
1976	34	0.012	0.05	0.25	4.70%	16%		
1977	21	0.008	0.06	0.24	3.04%	19%		
1978	20	0.007	0.06	0.24	2.99%	21%		
1979	15	0.005	0.07	0.23	2.31%	23%		
1980	25	0.009	0.08	0.22	3.94%	26%		
1981	24	0.009	0.09	0.21	3.94%	29%		
1982	30	0.011	0.10	0.20	5.13%	33%		
1983	59	0.022	0.12	0.18	10.64%	40%		1st PEAK
1984	45	0.016	0.14	0.16	9.08%	45%		
1985	45	0.016	0.15	0.15	9.98%	51%		50% PRODUCED
1986	37	0.014	0.17	0.13	9.17%	55%		
1987	33	0.012	0.18	0.12	8.90%	59%		
1988	30	0.011	0.19	0.11	9.06%	63%		

1989	18	0.007	0.20	0.10	5.90%	65%
1990	17	0.006	0.20	0.10	5.90%	67%
1991	16	0.006	0.21	0.09	5.90%	69%
1992	15	0.005	0.21	0.09	5.90%	71%
1993	14	0.005	0.22	0.08	5.90%	73%
1994	13	0.005	0.22	0.08	5.90%	74%
1995	13	0.005	0.23	0.07	5.90%	76%
1996	12	0.004	0.23	0.07	5.90%	77%
1997	11	0.004	0.24	0.06	5.90%	79%
1998	10	0.004	0.24	0.06	5.90%	80%
1999	10	0.004	0.24	0.06	5.90%	81%
2000	9	0.003	0.25	0.05	5.90%	82%
2001	9	0.003	0.25	0.05	5.90%	83%
2002	8	0.003	0.25	0.05	5.90%	84%
2003	8	0.003	0.26	0.04	5.90%	85%
2004	7	0.003	0.26	0.04	5.90%	86%
2005	7	0.002	0.26	0.04	5.90%	87%
2006	6	0.002	0.26	0.04	5.90%	88%
2007	6	0.002	0.26	0.04	5.90%	88%
2008	6	0.002	0.27	0.03	5.90%	89%
2009	5	0.002	0.27	0.03	5.90%	90%
2010	5	0.002	0.27	0.03	5.90%	90%
2011	5	0.002	0.27	0.03	5.90%	91%
2012	4	0.002	0.27	0.03	5.90%	91%
2013	4	0.002	0.28	0.02	5.90%	92%
2014	4	0.001	0.28	0.02	5.90%	92%
2015	4	0.001	0.28	0.02	5.90%	93%
2016	3	0.001	0.28	0.02	5.90%	93%
2017	3	0.001	0.28	0.02	5.90%	94%
2018	3	0.001	0.28	0.02	5.90%	94%
2019	3	0.001	0.28	0.02	5.90%	94%
2020	3	0.001	0.28	0.02	5.90%	95%
2021	3	0.001	0.29	0.01	5.90%	95%
2022	2	0.001	0.29	0.01	5.90%	95%
2023	2	0.001	0.29	0.01	5.90%	96%
2024	2	0.001	0.29	0.01	5.90%	96%
2025	2	0.001	0.29	0.01	5.90%	96%
2026	2	0.001	0.29	0.01	5.90%	96%
2027	2	0.001	0.29	0.01	5.90%	97%
2028	2	0.001	0.29	0.01	5.90%	97%
2029	2	0.001	0.29	0.01	5.90%	97%
2030	1	0.001	0.29	0.01	5.90%	97%
2031	1	0.001	0.29	0.01	5.90%	97%
2032	1	0.000	0.29	0.01	5.90%	97%
2033	1	0.000	0.29	0.01	5.90%	98%
2034	1	0.000	0.29	0.01	5.90%	98%
2035	1	0.000	0.29	0.01	5.90%	98%
2036	1	0.000	0.29	0.01	5.90%	98%
2037	1	0.000	0.29	0.01	5.90%	98%
2038	1	0.000	0.29	0.01	5.90%	98%
2039	1	0.000	0.29	0.01	5.90%	98%
2040	1	0.000	0.30	0.00	5.90%	98%
2041	1	0.000	0.30	0.00	5.90%	99%
2042	1	0.000	0.30	0.00	5.90%	99%
2043	1	0.000	0.30	0.00	5.90%	99%
2044	1	0.000	0.30	0.00	5.90%	99%
2045	1	0.000	0.30	0.00	5.90%	99%
2046	1	0.000	0.30	0.00	5.90%	99%
2047	1	0.000	0.30	0.00	5.90%	99%
2048	0	0.000	0.30	0.00	5.90%	99%
2049	0	0.000	0.30	0.00	5.90%	99%
2050	0	0.000	0.30	0.00	5.90%	99%

CUM.PROD	0.000 BBO	GIANT FIELDS			SUDAN
RESERVES	0.300 BBO	TOTAL BBO	0.0		AFRICA
DISCOVERED	0.300 BBO	% OF DISC.	0%		
UNDISCOVERED	0.200 BBO	Last Giant			
REMAINING	0.500 BBO				Variance Trend
ULTIMATE	0.500 BBO				1990 N/A
Rept.Reserves	0.300 BBO	Ref.Date (end)		1989	Future UP

Milestones	Date	Prod Mb/d	Dep Rate	% Produced	Forecast "C"
1st Peak	2029	68	8.01%	57%	A=Exponential decline
2nd Peak					B=5% inc. to Mid-Point
					C=Other
Mid-Point	2028	62	8.01%	53%	Pre-1930 Cum 0
Final Decline	2028		8.01%		(Added to 1930 Prod)

Date	Prod. M.b/d	Prod. BBO/yr	Cum Prod	Remain	Dep. Rate	% Produced	Giant Names & Notes
1930		0.000	0.00	0.50	ERR	0%	
1931		0.000	0.00	0.50	0.00%	0%	
1932		0.000	0.00	0.50	0.00%	0%	
1933		0.000	0.00	0.50	0.00%	0%	
1934		0.000	0.00	0.50	0.00%	0%	
1935		0.000	0.00	0.50	0.00%	0%	
1936		0.000	0.00	0.50	0.00%	0%	
1937		0.000	0.00	0.50	0.00%	0%	
1938		0.000	0.00	0.50	0.00%	0%	
1939		0.000	0.00	0.50	0.00%	0%	
1940		0.000	0.00	0.50	0.00%	0%	
1941		0.000	0.00	0.50	0.00%	0%	
1942		0.000	0.00	0.50	0.00%	0%	
1943		0.000	0.00	0.50	0.00%	0%	
1944		0.000	0.00	0.50	0.00%	0%	
1945		0.000	0.00	0.50	0.00%	0%	
1946		0.000	0.00	0.50	0.00%	0%	
1947		0.000	0.00	0.50	0.00%	0%	
1948		0.000	0.00	0.50	0.00%	0%	
1949		0.000	0.00	0.50	0.00%	0%	
1950		0.000	0.00	0.50	0.00%	0%	
1951		0.000	0.00	0.50	0.00%	0%	
1952		0.000	0.00	0.50	0.00%	0%	
1953		0.000	0.00	0.50	0.00%	0%	
1954		0.000	0.00	0.50	0.00%	0%	
1955		0.000	0.00	0.50	0.00%	0%	
1956		0.000	0.00	0.50	0.00%	0%	
1957		0.000	0.00	0.50	0.00%	0%	
1958		0.000	0.00	0.50	0.00%	0%	
1959		0.000	0.00	0.50	0.00%	0%	
1960		0.000	0.00	0.50	0.00%	0%	
1961		0.000	0.00	0.50	0.00%	0%	
1962		0.000	0.00	0.50	0.00%	0%	
1963		0.000	0.00	0.50	0.00%	0%	
1964		0.000	0.00	0.50	0.00%	0%	
1965		0.000	0.00	0.50	0.00%	0%	
1966		0.000	0.00	0.50	0.00%	0%	
1967		0.000	0.00	0.50	0.00%	0%	
1968		0.000	0.00	0.50	0.00%	0%	
1969		0.000	0.00	0.50	0.00%	0%	
1970		0.000	0.00	0.50	0.00%	0%	
1971		0.000	0.00	0.50	0.00%	0%	
1972		0.000	0.00	0.50	0.00%	0%	
1973		0.000	0.00	0.50	0.00%	0%	
1974		0.000	0.00	0.50	0.00%	0%	
1975		0.000	0.00	0.50	0.00%	0%	
1976		0.000	0.00	0.50	0.00%	0%	
1977		0.000	0.00	0.50	0.00%	0%	
1978		0.000	0.00	0.50	0.00%	0%	
1979		0.000	0.00	0.50	0.00%	0%	
1980		0.000	0.00	0.50	0.00%	0%	
1981		0.000	0.00	0.50	0.00%	0%	
1982		0.000	0.00	0.50	0.00%	0%	
1983		0.000	0.00	0.50	0.00%	0%	
1984		0.000	0.00	0.50	0.00%	0%	
1985		0.000	0.00	0.50	0.00%	0%	
1986		0.000	0.00	0.50	0.00%	0%	
1987		0.000	0.00	0.50	0.00%	0%	
1988		0.000	0.00	0.50	0.00%	0%	

1989		0.000	0.00	0.50	0.00%	0%	
1990		0.000	0.00	0.50	0.00%	0%	
1991	2.0	0.001	0.00	0.50	0.15%	0%	Prod. to commence & increase at 10%/yr
1992	2.0	0.001	0.00	0.50	0.16%	0%	
1993	2.2	0.001	0.00	0.50	0.18%	0%	
1994	2.4	0.001	0.00	0.50	0.20%	1%	
1995	2.7	0.001	0.00	0.50	0.22%	1%	
1996	2.9	0.001	0.00	0.49	0.24%	1%	
1997	3.2	0.001	0.01	0.49	0.26%	1%	
1998	3.5	0.001	0.01	0.49	0.29%	2%	
1999	3.9	0.002	0.01	0.49	0.32%	2%	
2000	4.3	0.002	0.01	0.49	0.35%	2%	
2001	4.7	0.002	0.01	0.49	0.39%	3%	
2002	5.2	0.002	0.01	0.48	0.43%	3%	
2003	5.7	0.002	0.02	0.48	0.47%	3%	
2004	6.3	0.003	0.02	0.48	0.52%	4%	
2005	6.9	0.003	0.02	0.48	0.58%	4%	
2006	7.6	0.003	0.03	0.47	0.64%	5%	
2007	8.4	0.003	0.03	0.47	0.71%	6%	
2008	9.2	0.004	0.03	0.47	0.78%	7%	
2009	10.1	0.004	0.04	0.46	0.87%	7%	
2010	11.1	0.004	0.04	0.46	0.97%	8%	
2011	12.2	0.005	0.05	0.45	1.07%	9%	
2012	13.5	0.005	0.05	0.45	1.19%	10%	
2013	14.8	0.006	0.06	0.44	1.33%	11%	
2014	16.3	0.007	0.06	0.44	1.48%	13%	
2015	17.9	0.007	0.07	0.43	1.65%	14%	
2016	19.7	0.008	0.08	0.42	1.85%	16%	
2017	21.7	0.009	0.09	0.41	2.07%	18%	
2018	23.8	0.010	0.10	0.40	2.33%	19%	
2019	26.2	0.011	0.11	0.39	2.62%	22%	
2020	28.8	0.012	0.12	0.38	2.96%	24%	
2021	31.7	0.013	0.13	0.37	3.35%	26%	
2022	34.9	0.014	0.15	0.35	3.82%	29%	
2023	38.4	0.015	0.16	0.34	4.36%	32%	
2024	42.2	0.017	0.18	0.32	5.02%	36%	
2025	46.5	0.019	0.20	0.30	5.81%	39%	
2026	51.1	0.021	0.22	0.28	6.79%	44%	
2027	56.2	0.023	0.24	0.26	8.01%	48%	
2028	61.8	0.025	0.27	0.23	8.01%	53%	50% PRODUCED
2029	68.0	0.019	0.28	0.22	8.01%	57%	1ST PEAK
2030	51.4	0.017	0.30	0.20	8.01%	60%	
2031	47.3	0.016	0.32	0.18	8.01%	63%	
2032	43.5	0.015	0.33	0.17	8.01%	66%	
2033	40.0	0.013	0.34	0.15	8.01%	69%	
2034	36.8	0.012	0.36	0.14	8.01%	71%	
2035	33.9	0.011	0.37	0.13	8.01%	74%	
2036	31.2	0.010	0.38	0.12	8.01%	76%	
2037	28.7	0.010	0.39	0.11	8.01%	78%	
2038	26.4	0.009	0.40	0.10	8.01%	80%	
2039	24.2	0.008	0.41	0.09	8.01%	81%	
2040	22.3	0.007	0.41	0.09	8.01%	83%	
2041	20.5	0.007	0.42	0.08	8.01%	84%	
2042	18.9	0.006	0.43	0.07	8.01%	85%	
2043	17.4	0.006	0.43	0.07	8.01%	86%	
2044	16.0	0.005	0.44	0.06	8.01%	88%	
2045	14.7	0.005	0.44	0.06	8.01%	89%	
2046	13.5	0.005	0.45	0.05	8.01%	89%	
2047	12.4	0.004	0.45	0.05	8.01%	90%	
2048	11.4	0.004	0.46	0.04	8.01%	91%	
2049	10.5	0.004	0.46	0.04	8.01%	92%	
2050	9.7	0.003	0.46	0.04	8.01%	92%	

CUM.PROD	0.004 BBO	GIANT FIELDS			SURINAME	
RESERVES	0.028 BBO	TOTAL BBO	0		L.AMERICA	
DISCOVERED	0.032 BBO	% OF DISC.	0%			
UNDISCOVERED	0.018 BBO	Last Giant	1868			
REMAINING	0.046 BBO				Variance	Trend
ULTIMATE	0.050 BBO				1990	+2%
Rept.Reserves	0.028 BBO	Ref. Date (end)		1989	Future	DOWN

Milestones	Date	Prod Mb/d	Dep. Rate	% Produced	Forecast "B"
1st Peak	2000	6	8.29%	49%	A=Exponential decline
2nd Peak					B=5% inc. to Mid-Point
					C=Other
Mid-Point	2001	6	8.29%	53%	Pre-1930 Cum 0
Final Decline	2001		8.29%		(added to 1930 Prod)

Date	Prod. M.b/d	Prod. BBO/yr	Cum Prod	Remain	Dep. Rate	% Produced	Giant Names & Notes
1930		0.000	0.00	0.05	ERR	0%	
1931		0.000	0.00	0.05	0.00%	0%	
1932		0.000	0.00	0.05	0.00%	0%	
1933		0.000	0.00	0.05	0.00%	0%	
1934		0.000	0.00	0.05	0.00%	0%	
1935		0.000	0.00	0.05	0.00%	0%	
1936		0.000	0.00	0.05	0.00%	0%	
1937		0.000	0.00	0.05	0.00%	0%	
1938		0.000	0.00	0.05	0.00%	0%	
1939		0.000	0.00	0.05	0.00%	0%	
1940		0.000	0.00	0.05	0.00%	0%	
1941		0.000	0.00	0.05	0.00%	0%	
1942		0.000	0.00	0.05	0.00%	0%	
1943		0.000	0.00	0.05	0.00%	0%	
1944		0.000	0.00	0.05	0.00%	0%	
1945		0.000	0.00	0.05	0.00%	0%	
1946		0.000	0.00	0.05	0.00%	0%	
1947		0.000	0.00	0.05	0.00%	0%	
1948		0.000	0.00	0.05	0.00%	0%	
1949		0.000	0.00	0.05	0.00%	0%	
1950		0.000	0.00	0.05	0.00%	0%	
1951		0.000	0.00	0.05	0.00%	0%	
1952		0.000	0.00	0.05	0.00%	0%	
1953		0.000	0.00	0.05	0.00%	0%	
1954		0.000	0.00	0.05	0.00%	0%	
1955		0.000	0.00	0.05	0.00%	0%	
1956		0.000	0.00	0.05	0.00%	0%	
1957		0.000	0.00	0.05	0.00%	0%	
1958		0.000	0.00	0.05	0.00%	0%	
1959		0.000	0.00	0.05	0.00%	0%	
1960		0.000	0.00	0.05	0.00%	0%	
1961		0.000	0.00	0.05	0.00%	0%	
1962		0.000	0.00	0.05	0.00%	0%	
1963		0.000	0.00	0.05	0.00%	0%	
1964		0.000	0.00	0.05	0.00%	0%	
1965		0.000	0.00	0.05	0.00%	0%	
1966		0.000	0.00	0.05	0.00%	0%	
1967		0.000	0.00	0.05	0.00%	0%	
1968		0.000	0.00	0.05	0.00%	0%	
1969		0.000	0.00	0.05	0.00%	0%	
1970		0.000	0.00	0.05	0.00%	0%	
1971		0.000	0.00	0.05	0.00%	0%	
1972		0.000	0.00	0.05	0.00%	0%	
1973		0.000	0.00	0.05	0.00%	0%	
1974		0.000	0.00	0.05	0.00%	0%	
1975		0.000	0.00	0.05	0.00%	0%	
1976		0.000	0.00	0.05	0.00%	0%	
1977		0.000	0.00	0.05	0.00%	0%	
1978		0.000	0.00	0.05	0.00%	0%	
1979		0.000	0.00	0.05	0.00%	0%	
1980		0.000	0.00	0.05	0.00%	0%	
1981		0.000	0.00	0.05	0.00%	0%	
1982		0.000	0.00	0.05	0.00%	0%	
1983		0.000	0.00	0.05	0.00%	0%	
1984	0.5	0.000	0.00	0.05	0.37%	0%	
1985	1.0	0.000	0.00	0.05	0.73%	1%	
1986	1.8	0.001	0.00	0.05	1.33%	2%	
1987	2.2	0.001	0.00	0.05	1.65%	4%	
1988	2.6	0.001	0.00	0.05	1.98%	6%	

1989	3.7	0.001	0.00	0.05	2.87%	9%	
1990	3.9	0.001	0.01	0.04	3.10%	11%	
1991	4.1	0.001	0.01	0.04	3.36%	14%	
1992	4.3	0.002	0.01	0.04	3.65%	18%	
1993	4.5	0.002	0.01	0.04	3.98%	21%	
1994	4.7	0.002	0.01	0.04	4.35%	24%	
1995	5.0	0.002	0.01	0.04	4.78%	28%	
1996	5.2	0.002	0.02	0.03	5.27%	32%	
1997	5.5	0.002	0.02	0.03	5.84%	36%	
1998	5.7	0.002	0.02	0.03	6.52%	40%	
1999	6.0	0.002	0.02	0.03	7.32%	44%	
2000	6.3	0.002	0.02	0.03	8.29%	49%	1ST PEAK
2001	5.8	0.002	0.03	0.02	8.29%	53%	50% PRODUCED
2002	5.3	0.002	0.03	0.02	8.29%	57%	
2003	4.9	0.002	0.03	0.02	8.29%	61%	
2004	4.5	0.002	0.03	0.02	8.29%	64%	
2005	4.1	0.001	0.03	0.02	8.29%	67%	
2006	3.8	0.001	0.03	0.02	8.29%	70%	
2007	3.5	0.001	0.04	0.01	8.29%	72%	
2008	3.2	0.001	0.04	0.01	8.29%	74%	
2009	2.9	0.001	0.04	0.01	8.29%	77%	
2010	2.7	0.001	0.04	0.01	8.29%	78%	
2011	2.4	0.001	0.04	0.01	8.29%	80%	
2012	2.2	0.001	0.04	0.01	8.29%	82%	
2013	2.1	0.001	0.04	0.01	8.29%	83%	
2014	1.9	0.001	0.04	0.01	8.29%	85%	
2015	1.7	0.001	0.04	0.01	8.29%	86%	
2016	1.6	0.001	0.04	0.01	8.29%	87%	
2017	1.5	0.001	0.04	0.01	8.29%	88%	
2018	1.3	0.000	0.04	0.01	8.29%	89%	
2019	1.2	0.000	0.05	0.00	8.29%	90%	
2020	1.1	0.000	0.05	0.00	8.29%	91%	
2021	1.0	0.000	0.05	0.00	8.29%	92%	
2022	0.9	0.000	0.05	0.00	8.29%	92%	
2023	0.9	0.000	0.05	0.00	8.29%	93%	
2024	0.8	0.000	0.05	0.00	8.29%	94%	
2025	0.7	0.000	0.05	0.00	8.29%	94%	
2026	0.7	0.000	0.05	0.00	8.29%	95%	
2027	0.6	0.000	0.05	0.00	8.29%	95%	
2028	0.6	0.000	0.05	0.00	8.29%	95%	
2029	0.5	0.000	0.05	0.00	8.29%	96%	
2030	0.5	0.000	0.05	0.00	8.29%	96%	
2031	0.4	0.000	0.05	0.00	8.29%	97%	
2032	0.4	0.000	0.05	0.00	8.29%	97%	
2033	0.4	0.000	0.05	0.00	8.29%	97%	
2034	0.3	0.000	0.05	0.00	8.29%	97%	
2035	0.3	0.000	0.05	0.00	8.29%	98%	
2036	0.3	0.000	0.05	0.00	8.29%	98%	
2037	0.3	0.000	0.05	0.00	8.29%	98%	
2038	0.2	0.000	0.05	0.00	8.29%	98%	
2039	0.2	0.000	0.05	0.00	8.29%	98%	
2040	0.2	0.000	0.05	0.00	8.29%	98%	
2041	0.2	0.000	0.05	0.00	8.29%	99%	
2042	0.2	0.000	0.05	0.00	8.29%	99%	
2043	0.2	0.000	0.05	0.00	8.29%	99%	
2044	0.1	0.000	0.05	0.00	8.29%	99%	
2045	0.1	0.000	0.05	0.00	8.29%	99%	
2046	0.1	0.000	0.05	0.00	8.29%	99%	
2047	0.1	0.000	0.05	0.00	8.29%	99%	
2048	0.1	0.000	0.05	0.00	8.29%	99%	
2049	0.1	0.000	0.05	0.00	8.29%	99%	
2050	0.1	0.000	0.05	0.00	8.29%	99%	

CUM.PROD	1.269 BBO	GIANT FIELDS		SYRIA	
RESERVES	1.730 BBO	TOTAL BBO	2.42	M.EAST (MINOR)	
DISCOVERED	2.999 BBO	% OF DISC.	81%		
UNDISCOVERED	1.001 BBO	Last Giant	1959		
REMAINING	2.731 BBO			Variance	Trend
ULTIMATE	4.000 BBO			1990	+21%
Rept.Reserve	1.730	Ref.date(end)	1989	Future	UP

Milestones	Date	Prod Mb/d	Dep. Rate	% Produced
1st Peak	1978	190	1.94%	12%
2nd Peak	1995	403	7.02%	51%
Mid-Point	1995	403	7.02%	51%
Final Decline	1995		7.02%	51%

Forecast "B"
A=Exponential decline
B=5% incr. to Mid-Point
C=Other
Pre-1930 Cum. 0

Date	Prod. M.b/d	Prod. BBO/yr	Cum. Prod.	Remain	Deplet. Rate	% Produced	Giant	Notes & Names
1930		0.000	0.00	4.00	ERR	0%		
1931		0.000	0.00	4.00	0.00%	0%		
1932		0.000	0.00	4.00	0.00%	0%		
1933		0.000	0.00	4.00	0.00%	0%		
1934		0.000	0.00	4.00	0.00%	0%		
1935		0.000	0.00	4.00	0.00%	0%		
1936		0.000	0.00	4.00	0.00%	0%		
1937		0.000	0.00	4.00	0.00%	0%		
1938		0.000	0.00	4.00	0.00%	0%		
1939		0.000	0.00	4.00	0.00%	0%		
1940		0.000	0.00	4.00	0.00%	0%		
1941		0.000	0.00	4.00	0.00%	0%		
1942		0.000	0.00	4.00	0.00%	0%		
1943		0.000	0.00	4.00	0.00%	0%		
1944		0.000	0.00	4.00	0.00%	0%		
1945		0.000	0.00	4.00	0.00%	0%		
1946		0.000	0.00	4.00	0.00%	0%		
1947		0.000	0.00	4.00	0.00%	0%		
1948		0.000	0.00	4.00	0.00%	0%		
1949		0.000	0.00	4.00	0.00%	0%		
1950		0.000	0.00	4.00	0.00%	0%		
1951		0.000	0.00	4.00	0.00%	0%		
1952		0.000	0.00	4.00	0.00%	0%		
1953		0.000	0.00	4.00	0.00%	0%		
1954		0.000	0.00	4.00	0.00%	0%		
1955		0.000	0.00	4.00	0.00%	0%		
1956		0.000	0.00	4.00	0.00%	0%	0.90	Karatchouk
1957		0.000	0.00	4.00	0.00%	0%		
1958		0.000	0.00	4.00	0.00%	0%		
1959		0.000	0.00	4.00	0.00%	0%	1.52	Souedie
1960		0.000	0.00	4.00	0.00%	0%		
1961		0.000	0.00	4.00	0.00%	0%		
1962		0.000	0.00	4.00	0.00%	0%		
1963		0.000	0.00	4.00	0.00%	0%		
1964		0.000	0.00	4.00	0.00%	0%		
1965		0.000	0.00	4.00	0.00%	0%		
1966		0.000	0.00	4.00	0.00%	0%		
1967		0.000	0.00	4.00	0.00%	0%		
1968	27	0.010	0.01	3.99	0.25%	0%		
1969	46	0.017	0.03	3.97	0.42%	1%		
1970	80	0.029	0.06	3.94	0.74%	1%		
1971	100	0.036	0.09	3.91	0.92%	2%		
1972	124	0.045	0.14	3.86	1.16%	3%		
1973	105	0.038	0.18	3.82	0.99%	4%		
1974	124	0.045	0.22	3.78	1.19%	6%		
1975	181	0.066	0.29	3.71	1.74%	7%		
1976	191	0.070	0.36	3.64	1.88%	9%		
1977	174	0.064	0.42	3.58	1.75%	11%		
1978	190	0.069	0.49	3.51	1.94%	12%		1ST PEAK
1979	165	0.060	0.55	3.45	1.72%	14%		
1980	172	0.063	0.61	3.39	1.82%	15%		
1981	172	0.063	0.68	3.32	1.85%	17%		
1982	156	0.057	0.73	3.27	1.71%	18%		
1983	168	0.061	0.79	3.21	1.88%	20%		
1984	171	0.062	0.86	3.14	1.94%	21%		
1985	178	0.065	0.92	3.08	2.07%	23%		
1986	194	0.071	0.99	3.01	2.30%	25%		
1987	230	0.084	1.08	2.92	2.79%	27%		
1988	230	0.084	1.16	2.84	2.87%	29%		

1989	301	0.110	1.27	2.73	3.86%	32%	
1990	316	0.115	1.38	2.62	4.22%	35%	
1991	331	0.121	1.51	2.49	4.63%	38%	
1992	348	0.127	1.63	2.37	5.09%	41%	
1993	365	0.133	1.77	2.23	5.63%	44%	
1994	384	0.140	1.91	2.09	6.27%	48%	
1995	403	0.147	2.05	1.95	7.02%	51%	50% PRODUCED & 2ND PEAK
1996	374	0.137	2.19	1.81	7.02%	55%	
1997	348	0.127	2.32	1.68	7.02%	58%	
1998	324	0.118	2.43	1.57	7.02%	61%	
1999	301	0.110	2.54	1.46	7.02%	64%	
2000	280	0.102	2.65	1.35	7.02%	66%	
2001	260	0.095	2.74	1.26	7.02%	69%	
2002	242	0.088	2.83	1.17	7.02%	71%	
2003	225	0.082	2.91	1.09	7.02%	73%	
2004	209	0.076	2.99	1.01	7.02%	75%	
2005	194	0.071	3.06	0.94	7.02%	76%	
2006	181	0.066	3.13	0.87	7.02%	78%	
2007	168	0.061	3.19	0.81	7.02%	80%	
2008	156	0.057	3.24	0.76	7.02%	81%	
2009	145	0.053	3.30	0.70	7.02%	82%	
2010	135	0.049	3.35	0.65	7.02%	84%	
2011	126	0.046	3.39	0.61	7.02%	85%	
2012	117	0.043	3.44	0.56	7.02%	86%	
2013	109	0.040	3.47	0.53	7.02%	87%	
2014	101	0.037	3.51	0.49	7.02%	88%	
2015	94	0.034	3.55	0.45	7.02%	89%	
2016	87	0.032	3.58	0.42	7.02%	89%	
2017	81	0.030	3.61	0.39	7.02%	90%	
2018	76	0.028	3.63	0.37	7.02%	91%	
2019	70	0.026	3.66	0.34	7.02%	92%	
2020	65	0.024	3.68	0.32	7.02%	92%	
2021	61	0.022	3.71	0.29	7.02%	93%	
2022	56	0.021	3.73	0.27	7.02%	93%	
2023	52	0.019	3.75	0.25	7.02%	94%	
2024	49	0.018	3.76	0.24	7.02%	94%	
2025	45	0.017	3.78	0.22	7.02%	95%	
2026	42	0.015	3.80	0.20	7.02%	95%	
2027	39	0.014	3.81	0.19	7.02%	95%	
2028	36	0.013	3.82	0.18	7.02%	96%	
2029	34	0.012	3.84	0.16	7.02%	96%	
2030	32	0.012	3.85	0.15	7.02%	96%	
2031	29	0.011	3.86	0.14	7.02%	96%	
2032	27	0.010	3.87	0.13	7.02%	97%	
2033	25	0.009	3.88	0.12	7.02%	97%	
2034	24	0.009	3.89	0.11	7.02%	97%	
2035	22	0.008	3.89	0.11	7.02%	97%	
2036	20	0.007	3.90	0.10	7.02%	98%	
2037	19	0.007	3.91	0.09	7.02%	98%	
2038	18	0.006	3.91	0.09	7.02%	98%	
2039	16	0.006	3.92	0.08	7.02%	98%	
2040	15	0.006	3.93	0.07	7.02%	98%	
2041	14	0.005	3.93	0.07	7.02%	98%	
2042	13	0.005	3.94	0.06	7.02%	98%	
2043	12	0.004	3.94	0.06	7.02%	99%	
2044	11	0.004	3.94	0.06	7.02%	99%	
2045	11	0.004	3.95	0.05	7.02%	99%	
2046	10	0.004	3.95	0.05	7.02%	99%	
2047	9	0.003	3.96	0.04	7.02%	99%	
2048	9	0.003	3.96	0.04	7.02%	99%	
2049	8	0.003	3.96	0.04	7.02%	99%	
2050	7	0.003	3.96	0.04	7.02%	99%	

CUM.PROD	0.024 BBO	GIANT FIELDS		TAIWAN	
RESERVES	0.005 BBO	TOTAL BBO	0	THE EAST	
DISCOVERED	0.029 BBO	% OF DISC.	0%		
UNDISCOVERED	0.021 BBO	Last Giant	-		
REMAINING	0.026 BBO			Variance	Trend
ULTIMATE	0.050 BBO			1990	n/a
Rept.Reserve	0.005	Ref.date(end)	1989	Future	DOWN

Milestones	Date	Prod Mb/d	Dep. Rate	% Produced
1st Peak	1979	5	4.83%	28%
2nd Peak				
Mid-Point	1990	3	3.52%	50%
Final Decline	1990		3.52%	50%

Forecast "A"
A=Exponential decline
B=5% incr. to Mid-Point
C=Other
Pre-1930 Cum. 0

Date	Prod. M.b/d	Prod. BBO/yr	Cum. Prod.	Remain	Deplet. Rate	% Produced	Giant	Notes & Names
1930		0.000	0.00	0.05	ERR	0%		
1931		0.000	0.00	0.05	0.00%	0%		
1932		0.000	0.00	0.05	0.00%	0%		
1933		0.000	0.00	0.05	0.00%	0%		
1934		0.000	0.00	0.05	0.00%	0%		
1935		0.000	0.00	0.05	0.00%	0%		
1936		0.000	0.00	0.05	0.00%	0%		
1937		0.000	0.00	0.05	0.00%	0%		
1938		0.000	0.00	0.05	0.00%	0%		
1939		0.000	0.00	0.05	0.00%	0%		
1940		0.000	0.00	0.05	0.00%	0%		
1941		0.000	0.00	0.05	0.00%	0%		
1942		0.000	0.00	0.05	0.00%	0%		
1943		0.000	0.00	0.05	0.00%	0%		
1944		0.000	0.00	0.05	0.00%	0%		
1945		0.000	0.00	0.05	0.00%	0%		
1946		0.000	0.00	0.05	0.00%	0%		
1947		0.000	0.00	0.05	0.00%	0%		
1948		0.000	0.00	0.05	0.00%	0%		
1949		0.000	0.00	0.05	0.00%	0%		
1950		0.000	0.00	0.05	0.00%	0%		
1951		0.000	0.00	0.05	0.00%	0%		
1952		0.000	0.00	0.05	0.00%	0%		
1953		0.000	0.00	0.05	0.00%	0%		
1954		0.000	0.00	0.05	0.00%	0%		
1955		0.000	0.00	0.05	0.00%	0%		
1956		0.000	0.00	0.05	0.00%	0%		
1957		0.000	0.00	0.05	0.00%	0%		
1958	0	0.000	0.00	0.05	0.03%	0%		
1959	0	0.000	0.00	0.05	0.03%	0%		
1960	0	0.000	0.00	0.05	0.05%	0%		
1961	0	0.000	0.00	0.05	0.00%	0%		
1962	0	0.000	0.00	0.05	0.00%	0%		
1963	0	0.000	0.00	0.05	0.00%	0%		
1964	0	0.000	0.00	0.05	0.04%	0%		
1965	0	0.000	0.00	0.05	0.22%	0%		
1966	1	0.000	0.00	0.05	0.44%	1%		
1967	1	0.000	0.00	0.05	0.44%	1%		
1968	1	0.000	0.00	0.05	0.44%	2%		
1969	2	0.001	0.00	0.05	1.19%	3%		
1970	2	0.001	0.00	0.05	1.28%	4%		
1971	2	0.001	0.00	0.05	1.52%	6%		
1972	2	0.001	0.00	0.05	1.55%	7%		
1973	2	0.001	0.00	0.05	1.57%	8%		
1974	3	0.001	0.01	0.04	2.63%	11%		
1975	5	0.002	0.01	0.04	4.10%	15%		
1976	4	0.002	0.01	0.04	3.59%	18%		
1977	5	0.002	0.01	0.04	4.43%	21%		
1978	4	0.002	0.01	0.04	3.99%	24%		
1979	5	0.002	0.01	0.04	4.83%	28%		1ST PEAK
1980	4	0.001	0.02	0.03	4.06%	31%		
1981	3	0.001	0.02	0.03	3.17%	33%		
1982	3	0.001	0.02	0.03	3.28%	35%		
1983	2	0.001	0.02	0.03	2.60%	37%		
1984	3	0.001	0.02	0.03	3.01%	39%		
1985	3	0.001	0.02	0.03	3.23%	41%		
1986	2	0.001	0.02	0.03	2.35%	42%		
1987	3	0.001	0.02	0.03	3.29%	44%		
1988	3	0.001	0.02	0.03	3.40%	46%		

1989	3	0.001	0.02	0.03	3.52%	48%	
1990	3	0.001	0.02	0.03	3.52%	50%	50% PRODUCED
1991	2	0.001	0.03	0.02	3.52%	52%	
1992	2	0.001	0.03	0.02	3.52%	53%	
1993	2	0.001	0.03	0.02	3.52%	55%	
1994	2	0.001	0.03	0.02	3.52%	56%	
1995	2	0.001	0.03	0.02	3.52%	58%	
1996	2	0.001	0.03	0.02	3.52%	60%	
1997	2	0.001	0.03	0.02	3.52%	61%	
1998	2	0.001	0.03	0.02	3.52%	62%	
1999	2	0.001	0.03	0.02	3.52%	64%	
2000	2	0.001	0.03	0.02	3.52%	65%	
2001	2	0.001	0.03	0.02	3.52%	66%	
2002	2	0.001	0.03	0.02	3.52%	67%	
2003	2	0.001	0.03	0.02	3.52%	68%	
2004	2	0.001	0.03	0.02	3.52%	70%	
2005	1	0.001	0.04	0.01	3.52%	71%	
2006	1	0.001	0.04	0.01	3.52%	72%	
2007	1	0.000	0.04	0.01	3.52%	73%	
2008	1	0.000	0.04	0.01	3.52%	74%	
2009	1	0.000	0.04	0.01	3.52%	75%	
2010	1	0.000	0.04	0.01	3.52%	75%	
2011	1	0.000	0.04	0.01	3.52%	76%	
2012	1	0.000	0.04	0.01	3.52%	77%	
2013	1	0.000	0.04	0.01	3.52%	78%	
2014	1	0.000	0.04	0.01	3.52%	79%	
2015	1	0.000	0.04	0.01	3.52%	80%	
2016	1	0.000	0.04	0.01	3.52%	80%	
2017	1	0.000	0.04	0.01	3.52%	81%	
2018	1	0.000	0.04	0.01	3.52%	82%	
2019	1	0.000	0.04	0.01	3.52%	82%	
2020	1	0.000	0.04	0.01	3.52%	83%	
2021	1	0.000	0.04	0.01	3.52%	83%	
2022	1	0.000	0.04	0.01	3.52%	84%	
2023	1	0.000	0.04	0.01	3.52%	85%	
2024	1	0.000	0.04	0.01	3.52%	85%	
2025	1	0.000	0.04	0.01	3.52%	86%	
2026	1	0.000	0.04	0.01	3.52%	86%	
2027	1	0.000	0.04	0.01	3.52%	87%	
2028	1	0.000	0.04	0.01	3.52%	87%	
2029	1	0.000	0.04	0.01	3.52%	88%	
2030	1	0.000	0.04	0.01	3.52%	88%	
2031	1	0.000	0.04	0.01	3.52%	88%	
2032	1	0.000	0.04	0.01	3.52%	89%	
2033	1	0.000	0.04	0.01	3.52%	89%	
2034	1	0.000	0.04	0.01	3.52%	90%	
2035	1	0.000	0.04	0.01	3.52%	90%	
2036	0	0.000	0.05	0.00	3.52%	90%	
2037	0	0.000	0.05	0.00	3.52%	91%	
2038	0	0.000	0.05	0.00	3.52%	91%	
2039	0	0.000	0.05	0.00	3.52%	91%	
2040	0	0.000	0.05	0.00	3.52%	92%	
2041	0	0.000	0.05	0.00	3.52%	92%	
2042	0	0.000	0.05	0.00	3.52%	92%	
2043	0	0.000	0.05	0.00	3.52%	92%	
2044	0	0.000	0.05	0.00	3.52%	93%	
2045	0	0.000	0.05	0.00	3.52%	93%	
2046	0	0.000	0.05	0.00	3.52%	93%	
2047	0	0.000	0.05	0.00	3.52%	93%	
2048	0	0.000	0.05	0.00	3.52%	94%	
2049	0	0.000	0.05	0.00	3.52%	94%	
2050	0	0.000	0.05	0.00	3.52%	94%	

CUM.PROD	0.079 BBO	GIANT FIELDS		THAILAND	
RESERVES	0.231 BBO	TOTAL BBO	0	THE EAST	
DISCOVERED	0.310 BBO	% OF DISC.	0%		
UNDISCOVERED	0.190 BBO	Last Giant	-		
REMAINING	0.421 BBO			Variance Trend	
ULTIMATE	0.500 BBO			1990	zero
Rept.Reserve	0.231	Ref.date(end)	1989	Future	UP

Milestones	Date	Prod Mb/d	Dep. Rate	% Produced	Forecast "B"
1st Peak	1998	61	7.95%	49%	A=Exponential decline
2nd Peak					B=5% incr. to Mid-Point
					C=Other
Mid-Point	1999	56	7.95%	53%	Pre-1930 Cum. 0
Final Decline	1999		7.95%	53%	

Date	Prod. M.b/d	Prod. BBO/yr	Cum. Prod.	Remain	Deplet. Rate	% Produced	Giant	Notes & Names
1930		0.000	0.00	0.50	ERR	0%		
1931		0.000	0.00	0.50	0.00%	0%		
1932		0.000	0.00	0.50	0.00%	0%		
1933		0.000	0.00	0.50	0.00%	0%		
1934		0.000	0.00	0.50	0.00%	0%		
1935		0.000	0.00	0.50	0.00%	0%		
1936		0.000	0.00	0.50	0.00%	0%		
1937		0.000	0.00	0.50	0.00%	0%		
1938		0.000	0.00	0.50	0.00%	0%		
1939		0.000	0.00	0.50	0.00%	0%		
1940		0.000	0.00	0.50	0.00%	0%		
1941		0.000	0.00	0.50	0.00%	0%		
1942		0.000	0.00	0.50	0.00%	0%		
1943		0.000	0.00	0.50	0.00%	0%		
1944		0.000	0.00	0.50	0.00%	0%		
1945		0.000	0.00	0.50	0.00%	0%		
1946		0.000	0.00	0.50	0.00%	0%		
1947		0.000	0.00	0.50	0.00%	0%		
1948		0.000	0.00	0.50	0.00%	0%		
1949		0.000	0.00	0.50	0.00%	0%		
1950		0.000	0.00	0.50	0.00%	0%		
1951		0.000	0.00	0.50	0.00%	0%		
1952		0.000	0.00	0.50	0.00%	0%		
1953		0.000	0.00	0.50	0.00%	0%		
1954		0.000	0.00	0.50	0.00%	0%		
1955		0.000	0.00	0.50	0.00%	0%		
1956		0.000	0.00	0.50	0.00%	0%		
1957		0.000	0.00	0.50	0.00%	0%		
1958		0.000	0.00	0.50	0.00%	0%		
1959		0.000	0.00	0.50	0.00%	0%		
1960		0.000	0.00	0.50	0.00%	0%		
1961		0.000	0.00	0.50	0.00%	0%		
1962		0.000	0.00	0.50	0.00%	0%		
1963		0.000	0.00	0.50	0.00%	0%		
1964		0.000	0.00	0.50	0.00%	0%		
1965	0	0.000	0.00	0.50	0.00%	0%		
1966	0	0.000	0.00	0.50	0.01%	0%		
1967	0	0.000	0.00	0.50	0.01%	0%		
1968	0	0.000	0.00	0.50	0.01%	0%		
1969	0	0.000	0.00	0.50	0.01%	0%		
1970	0	0.000	0.00	0.50	0.02%	0%		
1971	0	0.000	0.00	0.50	0.02%	0%		
1972	0	0.000	0.00	0.50	0.01%	0%		
1973	0	0.000	0.00	0.50	0.01%	0%		
1974	0	0.000	0.00	0.50	0.01%	0%		
1975	0	0.000	0.00	0.50	0.01%	0%		
1976	0	0.000	0.00	0.50	0.02%	0%		
1977	0	0.000	0.00	0.50	0.01%	0%		
1978	0	0.000	0.00	0.50	0.01%	0%		
1979	0	0.000	0.00	0.50	0.00%	0%		
1980	0	0.000	0.00	0.50	0.00%	0%		
1981	0	0.000	0.00	0.50	0.00%	0%		
1982	6	0.002	0.00	0.50	0.44%	1%		
1983	11	0.004	0.01	0.49	0.81%	1%		
1984	19	0.007	0.01	0.49	1.41%	3%		
1985	34	0.013	0.03	0.47	2.58%	5%		
1986	36	0.013	0.04	0.46	2.75%	8%		
1987	31	0.011	0.05	0.45	2.48%	10%		
1988	38	0.014	0.07	0.43	3.12%	13%		

1989	39	0.014	0.08	0.42	3.27%	16%	
1990	41	0.015	0.09	0.41	3.55%	19%	
1991	43	0.016	0.11	0.39	3.87%	22%	
1992	45	0.016	0.13	0.37	4.23%	25%	
1993	47	0.017	0.14	0.36	4.63%	29%	
1994	50	0.018	0.16	0.34	5.10%	32%	
1995	52	0.019	0.18	0.32	5.64%	36%	
1996	55	0.020	0.20	0.30	6.28%	40%	
1997	58	0.021	0.22	0.28	7.04%	44%	
1998	61	0.022	0.24	0.26	7.95%	49%	1ST PEAK
1999	56	0.020	0.26	0.24	7.95%	53%	50% PRODUCED
2000	51	0.019	0.28	0.22	7.95%	57%	
2001	47	0.017	0.30	0.20	7.95%	60%	
2002	43	0.016	0.32	0.18	7.95%	63%	
2003	40	0.015	0.33	0.17	7.95%	66%	
2004	37	0.013	0.34	0.16	7.95%	69%	
2005	34	0.012	0.36	0.14	7.95%	71%	
2006	31	0.011	0.37	0.13	7.95%	74%	
2007	29	0.010	0.38	0.12	7.95%	76%	
2008	26	0.010	0.39	0.11	7.95%	78%	
2009	24	0.009	0.40	0.10	7.95%	79%	
2010	22	0.008	0.41	0.09	7.95%	81%	
2011	21	0.008	0.41	0.09	7.95%	83%	
2012	19	0.007	0.42	0.08	7.95%	84%	
2013	17	0.006	0.43	0.07	7.95%	85%	
2014	16	0.006	0.43	0.07	7.95%	86%	
2015	15	0.005	0.44	0.06	7.95%	87%	
2016	14	0.005	0.44	0.06	7.95%	88%	
2017	13	0.005	0.45	0.05	7.95%	89%	
2018	12	0.004	0.45	0.05	7.95%	90%	
2019	11	0.004	0.46	0.04	7.95%	91%	
2020	10	0.004	0.46	0.04	7.95%	92%	
2021	9	0.003	0.46	0.04	7.95%	92%	
2022	8	0.003	0.46	0.04	7.95%	93%	
2023	8	0.003	0.47	0.03	7.95%	94%	
2024	7	0.003	0.47	0.03	7.95%	94%	
2025	6	0.002	0.47	0.03	7.95%	95%	
2026	6	0.002	0.47	0.03	7.95%	95%	
2027	5	0.002	0.48	0.02	7.95%	95%	
2028	5	0.002	0.48	0.02	7.95%	96%	
2029	5	0.002	0.48	0.02	7.95%	96%	
2030	4	0.002	0.48	0.02	7.95%	96%	
2031	4	0.001	0.48	0.02	7.95%	97%	
2032	4	0.001	0.48	0.02	7.95%	97%	
2033	3	0.001	0.49	0.01	7.95%	97%	
2034	3	0.001	0.49	0.01	7.95%	97%	
2035	3	0.001	0.49	0.01	7.95%	98%	
2036	3	0.001	0.49	0.01	7.95%	98%	
2037	2	0.001	0.49	0.01	7.95%	98%	
2038	2	0.001	0.49	0.01	7.95%	98%	
2039	2	0.001	0.49	0.01	7.95%	98%	
2040	2	0.001	0.49	0.01	7.95%	98%	
2041	2	0.001	0.49	0.01	7.95%	99%	
2042	2	0.001	0.49	0.01	7.95%	99%	
2043	1	0.001	0.49	0.01	7.95%	99%	
2044	1	0.000	0.49	0.01	7.95%	99%	
2045	1	0.000	0.49	0.01	7.95%	99%	
2046	1	0.000	0.50	0.00	7.95%	99%	
2047	1	0.000	0.50	0.00	7.95%	99%	
2048	1	0.000	0.50	0.00	7.95%	99%	
2049	1	0.000	0.50	0.00	7.95%	99%	
2050	1	0.000	0.50	0.00	7.95%	99%	

CUM.PROD	2.588 BBO	GIANT FIELDS			TRINIDAD	
RESERVES	0.521 BBO	TOTAL BBO	1.27		L.AMERICA	
DISCOVERED	3.109 BBO	% OF DISC.	41%			
UNDISCOVERED	0.141 BBO	Last Giant	1956			
REMAINING	0.662 BBO				Variance	Trend
ULTIMATE	3.250 BBO				1990	+7%
Rept.Reserves	0.521 BBO	Ref. Date (end)		1989	Future	UP

Milestones	Date	Prod Mb/d	Dep. Rate	% Produced
1st Peak	1968	187	3.21%	37%
2nd Peak	1978	240	5.93%	57%
Mid-Point	1976	224	4.98%	52%
Final Decline	1990		7.81%	

Forecast "A"
A=Exponential decline
B=5% inc. to Mid-Point
C=Other
Pre-1930 Cum 0.1
(added to 1930 Prod)

Date	Prod. M.b/d	Prod. BBO/yr	Cum Prod	Remain	Dep. Rate	% Produced	Giant	Names & Notes
1930	25	0.009	0.11	3.14	ERR	3%	0.75	Forest Res.
1931	28	0.010	0.12	3.13	0.33%	4%		
1932	31	0.011	0.13	3.12	0.37%	4%		
1933	34	0.013	0.14	3.11	0.40%	4%		
1934	37	0.014	0.16	3.09	0.44%	5%		
1935	40	0.015	0.17	3.08	0.48%	5%		
1936	43	0.016	0.19	3.06	0.52%	6%		
1937	46	0.017	0.21	3.04	0.55%	6%		
1938	49	0.018	0.22	3.03	0.58%	7%		
1939	53	0.019	0.24	3.01	0.64%	7%		
1940	61	0.022	0.26	2.99	0.74%	8%		
1941	60	0.022	0.29	2.96	0.73%	9%		
1942	60	0.022	0.31	2.94	0.74%	9%		
1943	59	0.021	0.33	2.92	0.73%	10%		
1944	61	0.022	0.35	2.90	0.76%	11%		
1945	58	0.021	0.37	2.88	0.73%	11%		
1946	55	0.020	0.39	2.86	0.70%	12%		
1947	56	0.020	0.41	2.84	0.71%	13%		
1948	55	0.020	0.43	2.82	0.71%	13%		
1949	55	0.020	0.45	2.80	0.71%	14%		
1950	56	0.020	0.47	2.78	0.73%	15%		
1951	57	0.021	0.49	2.76	0.75%	15%		
1952	58	0.021	0.52	2.73	0.77%	16%		
1953	57	0.021	0.54	2.71	0.76%	17%		
1954	65	0.024	0.56	2.69	0.87%	17%		
1955	68	0.025	0.58	2.67	0.92%	18%		
1956	79	0.029	0.61	2.64	1.08%	19%	0.52	Soldado
1957	93	0.034	0.65	2.60	1.29%	20%		
1958	99	0.036	0.68	2.57	1.39%	21%		
1959	112	0.041	0.72	2.53	1.59%	22%		
1960	116	0.042	0.77	2.48	1.68%	24%		
1961	125	0.045	0.81	2.44	1.83%	25%		
1962	133	0.049	0.86	2.39	1.99%	26%		
1963	134	0.049	0.91	2.34	2.05%	28%		
1964	135	0.049	0.96	2.29	2.10%	30%		
1965	134	0.049	1.01	2.24	2.13%	31%		
1966	149	0.054	1.06	2.19	2.43%	33%		
1967	173	0.063	1.13	2.12	2.88%	35%		
1968	187	0.068	1.19	2.06	3.21%	37%		1ST PEAK
1969	162	0.059	1.25	2.00	2.88%	39%		
1970	141	0.051	1.30	1.95	2.58%	40%		
1971	138	0.051	1.35	1.90	2.60%	42%		
1972	143	0.052	1.41	1.84	2.75%	43%		
1973	159	0.058	1.46	1.79	3.15%	45%		
1974	181	0.066	1.53	1.72	3.71%	47%		
1975	210	0.077	1.61	1.64	4.46%	49%		
1976	224	0.082	1.69	1.56	4.98%	52%		50% PRODUCED
1977	230	0.084	1.77	1.48	5.38%	55%		
1978	240	0.088	1.86	1.39	5.93%	57%		2ND PEAK
1979	215	0.078	1.94	1.31	5.65%	60%		
1980	215	0.078	2.02	1.23	5.99%	62%		
1981	240	0.088	2.11	1.14	7.11%	65%		
1982	182	0.066	2.17	1.08	5.80%	67%		
1983	158	0.058	2.23	1.02	5.35%	69%		
1984	169	0.062	2.29	0.96	6.05%	71%		
1985	179	0.065	2.36	0.89	6.81%	73%		
1986	168	0.061	2.42	0.83	6.86%	74%		
1987	163	0.060	2.48	0.77	7.15%	76%		
1988	149	0.054	2.53	0.72	7.05%	78%		

1989	154	0.056	2.59	0.66	7.81%	80%
1990	142	0.052	2.64	0.61	7.81%	81%
1991	131	0.048	2.69	0.56	7.81%	83%
1992	120	0.044	2.73	0.52	7.81%	84%
1993	111	0.041	2.77	0.48	7.81%	85%
1994	102	0.037	2.81	0.44	7.81%	86%
1995	94	0.034	2.84	0.41	7.81%	87%
1996	87	0.032	2.88	0.37	7.81%	88%
1997	80	0.029	2.90	0.35	7.81%	89%
1998	74	0.027	2.93	0.32	7.81%	90%
1999	68	0.025	2.96	0.29	7.81%	91%
2000	63	0.023	2.98	0.27	7.81%	92%
2001	58	0.021	3.00	0.25	7.81%	92%
2002	53	0.019	3.02	0.23	7.81%	93%
2003	49	0.018	3.04	0.21	7.81%	93%
2004	45	0.017	3.05	0.20	7.81%	94%
2005	42	0.015	3.07	0.18	7.81%	94%
2006	39	0.014	3.08	0.17	7.81%	95%
2007	36	0.013	3.10	0.15	7.81%	95%
2008	33	0.012	3.11	0.14	7.81%	96%
2009	30	0.011	3.12	0.13	7.81%	96%
2010	28	0.010	3.13	0.12	7.81%	96%
2011	26	0.009	3.14	0.11	7.81%	97%
2012	24	0.009	3.15	0.10	7.81%	97%
2013	22	0.008	3.16	0.09	7.81%	97%
2014	20	0.007	3.16	0.09	7.81%	97%
2015	19	0.007	3.17	0.08	7.81%	98%
2016	17	0.006	3.18	0.07	7.81%	98%
2017	16	0.006	3.18	0.07	7.81%	98%
2018	15	0.005	3.19	0.06	7.81%	98%
2019	13	0.005	3.19	0.06	7.81%	98%
2020	12	0.005	3.20	0.05	7.81%	98%
2021	11	0.004	3.20	0.05	7.81%	98%
2022	10	0.004	3.20	0.05	7.81%	99%
2023	10	0.004	3.21	0.04	7.81%	99%
2024	9	0.003	3.21	0.04	7.81%	99%
2025	8	0.003	3.21	0.04	7.81%	99%
2026	8	0.003	3.22	0.03	7.81%	99%
2027	7	0.003	3.22	0.03	7.81%	99%
2028	6	0.002	3.22	0.03	7.81%	99%
2029	6	0.002	3.22	0.03	7.81%	99%
2030	5	0.002	3.23	0.02	7.81%	99%
2031	5	0.002	3.23	0.02	7.81%	99%
2032	5	0.002	3.23	0.02	7.81%	99%
2033	4	0.002	3.23	0.02	7.81%	99%
2034	4	0.001	3.23	0.02	7.81%	99%
2035	4	0.001	3.23	0.02	7.81%	100%
2036	3	0.001	3.24	0.01	7.81%	100%
2037	3	0.001	3.24	0.01	7.81%	100%
2038	3	0.001	3.24	0.01	7.81%	100%
2039	3	0.001	3.24	0.01	7.81%	100%
2040	2	0.001	3.24	0.01	7.81%	100%
2041	2	0.001	3.24	0.01	7.81%	100%
2042	2	0.001	3.24	0.01	7.81%	100%
2043	2	0.001	3.24	0.01	7.81%	100%
2044	2	0.001	3.24	0.01	7.81%	100%
2045	2	0.001	3.24	0.01	7.81%	100%
2046	1	0.001	3.24	0.01	7.81%	100%
2047	1	0.001	3.24	0.01	7.81%	100%
2048	1	0.000	3.24	0.01	7.81%	100%
2049	1	0.000	3.24	0.01	7.81%	100%
2050	1	0.000	3.25	0.00	7.81%	100%

CUM.PROD	0.786 BBO	GIANT FIELDS			TUNISIA
RESERVES	1.750 BBO	TOTAL BBO	0.8		AFRICA
DISCOVERED	2.536 BBO	% OF DISC.	30%		
UNDISCOVERED	0.464 BBO	Last Giant	1964		
REMAINING	2.214 BBO				Variance Trend
ULTIMATE	3.000 BBO				1990 -9%
Rept.Reserves	1.750 BBO	Ref.Date (end)		1989	Future DOWN

Milestones	Date	Prod Mb/d	Dep Rate	% Produced
1st Peak	1985	125	1.89%	21%
2nd Peak	2003	184	4.41%	52%
Mid-Point	2003	184	4.41%	52%
Final Decline	2003		4.41%	

Forecast "B"
A=Exponential decline
B=5% inc. to Mid-Point
C=Other
Pre-1930 Cum 0
(Added to 1930 Prod)

Date	Prod. M.b/d	Prod. BBO/yr	Cum Prod	Remain	Dep. Rate	% Produced	Giant	Names & Notes
1930		0.000	0.00	3.00	ERR	0%		
1931		0.000	0.00	3.00	0.00%	0%		
1932		0.000	0.00	3.00	0.00%	0%		
1933		0.000	0.00	3.00	0.00%	0%		
1934		0.000	0.00	3.00	0.00%	0%		
1935		0.000	0.00	3.00	0.00%	0%		
1936		0.000	0.00	3.00	0.00%	0%		
1937		0.000	0.00	3.00	0.00%	0%		
1938		0.000	0.00	3.00	0.00%	0%		
1939		0.000	0.00	3.00	0.00%	0%		
1940		0.000	0.00	3.00	0.00%	0%		
1941		0.000	0.00	3.00	0.00%	0%		
1942		0.000	0.00	3.00	0.00%	0%		
1943		0.000	0.00	3.00	0.00%	0%		
1944		0.000	0.00	3.00	0.00%	0%		
1945		0.000	0.00	3.00	0.00%	0%		
1946		0.000	0.00	3.00	0.00%	0%		
1947		0.000	0.00	3.00	0.00%	0%		
1948		0.000	0.00	3.00	0.00%	0%		
1949		0.000	0.00	3.00	0.00%	0%		
1950		0.000	0.00	3.00	0.00%	0%		
1951		0.000	0.00	3.00	0.00%	0%		
1952		0.000	0.00	3.00	0.00%	0%		
1953		0.000	0.00	3.00	0.00%	0%		
1954		0.000	0.00	3.00	0.00%	0%		
1955		0.000	0.00	3.00	0.00%	0%		
1956		0.000	0.00	3.00	0.00%	0%		
1957		0.000	0.00	3.00	0.00%	0%		
1958		0.000	0.00	3.00	0.00%	0%		
1959		0.000	0.00	3.00	0.00%	0%		
1960		0.000	0.00	3.00	0.00%	0%		
1961		0.000	0.00	3.00	0.00%	0%		
1962		0.000	0.00	3.00	0.00%	0%		
1963		0.000	0.00	3.00	0.00%	0%		
1964		0.000	0.00	3.00	0.00%	0%	0.75	El Borma
1965	0.0	0.000	0.00	3.00	0.00%	0%		
1966	10.5	0.004	0.00	3.00	0.13%	0%		
1967	31.2	0.011	0.02	2.98	0.38%	1%		
1968	47.8	0.017	0.03	2.97	0.58%	1%		
1969	79.6	0.029	0.06	2.94	0.98%	2%		
1970	88.0	0.032	0.09	2.91	1.09%	3%		
1971	86.5	0.032	0.13	2.87	1.09%	4%		
1972	80.0	0.029	0.15	2.85	1.02%	5%		
1973	87.0	0.032	0.19	2.81	1.12%	6%		
1974	85.0	0.031	0.22	2.78	1.10%	7%		
1975	95.0	0.035	0.25	2.75	1.25%	8%		
1976	77.0	0.028	0.28	2.72	1.02%	9%		
1977	87.0	0.032	0.31	2.69	1.17%	10%		
1978	100.0	0.037	0.35	2.65	1.36%	12%		
1979	105.0	0.038	0.39	2.61	1.45%	13%		
1980	100.0	0.037	0.42	2.58	1.40%	14%		
1981	118.0	0.043	0.47	2.53	1.67%	16%		
1982	106.0	0.039	0.51	2.49	1.53%	17%		
1983	115.0	0.042	0.55	2.45	1.68%	18%		
1984	114.0	0.042	0.59	2.41	1.70%	20%		
1985	125.0	0.046	0.63	2.37	1.89%	21%		1ST PEAK
1986	106.0	0.039	0.67	2.33	1.64%	22%		
1987	104.3	0.038	0.71	2.29	1.64%	24%		
1988	102.8	0.038	0.75	2.25	1.64%	25%		

1989	102.0	0.037	0.79	2.21	1.65%	26%	
1990	107.1	0.039	0.82	2.18	1.77%	27%	
1991	112.5	0.041	0.87	2.13	1.89%	29%	
1992	118.1	0.043	0.91	2.09	2.02%	30%	
1993	124.0	0.045	0.95	2.05	2.16%	32%	
1994	130.2	0.048	1.00	2.00	2.32%	33%	
1995	136.7	0.050	1.05	1.95	2.50%	35%	
1996	143.5	0.052	1.10	1.90	2.69%	37%	
1997	150.7	0.055	1.16	1.84	2.90%	39%	
1998	158.2	0.058	1.22	1.78	3.14%	41%	
1999	166.1	0.061	1.28	1.72	3.40%	43%	
2000	174.5	0.064	1.34	1.66	3.70%	45%	
2001	183.2	0.067	1.41	1.59	4.03%	47%	
2002	192.3	0.070	1.48	1.52	4.41%	49%	
2003	183.9	0.067	1.55	1.45	4.41%	52%	2ND PEAK
2004	175.8	0.064	1.61	1.39	4.41%	54%	& 50% PRODUCED
2005	168.0	0.061	1.67	1.33	4.41%	56%	
2006	160.6	0.059	1.73	1.27	4.41%	58%	
2007	153.5	0.056	1.79	1.21	4.41%	60%	
2008	146.7	0.054	1.84	1.16	4.41%	61%	
2009	140.3	0.051	1.89	1.11	4.41%	63%	
2010	134.1	0.049	1.94	1.06	4.41%	65%	
2011	128.2	0.047	1.99	1.01	4.41%	66%	
2012	122.5	0.045	2.03	0.97	4.41%	68%	
2013	117.1	0.043	2.07	0.93	4.41%	69%	
2014	112.0	0.041	2.11	0.89	4.41%	70%	
2015	107.0	0.039	2.15	0.85	4.41%	72%	
2016	102.3	0.037	2.19	0.81	4.41%	73%	
2017	97.8	0.036	2.23	0.77	4.41%	74%	
2018	93.5	0.034	2.26	0.74	4.41%	75%	
2019	89.4	0.033	2.29	0.71	4.41%	76%	
2020	85.4	0.031	2.32	0.68	4.41%	77%	
2021	81.6	0.030	2.35	0.65	4.41%	78%	
2022	78.0	0.028	2.38	0.62	4.41%	79%	
2023	74.6	0.027	2.41	0.59	4.41%	80%	
2024	71.3	0.026	2.44	0.56	4.41%	81%	
2025	68.2	0.025	2.46	0.54	4.41%	82%	
2026	65.2	0.024	2.48	0.52	4.41%	83%	
2027	62.3	0.023	2.51	0.49	4.41%	84%	
2028	59.5	0.022	2.53	0.47	4.41%	84%	
2029	56.9	0.021	2.55	0.45	4.41%	85%	
2030	54.4	0.020	2.57	0.43	4.41%	86%	
2031	52.0	0.019	2.59	0.41	4.41%	86%	
2032	49.7	0.018	2.61	0.39	4.41%	87%	
2033	47.5	0.017	2.62	0.38	4.41%	87%	
2034	45.4	0.017	2.64	0.36	4.41%	88%	
2035	43.4	0.016	2.66	0.34	4.41%	89%	
2036	41.5	0.015	2.67	0.33	4.41%	89%	
2037	39.7	0.014	2.69	0.31	4.41%	90%	
2038	37.9	0.014	2.70	0.30	4.41%	90%	
2039	36.3	0.013	2.71	0.29	4.41%	90%	
2040	34.7	0.013	2.73	0.27	4.41%	91%	
2041	33.1	0.012	2.74	0.26	4.41%	91%	
2042	31.7	0.012	2.75	0.25	4.41%	92%	
2043	30.3	0.011	2.76	0.24	4.41%	92%	
2044	28.9	0.011	2.77	0.23	4.41%	92%	
2045	27.7	0.010	2.78	0.22	4.41%	93%	
2046	26.4	0.010	2.79	0.21	4.41%	93%	
2047	25.3	0.009	2.80	0.20	4.41%	93%	
2048	24.2	0.009	2.81	0.19	4.41%	94%	
2049	23.1	0.008	2.82	0.18	4.41%	94%	
2050	22.1	0.008	2.83	0.17	4.41%	94%	

CUM.PROD	0.511 BBO	GIANT FIELDS		TURKEY		
RESERVES	0.521 BBO	TOTAL BBO	0	W.EUROPE		
DISCOVERED	1.032 BBO	% OF DISC.	0%			
UNDISCOVERED	0.218 BBO	Last Giant	-			
REMAINING	0.739 BBO			Variance	Trend	
ULTIMATE	1.250 BBO			1990	+20%	
Rept.Reserve	0.521 BBO	Ref.Date(end)	1989	Future	UP	

Milestones	Date	Prod Mb/d	Dep. Rate	% Produced	Forecast "B"
1st Peak	1969	69	2.16%	9%	A=Exponential decline
2nd Peak	1994	71	4.00%	50%	B=5% incr. to Mid-Point
Mid-Point	1994	71	4.00%	50%	C=Other
Final Decline	1994		4.00%		Pre-1930 Cum. 0 (Added to 1930 Prod)

Date	Prod. M.b/d	Prod. BBO/yr	Cum Prod	Remain	Dep. Rate	% Produced	Giant	Names & Notes
1930		0.000	0.00	1.25	ERR	0%		
1931		0.000	0.00	1.25	0.00%	0%		
1932		0.000	0.00	1.25	0.00%	0%		
1933		0.000	0.00	1.25	0.00%	0%		
1934		0.000	0.00	1.25	0.00%	0%		
1935		0.000	0.00	1.25	0.00%	0%		
1936		0.000	0.00	1.25	0.00%	0%		
1937		0.000	0.00	1.25	0.00%	0%		
1938		0.000	0.00	1.25	0.00%	0%		
1939		0.000	0.00	1.25	0.00%	0%		
1940		0.000	0.00	1.25	0.00%	0%		
1941		0.000	0.00	1.25	0.00%	0%		
1942		0.000	0.00	1.25	0.00%	0%		
1943		0.000	0.00	1.25	0.00%	0%		
1944		0.000	0.00	1.25	0.00%	0%		
1945		0.000	0.00	1.25	0.00%	0%		
1946		0.000	0.00	1.25	0.00%	0%		
1947		0.000	0.00	1.25	0.00%	0%		
1948		0.000	0.00	1.25	0.00%	0%		
1949	0	0.000	0.00	1.25	0.01%	0%		
1950	0	0.000	0.00	1.25	0.01%	0%		
1951	0	0.000	0.00	1.25	0.01%	0%		
1952	0	0.000	0.00	1.25	0.01%	0%		
1953	0	0.000	0.00	1.25	0.01%	0%		
1954	1	0.000	0.00	1.25	0.03%	0%		
1955	5	0.002	0.00	1.25	0.13%	0%		
1956	6	0.002	0.01	1.24	0.19%	0%		
1957	8	0.003	0.01	1.24	0.24%	1%		
1958	7	0.002	0.01	1.24	0.19%	1%		
1959	7	0.003	0.01	1.24	0.21%	1%		
1960	7	0.003	0.02	1.23	0.22%	1%		
1961	8	0.003	0.02	1.23	0.22%	1%		
1962	10	0.004	0.02	1.23	0.29%	2%		
1963	14	0.005	0.03	1.22	0.42%	2%		
1964	16	0.006	0.03	1.22	0.49%	3%		
1965	27	0.010	0.04	1.21	0.80%	3%		
1966	37	0.014	0.06	1.19	1.12%	5%		
1967	38	0.014	0.07	1.18	1.15%	6%		
1968	49	0.018	0.09	1.16	1.52%	7%		
1969	69	0.025	0.11	1.14	2.16%	9%		1st PEAK
1970	68	0.025	0.14	1.11	2.18%	11%		
1971	65	0.024	0.16	1.09	2.13%	13%		
1972	65	0.024	0.19	1.06	2.18%	15%		
1973	65	0.024	0.21	1.04	2.23%	17%		
1974	65	0.024	0.23	1.02	2.29%	19%		
1975	60	0.022	0.25	1.00	2.15%	20%		
1976	67	0.024	0.28	0.97	2.46%	22%		
1977	65	0.024	0.30	0.95	2.44%	24%		
1978	50	0.018	0.32	0.93	1.93%	26%		
1979	56	0.020	0.34	0.91	2.20%	27%		
1980	42	0.015	0.36	0.89	1.69%	29%		
1981	46	0.017	0.37	0.88	1.88%	30%		
1982	45	0.016	0.39	0.86	1.87%	31%		
1983	45	0.016	0.41	0.84	1.91%	33%		
1984	41	0.015	0.42	0.83	1.77%	34%		
1985	42	0.015	0.44	0.81	1.83%	35%		
1986	47	0.017	0.45	0.80	2.10%	36%		
1987	52	0.019	0.47	0.78	2.38%	38%		
1988	50	0.018	0.49	0.76	2.35%	39%		

1989	56	0.020	0.51	0.74	2.67%	41%	
1990	58	0.021	0.53	0.72	2.88%	43%	
1991	61	0.022	0.55	0.70	3.11%	44%	
1992	64	0.023	0.58	0.67	3.37%	46%	
1993	67	0.025	0.60	0.65	3.67%	48%	
1994	71	0.026	0.63	0.62	4.00%	50%	50% PRODUCED
1995	68	0.025	0.65	0.60	4.00%	52%	& 2nd PEAK
1996	65	0.024	0.68	0.57	4.00%	54%	
1997	63	0.023	0.70	0.55	4.00%	56%	
1998	60	0.022	0.72	0.53	4.00%	58%	
1999	58	0.021	0.74	0.51	4.00%	59%	
2000	55	0.020	0.76	0.49	4.00%	61%	
2001	53	0.019	0.78	0.47	4.00%	63%	
2002	51	0.019	0.80	0.45	4.00%	64%	
2003	49	0.018	0.82	0.43	4.00%	66%	
2004	47	0.017	0.84	0.41	4.00%	67%	
2005	45	0.017	0.85	0.40	4.00%	68%	
2006	43	0.016	0.87	0.38	4.00%	70%	
2007	42	0.015	0.88	0.37	4.00%	71%	
2008	40	0.015	0.90	0.35	4.00%	72%	
2009	38	0.014	0.91	0.34	4.00%	73%	
2010	37	0.013	0.93	0.32	4.00%	74%	
2011	35	0.013	0.94	0.31	4.00%	75%	
2012	34	0.012	0.95	0.30	4.00%	76%	
2013	33	0.012	0.96	0.29	4.00%	77%	
2014	31	0.011	0.98	0.27	4.00%	78%	
2015	30	0.011	0.99	0.26	4.00%	79%	
2016	29	0.011	1.00	0.25	4.00%	80%	
2017	28	0.010	1.01	0.24	4.00%	81%	
2018	27	0.010	1.02	0.23	4.00%	81%	
2019	26	0.009	1.03	0.22	4.00%	82%	
2020	25	0.009	1.04	0.21	4.00%	83%	
2021	24	0.009	1.04	0.21	4.00%	83%	
2022	23	0.008	1.05	0.20	4.00%	84%	
2023	22	0.008	1.06	0.19	4.00%	85%	
2024	21	0.008	1.07	0.18	4.00%	85%	
2025	20	0.007	1.07	0.18	4.00%	86%	
2026	19	0.007	1.08	0.17	4.00%	87%	
2027	18	0.007	1.09	0.16	4.00%	87%	
2028	18	0.006	1.09	0.16	4.00%	88%	
2029	17	0.006	1.10	0.15	4.00%	88%	
2030	16	0.006	1.11	0.14	4.00%	89%	
2031	16	0.006	1.11	0.14	4.00%	89%	
2032	15	0.005	1.12	0.13	4.00%	89%	
2033	14	0.005	1.12	0.13	4.00%	90%	
2034	14	0.005	1.13	0.12	4.00%	90%	
2035	13	0.005	1.13	0.12	4.00%	91%	
2036	13	0.005	1.14	0.11	4.00%	91%	
2037	12	0.004	1.14	0.11	4.00%	91%	
2038	12	0.004	1.15	0.10	4.00%	92%	
2039	11	0.004	1.15	0.10	4.00%	92%	
2040	11	0.004	1.16	0.09	4.00%	92%	
2041	10	0.004	1.16	0.09	4.00%	93%	
2042	10	0.004	1.16	0.09	4.00%	93%	
2043	10	0.004	1.17	0.08	4.00%	93%	
2044	9	0.003	1.17	0.08	4.00%	94%	
2045	9	0.003	1.17	0.08	4.00%	94%	
2046	8	0.003	1.18	0.07	4.00%	94%	
2047	8	0.003	1.18	0.07	4.00%	94%	
2048	8	0.003	1.18	0.07	4.00%	95%	
2049	8	0.003	1.18	0.07	4.00%	95%	
2050	7	0.003	1.19	0.06	4.00%	95%	

CUM.PROD	9.347	BBO	GIANT FIELDS		UNITED KINGDOM	
RESERVES	4.256	BBO	TOTAL BBO	11.5	W.EUROPE	
DISCOVERED	13.603	BBO	% OF DISC.	85%		
UNDISCOVERED	1.397	BBO	Last Giant	-		
REMAINING	5.653	BBO			Variance Trend	
ULTIMATE	15.000	BBO			1990	19%
Rept.Reserve	4.256	BBO	Ref.Date(end) 1989		Future	UP

Milestones	Date	Prod Mb/d	Dep. Rate	% Produced
1st Peak	1986	2602	10.55%	46%
2nd Peak				
Mid-Point	1987	2446	11.09%	52%
Final Decline	1990		10.12%	66%

Forecast "A"
A=Exponential decline
B=5% incr. to Mid-Point
C=Other
Pre-1930 Cum. 0
(Added to 1930 Prod)

Date	Prod. M.b/d	Prod. BBO/yr	Cum Prod	Remain	Dep. Rate	% Produced	Giant	Names & Notes
1930		0.000	0.00	15.00	ERR	0%		
1931		0.000	0.00	15.00	0.00%	0%		
1932		0.000	0.00	15.00	0.00%	0%		
1933		0.000	0.00	15.00	0.00%	0%		
1934		0.000	0.00	15.00	0.00%	0%		
1935		0.000	0.00	15.00	0.00%	0%		
1936		0.000	0.00	15.00	0.00%	0%		
1937		0.000	0.00	15.00	0.00%	0%		
1938		0.000	0.00	15.00	0.00%	0%		
1939		0.000	0.00	15.00	0.00%	0%		
1940	0	0.000	0.00	15.00	0.00%	0%		
1941	1	0.000	0.00	15.00	0.00%	0%		
1942	2	0.001	0.00	15.00	0.00%	0%		
1943	2	0.001	0.00	15.00	0.01%	0%		
1944	2	0.001	0.00	15.00	0.00%	0%		
1945	2	0.001	0.00	15.00	0.00%	0%		
1946	1	0.000	0.00	15.00	0.00%	0%		
1947	1	0.000	0.00	15.00	0.00%	0%		
1948	1	0.000	0.00	15.00	0.00%	0%		
1949	1	0.000	0.00	15.00	0.00%	0%		
1950	1	0.000	0.00	15.00	0.00%	0%		
1951	1	0.000	0.01	14.99	0.00%	0%		
1952	1	0.000	0.01	14.99	0.00%	0%		
1953	1	0.000	0.01	14.99	0.00%	0%		
1954	1	0.000	0.01	14.99	0.00%	0%		
1955	1	0.000	0.01	14.99	0.00%	0%		
1956	1	0.000	0.01	14.99	0.00%	0%		
1957	2	0.001	0.01	14.99	0.00%	0%		
1958	2	0.001	0.01	14.99	0.00%	0%		
1959	2	0.001	0.01	14.99	0.00%	0%		
1960	2	0.001	0.01	14.99	0.00%	0%		
1961	2	0.001	0.01	14.99	0.00%	0%		
1962	2	0.001	0.01	14.99	0.01%	0%		
1963	3	0.001	0.01	14.99	0.01%	0%		
1964	3	0.001	0.01	14.99	0.01%	0%		
1965	3	0.001	0.01	14.99	0.01%	0%		
1966	2	0.001	0.01	14.99	0.00%	0%		
1967	2	0.001	0.02	14.98	0.00%	0%		
1968	2	0.001	0.02	14.98	0.00%	0%		
1969	2	0.001	0.02	14.98	0.00%	0%		
1970	2	0.001	0.02	14.98	0.00%	0%	2.47	Forties
1971	2	0.001	0.02	14.98	0.00%	0%	1.73	Brent
1972	2	0.001	0.02	14.98	0.00%	0%	1	Beryl/Cormorant
1973	2	0.001	0.02	14.98	0.00%	0%	2.086	
1974	2	0.001	0.02	14.98	0.00%	0%	2.165	#1
1975	40	0.015	0.03	14.97	0.10%	0%	2.064	#3
1976	230	0.084	0.12	14.88	0.56%	1%		
1977	775	0.283	0.40	14.60	1.90%	3%		
1978	1100	0.402	0.80	14.20	2.75%	5%		
1979	1570	0.573	1.38	13.62	4.04%	9%		
1980	1600	0.584	1.96	13.04	4.29%	13%		
1981	1790	0.653	2.61	12.39	5.01%	17%		
1982	2050	0.748	3.36	11.64	6.04%	22%		
1983	2260	0.825	4.19	10.81	7.09%	28%		
1984	2452	0.895	5.08	9.92	8.28%	34%		
1985	2520	0.920	6.00	9.00	9.27%	40%		
1986	2602	0.950	6.95	8.05	10.55%	46%		1st PEAK
1987	2446	0.893	7.84	7.16	11.09%	52%		50% PRODUCED
1988	2376	0.867	8.71	6.29	12.12%	58%		

1989	1743	0.636	9.35	5.65	10.12%	62%
1990	1567	0.572	9.92	5.08	10.12%	66%
1991	1409	0.514	10.43	4.57	10.12%	70%
1992	1266	0.462	10.90	4.10	10.12%	73%
1993	1138	0.415	11.31	3.69	10.12%	75%
1994	1023	0.373	11.68	3.32	10.12%	78%
1995	919	0.336	12.02	2.98	10.12%	80%
1996	826	0.302	12.32	2.68	10.12%	82%
1997	743	0.271	12.59	2.41	10.12%	84%
1998	668	0.244	12.84	2.16	10.12%	86%
1999	600	0.219	13.06	1.94	10.12%	87%
2000	539	0.197	13.25	1.75	10.12%	88%
2001	485	0.177	13.43	1.57	10.12%	90%
2002	436	0.159	13.59	1.41	10.12%	91%
2003	392	0.143	13.73	1.27	10.12%	92%
2004	352	0.128	13.86	1.14	10.12%	92%
2005	316	0.115	13.97	1.03	10.12%	93%
2006	284	0.104	14.08	0.92	10.12%	94%
2007	256	0.093	14.17	0.83	10.12%	94%
2008	230	0.084	14.26	0.74	10.12%	95%
2009	206	0.075	14.33	0.67	10.12%	96%
2010	186	0.068	14.40	0.60	10.12%	96%
2011	167	0.061	14.46	0.54	10.12%	96%
2012	150	0.055	14.51	0.49	10.12%	97%
2013	135	0.049	14.56	0.44	10.12%	97%
2014	121	0.044	14.61	0.39	10.12%	97%
2015	109	0.040	14.65	0.35	10.12%	98%
2016	98	0.036	14.68	0.32	10.12%	98%
2017	88	0.032	14.72	0.28	10.12%	98%
2018	79	0.029	14.74	0.26	10.12%	98%
2019	71	0.026	14.77	0.23	10.12%	98%
2020	64	0.023	14.79	0.21	10.12%	99%
2021	57	0.021	14.81	0.19	10.12%	99%
2022	52	0.019	14.83	0.17	10.12%	99%
2023	46	0.017	14.85	0.15	10.12%	99%
2024	42	0.015	14.86	0.14	10.12%	99%
2025	37	0.014	14.88	0.12	10.12%	99%
2026	34	0.012	14.89	0.11	10.12%	99%
2027	30	0.011	14.90	0.10	10.12%	99%
2028	27	0.010	14.91	0.09	10.12%	99%
2029	24	0.009	14.92	0.08	10.12%	99%
2030	22	0.008	14.93	0.07	10.12%	100%
2031	20	0.007	14.94	0.06	10.12%	100%
2032	18	0.006	14.94	0.06	10.12%	100%
2033	16	0.006	14.95	0.05	10.12%	100%
2034	14	0.005	14.95	0.05	10.12%	100%
2035	13	0.005	14.96	0.04	10.12%	100%
2036	12	0.004	14.96	0.04	10.12%	100%
2037	10	0.004	14.97	0.03	10.12%	100%
2038	9	0.003	14.97	0.03	10.12%	100%
2039	8	0.003	14.97	0.03	10.12%	100%
2040	8	0.003	14.98	0.02	10.12%	100%
2041	7	0.002	14.98	0.02	10.12%	100%
2042	6	0.002	14.98	0.02	10.12%	100%
2043	5	0.002	14.98	0.02	10.12%	100%
2044	5	0.002	14.98	0.02	10.12%	100%
2045	4	0.002	14.99	0.01	10.12%	100%
2046	4	0.001	14.99	0.01	10.12%	100%
2047	4	0.001	14.99	0.01	10.12%	100%
2048	3	0.001	14.99	0.01	10.12%	100%
2049	3	0.001	14.99	0.01	10.12%	100%
2050	3	0.001	14.99	0.01	10.12%	100%

Notes
#1 Claymore, Magnus, Ninian
#2 Piper, Dunlin, Thistle
#3 Brae, Murchison, Fulmar

CUM.PROD	156.97	BBO	GIANT FIELDS		U.S.A.	
RESERVES	25.86	BBO	TOTAL BBO	56.7	N.AMERICA	
DISCOVERED	182.83	BBO	% OF DISC.	31%		
UNDISCOVERED	17.17	BBO	Last Giant	1982		
REMAINING	43.03	BBO			Variance	Trend
ULTIMATE	200.00	BBO			1990	-0.24%
Rept.Reserve	25.86		Ref. date	1989	Future	DOWN

Milestones	Date	Prod Mb/d	Dep. Rate	% Produced	Forecast "A"
1st Peak	1971	9650	3.42%	50%	A=Exponential decline
2nd Peak	1985	8919	5.58%	72%	B=5% incr. to Mid-Point
Mid-Point	1971	9650	3.42%	50%	C=Other
Final Decline	1990		6.11%	80%	Pre-1930 Cum. 13.31 (Added to 1930 prod)

Date	Prod. M.b/d	Prod. BBO/yr	Cum. Prod.	Remain	Deplet. Rate	% Produced	Giant	Notes & Names
1930	2347	0.857	14.17	185.83	ERR	7%	28.93	Pre-1930
1931	2477	0.904	15.07	184.93	0.49%	8%	0.76	Conroe
1932	2609	0.952	16.03	183.97	0.51%	8%	2.79	Wilmington
1933	2743	1.001	17.03	182.97	0.54%	9%	0.81	Greta
1934	2878	1.051	18.08	181.92	0.57%	9%	0.77	Hastings
1935	3015	1.100	19.18	180.82	0.60%	10%	0.79	Goldsmith
1936	3152	1.151	20.33	179.67	0.64%	10%	3.47	#1
1937	3289	1.201	21.53	178.47	0.67%	11%		
1938	3327	1.214	22.74	177.26	0.68%	11%	1.04	#2
1939	3466	1.265	24.01	175.99	0.71%	12%		
1940	3697	1.350	25.36	174.64	0.77%	13%	0.86	Hawkins
1941	3842	1.402	26.76	173.24	0.80%	13%		
1942	3799	1.387	28.15	171.85	0.80%	14%		
1943	4125	1.506	29.65	170.35	0.88%	15%		
1944	4584	1.673	31.33	168.67	0.98%	16%		
1945	4695	1.714	33.04	166.96	1.02%	17%	0.50	Golden Trnd
1946	4751	1.734	34.77	165.23	1.04%	17%		
1947	5088	1.857	36.63	163.37	1.12%	18%	0.53	S.Ardo
1948	5520	2.015	38.65	161.35	1.23%	19%	1.35	Kelly
1949	5046	1.842	40.49	159.51	1.14%	20%		
1950	5402	1.972	42.46	157.54	1.24%	21%		
1951	6149	2.245	44.70	155.30	1.42%	22%		
1952	6268	2.288	46.99	153.01	1.47%	23%		
1953	6512	2.377	49.37	150.63	1.55%	25%		
1954	6340	2.314	51.68	148.32	1.54%	26%		
1955	6763	2.468	54.15	145.85	1.66%	27%		
1956	7150	2.610	56.76	143.24	1.79%	28%		
1957	7160	2.613	59.37	140.63	1.82%	30%		
1958	6459	2.358	61.73	138.27	1.68%	31%		
1959	7043	2.571	64.30	135.70	1.86%	32%		
1960	7019	2.562	66.86	133.14	1.89%	33%		
1961	7181	2.621	69.49	130.51	1.97%	35%		
1962	7337	2.678	72.16	127.84	2.05%	36%		
1963	7537	2.751	74.91	125.09	2.15%	37%		
1964	7640	2.789	77.70	122.30	2.23%	39%		
1965	7753	2.830	80.53	119.47	2.31%	40%	0.58	McArthur R.
1966	8301	3.030	83.56	116.44	2.54%	42%		
1967	8801	3.212	86.78	113.22	2.76%	43%	9.92	Prudhoe Bay
1968	9153	3.341	90.12	109.88	2.95%	45%		
1969	9178	3.350	93.47	106.53	3.05%	47%	1.50	Kuparuk
1970	9507	3.470	96.94	103.06	3.26%	48%		
1971	9650	3.522	100.46	99.54	3.42%	50%		50% PRODUCED
1972	9500	3.468	103.93	96.07	3.48%	52%		& 1st PEAK
1973	9225	3.367	107.29	92.71	3.50%	54%		
1974	8945	3.265	110.56	89.44	3.52%	55%		
1975	8370	3.055	113.61	86.39	3.42%	57%		
1976	8105	2.958	116.57	83.43	3.42%	58%		
1977	8240	3.008	119.58	80.42	3.60%	60%		
1978	8660	3.161	122.74	77.26	3.93%	61%		
1979	8650	3.157	125.90	74.10	4.09%	63%	0.81	E.Anschutz
1980	8650	3.157	129.05	70.95	4.26%	65%		
1981	8588	3.135	132.19	67.81	4.42%	66%	0.55	Pt.Arguello
1982	8655	3.159	135.35	64.65	4.66%	68%	0.71	Sprayberry
1983	8669	3.164	138.51	61.49	4.89%	69%		
1984	8750	3.194	141.71	58.29	5.19%	71%		
1985	8919	3.255	144.96	55.04	5.58%	72%		2ND PEAK
1986	8790	3.208	148.17	51.83	5.83%	74%		
1987	8277	3.021	151.19	48.81	5.83%	76%		
1988	8166	2.981	154.17	45.83	6.11%	77%		

1989	7676	2.802	156.97	43.03	6.11%	78%
1990	7203	2.629	159.60	40.40	6.11%	80%
1991	6763	2.468	162.07	37.93	6.11%	81%
1992	6349	2.318	164.39	35.61	6.11%	82%
1993	5961	2.176	166.56	33.44	6.11%	83%
1994	5597	2.043	168.61	31.39	6.11%	84%
1995	5255	1.918	170.52	29.48	6.11%	85%
1996	4934	1.801	172.33	27.67	6.11%	86%
1997	4633	1.691	174.02	25.98	6.11%	87%
1998	4350	1.588	175.60	24.40	6.11%	88%
1999	4084	1.491	177.09	22.91	6.11%	89%
2000	3834	1.400	178.49	21.51	6.11%	89%
2001	3600	1.314	179.81	20.19	6.11%	90%
2002	3380	1.234	181.04	18.96	6.11%	91%
2003	3174	1.158	182.20	17.80	6.11%	91%
2004	2980	1.088	183.29	16.71	6.11%	92%
2005	2798	1.021	184.31	15.69	6.11%	92%
2006	2627	0.959	185.27	14.73	6.11%	93%
2007	2466	0.900	186.17	13.83	6.11%	93%
2008	2315	0.845	187.01	12.99	6.11%	94%
2009	2174	0.794	187.81	12.19	6.11%	94%
2010	2041	0.745	188.55	11.45	6.11%	94%
2011	1916	0.700	189.25	10.75	6.11%	95%
2012	1799	0.657	189.91	10.09	6.11%	95%
2013	1689	0.617	190.52	9.48	6.11%	95%
2014	1586	0.579	191.10	8.90	6.11%	96%
2015	1489	0.544	191.65	8.35	6.11%	96%
2016	1398	0.510	192.16	7.84	6.11%	96%
2017	1313	0.479	192.64	7.36	6.11%	96%
2018	1233	0.450	193.09	6.91	6.11%	97%
2019	1157	0.422	193.51	6.49	6.11%	97%
2020	1087	0.397	193.91	6.09	6.11%	97%
2021	1020	0.372	194.28	5.72	6.11%	97%
2022	958	0.350	194.63	5.37	6.11%	97%
2023	899	0.328	194.96	5.04	6.11%	97%
2024	844	0.308	195.26	4.74	6.11%	98%
2025	793	0.289	195.55	4.45	6.11%	98%
2026	744	0.272	195.82	4.18	6.11%	98%
2027	699	0.255	196.08	3.92	6.11%	98%
2028	656	0.240	196.32	3.68	6.11%	98%
2029	616	0.225	196.54	3.46	6.11%	98%
2030	578	0.211	196.76	3.24	6.11%	98%
2031	543	0.198	196.95	3.05	6.11%	98%
2032	510	0.186	197.14	2.86	6.11%	99%
2033	479	0.175	197.31	2.69	6.11%	99%
2034	450	0.164	197.48	2.52	6.11%	99%
2035	422	0.154	197.63	2.37	6.11%	99%
2036	396	0.145	197.78	2.22	6.11%	99%
2037	372	0.136	197.91	2.09	6.11%	99%
2038	349	0.128	198.04	1.96	6.11%	99%
2039	328	0.120	198.16	1.84	6.11%	99%
2040	308	0.112	198.27	1.73	6.11%	99%
2041	289	0.106	198.38	1.62	6.11%	99%
2042	271	0.099	198.48	1.52	6.11%	99%
2043	255	0.093	198.57	1.43	6.11%	99%
2044	239	0.087	198.66	1.34	6.11%	99%
2045	225	0.082	198.74	1.26	6.11%	99%
2046	211	0.077	198.82	1.18	6.11%	99%
2047	198	0.072	198.89	1.11	6.11%	99%
2048	186	0.068	198.96	1.04	6.11%	99%
2049	175	0.064	199.02	0.98	6.11%	100%
2050	164	0.060	199.08	0.92	6.11%	100%

#1 Slaughter 1.08, Wasson 1.798, Webster .593
#2 Coalinga Nose .508, Levelland .534

CUM.PROD	106.33	BBO	GIANT FIELDS		U.S.S.R.	
RESERVES	58.40	BBO	TOTAL BBO	97.9	SINO-SOVIET	
DISCOVERED	164.73	BBO	% OF DISC.	59%		
UNDISCOVERED	35.27	BBO	Last Giant	1979	----------------	
REMAINING	93.67	BBO	------------------		Variance	Trend
ULTIMATE	200.00	BBO			1990	-8%
Rept.Reserve	58.400		Ref.date(end)	1989	Future	DOWN

Milestones	Date	Prod Mb/d	Dep. Rate	% Produced	Forecast	"A"
					A=Exponential decline	
1st Peak	1988	12531	4.45%	51%	B=5% incr. to Mid-Point	
2nd Peak	None				C=Other	
Mid-Point	1988	12531	4.45%	51%	Pre-1930 Cum.	0.6155
Final Decline	1990		4.52%	55%	(Added to 1930)	

Date	Prod. M.b/d	Prod. BBO/yr	Cum. Prod.	Remain	Deplet. Rate	% Produced	Giant	Notes & Names
1930	349	0.127	0.74	199.26	ERR	0%	10.37	Pre-1930
1931	454	0.166	0.91	199.09	0.08%	0%		
1932	431	0.157	1.07	198.93	0.08%	1%		See Giant Field
1933	431	0.157	1.22	198.78	0.08%	1%		List for names
1934	485	0.177	1.40	198.60	0.09%	1%	0.50	
1935	507	0.185	1.59	198.41	0.09%	1%		
1936	519	0.189	1.77	198.23	0.10%	1%		
1937	539	0.197	1.97	198.03	0.10%	1%	1.00	
1938	572	0.209	2.18	197.82	0.11%	1%		
1939	605	0.221	2.40	197.60	0.11%	1%		
1940	610	0.223	2.62	197.38	0.11%	1%		
1941	663	0.242	2.87	197.13	0.12%	1%		
1942	634	0.231	3.10	196.90	0.12%	2%		
1943	564	0.206	3.30	196.70	0.10%	2%		
1944	767	0.280	3.58	196.42	0.14%	2%	1.25	
1945	425	0.155	3.74	196.26	0.08%	2%	1.53	
1946	448	0.164	3.90	196.10	0.08%	2%	0.50	
1947	533	0.194	4.10	195.90	0.10%	2%		
1948	616	0.225	4.32	195.68	0.11%	2%	14.00	
1949	670	0.245	4.57	195.43	0.13%	2%	1.23	
1950	748	0.273	4.84	195.16	0.14%	2%		
1951	800	0.292	5.13	194.87	0.15%	3%		
1952	903	0.329	5.46	194.54	0.17%	3%		
1953	1042	0.380	5.84	194.16	0.20%	3%		
1954	1170	0.427	6.27	193.73	0.22%	3%	0.51	
1955	1397	0.510	6.78	193.22	0.26%	3%	4.12	
1956	1676	0.612	7.39	192.61	0.32%	4%	2.90	
1957	1967	0.718	8.11	191.89	0.37%	4%		
1958	2264	0.826	8.93	191.07	0.43%	4%	0.76	
1959	2591	0.946	9.88	190.12	0.49%	5%	0.50	
1960	2957	1.079	10.96	189.04	0.57%	5%	0.60	
1961	3321	1.212	12.17	187.83	0.64%	6%	5.06	
1962	3725	1.360	13.53	186.47	0.72%	7%	3.08	
1963	4121	1.504	15.03	184.97	0.81%	8%	5.43	
1964	4503	1.643	16.68	183.32	0.89%	8%	1.45	
1965	4901	1.789	18.47	181.53	0.98%	9%	25.90	
1966	5337	1.948	20.42	179.58	1.07%	10%	0.60	
1967	5733	2.093	22.51	177.49	1.17%	11%		
1968	6168	2.251	24.76	175.24	1.27%	12%		
1969	6545	2.389	27.15	172.85	1.36%	14%	1.44	
1970	6985	2.550	29.70	170.30	1.47%	15%	2.10	
1971	7612	2.778	32.48	167.52	1.63%	16%	1.10	
1972	7934	2.896	35.37	164.63	1.73%	18%	1.00	
1973	8478	3.094	38.47	161.53	1.88%	19%	0.5	
1974	9243	3.374	41.84	158.16	2.09%	21%	0.50	
1975	9887	3.609	45.45	154.55	2.28%	23%		
1976	10471	3.822	49.27	150.73	2.47%	25%		
1977	10934	3.991	53.26	146.74	2.65%	27%		
1978	11215	4.093	57.36	142.64	2.79%	29%		
1979	11700	4.271	61.63	138.37	2.99%	31%	10.00	
1980	12063	4.403	66.03	133.97	3.18%	33%		
1981	12176	4.444	70.47	129.53	3.32%	35%		
1982	12251	4.472	74.94	125.06	3.45%	37%		
1983	12327	4.499	79.44	120.56	3.60%	40%		
1984	12338	4.503	83.95	116.05	3.74%	42%		
1985	11882	4.337	88.28	111.72	3.74%	44%		
1986	12289	4.485	92.77	107.23	4.02%	46%		
1987	12478	4.554	97.32	102.68	4.25%	49%		
1988	12531	4.574	101.90	98.10	4.45%	51%		1ST PEAK

1989	12140	4.431	106.33	93.67	4.52%	53%	& 50% PRODUCED
1990	11600	4.234	110.56	89.44	4.52%	55%	
1991	11075	4.043	114.61	85.39	4.52%	57%	
1992	10575	3.860	118.47	81.53	4.52%	59%	
1993	10097	3.685	122.15	77.85	4.52%	61%	
1994	9641	3.519	125.67	74.33	4.52%	63%	
1995	9205	3.360	129.03	70.97	4.52%	65%	
1996	8789	3.208	132.24	67.76	4.52%	66%	
1997	8391	3.063	135.30	64.70	4.52%	68%	
1998	8012	2.924	138.22	61.78	4.52%	69%	
1999	7650	2.792	141.02	58.98	4.52%	71%	
2000	7304	2.666	143.68	56.32	4.52%	72%	
2001	6974	2.546	146.23	53.77	4.52%	73%	
2002	6659	2.430	148.66	51.34	4.52%	74%	
2003	6358	2.321	150.98	49.02	4.52%	75%	
2004	6070	2.216	153.20	46.80	4.52%	77%	
2005	5796	2.116	155.31	44.69	4.52%	78%	
2006	5534	2.020	157.33	42.67	4.52%	79%	
2007	5284	1.929	159.26	40.74	4.52%	80%	
2008	5045	1.841	161.10	38.90	4.52%	81%	
2009	4817	1.758	162.86	37.14	4.52%	81%	
2010	4599	1.679	164.54	35.46	4.52%	82%	
2011	4391	1.603	166.14	33.86	4.52%	83%	
2012	4193	1.530	167.67	32.33	4.52%	84%	
2013	4003	1.461	169.13	30.87	4.52%	85%	
2014	3823	1.395	170.53	29.47	4.52%	85%	
2015	3650	1.332	171.86	28.14	4.52%	86%	
2016	3485	1.272	173.13	26.87	4.52%	87%	
2017	3327	1.214	174.35	25.65	4.52%	87%	
2018	3177	1.160	175.51	24.49	4.52%	88%	
2019	3033	1.107	176.61	23.39	4.52%	88%	
2020	2896	1.057	177.67	22.33	4.52%	89%	
2021	2765	1.009	178.68	21.32	4.52%	89%	
2022	2640	0.964	179.64	20.36	4.52%	90%	
2023	2521	0.920	180.56	19.44	4.52%	90%	
2024	2407	0.879	181.44	18.56	4.52%	91%	
2025	2298	0.839	182.28	17.72	4.52%	91%	
2026	2194	0.801	183.08	16.92	4.52%	92%	
2027	2095	0.765	183.85	16.15	4.52%	92%	
2028	2000	0.730	184.58	15.42	4.52%	92%	
2029	1910	0.697	185.27	14.73	4.52%	93%	
2030	1824	0.666	185.94	14.06	4.52%	93%	
2031	1741	0.636	186.57	13.43	4.52%	93%	
2032	1663	0.607	187.18	12.82	4.52%	94%	
2033	1587	0.579	187.76	12.24	4.52%	94%	
2034	1516	0.553	188.31	11.69	4.52%	94%	
2035	1447	0.528	188.84	11.16	4.52%	94%	
2036	1382	0.504	189.35	10.65	4.52%	95%	
2037	1319	0.482	189.83	10.17	4.52%	95%	
2038	1260	0.460	190.29	9.71	4.52%	95%	
2039	1203	0.439	190.73	9.27	4.52%	95%	
2040	1148	0.419	191.15	8.85	4.52%	96%	
2041	1096	0.400	191.55	8.45	4.52%	96%	
2042	1047	0.382	191.93	8.07	4.52%	96%	
2043	1000	0.365	192.29	7.71	4.52%	96%	
2044	954	0.348	192.64	7.36	4.52%	96%	
2045	911	0.333	192.97	7.03	4.52%	96%	
2046	870	0.318	193.29	6.71	4.52%	97%	
2047	831	0.303	193.59	6.41	4.52%	97%	
2048	793	0.290	193.88	6.12	4.52%	97%	
2049	757	0.276	194.16	5.84	4.52%	97%	
2050	723	0.264	194.42	5.58	4.52%	97%	

CUM.PROD	16.357 BBO	GIANT FIELDS			MEXICO	
RESERVES	40.000 BBO	TOTAL BBO	40.3		L.AMERICA	
DISCOVERED	56.357 BBO	% OF DISC.	72%			
UNDISCOVERED	3.643 BBO	Last Giant	1978			
REMAINING	43.643 BBO				Variance Trend	
ULTIMATE	60.000 BBO				1990	-4%
Rept.Reserves	56.365 BBO	Ref. Date (end)		1989	Future	DOWN

Milestones	Date	Prod Mb/d	Dep. Rate	% Produced
1st Peak	1985	2797	2.11%	21%
2nd Peak	2000	4063	4.78%	51%
Mid-Point	2000	4063	4.78%	51%
Final Decline	2000		4.78%	

Forecast "B"
A=Exponential decline
B=5% inc. to Mid-Point
C=Other
Pre-1930 Cum 1.374
(added to 1930 Prod)

Date	Prod. M.b/d	Prod. BBO/yr	Cum Prod	Remain	Dep. Rate	% Produced	Giant	Names & Notes
1930	112.3	0.041	1.42	58.59	ERR	2%	11.7	#1
1931	110.0	0.040	1.46	58.54	0.07%	2%		
1932	110.0	0.040	1.50	58.50	0.07%	2%		
1933	110.0	0.040	1.54	58.46	0.07%	3%		
1934	110.0	0.040	1.58	58.42	0.07%	3%		
1935	110.0	0.040	1.62	58.38	0.07%	3%		
1936	110.0	0.040	1.66	58.34	0.07%	3%		
1937	110.0	0.040	1.70	58.30	0.07%	3%		
1938	105.6	0.039	1.73	58.27	0.07%	3%		
1939	117.7	0.043	1.78	58.22	0.07%	3%		
1940	120.4	0.044	1.82	58.18	0.08%	3%		
1941	115.8	0.042	1.86	58.14	0.07%	3%		
1942	95.5	0.035	1.90	58.10	0.06%	3%		
1943	96.4	0.035	1.93	58.07	0.06%	3%		
1944	104.7	0.038	1.97	58.03	0.07%	3%		
1945	119.4	0.044	2.02	57.98	0.08%	3%		
1946	135.0	0.049	2.06	57.94	0.08%	3%		
1947	153.8	0.056	2.12	57.88	0.10%	4%		
1948	159.5	0.058	2.18	57.82	0.10%	4%		
1949	166.4	0.061	2.24	57.76	0.11%	4%		
1950	197.6	0.072	2.31	57.69	0.12%	4%		
1951	211.2	0.077	2.39	57.61	0.13%	4%		
1952	218.0	0.080	2.47	57.53	0.14%	4%		
1953	201.0	0.073	2.54	57.46	0.13%	4%		
1954	232.0	0.085	2.63	57.37	0.15%	4%		
1955	250.0	0.091	2.72	57.28	0.16%	5%		
1956	253.0	0.092	2.81	57.19	0.16%	5%		
1957	241.0	0.088	2.90	57.10	0.15%	5%		
1958	243.1	0.089	2.99	57.01	0.16%	5%		
1959	264.0	0.096	3.08	56.92	0.17%	5%		
1960	275.0	0.100	3.18	56.82	0.18%	5%	4.5	#2
1961	283.0	0.103	3.29	56.71	0.18%	5%		
1962	305.0	0.111	3.40	56.60	0.20%	6%		
1963	320.0	0.117	3.52	56.48	0.21%	6%		
1964	310.0	0.113	3.63	56.37	0.20%	6%		
1965	333.2	0.122	3.75	56.25	0.22%	6%		
1966	322.2	0.118	3.87	56.13	0.21%	6%		
1967	357.6	0.131	4.00	56.00	0.23%	7%	1	#3
1968	383.5	0.140	4.14	55.86	0.25%	7%		
1969	405.1	0.148	4.29	55.71	0.26%	7%		
1970	427.4	0.156	4.44	55.56	0.28%	7%		
1971	427.0	0.156	4.60	55.40	0.28%	8%		
1972	440.0	0.161	4.76	55.24	0.29%	8%	1.85	#4
1973	478.0	0.174	4.93	55.07	0.32%	8%		
1974	513.0	0.187	5.12	54.88	0.34%	9%		
1975	710.0	0.259	5.38	54.62	0.47%	9%	0.5	#5
1976	850.0	0.310	5.69	54.31	0.57%	9%		
1977	990.0	0.361	6.05	53.95	0.67%	10%	8.5	#6
1978	1270.0	0.464	6.51	53.49	0.86%	11%	12.285	#7
1979	1490.0	0.544	7.06	52.94	1.02%	12%		
1980	1960.0	0.715	7.77	52.23	1.35%	13%		
1981	2390.0	0.872	8.65	51.35	1.67%	14%		
1982	2734.0	0.998	9.64	50.36	1.94%	16%		
1983	2702.0	0.986	10.63	49.37	1.96%	18%		
1984	2743.0	1.001	11.63	48.37	2.03%	19%		
1985	2797.0	1.021	12.65	47.35	2.11%	21%		1ST PEAK
1986	2468.0	0.901	13.55	46.45	1.90%	23%		
1987	2537.0	0.926	14.48	45.52	1.99%	24%		
1988	2527.0	0.922	15.40	44.60	2.03%	26%		

1989	2618.0	0.956	16.36	43.64	2.14%	27%	
1990	2748.9	1.003	17.36	42.64	2.30%	29%	
1991	2886.3	1.054	18.41	41.59	2.47%	31%	
1992	3030.7	1.106	19.52	40.48	2.66%	33%	
1993	3182.2	1.162	20.68	39.32	2.87%	34%	
1994	3341.3	1.220	21.90	38.10	3.10%	37%	
1995	3508.4	1.281	23.18	36.82	3.36%	39%	
1996	3683.8	1.345	24.53	35.47	3.65%	41%	
1997	3868.0	1.412	25.94	34.06	3.98%	43%	
1998	4061.4	1.482	27.42	32.58	4.35%	46%	
1999	4264.4	1.557	28.98	31.02	4.78%	48%	
2000	4062.7	1.483	30.46	29.54	4.78%	51%	2ND PEAK
2001	3868.5	1.412	31.87	28.13	4.78%	53%	& 50% PRODUCED
2002	3683.6	1.345	33.22	26.78	4.78%	55%	
2003	3507.5	1.280	34.50	25.50	4.78%	57%	
2004	3339.9	1.219	35.72	24.28	4.78%	60%	
2005	3180.2	1.161	36.88	23.12	4.78%	61%	
2006	3028.2	1.105	37.98	22.02	4.78%	63%	
2007	2883.5	1.052	39.03	20.97	4.78%	65%	
2008	2745.6	1.002	40.04	19.96	4.78%	67%	
2009	2614.4	0.954	40.99	19.01	4.78%	68%	
2010	2489.4	0.909	41.90	18.10	4.78%	70%	
2011	2370.4	0.865	42.76	17.24	4.78%	71%	
2012	2257.1	0.824	43.59	16.41	4.78%	73%	
2013	2149.2	0.784	44.37	15.63	4.78%	74%	
2014	2046.5	0.747	45.12	14.88	4.78%	75%	
2015	1948.7	0.711	45.83	14.17	4.78%	76%	
2016	1855.5	0.677	46.51	13.49	4.78%	78%	
2017	1766.8	0.645	47.15	12.85	4.78%	79%	
2018	1682.4	0.614	47.77	12.23	4.78%	80%	
2019	1602.0	0.585	48.35	11.65	4.78%	81%	
2020	1525.4	0.557	48.91	11.09	4.78%	82%	
2021	1452.5	0.530	49.44	10.56	4.78%	82%	
2022	1383.1	0.505	49.94	10.06	4.78%	83%	
2023	1316.9	0.481	50.42	9.58	4.78%	84%	
2024	1254.0	0.458	50.88	9.12	4.78%	85%	
2025	1194.1	0.436	51.32	8.68	4.78%	86%	
2026	1137.0	0.415	51.73	8.27	4.78%	86%	
2027	1082.6	0.395	52.13	7.87	4.78%	87%	
2028	1030.9	0.376	52.50	7.50	4.78%	88%	
2029	981.6	0.358	52.86	7.14	4.78%	88%	
2030	934.7	0.341	53.20	6.80	4.78%	89%	
2031	890.0	0.325	53.53	6.47	4.78%	89%	
2032	847.5	0.309	53.84	6.16	4.78%	90%	
2033	807.0	0.295	54.13	5.87	4.78%	90%	
2034	768.4	0.280	54.41	5.59	4.78%	91%	
2035	731.7	0.267	54.68	5.32	4.78%	91%	
2036	696.7	0.254	54.93	5.07	4.78%	92%	
2037	663.4	0.242	55.18	4.82	4.78%	92%	
2038	631.7	0.231	55.41	4.59	4.78%	92%	
2039	601.5	0.220	55.63	4.37	4.78%	93%	
2040	572.7	0.209	55.84	4.16	4.78%	93%	
2041	545.3	0.199	56.03	3.97	4.78%	93%	
2042	519.3	0.190	56.22	3.78	4.78%	94%	
2043	494.5	0.180	56.40	3.60	4.78%	94%	
2044	470.8	0.172	56.58	3.42	4.78%	94%	
2045	448.3	0.164	56.74	3.26	4.78%	95%	
2046	426.9	0.156	56.90	3.10	4.78%	95%	
2047	406.5	0.148	57.04	2.96	4.78%	95%	
2048	387.1	0.141	57.19	2.81	4.78%	95%	
2049	368.6	0.135	57.32	2.68	4.78%	96%	
2050	350.9	0.128	57.45	2.55	4.78%	96%	

CUM.PROD	0.007	BBO	GIANT FIELDS		VIET NAM	
RESERVES	0.500	BBO	TOTAL BBO	0	THE EAST	
DISCOVERED	0.507	BBO	% OF DISC.	0%		
UNDISCOVERED	0.243	BBO	Last Giant	-		
REMAINING	0.743	BBO			Variance Trend	
ULTIMATE	0.750	BBO			1990	zero
Rept.Reserve	0.500		Ref.date(end) 1989		Future	UP

Milestones	Date	Prod Mb/d	Dep. Rate	% Produced	Forecast "C"
1st Peak	2005	83	7.10%	47%	A=Exponential decline B=5% incr. to Mid-Point
2nd Peak					C=Other
Mid-Point	2006	77	7.10%	51%	Pre-1930 Cum. 0
Final Decline	2007		7.10%		

Date	Prod. M.b/d	Prod. BBO/yr	Cum. Prod.	Remain	Deplet. Rate	% Produced	Giant	Notes & Names
1930		0.000	0.00	0.75	ERR	0%		
1931		0.000	0.00	0.75	0.00%	0%		
1932		0.000	0.00	0.75	0.00%	0%		
1933		0.000	0.00	0.75	0.00%	0%		
1934		0.000	0.00	0.75	0.00%	0%		
1935		0.000	0.00	0.75	0.00%	0%		
1936		0.000	0.00	0.75	0.00%	0%		
1937		0.000	0.00	0.75	0.00%	0%		
1938		0.000	0.00	0.75	0.00%	0%		
1939		0.000	0.00	0.75	0.00%	0%		
1940		0.000	0.00	0.75	0.00%	0%		
1941		0.000	0.00	0.75	0.00%	0%		
1942		0.000	0.00	0.75	0.00%	0%		
1943		0.000	0.00	0.75	0.00%	0%		
1944		0.000	0.00	0.75	0.00%	0%		
1945		0.000	0.00	0.75	0.00%	0%		
1946		0.000	0.00	0.75	0.00%	0%		
1947		0.000	0.00	0.75	0.00%	0%		
1948		0.000	0.00	0.75	0.00%	0%		
1949		0.000	0.00	0.75	0.00%	0%		
1950		0.000	0.00	0.75	0.00%	0%		
1951		0.000	0.00	0.75	0.00%	0%		
1952		0.000	0.00	0.75	0.00%	0%		
1953		0.000	0.00	0.75	0.00%	0%		
1954		0.000	0.00	0.75	0.00%	0%		
1955		0.000	0.00	0.75	0.00%	0%		
1956		0.000	0.00	0.75	0.00%	0%		
1957		0.000	0.00	0.75	0.00%	0%		
1958		0.000	0.00	0.75	0.00%	0%		
1959		0.000	0.00	0.75	0.00%	0%		
1960		0.000	0.00	0.75	0.00%	0%		
1961		0.000	0.00	0.75	0.00%	0%		
1962		0.000	0.00	0.75	0.00%	0%		
1963		0.000	0.00	0.75	0.00%	0%		
1964		0.000	0.00	0.75	0.00%	0%		
1965		0.000	0.00	0.75	0.00%	0%		
1966		0.000	0.00	0.75	0.00%	0%		
1967		0.000	0.00	0.75	0.00%	0%		
1968		0.000	0.00	0.75	0.00%	0%		
1969		0.000	0.00	0.75	0.00%	0%		
1970		0.000	0.00	0.75	0.00%	0%		
1971		0.000	0.00	0.75	0.00%	0%		
1972		0.000	0.00	0.75	0.00%	0%		
1973		0.000	0.00	0.75	0.00%	0%		
1974		0.000	0.00	0.75	0.00%	0%		
1975		0.000	0.00	0.75	0.00%	0%		
1976		0.000	0.00	0.75	0.00%	0%		
1977		0.000	0.00	0.75	0.00%	0%		
1978		0.000	0.00	0.75	0.00%	0%		
1979		0.000	0.00	0.75	0.00%	0%		
1980		0.000	0.00	0.75	0.00%	0%		
1981		0.000	0.00	0.75	0.00%	0%		
1982		0.000	0.00	0.75	0.00%	0%		
1983		0.000	0.00	0.75	0.00%	0%		
1984		0.000	0.00	0.75	0.00%	0%		
1985		0.000	0.00	0.75	0.00%	0%		
1986		0.000	0.00	0.75	0.00%	0%		
1987		0.000	0.00	0.75	0.00%	0%		
1988		0.000	0.00	0.75	0.00%	0%		

1989	20	0.007	0.01	0.74	0.97%	1%	Prod starts
1990	40	0.015	0.02	0.73	1.97%	3%	
1991	42	0.015	0.04	0.71	2.11%	5%	
1992	44	0.016	0.05	0.70	2.26%	7%	
1993	46	0.017	0.07	0.68	2.43%	9%	
1994	49	0.018	0.09	0.66	2.61%	12%	
1995	51	0.019	0.11	0.64	2.81%	14%	
1996	54	0.020	0.13	0.62	3.04%	17%	
1997	56	0.021	0.15	0.60	3.29%	20%	
1998	59	0.022	0.17	0.58	3.58%	22%	
1999	62	0.023	0.19	0.56	3.89%	25%	
2000	65	0.024	0.21	0.54	4.25%	29%	
2001	68	0.025	0.24	0.51	4.67%	32%	
2002	72	0.026	0.27	0.48	5.14%	35%	
2003	75	0.028	0.29	0.46	5.69%	39%	
2004	79	0.029	0.32	0.43	6.33%	43%	
2005	83	0.030	0.35	0.40	7.10%	47%	1ST PEAK
2006	77	0.028	0.38	0.37	7.10%	51%	50% PRODUCED
2007	72	0.026	0.41	0.34	7.10%	54%	
2008	67	0.024	0.43	0.32	7.10%	58%	
2009	62	0.023	0.45	0.30	7.10%	61%	
2010	58	0.021	0.48	0.27	7.10%	63%	
2011	53	0.020	0.49	0.26	7.10%	66%	
2012	50	0.018	0.51	0.24	7.10%	68%	
2013	46	0.017	0.53	0.22	7.10%	71%	
2014	43	0.016	0.55	0.20	7.10%	73%	
2015	40	0.015	0.56	0.19	7.10%	75%	
2016	37	0.014	0.57	0.18	7.10%	76%	
2017	34	0.013	0.59	0.16	7.10%	78%	
2018	32	0.012	0.60	0.15	7.10%	80%	
2019	30	0.011	0.61	0.14	7.10%	81%	
2020	28	0.010	0.62	0.13	7.10%	82%	
2021	26	0.009	0.63	0.12	7.10%	84%	
2022	24	0.009	0.64	0.11	7.10%	85%	
2023	22	0.008	0.64	0.11	7.10%	86%	
2024	21	0.007	0.65	0.10	7.10%	87%	
2025	19	0.007	0.66	0.09	7.10%	88%	
2026	18	0.006	0.67	0.08	7.10%	89%	
2027	16	0.006	0.67	0.08	7.10%	90%	
2028	15	0.006	0.68	0.07	7.10%	90%	
2029	14	0.005	0.68	0.07	7.10%	91%	
2030	13	0.005	0.69	0.06	7.10%	92%	
2031	12	0.004	0.69	0.06	7.10%	92%	
2032	11	0.004	0.70	0.05	7.10%	93%	
2033	11	0.004	0.70	0.05	7.10%	93%	
2034	10	0.004	0.70	0.05	7.10%	94%	
2035	9	0.003	0.71	0.04	7.10%	94%	
2036	8	0.003	0.71	0.04	7.10%	95%	
2037	8	0.003	0.71	0.04	7.10%	95%	
2038	7	0.003	0.72	0.03	7.10%	95%	
2039	7	0.002	0.72	0.03	7.10%	96%	
2040	6	0.002	0.72	0.03	7.10%	96%	
2041	6	0.002	0.72	0.03	7.10%	96%	
2042	5	0.002	0.72	0.03	7.10%	97%	
2043	5	0.002	0.73	0.02	7.10%	97%	
2044	5	0.002	0.73	0.02	7.10%	97%	
2045	4	0.002	0.73	0.02	7.10%	97%	
2046	4	0.001	0.73	0.02	7.10%	97%	
2047	4	0.001	0.73	0.02	7.10%	98%	
2048	4	0.001	0.73	0.02	7.10%	98%	
2049	3	0.001	0.73	0.02	7.10%	98%	
2050	3	0.001	0.74	0.01	7.10%	98%	

CUM.PROD	0.130 BBO	GIANT FIELDS		YEMEN (NORTH)	
RESERVES	1.000 BBO	TOTAL BBO	0.5	M.EAST (MINOR)	
DISCOVERED	1.130 BBO	% OF DISC.	44%		
UNDISCOVERED	0.870 BBO	Last Giant	1984		
REMAINING	1.870 BBO			Variance	Trend
ULTIMATE	2.000 BBO			1990	n/a
Rept.Reserve	1.000	Ref.date(end)	1989	Future	UP

Milestones	Date	Prod Mb/d	Dep. Rate	% Produced	Forecast "A"
					A=Exponential decline
1st Peak	2000	302	10.82%	55%	B=5% incr. to Mid-Point
2nd Peak					C=Other
	2000	302	10.82%	55%	Pre-1930 Cum. 0
Final Decline	2001		10.82%		

Date	Prod. M.b/d	Prod. BBO/yr	Cum. Prod.	Remain	Deplet. Rate	% Produced	Giant	Notes & Names
1930		0.000	0.00	2.00	ERR	0%		
1931		0.000	0.00	2.00	0.00%	0%		
1932		0.000	0.00	2.00	0.00%	0%		
1933		0.000	0.00	2.00	0.00%	0%		
1934		0.000	0.00	2.00	0.00%	0%		
1935		0.000	0.00	2.00	0.00%	0%		
1936		0.000	0.00	2.00	0.00%	0%		
1937		0.000	0.00	2.00	0.00%	0%		
1938		0.000	0.00	2.00	0.00%	0%		
1939		0.000	0.00	2.00	0.00%	0%		
1940		0.000	0.00	2.00	0.00%	0%		
1941		0.000	0.00	2.00	0.00%	0%		
1942		0.000	0.00	2.00	0.00%	0%		
1943		0.000	0.00	2.00	0.00%	0%		
1944		0.000	0.00	2.00	0.00%	0%		
1945		0.000	0.00	2.00	0.00%	0%		
1946		0.000	0.00	2.00	0.00%	0%		
1947		0.000	0.00	2.00	0.00%	0%		
1948		0.000	0.00	2.00	0.00%	0%		
1949		0.000	0.00	2.00	0.00%	0%		
1950		0.000	0.00	2.00	0.00%	0%		
1951		0.000	0.00	2.00	0.00%	0%		
1952		0.000	0.00	2.00	0.00%	0%		
1953		0.000	0.00	2.00	0.00%	0%		
1954		0.000	0.00	2.00	0.00%	0%		
1955		0.000	0.00	2.00	0.00%	0%		
1956		0.000	0.00	2.00	0.00%	0%		
1957		0.000	0.00	2.00	0.00%	0%		
1958		0.000	0.00	2.00	0.00%	0%		
1959		0.000	0.00	2.00	0.00%	0%		
1960		0.000	0.00	2.00	0.00%	0%		
1961		0.000	0.00	2.00	0.00%	0%		
1962		0.000	0.00	2.00	0.00%	0%		
1963		0.000	0.00	2.00	0.00%	0%		
1964		0.000	0.00	2.00	0.00%	0%		
1965		0.000	0.00	2.00	0.00%	0%		
1966		0.000	0.00	2.00	0.00%	0%		
1967		0.000	0.00	2.00	0.00%	0%		
1968		0.000	0.00	2.00	0.00%	0%		
1969		0.000	0.00	2.00	0.00%	0%		
1970		0.000	0.00	2.00	0.00%	0%		
1971		0.000	0.00	2.00	0.00%	0%		
1972		0.000	0.00	2.00	0.00%	0%		
1973		0.000	0.00	2.00	0.00%	0%		
1974		0.000	0.00	2.00	0.00%	0%		
1975		0.000	0.00	2.00	0.00%	0%		
1976		0.000	0.00	2.00	0.00%	0%		
1977		0.000	0.00	2.00	0.00%	0%		
1978		0.000	0.00	2.00	0.00%	0%		
1979		0.000	0.00	2.00	0.00%	0%		
1980		0.000	0.00	2.00	0.00%	0%		
1981		0.000	0.00	2.00	0.00%	0%		
1982		0.000	0.00	2.00	0.00%	0%		
1983		0.000	0.00	2.00	0.00%	0%		
1984		0.000	0.00	2.00	0.00%	0%	0.50	Alif
1985		0.000	0.00	2.00	0.00%	0%		
1986	10	0.004	0.00	2.00	0.18%	0%		
1987	10	0.004	0.01	1.99	0.18%	0%		
1988	160	0.058	0.07	1.93	2.93%	3%		

1989	177	0.064	0.13	1.87	3.33%	7%	
1990	185	0.068	0.20	1.80	3.62%	10%	
1991	195	0.071	0.27	1.73	3.94%	13%	
1992	204	0.075	0.34	1.66	4.31%	17%	
1993	215	0.078	0.42	1.58	4.73%	21%	
1994	225	0.082	0.50	1.50	5.21%	25%	
1995	237	0.086	0.59	1.41	5.77%	30%	
1996	248	0.091	0.68	1.32	6.43%	34%	
1997	261	0.095	0.78	1.22	7.22%	39%	
1998	274	0.100	0.88	1.12	8.17%	44%	
1999	288	0.105	0.98	1.02	9.34%	49%	
2000	302	0.110	1.09	0.91	10.82%	55%	1ST PEAK
2001	269	0.098	1.19	0.81	10.82%	59%	50% PRODUCED
2002	240	0.088	1.28	0.72	10.82%	64%	
2003	214	0.078	1.36	0.64	10.82%	68%	
2004	191	0.070	1.43	0.57	10.82%	71%	
2005	170	0.062	1.49	0.51	10.82%	74%	
2006	152	0.055	1.54	0.46	10.82%	77%	
2007	135	0.049	1.59	0.41	10.82%	80%	
2008	121	0.044	1.64	0.36	10.82%	82%	
2009	108	0.039	1.68	0.32	10.82%	84%	
2010	96	0.035	1.71	0.29	10.82%	86%	
2011	86	0.031	1.74	0.26	10.82%	87%	
2012	76	0.028	1.77	0.23	10.82%	89%	
2013	68	0.025	1.80	0.20	10.82%	90%	
2014	61	0.022	1.82	0.18	10.82%	91%	
2015	54	0.020	1.84	0.16	10.82%	92%	
2016	48	0.018	1.85	0.15	10.82%	93%	
2017	43	0.016	1.87	0.13	10.82%	94%	
2018	38	0.014	1.88	0.12	10.82%	94%	
2019	34	0.013	1.90	0.10	10.82%	95%	
2020	31	0.011	1.91	0.09	10.82%	95%	
2021	27	0.010	1.92	0.08	10.82%	96%	
2022	24	0.009	1.93	0.07	10.82%	96%	
2023	22	0.008	1.93	0.07	10.82%	97%	
2024	19	0.007	1.94	0.06	10.82%	97%	
2025	17	0.006	1.95	0.05	10.82%	97%	
2026	15	0.006	1.95	0.05	10.82%	98%	
2027	14	0.005	1.96	0.04	10.82%	98%	
2028	12	0.004	1.96	0.04	10.82%	98%	
2029	11	0.004	1.97	0.03	10.82%	98%	
2030	10	0.004	1.97	0.03	10.82%	99%	
2031	9	0.003	1.97	0.03	10.82%	99%	
2032	8	0.003	1.98	0.02	10.82%	99%	
2033	7	0.003	1.98	0.02	10.82%	99%	
2034	6	0.002	1.98	0.02	10.82%	99%	
2035	5	0.002	1.98	0.02	10.82%	99%	
2036	5	0.002	1.99	0.01	10.82%	99%	
2037	4	0.002	1.99	0.01	10.82%	99%	
2038	4	0.001	1.99	0.01	10.82%	99%	
2039	3	0.001	1.99	0.01	10.82%	99%	
2040	3	0.001	1.99	0.01	10.82%	100%	
2041	3	0.001	1.99	0.01	10.82%	100%	
2042	2	0.001	1.99	0.01	10.82%	100%	
2043	2	0.001	1.99	0.01	10.82%	100%	
2044	2	0.001	1.99	0.01	10.82%	100%	
2045	2	0.001	1.99	0.01	10.82%	100%	
2046	2	0.001	2.00	0.00	10.82%	100%	
2047	1	0.001	2.00	0.00	10.82%	100%	
2048	1	0.000	2.00	0.00	10.82%	100%	
2049	1	0.000	2.00	0.00	10.82%	100%	
2050	1	0.000	2.00	0.00	10.82%	100%	

CUM.PROD	0.008 BBO	GIANT FIELDS		YEMEN (SOUTH)	
RESERVES	3.000 BBO	TOTAL BBO	0	M.EAST (MINOR)	
DISCOVERED	3.008 BBO	% OF DISC.	0%		
UNDISCOVERED	1.492 BBO	Last Giant	-		
REMAINING	4.492 BBO			Variance Trend	
ULTIMATE	4.500 BBO			1990	n/a
Rept.Reserve	3.000	Ref.date(end) 1989		Future	UP

Milestones	Date	Prod Mb/d	Dep. Rate	% Produced	Forecast "C"
1st Peak	2031	548	7.89%	48%	A=Exponential decline
2nd Peak					B=5% incr. to Mid-Point
					C=Other +10%
Mid-Point	2032	504	7.89%	52%	Pre-1930 Cum. 0
Final Decline	2032		7.89%	52%	

Date	Prod. M.b/d	Prod. BBO/yr	Cum. Prod.	Remain	Deplet. Rate	% Produced	Giant	Notes & Names
1930		0.000	0.00	4.50	ERR	0%		
1931		0.000	0.00	4.50	0.00%	0%		
1932		0.000	0.00	4.50	0.00%	0%		
1933		0.000	0.00	4.50	0.00%	0%		
1934		0.000	0.00	4.50	0.00%	0%		
1935		0.000	0.00	4.50	0.00%	0%		
1936		0.000	0.00	4.50	0.00%	0%		
1937		0.000	0.00	4.50	0.00%	0%		
1938		0.000	0.00	4.50	0.00%	0%		
1939		0.000	0.00	4.50	0.00%	0%		
1940		0.000	0.00	4.50	0.00%	0%		
1941		0.000	0.00	4.50	0.00%	0%		
1942		0.000	0.00	4.50	0.00%	0%		
1943		0.000	0.00	4.50	0.00%	0%		
1944		0.000	0.00	4.50	0.00%	0%		
1945		0.000	0.00	4.50	0.00%	0%		
1946		0.000	0.00	4.50	0.00%	0%		
1947		0.000	0.00	4.50	0.00%	0%		
1948		0.000	0.00	4.50	0.00%	0%		
1949		0.000	0.00	4.50	0.00%	0%		
1950		0.000	0.00	4.50	0.00%	0%		
1951		0.000	0.00	4.50	0.00%	0%		
1952		0.000	0.00	4.50	0.00%	0%		
1953		0.000	0.00	4.50	0.00%	0%		
1954		0.000	0.00	4.50	0.00%	0%		
1955		0.000	0.00	4.50	0.00%	0%		
1956		0.000	0.00	4.50	0.00%	0%		
1957		0.000	0.00	4.50	0.00%	0%		
1958		0.000	0.00	4.50	0.00%	0%		
1959		0.000	0.00	4.50	0.00%	0%		
1960		0.000	0.00	4.50	0.00%	0%		
1961		0.000	0.00	4.50	0.00%	0%		
1962		0.000	0.00	4.50	0.00%	0%		
1963		0.000	0.00	4.50	0.00%	0%		
1964		0.000	0.00	4.50	0.00%	0%		
1965		0.000	0.00	4.50	0.00%	0%		
1966		0.000	0.00	4.50	0.00%	0%		
1967		0.000	0.00	4.50	0.00%	0%		
1968		0.000	0.00	4.50	0.00%	0%		
1969		0.000	0.00	4.50	0.00%	0%		
1970		0.000	0.00	4.50	0.00%	0%		
1971		0.000	0.00	4.50	0.00%	0%		
1972		0.000	0.00	4.50	0.00%	0%		
1973		0.000	0.00	4.50	0.00%	0%		
1974		0.000	0.00	4.50	0.00%	0%		
1975		0.000	0.00	4.50	0.00%	0%		
1976		0.000	0.00	4.50	0.00%	0%		
1977		0.000	0.00	4.50	0.00%	0%		
1978		0.000	0.00	4.50	0.00%	0%		
1979		0.000	0.00	4.50	0.00%	0%		
1980		0.000	0.00	4.50	0.00%	0%		
1981		0.000	0.00	4.50	0.00%	0%		
1982		0.000	0.00	4.50	0.00%	0%		
1983		0.000	0.00	4.50	0.00%	0%		
1984		0.000	0.00	4.50	0.00%	0%		
1985		0.000	0.00	4.50	0.00%	0%		
1986		0.000	0.00	4.50	0.00%	0%		
1987		0.000	0.00	4.50	0.00%	0%		
1988	13	0.005	0.00	4.50	0.10%	0%		

1989	10	0.004	0.01	4.49	0.08%	0%	
1990	11	0.004	0.01	4.49	0.09%	0%	
1991	12	0.004	0.02	4.48	0.10%	0%	
1992	13	0.005	0.02	4.48	0.11%	0%	
1993	15	0.005	0.03	4.47	0.12%	1%	
1994	16	0.006	0.03	4.47	0.13%	1%	
1995	18	0.006	0.04	4.46	0.14%	1%	
1996	19	0.007	0.05	4.45	0.16%	1%	
1997	21	0.008	0.05	4.45	0.18%	1%	
1998	24	0.009	0.06	4.44	0.19%	1%	
1999	26	0.009	0.07	4.43	0.21%	2%	
2000	29	0.010	0.08	4.42	0.24%	2%	
2001	31	0.011	0.09	4.41	0.26%	2%	
2002	35	0.013	0.11	4.39	0.29%	2%	
2003	38	0.014	0.12	4.38	0.32%	3%	
2004	42	0.015	0.14	4.36	0.35%	3%	
2005	46	0.017	0.15	4.35	0.38%	3%	
2006	51	0.018	0.17	4.33	0.42%	4%	
2007	56	0.020	0.19	4.31	0.47%	4%	
2008	61	0.022	0.21	4.29	0.52%	5%	
2009	67	0.025	0.24	4.26	0.57%	5%	
2010	74	0.027	0.27	4.23	0.63%	6%	
2011	81	0.030	0.29	4.21	0.70%	7%	
2012	90	0.033	0.33	4.17	0.78%	7%	
2013	98	0.036	0.36	4.14	0.86%	8%	
2014	108	0.040	0.40	4.10	0.96%	9%	
2015	119	0.044	0.45	4.05	1.06%	10%	
2016	131	0.048	0.49	4.01	1.18%	11%	
2017	144	0.053	0.55	3.95	1.31%	12%	
2018	159	0.058	0.60	3.90	1.46%	13%	
2019	174	0.064	0.67	3.83	1.64%	15%	
2020	192	0.070	0.74	3.76	1.83%	16%	
2021	211	0.077	0.82	3.68	2.05%	18%	
2022	232	0.085	0.90	3.60	2.30%	20%	
2023	255	0.093	0.99	3.51	2.59%	22%	
2024	281	0.103	1.10	3.40	2.93%	24%	
2025	309	0.113	1.21	3.29	3.32%	27%	
2026	340	0.124	1.33	3.17	3.77%	30%	
2027	374	0.137	1.47	3.03	4.31%	33%	
2028	411	0.150	1.62	2.88	4.96%	36%	
2029	453	0.165	1.79	2.71	5.74%	40%	
2030	498	0.182	1.97	2.53	6.69%	44%	
2031	548	0.200	2.17	2.33	7.89%	48%	1ST PEAK
2032	504	0.184	2.35	2.15	7.89%	52%	50% PRODUCED
2033	465	0.170	2.52	1.98	7.89%	56%	
2034	428	0.156	2.68	1.82	7.89%	59%	
2035	394	0.144	2.82	1.68	7.89%	63%	
2036	363	0.133	2.95	1.55	7.89%	66%	
2037	334	0.122	3.08	1.42	7.89%	68%	
2038	308	0.112	3.19	1.31	7.89%	71%	
2039	284	0.104	3.29	1.21	7.89%	73%	
2040	261	0.095	3.39	1.11	7.89%	75%	
2041	241	0.088	3.47	1.03	7.89%	77%	
2042	222	0.081	3.56	0.94	7.89%	79%	
2043	204	0.075	3.63	0.87	7.89%	81%	
2044	188	0.069	3.70	0.80	7.89%	82%	
2045	173	0.063	3.76	0.74	7.89%	84%	
2046	160	0.058	3.82	0.68	7.89%	85%	
2047	147	0.054	3.87	0.63	7.89%	86%	
2048	135	0.049	3.92	0.58	7.89%	87%	
2049	125	0.046	3.97	0.53	7.89%	88%	
2050	115	0.042	4.01	0.49	7.89%	89%	

Note : Production assumed to start in 1993 following construction of a pipeline at 55,000 b/d and rise at 5% until 50% depleted

CUM.PROD	0.662 BBO	GIANT FIELDS		YUGOSLAVIA	
RESERVES	0.247 BBO	TOTAL BBO	0	SINO-SOVIET	
DISCOVERED	0.909 BBO	% OF DISC.	0%		
UNDISCOVERED	0.341 BBO	Last Giant	-		
REMAINING	0.588 BBO			Variance Trend	
ULTIMATE	1.250 BBO			1990	-14%
Rept.Reserve	0.230 #1	Ref.date(end)	1989	Future	UP

Milestones	Date	Prod Mb/d	Dep. Rate	% Produced	Forecast "A"
1st Peak	1981	88	4.03%	34%	A=Exponential decline
2nd Peak	None				B=5% incr. to Mid-Point
Mid-Point	1988	83	5.21%	51%	C=Other
Final Decline	1989		5.29%		Pre-1930 Cum. 0 (Added to 1930)

Date	Prod. M.b/d	Prod. BBO/yr	Cum. Prod.	Remain	Deplet. Rate	% Produced	Giant	Notes & Names
1930		0.000	0.00	1.25	ERR	0%		Data-base weak
1931		0.000	0.00	1.25	0.00%	0%		
1932		0.000	0.00	1.25	0.00%	0%		
1933		0.000	0.00	1.25	0.00%	0%		
1934		0.000	0.00	1.25	0.00%	0%		
1935		0.000	0.00	1.25	0.00%	0%		
1936		0.000	0.00	1.25	0.00%	0%		
1937		0.000	0.00	1.25	0.00%	0%		
1938		0.000	0.00	1.25	0.00%	0%		
1939	0	0.000	0.00	1.25	0.00%	0%		
1940	0	0.000	0.00	1.25	0.00%	0%		
1941	0	0.000	0.00	1.25	0.00%	0%		
1942	0	0.000	0.00	1.25	0.00%	0%		
1943	0	0.000	0.00	1.25	0.00%	0%		
1944	0	0.000	0.00	1.25	0.00%	0%		
1945	1	0.000	0.00	1.25	0.02%	0%		
1946	1	0.000	0.00	1.25	0.01%	0%		
1947	0	0.000	0.00	1.25	0.01%	0%		
1948	1	0.000	0.00	1.25	0.02%	0%		
1949	1	0.000	0.00	1.25	0.02%	0%		
1950	1	0.000	0.00	1.25	0.04%	0%		
1951	2	0.001	0.00	1.25	0.06%	0%		
1952	3	0.001	0.00	1.25	0.08%	0%		
1953	3	0.001	0.00	1.25	0.09%	0%		
1954	3	0.001	0.01	1.24	0.10%	0%		
1955	4	0.002	0.01	1.24	0.13%	1%		
1956	6	0.002	0.01	1.24	0.16%	1%		
1957	8	0.003	0.01	1.24	0.23%	1%		
1958	9	0.003	0.02	1.23	0.27%	1%		
1959	12	0.004	0.02	1.23	0.34%	2%		
1960	18	0.007	0.03	1.22	0.54%	2%		
1961	26	0.009	0.04	1.21	0.78%	3%		
1962	31	0.011	0.05	1.20	0.93%	4%		
1963	33	0.012	0.06	1.19	0.99%	5%		
1964	36	0.013	0.07	1.18	1.11%	6%		
1965	42	0.015	0.09	1.16	1.30%	7%		
1966	45	0.016	0.10	1.15	1.42%	8%		
1967	48	0.018	0.12	1.13	1.54%	10%		
1968	51	0.018	0.14	1.11	1.64%	11%		
1969	55	0.020	0.16	1.09	1.80%	13%		
1970	58	0.021	0.18	1.07	1.94%	15%		
1971	60	0.022	0.20	1.05	2.05%	16%		
1972	65	0.024	0.23	1.02	2.27%	18%		
1973	68	0.025	0.25	1.00	2.41%	20%		
1974	70	0.026	0.21	0.97	2.57%	17%		
1975	75	0.027	0.24	0.95	2.81%	19%		
1976	79	0.029	0.27	0.92	3.04%	21%		
1977	80	0.029	0.30	0.89	3.19%	24%		
1978	83	0.030	0.33	0.86	3.40%	26%		
1979	81	0.030	0.36	0.83	3.46%	29%		
1980	86	0.031	0.39	0.80	3.79%	31%		
1981	88	0.032	0.42	0.76	4.03%	34%		1ST PEAK
1982	86	0.032	0.45	0.73	4.13%	36%		
1983	86	0.032	0.48	0.70	4.30%	39%		
1984	82	0.030	0.51	0.67	4.27%	41%		
1985	84	0.031	0.54	0.64	4.59%	44%		
1986	83	0.030	0.57	0.61	4.73%	46%		
1987	77	0.028	0.60	0.58	4.63%	48%		
1988	83	0.030	0.63	0.55	5.21%	51%		50% PRODUCED

1989	80	0.029	0.66	0.52	5.29%	53%
1990	76	0.028	0.69	0.49	5.29%	55%
1991	72	0.026	0.72	0.47	5.29%	57%
1992	68	0.025	0.74	0.44	5.29%	59%
1993	64	0.023	0.76	0.42	5.29%	61%
1994	61	0.022	0.79	0.40	5.29%	63%
1995	58	0.021	0.81	0.38	5.29%	65%
1996	55	0.020	0.83	0.36	5.29%	66%
1997	52	0.019	0.85	0.34	5.29%	68%
1998	49	0.018	0.86	0.32	5.29%	69%
1999	46	0.017	0.88	0.30	5.29%	70%
2000	44	0.016	0.90	0.29	5.29%	72%
2001	42	0.015	0.91	0.27	5.29%	73%
2002	39	0.014	0.93	0.26	5.29%	74%
2003	37	0.014	0.94	0.24	5.29%	75%
2004	35	0.013	0.95	0.23	5.29%	76%
2005	33	0.012	0.97	0.22	5.29%	77%
2006	32	0.012	0.98	0.21	5.29%	78%
2007	30	0.011	0.99	0.20	5.29%	79%
2008	28	0.010	1.00	0.19	5.29%	80%
2009	27	0.010	1.01	0.18	5.29%	81%
2010	26	0.009	1.02	0.17	5.29%	81%
2011	24	0.009	1.03	0.16	5.29%	82%
2012	23	0.008	1.03	0.15	5.29%	83%
2013	22	0.008	1.04	0.14	5.29%	83%
2014	21	0.007	1.05	0.13	5.29%	84%
2015	19	0.007	1.06	0.13	5.29%	85%
2016	18	0.007	1.06	0.12	5.29%	85%
2017	17	0.006	1.07	0.11	5.29%	86%
2018	17	0.006	1.08	0.11	5.29%	86%
2019	16	0.006	1.08	0.10	5.29%	87%
2020	15	0.005	1.09	0.10	5.29%	87%
2021	14	0.005	1.09	0.09	5.29%	87%
2022	13	0.005	1.10	0.09	5.29%	88%
2023	13	0.005	1.10	0.08	5.29%	88%
2024	12	0.004	1.11	0.08	5.29%	88%
2025	11	0.004	1.11	0.07	5.29%	89%
2026	11	0.004	1.11	0.07	5.29%	89%
2027	10	0.004	1.12	0.07	5.29%	89%
2028	10	0.004	1.12	0.06	5.29%	90%
2029	9	0.003	1.12	0.06	5.29%	90%
2030	9	0.003	1.13	0.06	5.29%	90%
2031	8	0.003	1.13	0.05	5.29%	90%
2032	8	0.003	1.13	0.05	5.29%	91%
2033	7	0.003	1.14	0.05	5.29%	91%
2034	7	0.003	1.14	0.05	5.29%	91%
2035	7	0.002	1.14	0.04	5.29%	91%
2036	6	0.002	1.14	0.04	5.29%	91%
2037	6	0.002	1.15	0.04	5.29%	92%
2038	6	0.002	1.15	0.04	5.29%	92%
2039	5	0.002	1.15	0.03	5.29%	92%
2040	5	0.002	1.15	0.03	5.29%	92%
2041	5	0.002	1.15	0.03	5.29%	92%
2042	4	0.002	1.15	0.03	5.29%	92%
2043	4	0.002	1.16	0.03	5.29%	93%
2044	4	0.001	1.16	0.03	5.29%	93%
2045	4	0.001	1.16	0.02	5.29%	93%
2046	4	0.001	1.16	0.02	5.29%	93%
2047	3	0.001	1.16	0.02	5.29%	93%
2048	3	0.001	1.16	0.02	5.29%	93%
2049	3	0.001	1.16	0.02	5.29%	93%
2050	3	0.001	1.17	0.02	5.29%	93%

CUM.PROD	0.129 BBO	GIANT FIELDS		ZAIRE	
RESERVES	0.096 BBO	TOTAL BBO	0.0	AFRICA	
DISCOVERED	0.225 BBO	% OF DISC.	0%		
UNDISCOVERED	0.075 BBO	Last Giant			
REMAINING	0.171 BBO			Variance	Trend
ULTIMATE	0.300 BBO			1990	-3%
Rept.Reserves	0.096 BBO	Ref.Date (end)	1989	Future	DOWN

Milestones	Date	Prod Mb/d	Dep Rate	% Produced
1st Peak	1986	32	5.45%	32%
2nd Peak				
Mid-Point	1991	30	6.16%	50%
Final Decline	1991		6.16%	

Forecast "B"
A=Exponential decline
B=5% inc. to Mid-Point
C=Other
Pre-1930 Cum 0
(Added to 1930 Prod)

Date	Prod. M.b/d	Prod. BBO/yr	Cum Prod	Remain	Dep. Rate	% Produced	Giant	Names & Notes
1930		0.000	0.00	0.30	ERR	0%		
1931		0.000	0.00	0.30	0.00%	0%		
1932		0.000	0.00	0.30	0.00%	0%		
1933		0.000	0.00	0.30	0.00%	0%		
1934		0.000	0.00	0.30	0.00%	0%		
1935		0.000	0.00	0.30	0.00%	0%		
1936		0.000	0.00	0.30	0.00%	0%		
1937		0.000	0.00	0.30	0.00%	0%		
1938		0.000	0.00	0.30	0.00%	0%		
1939		0.000	0.00	0.30	0.00%	0%		
1940		0.000	0.00	0.30	0.00%	0%		
1941		0.000	0.00	0.30	0.00%	0%		
1942		0.000	0.00	0.30	0.00%	0%		
1943		0.000	0.00	0.30	0.00%	0%		
1944		0.000	0.00	0.30	0.00%	0%		
1945		0.000	0.00	0.30	0.00%	0%		
1946		0.000	0.00	0.30	0.00%	0%		
1947		0.000	0.00	0.30	0.00%	0%		
1948		0.000	0.00	0.30	0.00%	0%		
1949		0.000	0.00	0.30	0.00%	0%		
1950		0.000	0.00	0.30	0.00%	0%		
1951		0.000	0.00	0.30	0.00%	0%		
1952		0.000	0.00	0.30	0.00%	0%		
1953		0.000	0.00	0.30	0.00%	0%		
1954		0.000	0.00	0.30	0.00%	0%		
1955		0.000	0.00	0.30	0.00%	0%		
1956		0.000	0.00	0.30	0.00%	0%		
1957		0.000	0.00	0.30	0.00%	0%		
1958		0.000	0.00	0.30	0.00%	0%		
1959		0.000	0.00	0.30	0.00%	0%		
1960		0.000	0.00	0.30	0.00%	0%		
1961		0.000	0.00	0.30	0.00%	0%		
1962		0.000	0.00	0.30	0.00%	0%		
1963		0.000	0.00	0.30	0.00%	0%		
1964		0.000	0.00	0.30	0.00%	0%		
1965		0.000	0.00	0.30	0.00%	0%		
1966		0.000	0.00	0.30	0.00%	0%		
1967		0.000	0.00	0.30	0.00%	0%		
1968		0.000	0.00	0.30	0.00%	0%		
1969		0.000	0.00	0.30	0.00%	0%		
1970		0.000	0.00	0.30	0.00%	0%		
1971		0.000	0.00	0.30	0.00%	0%		
1972		0.000	0.00	0.30	0.00%	0%		
1973		0.000	0.00	0.30	0.00%	0%		
1974		0.000	0.00	0.30	0.00%	0%		
1975		0.000	0.00	0.30	0.00%	0%		
1976	20	0.007	0.01	0.29	2.43%	2%		
1977	24	0.009	0.02	0.28	2.99%	5%		
1978	20	0.007	0.02	0.28	2.57%	8%		
1979	21	0.008	0.03	0.27	2.77%	10%		
1980	22	0.008	0.04	0.26	2.99%	13%		
1981	20	0.007	0.05	0.25	2.80%	15%		
1982	22	0.008	0.05	0.25	3.17%	18%		
1983	26	0.009	0.06	0.24	3.86%	21%		
1984	27	0.010	0.07	0.23	4.17%	25%		
1985	31	0.011	0.08	0.22	4.92%	28%		
1986	32	0.012	0.10	0.20	5.45%	32%		1ST PEAK
1987	32	0.012	0.11	0.19	5.71%	36%		
1988	30	0.011	0.12	0.18	5.61%	40%		

1989	28	0.010	0.13	0.17	5.54%	43%	
1990	29	0.011	0.14	0.16	6.16%	47%	
1991	30	0.011	0.15	0.15	6.16%	50%	50% PRODUCED
1992	25	0.009	0.16	0.14	6.16%	53%	
1993	24	0.009	0.17	0.13	6.16%	56%	
1994	22	0.008	0.18	0.12	6.16%	59%	
1995	21	0.008	0.18	0.12	6.16%	61%	
1996	20	0.007	0.19	0.11	6.16%	64%	
1997	18	0.007	0.20	0.10	6.16%	66%	
1998	17	0.006	0.20	0.10	6.16%	68%	
1999	16	0.006	0.21	0.09	6.16%	70%	
2000	15	0.006	0.22	0.08	6.16%	72%	
2001	14	0.005	0.22	0.08	6.16%	74%	
2002	13	0.005	0.23	0.07	6.16%	75%	
2003	13	0.005	0.23	0.07	6.16%	77%	
2004	12	0.004	0.23	0.07	6.16%	78%	
2005	11	0.004	0.24	0.06	6.16%	80%	
2006	10	0.004	0.24	0.06	6.16%	81%	
2007	10	0.004	0.25	0.05	6.16%	82%	
2008	9	0.003	0.25	0.05	6.16%	83%	
2009	9	0.003	0.25	0.05	6.16%	84%	
2010	8	0.003	0.26	0.04	6.16%	85%	
2011	8	0.003	0.26	0.04	6.16%	86%	
2012	7	0.003	0.26	0.04	6.16%	87%	
2013	7	0.002	0.26	0.04	6.16%	88%	
2014	6	0.002	0.27	0.03	6.16%	88%	
2015	6	0.002	0.27	0.03	6.16%	89%	
2016	5	0.002	0.27	0.03	6.16%	90%	
2017	5	0.002	0.27	0.03	6.16%	90%	
2018	5	0.002	0.27	0.03	6.16%	91%	
2019	5	0.002	0.27	0.03	6.16%	92%	
2020	4	0.002	0.28	0.02	6.16%	92%	
2021	4	0.001	0.28	0.02	6.16%	93%	
2022	4	0.001	0.28	0.02	6.16%	93%	
2023	4	0.001	0.28	0.02	6.16%	93%	
2024	3	0.001	0.28	0.02	6.16%	94%	
2025	3	0.001	0.28	0.02	6.16%	94%	
2026	3	0.001	0.28	0.02	6.16%	95%	
2027	3	0.001	0.28	0.02	6.16%	95%	
2028	3	0.001	0.29	0.01	6.16%	95%	
2029	2	0.001	0.29	0.01	6.16%	96%	
2030	2	0.001	0.29	0.01	6.16%	96%	
2031	2	0.001	0.29	0.01	6.16%	96%	
2032	2	0.001	0.29	0.01	6.16%	96%	
2033	2	0.001	0.29	0.01	6.16%	97%	
2034	2	0.001	0.29	0.01	6.16%	97%	
2035	2	0.001	0.29	0.01	6.16%	97%	
2036	2	0.001	0.29	0.01	6.16%	97%	
2037	1	0.001	0.29	0.01	6.16%	97%	
2038	1	0.000	0.29	0.01	6.16%	97%	
2039	1	0.000	0.29	0.01	6.16%	98%	
2040	1	0.000	0.29	0.01	6.16%	98%	
2041	1	0.000	0.29	0.01	6.16%	98%	
2042	1	0.000	0.29	0.01	6.16%	98%	
2043	1	0.000	0.29	0.01	6.16%	98%	
2044	1	0.000	0.29	0.01	6.16%	98%	
2045	1	0.000	0.30	0.00	6.16%	98%	
2046	1	0.000	0.30	0.00	6.16%	98%	
2047	1	0.000	0.30	0.00	6.16%	99%	
2048	1	0.000	0.30	0.00	6.16%	99%	
2049	1	0.000	0.30	0.00	6.16%	99%	
2050	1	0.000	0.30	0.00	6.16%	99%	

GIANT OILFIELDS

GIANT OIL FIELDS
(1990)
(*=uncertain/duplicated)

Name	Country	Date	Country Total	Ultimate Recovery BBO
Asab	Abu Dhabi	1965		6.500
Bab	Abu Dhabi	1954		8.000
Bu Hasa	Abu Dhabi	1962		11.000
Mubarras	Abu Dhabi	1969		2.000
Murban	Abu Dhabi	1962		3.000
Sahil	Abu Dhabi	1967		1.200
Sath Al Raazboot	Abu Dhabi	1970		0.900
Shah	Abu Dhabi	1966		0.900
Umm Ad Dalkh	Abu Dhabi	1973		0.828
Umm Shaif	Abu Dhabi	1958		5.500
Zakum	Abu Dhabi	1964		21.000
Zubbaya	Abu Dhabi	1969	62.23	1.400
Gassi Touil	Algeria	1963		0.500
Hassi Messaoud	Algeria	1956		9.000
Rhourde El Baguel	Algeria	1962		0.500
Zarzaitine	Algeria	1957	10.90	0.900
Malongo N & S	Angola	1966		0.600
Malongo West	Angola	1969	1.50	0.900
Comodoro-Rivadavia	Argentina	1907	2.00	2.000
Barracouta	Australia	1965		0.500
Halibut	Australia	1967		0.800
Kingfish	Australia	1967	2.55	1.250
Greater Matzen	Austria	1949	0.50	0.500
Awali	Bahrain	1932	1.00	1.000
Albacora	Brazil	1984		1.100
Marlim	Brazil	1985		4.300
Miranga	Brazil		5.99	0.590
Seria	Brunei	1929		1.300
S.W.Ampa	Brunei	1963	2.35	1.050
Bonnie Glen	Canada			0.509
Hibernia	Canada	1979		0.525
Judy Creek	Canada	1959		0.536
Leduc	Canada	1947		0.500
Pembina	Canada	1953		1.800
Rainbow Complex	Canada	1965		0.667
Redwater	Canada	1948		0.809
Swan Hills	Canada	1957		1.005
S.Swan Hills	Canada	1960	7.01	0.656
Karamay	China	1955		1.730
Leng-hu	China	1958		1.100
Lung Nu	China	1958		0.600
Tach'ing	China	1959		6.000
Ya-Erh-Hsia	China	1950	9.96	0.525
Canon Limon	Colombia	1983		1.000
La Cira- Infantas	Colombia	1926	1.70	0.700
Emeraude	Congo	1969		0.500
Loango	Congo		1.00	0.500
Fateh	Dubai	1966		1.500
Fateh SW	Dubai	1970	2.60	1.100
Auca	Ecuador	1970		0.500
Sacha	Ecuador	1969		0.750
Shushufindi	Ecuador	1969	2.46	1.212
Belayim (Land)	Egypt	1955		0.500
Belayim (Marine)	Egypt	1961		1.000
El Morgan	Egypt	1965		1.500
July	Egypt	1967		0.720
Ramadan	Egypt	1974	4.72	1.000
Grondin	Gabon		0.50	0.500
Ankleshwar	India			0.520
Bassein N	India	1975		0.550
Bombay High	India	1974	3.47	2.400
Arjuna	Indonesia	1969		0.600
Arun	Indonesia	1971		0.700
Attaka	Indonesia	1970		0.550
Bangko	Indonesia	1970		0.500
Bekasap	Indonesia	1955		0.570
Duri	Indonesia	1941		0.500
Handil	Indonesia	1974		0.900
Minas	Indonesia	1944	8.62	4.300
Ab Teymur	Iran	1968		0.500
Agha Jari	Iran	1936		8.700

Ahwaz	Iran	1958		10.160
Ardenshir	Iran	1961		0.600
Bibi Hakimeh	Iran	1961		2.500
Binak	Iran	1959		0.500
Darius-Kharg	Iran	1961		0.820
Darood	Iran	1961		0.850
Foroozan	Iran	1966		2.000
Gash Seran	Iran	1928		8.500
Haft Kei	Iran	1927		1.750
Karanj	Iran	1963		1.650
Kupal	Iran	1965		1.200
Lab-E-Safid	Iran	1968		0.550
Mansun	Iran			0.656
Mansuri	Iran	1963		1.500
Marun	Iran	1963		9.500
Masjid-i-Sulemein	Iran	1908		1.125
Naft-i-Safd	Iran	1935		1.000
Paris	Iran	1964		3.000
Pazanan	Iran	1961		1.130
Rag-e Safid	Iran	1964		2.400
Rostam	Iran	1967		1.000
Sarkham	Iran	1969		0.800
Sassan	Iran	1967		1.500
Shurom	Iran	1970	64.39	0.500
Bai Hassan	Iraq	1953		1.236
Buzurgan	Iraq	1969		2.500
East Baghdad	Iraq	1979		18.000
Fauqi	Iraq	1974		1.000
Hamrin	Iraq	1973		1.000
Jambur	Iraq	1954		1.000
Kirkuk	Iraq	1927		16.000
Luhais	Iraq	1961		0.500
Majoon	Iraq	1976		8.000
Nahr Umr	Iraq	1948		1.130
North Rumaila	Iraq			5.000
Quaiyarah	Iraq			0.500
Rumailia	Iraq	1953		14.141
Zubair	Iraq	1948		4.243
Abu Ghirab/Dehluran	Iraq/Iran	1971	75.75	1.500
Bahrah	Kuwait	1956		1.000
Burgan Complex	Kuwait	1938		66.500
Minagish	Kuwait	1959		2.100
Ratqa	Kuwait	1977		1.000
Raudhatain	Kuwait	1955		7.875
Sabriyah	Kuwait	1957		5.300
Umm Gudair	Kuwait/N.Z.	1962	84.88	1.100
Amal	Libya	1959		4.250
Bahi	Libya	1958		0.600
Bu Attifel	Libya	1968		1.000
Dahra-Hofra	Libya	1958		0.700
Defa	Libya	1960		1.800
Gialo	Libya	1961		4.000
Intisar A	Libya	1967		0.750
Intisar D	Libya	1967		1.500
Messia	Libya	1971		1.500
Nafoora-Augila	Libya	1966		1.500
Nasser	Libya	1959		2.200
Raguba	Libya	1961		1.000
Samah	Libya	1962		0.500
Sarir	Libya	1961		6.500
Waha	Libya	1959		1.200
Bouri	Libya/Tunis.	1976	34.00	5.000
Baronia	Malaysia			0.500
Smarang	Malaysia		1.00	0.500
Agave	Mexico			0.800
Arenque	Mexico	1967		1.000
Bermudez Complex	Mexico			6.500
Cactus-Nispero	Mexico	1972		1.850
Carntarell Complex	Mexico	1977		8.000
Chicontapec	Mexico	1978		12.285
Ebano-Panuco	Mexico	1901		1.304
Giraidas	Mexico			1.500
Naranjos-Cerro Azul	Mexico	1916		1.400
Paredon	Mexico	1977		0.500
Posa Rica	Mexico	1930		2.000
Rio Nuevo	Mexico	1975		0.500
Samaria-Cuncuacan	Mexico	1960		4.500

San Andres	Mexico		42.64	0.500
Hout	Neutral Z.	1963		0.500
Wafra	Neutral Z.	1953	5.50	5.000
Bomu	Nigeria	1958		0.500
Forcados-Yokri	Nigeria	1968		0.700
Imo River	Nigeria	1959		0.700
Jones Creek	Nigeria	1967		1.000
Kokon	Nigeria			0.500
Meren	Nigeria	1965		0.650
Nembe Creek	Nigeria	1973		0.645
Okan	Nigeria	1964		0.580
Olomoro-Afiesere	Nigeria		5.88	0.600
Ekofisk	Norway	1969		1.673
Gullfaks	Norway	1978		1.320
Heidrun	Norway	1985		0.547
Oseberg	Norway	1979		1.520
Snorre	Norway	1979		0.679
Statfjord	Norway/UK	1974	8.10	2.358
Fahud	Oman	1964		0.900
Marmul	Oman	1957		0.600
Natih	Oman	1960		0.600
Yibal	Oman	1962	2.70	0.600
La Brea Parinas	Peru	1868	1.39	1.385
Bul Hanine	Qatar	1970		1.250
Dukhan	Qatar	1940		2.850
Idd-El-Sharqi	Qatar	1960		2.100
Maydan/Mahzan	Qatar	1963	7.40	1.200
Moreni-Gura-Ocnitei	Rumania	1890	0.80	0.800
Abqaiq	Saudi	1941		12.800
Abu Hadriya	Saudi	1940		1.758
Abu Jiffan	Saudi	1973		0.500
Abu Sa'Fah	Saudi	1963		7.808
Bargan	Saudi	1969		0.500
Berri	Saudi	1964		7.300
Dammam	Saudi	1938		1.050
Fadhili	Saudi	1949		0.952
Ghawar	Saudi	1948		83.000
Harmaliya	Saudi	1972		1.025
Khurais	Saudi	1957		8.700
Khursaniyah	Saudi	1956		2.300
Lawnah	Saudi	1975		1.172
Maharah	Saudi	1973		1.100
Manifa	Saudi	1957		17.100
Mazalij	Saudi	1971		0.627
Qatif	Saudi	1945		3.200
Shaybah	Saudi	1968		5.710
Zulut	Saudi	1965		10.636
Fereidoon-Marjan	Saudi/Iran	1966		10.000
Safaniya-Khafji	Saudi/NZ	1951	209.5	32.300
Karatchouk	Syria	1956		0.900
Souedie	Syria	1959	2.42	1.519
Fyzabad Forest Res.	Trinidad	1913		0.750
Soldado	Trinidad	1956	1.27	0.520
El Borma	Tunisia/Alg.	1964	0.75	0.750
Alwyn	UK			0.500
Beryl	UK	1972		0.500
Brae	UK	1975		0.504
Brent	UK	1971		1.736
Claymore	UK	1974		0.500
Cormorant	UK	1972		0.500
Dunlin	UK	1973		0.585
Forties	UK	1970		2.470
Magnus	UK	1974		0.565
Murchison	UK	1975		1.060
Ninian	UK	1974		1.100
Piper	UK	1973		0.951
Thistle	UK	1973	11.52	0.550
Bay Marchand Bk 2	USA	1930		0.656
Belridge S.	USA	1911		1.111
Bradford	USA	1871		0.653
Buena Vista	USA	1909		0.686
Burbank	USA	1920		0.545
Caillou Island	USA	1930		0.676
Coalinga	USA	1887		0.915
Coalinga Nose	USA	1938		0.508
Conroe	USA	1931		0.761
Cowden Complex	USA	1930		0.538

East Anschutz Ranch	USA	1979		0.807
East Texas	USA	1930		5.998
Elk Hills	USA	1911		1.473
Golden Trend	USA	1945		0.503
Goldsmith-Andector	USA	1935		0.789
Greater Seminole	USA	1926		0.566
Greta/ Tom O'Connor	USA	1933		0.803
Hastings	USA	1934		0.766
Hawkins	USA	1940		0.864
Huntington Beach	USA	1920		1.138
Kelly/Sneid./Diamond	USA	1948		1.350
Kern River	USA	1899		1.947
Kuparuk	USA	1969		1.501
Levelland	USA	1938		0.534
Long Beach	USA	1921		0.927
McArthur River	USA	1965		0.576
McElroy-Dune	USA	1926		0.522
Midway-Sunset	USA	1894		2.253
Oklahoma City	USA	1928		0.822
Panhandle	USA	1918		1.443
Prudhoe Bay	USA	1967		9.918
Pt. Arguello	USA	1981		0.547
Rangely	USA	1902		0.787
Salt Creek	USA	1906		0.655
San Ardo	USA	1947		0.531
Santa Fe Springs	USA	1919		0.629
Sho-Vel-Tum	USA	1914		1.235
Slaughter	USA	1936		1.080
Smackover	USA	1922		0.568
Sprayberry	USA	1982		0.711
Van	USA	1928		0.537
Ventura-Rincon	USA	1916		0.992
Wasson	USA	1936		1.798
Webster	USA	1936		0.593
Wetz	USA	1921		0.655
Wilmington	USA	1932		2.788
Yates	USA	1926	59.21	1.955
Agan	USSR	1966		0.600
Arlan Complex	USSR	1955		2.800
Baku Archipelago	USSR	1963		0.860
Balakhany-Sab-Rom.	USSR	1896		2.400
Barsa	USSR	1962		0.500
Bavly	USSR	1946		0.500
Bibi Eybat	USSR	1871		1.200
Bolshoye-Chernogor	USSR	1970		0.800
Cheleken	USSR	1956		0.640
Fedorov	USSR	1971		0.600
Karachukhurzykh	USSR	1928		0.600
Kholomgor	USSR	1973		0.500
Kotur-Tepe	USSR	1956		1.460
Kuleshova	USSR	1958		0.760
Malgobek-V.-A.	USSR	1915		3.000
Mamontovo	USSR	1965		0.900
Megion	USSR	1961		0.630
Mokhanova	USSR	1945		1.530
Nebit Dag	USSR	1934		0.500
Neftyany/Kamni	USSR	1949		1.225
North Varyegan	USSR	1971		0.600
Novayekhov/Aktash	USSR	1955		1.320
Oktyabishoye	USSR	1913		0.520
Ostrov Bulla	USSR	1959		0.500
Pokachev	USSR	1970		0.800
Pravdinsk	USSR	1964		0.950
Prilukskoya	USSR	1960		0.600
Romashkino	USSR	1948		14.000
Salymskoya	USSR	1965		10.000
Samgori	USSR	1974		0.500
Samotlor	USSR	1965		15.000
Sangachaly-Duvannyy	USSR	1963		0.860
Severo Var.	USSR	1971		0.500
Shkapovo	USSR	1944		1.250
Sovetskoye	USSR	1962		1.250
Straroqroznyy	USSR	1893		0.650
Sverny Pokur	USSR	1964		0.500
Tengiz	USSR	1979		10.000
Tuymazy	USSR	1937		1.000
Usa	USSR	1963		0.614

Usin	USSR	1963		3.100
Usk-Balyk	USSR	1961		0.930
Uzen	USSR	1961		2.500
Var'yegen	USSR	1970		0.500
Vatyegan	USSR	1961		0.500
Vozey	USSR	1972		1.000
West Surgut	USSR	1962		0.730
Yarino	USSR	1956		0.800
Yarino-Kam	USSR	1954		0.510
Yuzhino-Caerenskaya	USSR	1969		1.440
Zapagno-Surgut	USSR	1962		0.600
Zhety Bay	USSR	1961		0.500
Zupando-Surquit	USSR		98.53	2.000
Bolivar Coastal	Venezuela	1957		32.000
Boscan	Venezuela	1946		1.000
Centro	Venezuela	1957		1.000
Dacion/Lobo/Leona	Venezuela			1.600
El Furrial	Venezuela	1986		2.900
Guara-Guico	Venezuela	1944		0.719
La Paz	Venezuela	1923		0.890
La Victoria	Venezuela	1984		0.500
Lama	Venezuela	1957		3.200
Lamar	Venezuela	1958		1.470
Mene Grande	Venezuela	1914		0.700
Oficina	Venezuela	1937		0.960
Quiriquire	Venezuela	1928		0.770
Urdaneta	Venezuela	1955		1.000
Zoro/Mapiri	Venezuela		49.81	1.100
Alif	Yemen N.	1984		0.500
Abu Jidu	*Abu Dhabi	1966		1.000
Bu Tini	*Abu Dhabi	1980		0.700
Murban Bab	*Abu Dhabi	1954		2.000
Nasr	*Abu Dhabi	1971		0.500
Zarrarah	*Abu Dhabi	1970		0.571
Campos	*Brazil			0.918
Champion	*Brunei	1970		0.950
Yenanma	*Burma	1920		3.100
Dagang	*China	1964		1.000
Gaoxi	*China	1954		0.500
Gudong	*China	1984		0.730
Karamai	*China	1955		0.730
Liaohe	*China	1964		4.380
Rengiu	*China	1975		2.300
Shengli	*China	1962		3.000
Songfangtun	*China	1975		0.735
Suizhong-36	*China	1987		0.500
Yumen-Yaerhsia	*China	1938		0.700
"Amoco"	*Congo	1988		0.800
Pungarayacu	*Ecuador			1.550
Amal	*Egypt	1968		2.500
Nahorkatiya	*India	1953		0.630
Panna	*India	1976		0.733
Ahdab	*Iran	1979		0.500
Bushgan	*Iran	1963		1.000
Dehluran	*Iran	1972		0.600
Emam Hassan	*Iran	1963		0.550
Jufeyr	*Iran	1976		0.550
Kaki	*Iran	1977		0.550
Khaviz	*Iran	1977		0.550
Khayrabad	*Iran	1988		0.600
Namak-E-Kangan	*Iran	1977		0.550
Nargesi	*Iran	1975		0.550
Paydar	*Iran	1978		0.550
Rudak	*Iran	1979		0.550
Sarvestan	*Iran	1973		0.800
Shakeh	*Iran	1978		0.550
Shiah Makan	*Iran	1975		0.550
Tangu	*Iran	1978		0.550
Gharraf	*Iraq	1984		0.500
Halfaya	*Iraq	1977		3.000
Khabbaz	*Iraq	1983		0.500
Nasiryah	*Iraq	1979		0.500
Nur	*Iraq	1978		0.500
Rachi	*Iraq	1956		0.500
Ratavi	*Iraq	1948		1.000
Saddam	*Iraq	1983		2.000
Safwan	*Iraq	1983		0.500

Suba	*Iraq	1974	0.500
Tikrit	*Iraq	1986	1.000
Tuba	*Iraq	1960	0.800
West Qurna	*Iraq	1973	6.000
Khashman	*Kuwait	1963	0.520
Umm Gudair S	*Kuwait/N.Z.	1966	0.650
Khafji	*Kuwait/Saud	1959	8.870
Beda	*Libya	1959	0.900
Mabruk	*Libya	1959	0.500
NC120-A-001	*Libya	1984	0.765
Sarir (2)	*Libya	1966	0.700
Abkatum	*Mexico	1977	2.000
Agave	*Mexico	78-88	0.500
Batab	*Mexico	78-88	0.500
Bellota	*Mexico	78-88	0.500
Caan	*Mexico	78-88	0.500
Cardenas	*Mexico	1980	0.600
Chac	*Mexico	78-88	0.500
Ek	*Mexico	78-88	0.500
Jujo	*Mexico	1980	0.600
Ku	*Mexico	1979	1.400
Luna	*Mexico	78-88	0.500
Malob	*Mexico	78-88	0.500
Muspac	*Mexico	78-88	0.500
Pol	*Mexico	1980	0.900
Sen	*Mexico	78-88	0.500
Sitio Grande	*Mexico	1972	1.250
Tecominoacan	*Mexico	78-88	0.500
Uech	*Mexico	78-88	0.500
Dawl 1	*Saudi	1980	0.600
Dhib	*Saudi	1979	0.600
Dubaynah	*Saudi	1980	0.600
Farhah	*Saudi	1984	0.500
Faridah	*Saudi	1979	0.600
Habari	*Saudi	1976	0.600
Hamd	*Saudi	1982	0.600
Hamur	*Saudi	1979	0.600
Harqus	*Saudi	1978	0.600
Hasbah	*Saudi	1976	0.600
Hurayi	*Saudi	1981	0.600
Jaladi	*Saudi	1978	0.600
Jana	*Saudi	1967	0.500
Jauf	*Saudi	1980	0.600
Jawb	*Saudi	1979	0.600
Jurayd	*Saudi	1968	0.600
Kurayn	*Saudi	1974	0.600
Lughfah	*Saudi	1979	0.600
Magarib	*Saudi	1982	0.600
Marjan	*Saudi	1967	2.226
Nita	*Saudi	1981	0.600
Quhqub	*Saudi	1977	0.600
Rimthan	*Saudi	1974	0.600
Sadawi	*Saudi	1976	0.600
Sahba	*Saudi	1984	0.600
Salsal	*Saudi	1980	0.600
Samin	*Saudi	1979	0.600
Sharer	*Saudi	1976	0.600
Suban	*Saudi	1976	0.600
Tinat	*Saudi	1982	0.600
Wari'ah	*Saudi	1978	0.600
Watban	*Saudi	1975	0.600
Alba	*UK	1985	0.500
Borregas-Seeligson	*USA	1938	0.510
Elk Basin	*USA		0.457
Eunice	*USA	1927	0.136
Hondo	*USA		0.200
Kettleman N Dome	*USA		0.458
Lima-Indiana	*USA	1895	0.514
Old Illinois	*USA	1905	0.500
Scurry	*USA	1948	1.750
Seeligson	*USA	1937	0.327
South Pass	*USA	1950	0.750
S.Sand Belt	*USA	1926	0.560
S.Timbalier 21	*USA		0.263
Thompson	*USA	1921	0.500
West Delta	*USA	1962	0.493
Wichita Co. Reg	*USA		0.556

Bibieybatskoya	*USSR	1850	2.000
Gazlinskoya	*USSR	1956	10.000
Kalinskoya	*USSR	1931	0.500
Mamontovo	*USSR	1965	1.750
Ozeksuatskoya	*USSR	1953	1.000
Pezchanyy More	*USSR	1952	0.600
Salyukinskoya	*USSR	1971	0.870
Sosnino	*USSR	1962	4.200
Surakhanskoya	*USSR		0.900
Ust-Balyk	*USSR	1961	2.300
Ambrosio	*Venezuela	1925	4.000
Bachaquero	*Venezuela	1930	4.000
Cabimas	*Venezuela	1917	3.000
Carito	*Venezuela	1988	4.120
Cerro Negro	*Venezuela	1939	1.000
Cerro Negro -18	*Venezuela	1980	1.000
Cerro Negro -29	*Venezuela	1980	2.000
Ceuta	*Venezuela	1957	0.800
Ceuta S,SE	*Venezuela	1985	0.800
Greater Jusepin	*Venezuela	1938	0.575
Guafita	*Venezuela	1984	1.000
Jobo	*Venezuela	1953	0.850
Lago	*Venezuela	1958	1.350
Lagunillas	*Venezuela	1926	8.000
Mata	*Venezuela	1954	0.600
Musipan	*Venezuela	1987	0.890
Nipa	*Venezuela	1945	0.000
Pueblo Viejo	*Venezuela	1939	2.000
Tia Juana	*Venezuela	1928	13.000

The GeoJournal Library

1. B. Currey and G. Hugo (eds.): *Famine as Geographical Phenomenon.* 1984 ISBN 90-277-1762-1
2. S. H. U. Bowie, F.R.S. and I. Thornton (eds.): *Environmental Geochemistry and Health.* Report of the Royal Society's British National Committee for Problems of the Environment. 1985 ISBN 90-277-1879-2
3. L. A. Kosiński and K. M. Elahi (eds.): *Population Redistribution and Development in South Asia.* 1985 ISBN 90-277-1938-1
4. Y. Gradus (ed.): *Desert Development.* Man and Technology in Sparselands. 1985 ISBN 90-277-2043-6
5. F. J. Calzonetti and B. D. Solomon (eds.): *Geographical Dimensions of Energy.* 1985 ISBN 90-277-2061-4
6. J. Lundqvist, U. Lohm and M. Falkenmark (eds.): *Strategies for River Basin Management.* Environmental Integration of Land and Water in River Basin. 1985 ISBN 90-277-2111-4
7. A. Rogers and F. J. Willekens (eds.): *Migration and Settlement.* A Multiregional Comparative Study. 1986 ISBN 90-277-2119-X
8. R. Laulajainen: *Spatial Strategies in Retailing.* 1987 ISBN 90-277-2595-0
9. T. H. Lee, H. R. Linden, D. A. Dreyfus and T. Vasko (eds.): *The Methane Age.* 1988 ISBN 90-277-2745-7
10. H. J. Walker (ed.): *Artificial Structures and Shorelines.* 1988 ISBN 90-277-2746-5
11. A. Kellerman: *Time, Space, and Society.* Geographical Societal Perspectives. 1989 ISBN 0-7923-0123-4
12. P. Fabbri (ed.): *Recreational Uses of Coastal Areas.* A Research Project of the Commission on the Coastal Environment, International Geographical Union. 1990 ISBN 0-7923-0279-6
13. L. M. Brush, M. G. Wolman and Huang Bing-Wei (eds.): *Taming the Yellow River: Silt and Floods.* Proceedings of a Bilateral Seminar on Problems in the Lower Reaches of the Yellow River, China. 1989 ISBN 0-7923-0416-0
14. J. Stillwell and H. J. Scholten (eds.): *Contemporary Research in Population Geography.* A Comparison of the United Kingdom and the Netherlands. 1990 ISBN 0-7923-0431-4
15. M. S. Kenzer (ed.): *Applied Geography.* Issues, Questions, and Concerns. 1989 ISBN 0-7923-0438-1
16. D. Nir: *Region as a Socio-environmental System.* An Introduction to a Systemic Regional Geography. 1990 ISBN 0-7923-0516-7
17. H. J. Scholten and J. C. H. Stillwell (eds.): *Geographical Information Systems for Urban and Regional Planning.* 1990 ISBN 0-7923-0793-3
18. F. M. Brouwer, A. J. Thomas and M. J. Chadwick (eds.): *Land Use Changes in Europe.* Processes of Change, Environmental Transformations and Future Patterns. 1991 ISBN 0-7923-1099-3
19. C. J. Campbell: *The Golden Century of Oil 1950–2050.* The Depletion of a Resource ISBN 0-7923-1442-5

KLUWER ACADEMIC PUBLISHERS – DORDRECHT / BOSTON / LONDON

MIX
Papier aus verantwortungsvollen Quellen
Paper from responsible sources
FSC® C105338

If you have any concerns about our products,
you can contact us on
ProductSafety@springernature.com

In case Publisher is established outside the EU,
the EU authorized representative is:
Springer Nature Customer Service Center GmbH
Europaplatz 3, 69115 Heidelberg, Germany

Printed by Libri Plureos GmbH
in Hamburg, Germany